Al-Ghazālī's Philosophical Theology

Al-Ghazālī's Philosophical Theology

Frank Griffel

OXFORD
UNIVERSITY PRESS

OXFORD
UNIVERSITY PRESS

Oxford University Press, Inc., publishes works that further
Oxford University's objective of excellence
in research, scholarship, and education.

Oxford New York
Auckland Cape Town Dar es Salaam Hong Kong Karachi
Kuala Lumpur Madrid Melbourne Mexico City Nairobi
New Delhi Shanghai Taipei Toronto

With offices in
Argentina Austria Brazil Chile Czech Republic France Greece
Guatemala Hungary Italy Japan Poland Portugal Singapore
South Korea Switzerland Thailand Turkey Ukraine Vietnam

Published by Oxford University Press, Inc.
198 Madison Avenue, New York, New York 10016

www.oup.com

First issued as an Oxford University Press paperback, 2010

Library of Congress Cataloging-in-Publication Data

Griffel, Frank, 1965–
Al-Ghazali's philosophical theology / Frank Griffel.
p. cm.
Includes bibliographical references and index.
ISBN 978-0-19-977370-1
1. Ghazzali, 1058–1111. 2. Islamic cosmology. 3. Philosophy, Islamic.
I. Title.
B753.G34G75 2009
181'.5—dc22 2008030813

Printed in the United States of America
on acid-free paper

per Magda Lena e Tanya

La gloria di colui che tutto move
per l'universo penetra, e risplende
in una parte più e meno altrove.

—*Dante,* Paradiso, *1.1–3*

Contents

Timetable, xi

Introduction, 3

1. A Life between Public and Private Instruction: Al-Ghazālī's Biography, 19
    - The Main Sources for al-Ghazālī's Biography, 21
    - Al-Ghazālī's Date of Birth: Around 448/1056, 23
    - Al-Ghazālī's Early Years and His Education, 25
    - Becoming a Famous Jurist and Theologian, 31
    - Leaving Baghdad, Traveling in Syria and the Hijaz, and Returning to Khorasan, 40
    - The Ideal of a Secluded Life—His Last Years in Khorasan, 49
2. Al-Ghazālī's Most Influential Students and Early Followers, 61
    - Abū Bakr ibn al-ʿArabī (d. 543/1148), 62
    - Ibn al-ʿArabī's First Report of His Meeting with al-Ghazālī, 65
    - Ibn al-ʿArabī's Second Report of His Meeting with al-Ghazālī, 66
    - Asʿad al-Mayhanī (d. 523/1130 or 527/1132–33), 71
    - Muḥammad ibn Yaḥyā al-Janzī (d. 549/1154), 74
    - Ibn Tūmart (d. 524/1130), 77
    - ʿAyn al-Quḍāt al-Hamadhānī (d. 525/1131), 81
    - The Anonymous Author of *The Lion and the Diver* (*al-Asad wa-l-ghawwāṣ*), 87

3. Al-Ghazālī on the Role of *falsafa* in Islam, 97
The Refutation of the *falāsifa* in the *Incoherence* (*Tahāfut*), 98
Al-Ghazālī's *fatwā* against Three Teachings of the *falāsifa*, 101
Unbelief and Apostasy, 103
The Decisive Criterion (*Fayṣal al-tafriqa*), 105

4. The Reconciliation of Reason and Revelation through the "Rule of Interpretation" (*Qānūn al-taʾwīl*), 111
Three Different Types of Passages in Revelation, 115
A Dispute about al-Ghazālī's Approach: Ibn Ghaylān versus Fakhr al-Dīn al-Rāzī, 116
Demonstrative Knowledge (*burhān*) and Its Opposite: Emulation of Authorities (*taqlīd*), 120

5. Cosmology in Early Islam: Developments That Led to al-Ghazālī's *Incoherence of the Philosophers*, 123
Ashʿarite Occasionalism in the Generations before al-Ghazālī, 124
Secondary Causes in Ashʿarite Theology, 128
The *falāsifa*'s View of Creation by Means of Secondary Causality, 133
The *falāsifa*'s View That This World Is Necessary, 141
Al-Ghazālī's Treatment of Causality in MS London, Or. 3126, 143

6. The Seventeenth Discussion of *The Incoherence of the Philosophers*, 147
The First Position: Observation Does Not Establish Causal Connections, 150
The First Approach of the Second Position: How the Natural Sciences Are Possible Even in an Occasionalist Universe, 153
The Second Approach of the Second Position: An Immanent Explanation of Miracles, 156
Overcoming Occasionalism: The Third Position, 157
Julian Obermann's "Subjectivist" Interpretation of the Seventeenth Discussion, 160
Al-Ghazālī's Critique of Avicenna's Conception of the Modalities, 162
The Different Conceptions of the Modalities in *falsafa* and *kalām*, 167
What Does al-Ghazālī Mean When He Claims That Causal Connections Are Not Necessary?, 172

7. Knowledge of Causal Connection Is Necessary, 175
The Dispute over al-Ghazālī's Cosmology, 179
Five Conditions for Cosmological Explanations in the *Incoherence*, 183
Determination by an Unchanging Divine Foreknowledge, 187
Divine Foreknowledge in the *Revival of Religious Sciences*, 190
Prophetical Miracles and the Unchanging Nature of God's Habit, 194
Necessary Knowledge in an Occasionalist Universe, 201

Concomitant Events and Rational Judgments, 204
Experience (*tajriba*) in Avicenna and in al-Ghazālī, 208

8. Causes and Effects in *The Revival of the Religious Sciences*, 215
The Creation of Human Acts, 216
The Conditional Dependence of God's Actions, 222
The Conditions of a Creation That Is the Best of All Possible Creations, 225
The Necessity of the Conditions in God's Creation, 231

9. Cosmology in Works Written after *The Revival*, 235
God's Creation as an Apparatus: The Simile of the Water Clock, 236
Cosmology in *The Highest Goal in Explaining the Beautiful Names of God*, 242
The Niche of Lights: The Philosophers' God as the First Created Being, 245
The Cosmology of the "Fourth Group" in the Veil Section of *The Niche of Lights*, 253
An Ismāʿīlite Influence on the Cosmology in the Veil Section?, 260
Final Doubts about Cosmology: *Restraining the Ordinary People* (*Iljām al-ʿawāmm*), 265

Conclusion, 275

Notes, 287

Bibliography, 361

General Index, 395

Index of Works by al-Ghazālī, 403

Index of Manuscripts, 405

Index of Verses in the Qur'an, 407

Timetable

445/1053	The persecution of Ash'arites in Khorasan (in north-east Iran) begins. Al-Juwaynī emigrates to the Hijaz; Abū l-Qāsim al-Qushayrī is incarcerated in Nishapur, the capital of Khorasan.
Circa 448/1056	Al-Ghazālī is born in Ṭābarān, one of two major towns in the district of Ṭūs in northeast Iran.
455/1063	The death of Sultan Toghril-Bey ends the threat of persecution for the Ash'arites in Khorasan. Alp Arslan becomes his successor; his vizier Niẓām al-Mulk supports Ash'arism. Al-Juwaynī returns to Nishapur and becomes head teacher at the Niẓāmiyya madrasa.
Circa 461/1069	At age thirteen, al-Ghazālī begins his "plunge into the sea of religious sciences." After studying with local teachers in Ṭūs, he enters the Niẓāmiyya madrasa in Nishapur and studies with al-Juwaynī.
465/1072	Alp Arslan is assassinated; his son Malikshāh becomes sultan; Niẓām al-Mulk becomes an even more powerful vizier. According to a statement in one of al-Ghazālī letters, it was at this time that he joins the service of Malikshāh.

Circa 473/1080	Al-Ghazālī composes his first book, *The Sifted* (*al-Mankhūl*).
478/1085	Al-Juwaynī dies in Nishapur. Now or sometime earlier, al-Ghazālī joins the entourage (*muʿaskar*) of Niẓām al-Mulk. He spends most of his time in the Seljuq capital Isfahan.
Jumāda I 484 / July 1091	Al-Ghazālī arrives in Baghdad from Isfahan to take his post as head teacher of the Niẓāmiyya madrasa.
Ramaḍān 485 / October 1092	Assassination of Niẓām al-Mulk on the road between Isfahan and Baghdad.
Shawwāl 485 / November 1092	Death of Malikshāh in Baghdad. His succession is contested between the supporters of his two minor sons Berk-Yaruq and Maḥmūd, who have different mothers. Al-Ghazālī is involved in the negotiations with Terken Khātūn, a Qipchak princess and the mother of Maḥmūd, about the appointment of her son as sultan.
Muḥarram 487 / February 1094	After the death of Maḥmūd and his mother, Terken Khātūn, Berk-Yaruq is declared Sultan of the Seljuq Empire. A day after his appointment, the caliph al-Muqtadī dies in Baghdad. His fifteen-year-old son, al-Mustaẓhir (d. 512/1118), becomes his successor.
Muḥarram 488 / January 1095	Al-Ghazālī completes work on *The Incoherence of the Philosophers*.
Dhū l-Qaʿda 488 / November 1095	Al-Ghazālī suddenly gives up his post at the Niẓāmiyya madrasa, departs from Baghdad, and travels to Damascus.
Summer 489 / 1096	Al-Ghazālī travels from Damascus to Jerusalem.
Dhū l-Qaʿda 489 / October–November 1096	He visits Hebron and vows at the grave of Abraham no longer to serve state authorities. From Hebron, he joins the pilgrimage caravan to Mecca.
Dhū l-ḥijja 489 / November–December 1096	Pilgrimage to Mecca, later to Medina.
Muḥarram 490 / January 1097	Al-Ghazālī returns to Damascus. His host, Abū l-Fatḥ Naṣr, had just died. Al-Ghazālī stays only a few months and embarks on his return to Baghdad.
Jumāda II 490 / May–June 1097	Al-Ghazālī arrives in Baghdad. He reads from his *Revival of the Religious Sciences*.

490/1097	Berk-Yaruq appoints his half brother Sanjar as governor (*malik*) of Khorasan. He reigns in Khorasan until a few years before his death in 552/1157.
Fall of 490 / 1097	Al-Ghazālī leaves Baghdad for Khorasan.
Dhū l-ḥijja 490 / November 1097	Al-Ghazālī arrives in Khorasan. He establishes a madrasa and a *khānqāh* in Ṭābarān-Ṭūs where he teaches numerous students.
497/1104	Berk-Yaruq agrees to a division of power with his half brothers Muḥammad Tapar and Sanjar. Muḥammad Tapar becomes ruler of northwestern Persia, the Jazira, and Syria, while Sanjar remains in Khorasan, acknowledging Berk-Yaruk as the supreme sultan.
498/1105	Death of Berk-Yaruq. Muḥammad Tapar becomes the supreme sultan in Isfahan. Sanjar remains his governor (*malik*) in Khorasan.
Summer 499 / 1106	Under pressure from Sanjar and assurances from his vizier, Fakhr al-Mulk, al-Ghazālī begins to teach at the Niẓāmiyya madrasa in Nishapur.
Muḥarram 500 / September 1106	Ismāʿīlite agents asassinate Fakhr al-Mulk.
around 502/1109	Al-Ghazālī is summoned before Sanjar and confronted with accusations of his adversaries. A short time later, he composes the Persian mirror for princes *Council for Kings* (*Naṣīḥat al-mulūk*).
Fall of 504 / 1110	After the death of al-Kiyāʾ al-Harrāsī in Baghdad, al-Ghazālī is offered the same position at the Baghdad Niẓāmiyya that he left fifteen years earlier. He declines in a widely publicized letter.
14 Jumāda II 505 / 18 December 1111	Al-Ghazālī dies in his *khānqāh* in Ṭābarān-Ṭūs.

Al-Ghazālī's Philosophical Theology

Introduction

Today people both in the West and in the Muslim world think of Islamic civilization as a phenomenon of the past. We assume that like the ancient Egyptian or Roman civilizations, the Islamic civilization had a "Golden Age," a period of prosperity and discovery from the second/eighth to the sixth/twelfth centuries that was followed by decline and the rise of another, more innovative civilization. This later civilization is usually referred to as "the West," a vague term that includes the achievements of Galileo Galilei and Christopher Columbus just as much as the development of the personal computer and the Internet search engine. Since the eighteenth century, scholars in the West who have examined the reasons for the end of Islam's "Golden Age" often focus on the differing roles of philosophy in these two cultures.[1] In the West, philosophy and the production of rational arguments have always been regarded as motors that triggered and accelerated the development of new ideas and technologies. It was assumed, however, that although in the Islamic world philosophy grew tremendously during its Golden Age, later scholars in Muslim societies abandoned the study of philosophy and turned their attention toward religious scholarship. During the nineteenth century, Western researchers of Islam developed a by-now well-established account of the fate of philosophy in Islam, postulating that Islamic civilization became acquainted with the tradition of Greek philosophy during the second/eighth and third/ninth centuries, when many philosophical works—most important the writings of Aristotle and their ancient commentaries—were translated into Arabic. These translations triggered the development of a philosophic movement in Islam known in Arabic as *falsafa* (from the Greek word *philosophía*). This movement was not limited to Muslims, and included Christian,

Jewish, and even some pagan authors. It benefited from the open-mindedness and curiosity about other societies that characterizes the early Islamic period. Although *falsafa* developed in Islamic society, it quickly became subject to the harsh criticism of a conservative group of Muslims. Still, until the fifth/eleventh century, the philosophical movement in Islam was able to generate significant support among scholars, literates, and, most important, caliphs and local rulers who patronized their works. During these years, *falsafa* and its critics existed side-by-side among the numerous intellectual movements of classical Islam. According to the traditional understanding, which dominated Western Islamic studies through the nineteenth and most of the twentieth centuries, philosophy ceased to exist in Islam after the sixth/twelfth century. It was assumed that some of the instigating factors for its disappearance were political, such as a lack of patronage from local rulers; some of them economical, such as an assumed demise of the city economies after the arrival of nomadic Turks in the mid-fifth/eleventh century; and some of them educational, such as the beginning of a state-sponsored system of religious seminaries (*madrasas*) that supposedly favored traditionalist religious scholarship and put obstacles to the pursuit of the rational sciences. According to this account, these factors led to the demise of rational science and philosophy under Islam.

Tjitze J. de Boer's *History of Philosophy in Islam*, published in German in 1901 and in English two years later, was the first textbook on this subject. It ends its presentation—apart from an appendix on the thought of Ibn Khaldūn (d. 808/1406)—with Averroes (Ibn Rushd), who died in 595/1198. De Boer realized that, after Averroes, there were philosophical teachers and students "by hundreds and by thousands." Yet these were mostly epitomists—that is, authors who only commented on early works without themselves contributing original thoughts—he says, and after Averroes, "philosophy was not permitted to influence general culture or the condition of affairs."[2] In a widely read article of 1916, Ignaz Goldziher analyzed the attitude of Muslim theologians toward the rationalist sciences. He concluded that although there had always been opposition to rational science among the theologians of Islam, after the fifth/eleventh century, this opposition manifested itself much more forcefully. In the case of philosophical logics, for instance, he concluded that "from this period on, the study of logic was more or less decisively considered to be part of the category of *haram* (forbidden)."[3] In an earlier article, Goldziher had already said that with Averroes, the history of philosophy in Islam had come to an end.[4]

When new sources and fresh studies corrected Goldziher's view that the study of logic fell into decay after the fifth/eleventh century—there was indeed a blossoming of logics in the three subsequent centuries—this traditional view of the fate of philosophy in Islam did not change. In the 1960s, for instance, George Makdisi argued that the main current in Muslim theological thought after the fifth/eleventh century was represented by conservative traditionalists such as Ibn Taymiyya (d. 728/1328), who opposed *falsafa*.[5] In a well-received introductory textbook on Islam, Jonathan Berkey wrote in 2003 that between the fifth/eleventh and ninth/fifteenth centuries, the rational sciences such as philosophy and logics tended to become marginalized from what he calls "Sunni intellectual mainstream."[6]

The influential Muslim theologian al-Ghazālī (d. 505/1111) has always played a leading role in Western attempts to explain the assumed decline of philosophy in Islam. In his work *The Incoherence of the Philosophers* (*Tahāfut al-falāsifa*), al-Ghazālī criticizes twenty teachings of the Muslim philosophers. According to al-Ghazālī, three of those twenty teachings not only are unproven but also violate central tenets of Islam that all Muslims have agreed upon. For al-Ghazālī, these three teachings mark a departure from Islam. These are the views (1) that the word has no beginning in the past and is not created in time, (2) that God's knowledge includes only classes of beings (universals) and does not extend to individual beings and their circumstances (particulars), and (3) that after death the souls of humans will never again return into bodies. In these three cases, the teachings of Islam, which are based on revelation, suggest the opposite, al-Ghazālī says, and thus overrule the unfounded claims of the Muslim philosophers. Those people who actively propagate these three teachings cannot be regarded as Muslims, he says. Rather, they are apostates from Islam and—according to a ruling of Islamic law—subject to the death penalty.[7]

The fact that the alleged end of the philosophical tradition in Islam largely coincided with al-Ghazālī's condemnation of 487/1095, or happened within the next three generations, triggered the suggestion that his verdict contributed to or even caused the disappearance of philosophy in Islam. Solomon Munk, author of the first comprehensive history of Arabic and Islamic philosophy, set the tone of the debate when in 1844, he wrote that with his *Incoherence*, al-Ghazālī "struck a blow against philosophy from which it never recovered in the Orient."[8] Soon thereafter, Ernest Renan described al-Ghazālī as an enemy of philosophy who set off its persecution. According to Renan, a war was waged against philosophy in all lands of Islam during the century following al-Ghazālī's condemnation.[9] For Ignaz Goldziher, by the time of al-Ghazālī, the practice of philosophy in the heartlands of Islam had already weakened so much that the critique in his *Incoherence* was a mere *coup de grace* to an already ailing tradition. After al-Ghazālī, Goldziher continues, "we find the philosophical works every now and then on the pyre."[10] Goldziher was, of course, the most influential teacher in the formative period of Islamic studies in the West, and numerous statements of a similar kind appear there during the twentieth century. These comments still represent a good part of the more popular understanding of al-Ghazālī's position in Muslim intellectual history among contemporaries. William M. Watt, for instance, who shaped the historiography of Islamic thought for a whole generation of scholars studying Islam, acknowledged in 1962 that al-Ghazālī had brought together philosophy and theology. Watt, however, limited this fusion to al-Ghazālī's introduction of syllogistic logic into the Muslim theological discourse. In his *Incoherence of the Philosophers*, Watt wrote, al-Ghazālī argued powerfully against the philosophers, "and after this there was no further philosopher of note in the eastern Islamic world."[11]

Already in 1937, however, less than forty years after de Boer described Averroes as the last philosopher of note in Islam, Shlomo Pines remarked that it is a widespread but hasty generalization to assume that al-Ghazālī's polemics dealt a death blow to philosophy in Islam.[12] In Pines's view, there was no decline of the rational sciences and of philosophy after al-Ghazālī. There was no lack of new ideas under Islam, Pines wrote, although the tendency to maintain old systems of thought and the stability of the scientific environment led to a more gradual development of ideas than in Europe, where fundamental conceptions were periodically revised and sometimes discarded. Science in Islam included a large number of elements of diverse origin, Pines maintained, and it integrated Oriental, Persian, Indian, and Greek influences: "In its further development, it did not, as a rule, eliminate one of them; it led them to subsist side by side—or on different planes."[13] In Islam, there was a trend toward syncretism, in which elements of *kalām*, *falsafa*, and Sufism would appear within one and the same thinker. In 1974, Alessandro Bausani added an interesting observation: while in the medieval West, the mainstream of scientific discourse—the scientific orthodoxy, so to speak—was dominated by a systematic Aristotelian approach and the progressive trends, that is, the scientific heterodoxy, were often radical anti-Aristotelian, in Islam, these roles were reversed. Here, the orthodoxy included various trends of anti-Aristotelianism, and it developed a flexible and syncretistic approach to the methods of science. The much more static situation in the West was—Bausani adds: paradoxically—one of the reasons for its progress, since it was forced to change its approaches in the methods of science radically: "It is much more difficult to be radically revolutionary, if one is confronted by a comparatively more progressive establishment!"[14]

In 1987, Abdelhamid I. Sabra would give another vocal expression to the notion that mainstream Islam integrated the Greek philosophical tradition rather than excluded it. In a seminal article, Sabra argued that after a period of appropriation of the Greek sciences in their translation from Greek to Arabic and in the writings of the *falāsifa* up to Avicenna (Ibn Sīnā, d. 428/1037), philosophy and the Greek sciences were naturalized into the discourse of *kalām* and Islamic theology.[15] The discipline of *kalām*, that is, Muslim rationalist theology as it had been developed by the Muʿtazilites and continued by Sunni and Shiite schools of thought, most prominently the Ashʿarites, offered a new homeland to *falsafa*. The situation was in this respect similar to that in the Latin Middle Ages, during which the study of philosophy could not be distinguished from Christian theology. The discourse of Islamic theology integrated the tradition of *falsafa* so much so that Muslim theologians such as Fakhr al-Dīn al-Rāzī (d. 606/1210), Naṣīr al-Dīn al-Ṭūsī (d. 672/1274), and many other scholars of this period must be considered philosophers as well as theologians.[16] They studied the works of the philosophical tradition in Islam, most notably those of Avicenna; composed comments on these works; discussed the philosophers' teachings; and often adopted positions that were developed by one of the earlier *falāsifa*. With highly original theories, many of the later Islamic philosophers such as al-Suhrawardī (d. 587/1191) or Mullā Ṣadrā Shīrāzī (d. 1050/1640) founded their own philosophical schools.

Until now, scholars have been divided as to al-Ghazālī's place in this process of the naturalization of the Greek sciences into the discourse of Islamic theology. Given that he criticizes twenty teachings of the *falāsifa* in his *Incoherence of the Philosophers* and even condemns three of them as apostasy from Islam, must we say that the naturalization and effective integration of the philosophical discourse in Islam happened despite al-Ghazālī? Or, rather, should we think of al-Ghazālī as a thinker who stands at the center of developments in Islamic theology and whose *Incoherence* and subsequent works on Islamic theology are, in fact, a vital part of this process?

In this book I will explain why al-Ghazālī is indeed the first Muslim theologian who actively promotes the naturalization of the philosophical tradition into Islamic theology. His works document an attempt to integrate Aristotelian logics into the tradition of *kalām*, of rationalist Islamic theology. Al-Ghazālī tirelessly stresses the merits of syllogistic logics and urges his peers in Islamic theology to adopt this rational technique. He was quite outspoken about this project and propagates it, for instance, in his autobiography, *The Deliverer from Error* (*al-Munqidh min al-ḍalāl*) as well as in the *Incoherence* and this aspect of al-Ghazālī's relationship to *falsafa* is well known.[17] Some critics and interpreters of al-Ghazālī have questioned how he could make use of Aristotelian logics without also adopting Aristotelian ontology.[18] In the Aristotelian tradition, logic is so closely connected to the specific explanation of the world's most elementary constituents and their relations to one another that Aristotelian logic can hardly be adopted without Aristotelian ontology. Al-Ghazālī understood this connection very well, and while propagating learning logics from the *falāsifa*, he knew that he was also asking his peers to subscribe to fundamental assumptions that would change their positions on ontology and metaphysics. About this, however, al-Ghazālī was less open. When he summarized his views about the metaphysics of the *falāsifa* in such popular works as his autobiography, he turns his criticism of metaphysics to the fore and mentions his appreciation of their teachings only in passing.[19] Yet a thorough study of al-Ghazālī's works on theology leaves no doubt that his views on ontology, the human soul, and prophecy are particularly shaped by Avicenna.[20] Furthermore, even the aforementioned condemnation of three philosophical teachings in the *Incoherence of the Philosophers* was actually a part of the naturalization of Aristotelian philosophy into Muslim theology. With this condemnation, the book identifies those elements of Aristotelianism that were, according to al-Ghazālī, unfit to be integrated. By highlighting these three teachings, the great Muslim theologian opened the Muslim theological discourse to the *many other* important positions held by the *falāsifa*.

This book approaches the subject of al-Ghazālī's philosophical theology from two angles, offering a close study both of his life and of his teachings on cosmology. I have chosen these two subjects because I believe they currently pose the greatest obstacle for positioning al-Ghazālī as someone who contributed to the process of the naturalization of *falsafa* within the Islamic theological discourse. With regard to the study of al-Ghazālī's life, the currently prevailing views in Western scholarship are very much shaped by his own report in his autobiography, *The Deliverer from Error*. English scholarship does not adequately represent key additional sources on his life and work, such as reports from his students and the collection of his Persian letters. These additional sources became available during the mid-twentieth century, and they contain a wealth of information that settles many remaining uncertainties about the chronology of al-Ghazālī's actions and whereabouts. In particular, the collection of Persian letters illuminates many details about the circumstances surrounding the last fifteen years of his life. For example, as in his autobiography, al-Ghazālī often refers in his letters to the crisis that led to his departure from Baghdad in 488/1095. Yet in these Persian letters, al-Ghazālī also mentions another event, one that we must consider just as important as the departure from Baghdad: in Dhū l-Qaʿda 489 / October–November 1096, about a year after his departure from Baghdad, al-Ghazālī vowed at the tomb of Abraham in Hebron never again "to go to any ruler, to take a ruler's money, or to engage in one of his public disputations."[21] Although in his autobiography, he portrays the dramatic process that led to his departure from Baghdad in bright colors, he never mentions the vow at Hebron. This omission can be seen as connected to his contemporaries having accused him of breaking this vow, so he had little interest in reminding his readers of it. Leaving Baghdad and vowing not to cooperate with the representatives of state authority are, of course, two events that belong together, although a reader of al-Ghazālī's autobiography may not understand the connection. The distance of eleven years between al-Ghazālī's decision to leave Baghdad and his writing of the autobiography created this significant change in the representation of that event. Reading the letters and studying the comments of his students gives a much clearer picture of what triggered his decision to leave his post at the Niẓāmiyya *madrasa* in Baghdad.

Al-Ghazālī's change from one of the most successful and visible intellectuals in Baghdad to someone who shunned fame and lived withdrawn at his birthplace in the Iranian province has always captured the public imagination. It allowed the idea to emerge that there are two or even more al-Ghazālīs speaking in his works. Many interpreters also sensed that al-Ghazālī's relationship to *falsafa* was more ambiguous than he admitted in his autobiography. The idea that al-Ghazālī's teachings underwent a significant change during his life has been put forward so often that it has become part of the scholarly as well as more popular impression about his œuvre. In 1994, however, Richard M. Frank observed that there was no notable theoretical development or evolution in al-Ghazālī's theology between his earliest works, which were published before his departure from Baghdad in 488/1095, and his last.[22] Frank is right about this; there is hardly any evidence to support the widely held view that al-Ghazālī changed some of his positions after his departure from Baghdad and that he moved away from being a more *kalām*-oriented theologian toward being a Sufi. Although it is true that some motifs appear more prominently in al-Ghazālī's work after his departure from Baghdad—for instance, the concern about gaining salvation in the afterlife—none of them are absent from the early works, and to say that al-Ghazālī's theological teachings underwent a change cannot, in fact, be maintained.

One of my main interests in studying al-Ghazālī's life was to find out whether the current popular view about his changing from being a *mutakallim* (a Muslim rationalist theologian) and opponent of *falsafa* before departing Baghdad to being a Sufi, one who shunned *kalām* and worked to reconcile Sufism with Muslim orthodoxy and maybe even with *falsafa* can be supported by any of the most authoritative sources on his life. And although these sources do indeed talk about changes in al-Ghazālī's life, none of them reports that his teachings have significantly changed. ʿAbd al-Ghāfir al-Fārisī (d. 529/1134), one of his colleagues and contemporaries, reports eloquently how the intellectual arrogance of the young al-Ghazālī changed to a much more balanced personality.[23] Yet this subtle maturation is not the change from a dogmatic theologian to a mystic that many modern accounts talk about. Indeed, this same contemporary tells us that al-Ghazālī received a thorough introduction into Sufism from his master al-Fāramadhī (d. 477/1084) before he was thirty. This was at least ten—and probably many more—years before the dramatic crisis reported in his autobiography. In that work, al-Ghazālī pictures his departure from Baghdad as a more or less sudden effect of his discovery of Sufi literature. One of his students, Abū Bakr ibn al-ʿArabī (d. 543/1148), informs us that this process was not at all sudden. The student mentions that already two years before his departure from Baghdad, al-Ghazālī had "accepted the Sufi path and made himself free for what it requires."[24] All these accounts should lead us to reevaluate al-Ghazālī's own narrative of his crisis in 488/1095, which has thus far dominated all Western biographies of him.

Al-Ghazālī's teachings on cosmology are currently the biggest obstacle to a coherent understanding of his theology. The word "cosmology" refers to views about the most elementary constituents of the universe and how they interact with one another, if, in fact, they are assumed to do so. In the case of al-Ghazālī, who teaches that God creates every being and every event in the universe, cosmology refers to *how* God creates the world and *how* He relates to His creation. In Western scholarship, the problematic nature of al-Ghazālī's teaching on cosmology was raised soon after 1904, when his work *The Niche of Lights* (*Mishkāt al-anwār*) first appeared in print. In that book, al-Ghazālī chooses language reflecting and implying cosmological principles that were developed by philosophers and that had not appeared in any earlier work by a Sunni theologian. The teachings in *The Niche of Lights* also seem to be at odds with those in his other works, most significantly in his *Balanced Book on What-to-Believe* (*al-Iqtiṣād fī l-iʿtiqād*).[25] Within the next thirty years, scholars such as W. H. T. Gairdner, Arent J. Wensinck, and Miguel Asín Palacios documented these differences, yet they could not provide much of a reconciliation or an explanation. During the second half of the twentieth century, with the works of William M. Watt, Hava Lazarus-Yafeh, and others, Western scholars attempted to solve this puzzle by excluding the most problematic texts, those most at odds with the established teaching from an accepted corpus of al-Ghazālī. Lazarus-Yafeh argued that those works that use a distinctly philosophical language are spurious and should not be attributed to the great Muslim theologian, as Watt argued regarding a specific chapter in *The Niche of Lights.*[26] I find their arguments unpersuasive; it seems rather improbable that individual chapters could be added to a work of a prominent scholar such as al-Ghazālī if he had published this particular work during his lifetime. Classical Muslim scholarship greatly respected the textual tradition of an author's work; manuscripts were checked for their accuracy by comparing and collating them with other copies of the same work.[27] The author and many of his readers had an interest in safeguarding the integrity of his published works. Collectively, they would have been able to identify mistakes in the manuscript tradition even centuries after a book had been put on the market. First published in 1966, Hava Lazarus-Yafeh's argument that books that use philosophical terminology cannot have been authored by al-Ghazālī is methodologically problematic.[28] Lazarus-Yafeh observed that philosophical terms are absent from those works universally accepted as authored by al-Ghazālī, leading to her assumption that any usages of philosophical language are later and inauthentic additions to the Ghazalian corpus. Since many of al-Ghazālī's interpreters were reluctant to acknowledge that he may have occasionally used philosophical language, any use of such language, Lazarus-Yafeh argues, can be used to discredit the authenticity of his writings. Lazarus-Yafeh rejected al-Ghazālī's authorship of books that use philosophical language simply because there had always been scholars who had rejected those aspects of his thought, not because the passages were themselves problematic.

New controversy entered the study of al-Ghazālī in 1992 when Richard M. Frank suggested that al-Ghazālī had abandoned the cosmological system developed by the Ash'arite school of Muslim theology, the school tradition from whence he came, and that he had adopted the cosmology of Avicenna. Frank said that al-Ghazālī ceased to believe that God creates every event in the world directly and immediately, as Ash'arites believed before him. Rather, he subscribed to the philosophical explanation that God's creative power reaches the objects of creation through chains of intermediaries and secondary causes. Celestial intellects that reside in the nine heavenly spheres mediate the divine creative activity to the sublunar sphere, in which chains of secondary causes and their effects unfold. These causes create change according to their natures (*ṭabāʾiʿ*) and make God's performance of prophetical miracles impossible, at least in the way they were understood by Muslim theologians. According to Frank's analysis, al-Ghazālī no longer believed that God performs miracles to verify the claims of His prophets.[29] Yet the existence of prophetical miracles is one of the most fundamental elements of classical Ash'arite theology, and they are, at least according to Ash'arite theology before al-Ghazālī, a clear necessity of their theological system.[30]

In several articles published before and after 1992, Michael E. Marmura advanced the position that al-Ghazālī never broke with any fundamental principles of Ash'arite theology, remaining faithful to its cosmology. Based on a solid documentation, Marmura rejects Frank's results. Perhaps one could argue that al-Ghazālī wrote two types of works, one that supports Frank's analysis of a philosophical cosmology and one that provides evidence for Marmura's interpretation that he still applied the traditional Ash'arite cosmology. But indeed, the problem runs even deeper: Frank and Marmura use some of the same works to underline their theses. Apparently, the same texts by al-Ghazālī could be interpreted either as Frank or as Marmura interprets them.[31]

My own interest in al-Ghazālī's cosmology began in the summer of 1993 when Frank's work fell into my hands. I was doing my mandatory civil service in one of the academic backwaters of Germany and combed the local library for some interesting reading. Had Frank's study not been published by a German academic publisher, it would have never arrived there. Reading his *Creation and the Cosmic System* changed my academic interests. After studying Frank's book and returning to the Freie Universität Berlin, I wrote my master's thesis on Avicenna's influence on al-Ghazālī's *Decisive Criterion* (*Fayṣal al-tafriqa*). In turn, this research led me to focus on al-Ghazālī's condemnation of three philosophical teaching in his *Incoherence of the Philosophers* for my Ph.D. I was fascinated by the legal and theological development in Islam that had led to al-Ghazālī's harsh condemnation. In my book *Apostasie und Toleranz im Islam*, I present the development that led to al-Ghazālī's verdict on philosophy, and I document some of the reactions from the side of the philosophers.

In recent years, I have returned to the problem of cosmology, aiming to resolve the academic impasse between the different interpretations put forward by Frank and Marmura. Although I was first drawn to this subject through Frank's work, the reader will note that my current conclusions about al-Ghazālī's cosmology differ widely from Frank's conclusions. The path my results have taken from those of Frank and Marmura seems to me a fitting example of what G. W. Hegel called a dialectical progression. While Frank's and Marmura's works are the thesis and the antithesis (or the other way round), this book wishes to be considered a synthesis. In truly Hegelian fashion, it does not aim to reject any of their work or make it obsolete. Rather, its aim is the *Aufhebung* of these earlier contributions in all meanings of that German word: a synthesis that picks up the earlier theses, elevates them, dissolves their conflict, and leads to a new resolution and progress.

In this book, I try to offer a consistent interpretation of the different motifs in al-Ghazālī's thinking about how God creates the world and how He governs over it. Of course, this interpretation is not the only possible way to read al-Ghazālī, as we saw from Marmura and Frank. Yet I believe that these other readings do not give appropriate attention to all the motifs that al-Ghazālī considered important. Frank, for instance, accuses al-Ghazālī of being deceptive when he writes that God is a free agent who has a free choice in His actions. Marmura neglects to take full account of the handful of passages in which al-Ghazālī writes that God's actions are necessary. I present a reading that tries to reconcile these two apparently contradictory statements—and some other statements in al-Ghazālī's works that seem equally irreconcilable at first.

Perched between the Ashʿarite and the Avicennan poles, al-Ghazālī develops his own cosmology. Al-Ghazālī was a very systematic thinker, and given that the Avicennan system is much more systematic than the Ashʿarite one, it is unsurprising that his synthesis owes much more to Avicenna than to al-Ashʿarī. Through his analysis, he finds a very elegant path toward adopting

Avicenna's determinist cosmology while remaining a Muslim theologian who wishes to preserve God's free choice over His actions. Al-Ghazālī's solution as to how a theologian might adopt a deterministic cosmology is just as relevant today as it was at the turn of the sixth/twelfth century. Modern cosmology has become part of physics, yet contemporary cosmological systems leave room for the belief that given the existing laws of nature and an existing configuration of energy at the starting point of this universe—usually referred to as the Big Bang—all later developments of subatomic particles, atoms, galaxies, stars, planets, life on some planets, humanity, and even me, is, in fact, a necessary effect of the first moment and could not have been altered once the process started 14 billion years ago.[32] As a theologian, al-Ghazālī would have accepted this determinist statement. In fact, his view of the universe was quite similar, though defined by the parameters of Ptolemy's geocentric cosmos in which the beginning of the world is marked not by the Big Bang but by the *primum mobile* (*falak al-aflāk*), the outermost, starless sphere and the intellect that governs it.[33] Yet despite his determinist view of the universe, al-Ghazālī tirelessly maintained that God acts freely and that He is the only "maker" or efficient cause in the whole universe. Every event, even the beating of a gnat's wing, is willed and created by Him.

This book is divided into two main parts. The first part is a close study of the sources on al-Ghazālī's life, and the second offers an analysis of his cosmology. Al-Ghazālī's cosmology introduces us to a wider range of philosophical and theological subjects. Though the Ashʿarite and the Avicennan positions on cosmology have mutually exclusive views of how God relates to His creation, they share many similarities as to the consequences these two cosmologies have on God's creatures. It is this similarity that al-Ghazālī exploits when he develops something like a synthetic position between these two poles. His views on the conflict between human free will and divine predestination, on the generation of human acts, on prophecy, on the parallels between the human microcosm and the macrocosm of the universe, and on the question of whether God could have created a better world than this are all connected to the position he takes on cosmology. All these subjects will be discussed in the second part of this book.

The book is divided into nine chapters. Chapters one and two belong to the first part of the book, covering al-Ghazālī's life and his most important students and early followers. The second part starts with the third chapter. Relatively short, the third and fourth chapters lay the groundwork for a thorough analysis of al-Ghazālī's treatment of causality by explaining his position on the role of *falsafa* in Islam as well as his "Rule of Interpretation" (*qānūn al-taʾwīl*), the epistemological principle that al-Ghazālī applies to cases in which the results of a demonstrative argument clash with the literal wording of revelation. In these two chapters I summarize results of my previous studies, in particular, my German book *Apostasie und Toleranz im Islam*. The following chapters are again original, proceeding almost chronologically through the different texts that al-Ghazālī wrote on matters relating to cosmology. After explaining the relevance of cosmology in Muslim theology and the theological problems related to it in the fifth chapter, I discuss al-Ghazālī's *Incoherence of the Philosophers* in the sixth chapter. This is his first and most comprehensive treatment of the issue. Chapter seven discusses works stemming from the *Incoherence* that respond to questions left open in al-Ghazālī's very programmatic critique of philosophy. These are mostly his *Standard of Knowledge* (*Miʿyār al-ʿilm*) and his *Touchstone of Reasoning* (*Miḥakk al-naẓar*). Chapter eight takes a close look at the *Revival of the Religious Sciences* (*Iḥyāʾ ʿulūm al-dīn*), in which the most explicit statements on cosmology and the generation of human acts can be found in its thirty-fifth book, "Belief in God's Unity and Trust in God" (*Kitāb al-Tawḥīd wa-l-tawakkul*). The ninth chapter deals with works that were published after the *Revival*. Here I focus on three subjects: Al-Ghazālī's famous comparison between God's universe and a water clock, the equally famous "Veil-Section" in *The Niche of Lights*, and his last and probably most explicit statement about how God creates and how He acts upon His creation in *Restraining the Ordinary People from the Science of kalām* (*Iljām al-ʿawāmm ʿan ʿilm al-kalām*), a book finished only days before he died.

Initially, when I started writing this book, I planned to include an inventory of those texts of al-Ghazālī that are relevant to his theology. The file of that inventory kept expanding; and when I realized that any satisfactory treatment of such a list would make up more than half of this book, I postponed its publication and decided not to load this work with a heavily footnoted bibliographical study requiring detailed analysis of numerous manuscripts. I also realized that it would have taken much more time and effort than I first thought. Maurice Bouyges, who undertook a pioneering work on the cataloging and dating of al-Ghazālī's writings during the 1920s, did not think that he had ever finished the task, and his bibliographic study was published not until after his death in the 1950s. Since Bouyges's study, numerous new texts and manuscripts have become available that support many of his findings while challenging others. There was also a significant change of opinion among scholars of al-Ghazālī about the range of teachings his texts support. The canon of al-Ghazālī's works is a particularly contested subject. Although there is an acknowledged core of writings unequivocally ascribed to him, numerous texts attributed to him in manuscripts are not fully accepted as genuine. Carl Brockelmann, Maurice Bouyges, ʿAbd al-Raḥmān Badawī, and other researchers formed their opinions about the Ghazalian corpus largely on the basis of the work undertaken by the Muslim bibliographer Ḥājjī Khalīfa (d. 1067/1657) and by other, earlier bibliographers such as Tāj al-Dīn al-Subkī (d. 771/1370) or al-Wāsiṭī (d. 776/1374). Since the 1950s, new methods of determining which of the works attributed to all-Ghazālī were actually composed by him have been suggested, but most have not been very successful.

I believe this particular field of study has always suffered from a certain lack of understanding of what al-Ghazālī truly teaches in his core texts. The most reliable method of determining the authenticity of works that are not unanimously accepted as being those of al-Ghazālī is to develop a detailed understanding of the teachings in the core group and use this understanding as a yardstick to measure the ambiguous works. Given that most of the works of doubtful authenticity such as *The Epistle on Intimate Knowledge* (*Risāla Fī l-ʿilm al-ladunī*), *The Book to Be Withheld from Those for Whom It Is Not Written* (*al-Maḍnūn bihi ʿalā ghayri ahlihi*), or *Breathing of the Spirit and the Shaping* (*Nafkh al-rūḥ wa-l-taswiya*)—a work also known as the *Small Book to Be Withheld* (*al-Maḍnūn al-ṣaghīr*)—can be described as Avicennan texts,[34] it is important to understand the distinctive markers of al-Ghazālī's theology and philosophy and how they differ from those of Avicenna. This study argues that al-Ghazālī's theology and philosophy are a particular kind of Avicennism. Only a thorough understanding of its precise kind of Avicennism will allow us to determine the authenticity of the disputed works.

In order to start this first step and establish the teachings from the core group of al-Ghazālī's books, I have limited this study to those of his works unanimously regarded as genuine by the aforementioned bibliographical authorities. Al-Ghazālī refers to all of these works in his other writings, thus creating a network of authentic texts.[35] A further methodological question is how to obtain and verify reliable textual versions of these core works, a difficult task given that only one of al-Ghazālī's books, *The Incoherence of the Philosophers*, is critically edited, while a number of others, such as *The Balanced Book on What-to-Believe*, *The Highest Goal* (*al-Maqṣad al-asnā*), *The Choice Essentials* (*al-Mustaṣfā*), and *The Deliverer from Error* (*al-Munqidh min al-ḍālāl*) are available in reliable "semi-critical" editions that use many manuscripts but neglect to compare their importance relative to one another.[36] Other works have been edited uncritically, yet their editors made efforts to compare the text they print to more than just one manuscript source and to base it on a random sample of three or more manuscripts or earlier prints. In many cases, however, we simply have no idea how the text that we find in print has been established. We must trust the claims of the editors that they faithfully present one or more manuscripts. These claims are sometimes quite portentous, as in a 1910 print in which the meritorious editor asserts "that the manuscript on which this printing is based is among the most important ones, written by the hand of one of the great Muslim scholars during the seventh Islamic century (13th century CE)."[37]

Unfortunately, many of the prints of al-Ghazālī's works are not fully reliable when it comes to textual details. As an example, the most widely used edition of al-Ghazālī's *Decisive Criterion for Distinguishing Islam from Clandestine Unbelief* (*Fayṣal al-tafriqa bayna l-Islām wa-l-zandaqa*) is the one by the respected Azhar scholar Sulaymān Dunyā of 1961. That edition, however, is not based on an independent study of manuscript evidence but takes its text from an earlier edition of 1901 that is a collation of three manuscripts from Egypt and Damascus. This amalgamation can lead to ambiguities, as when Dunyā reproduces a passage that says the unbelief (*kufr*) of a Muslim scholar is established when he violates one of the "foundations of the rules" (*uṣūl al-qawāʿid*). This makes little sense, however, and another version of the text, which has "foundations of what-to-believe" (*uṣūl al-ʿaqāʾid*) in this passage, seems to express much better what al-Ghazālī may have had in mind.[38] Of course, without a critical edition that establishes a *stemma codicum*, one can only conjecture. But given that for al-Ghazālī, the unbelief of a Muslim implies the death penalty, finding out what exactly constitutes unbelief is no trivial matter. Among the available editions, the latter reading of the text is established only by Maḥmūd Bījū, who studied two manuscripts of the Ẓāhiriyya Collection in Damascus and published his edition in 1993 in his own small publishing house in Damascus' Ḥalbūnī quarter.[39] In this case, the less widespread edition seems to offer a better text and should be preferred. Realizing that few of my readers currently have access to the better edition, I refer in the footnotes to both Dunyā's and Bījū's texts and explain textual differences where they occur. The same applies to other works by al-Ghazālī such as his *Choice Essentials of the Methods of Jurisprudence (al-Mustaṣfā min ʿilm al-uṣūl)* in which case a recent edition by Ḥamza Ḥāfiẓ is established on the basis of two manuscripts from Istanbul and an early print. This book was published in Jeddah (Saudi Arabia), and because many readers may not have access to it, I also provide page references to the early print, which is more widely available and which, in principle, has become superfluous by the new edition.

Where no critical or semi-critical edition exists and where no thorough study of manuscripts has been undertaken, I prefer to use editions older than the ones that have appeared in recent years. At the beginning of the twentieth century, many of al-Ghazālī's smaller texts were edited for the first time, mostly in Cairo. The printers and scholars who prepared these editions often compared several manuscripts in order to establish the texts.[40] In the great majority of cases, later editions simply reprint these early editions and rely entirely on the manuscript studies undertaken by a group of early editors. Failing to make any improvements, some newer editions add punctuation, commas, paragraph breaks, and sometimes even textual emendations that distort the original. Finally, the new typesetting often introduces new mistakes. In addition to these scholarly concerns, there are two practical reasons why I chose to work with editions that are often almost a century old. First, these editions are no longer protected by copyright, which facilitates their future availability through new media such as the Internet. Second, it can be hoped that the sheer antiquarian value of these prints will guarantee their preservation for future generations, something less definite with more recent printings. Wherever possible, I compare the printed text to a manuscript that has not been used in the process of establishing the print. My preference for older prints implies that when the only edition listed in the bibliography of al-Ghazālī's works at the end of this book is a more recent one, the reader can assume that it has been established on the basis of an original study of manuscripts.

Although I try to work with a text directly established from manuscripts, that principle could not be applied in the case of *The Revival of the Religious Sciences*. Al-Ghazālī's major work on ethics and human behavior was one of the first books of classical Arabic literature printed at the Egyptian viceroy's press in Būlāq. Since 1269/1853, it has been continuously in print.[41] The textual history of the *Revival* is almost completely unknown and urgently needs to be researched. In 1912, Hans Bauer remarked that all available prints of the work seem to generate from the Egyptian *editio princeps*.[42] Its supervising editor,

Muḥammad ibn ʿAbd al-Raḥmān Quṭṭa al-ʿAdawī (d. 1281/1864), asserts that the Būlāq printers took the text "from the best testimonies at the Khedival library."[43] Since Bauer made his remark, however, the text has developed its own variants. Modern prints show small but sometimes significant variations from earlier prints. For instance, the word *mūjiduhu* ("the one who brings it into being") in one passage became *mūjibuhu* ("the one who makes it necessary"), or *ʿaql* ("intellect, rationality") in another became *naql* ("transmitted knowledge, revelation")—a quite considerable change of meaning.[44] Bauer had already discovered that the text included in the *matn*, the cited text, of al-Murtaḍā al-Zabīdī's (d. 1205/1791) commentary to the *Revival* offers a textual testimony independent of the other available prints.[45] Al-Murtaḍā al-Zabīdī collated this text from a number of manuscripts, and he notes their variants. This edition appears more reliable than any of the other available prints of the *Revival*.[46] It has since been used in the translations of Hans Wehr, Nabih Amin Faris, Richard Gramlich, and Timothy J. Winter and should be consulted whenever one attempts to establish the precise meaning of the *Revival*.

In order to encourage further research on the *Revival*, I refer to the text in a way that allows the reader to locate the passage in more than just a single edition. I expect that scholars will eventually adopt a future "standard" edition for ease of reference; in this book, I refer to two editions that are likely to achieve the status of such a standard. The first is a five-volume edition published in 1387/1967 by the Ḥalabī Firm, the successor of Muṣṭafā al-Bābī al-Ḥalabī in Cairo. In 1306/1888, three brothers of the al-Bābī al-Ḥalabī family—Muṣṭafā, Bakrī, and ʿĪsā—started to offer four-volume prints of the *Revival* under their Maymaniyya imprint. Their editions established the by-now canonical practice of printing supplementary texts by al-ʿIrāqī (d. 806/1404), al-ʿAydarūs (d. 1038/1628), Shihāb al-Dīn ʿUmar al-Suhrawardī (d. 632/1234), and al-Ghazālī's own *Dictation* (*Imlāʾ*) alongside the *Revival*.[47] The scholar, editor, and printer Muṣṭafā al-Bābī al-Ḥalabī, who took over the business in 1919, prepared a great number of print runs of the text through the end of the 1930s. He thus responded to the new demand for *Revival* printings created by the educational activity of the Muslim Brotherhood.[48] These four-volume editions have a similar, but unfortunately not identical pagination.[49] Given this possibility for confusion, I opted for the 1967 edition of the Ḥalabī Firm, available in many Western libraries.[50] The second edition I refer to is the sixteen-parts set—originally printed in four volumes—of the Committee for the Distribution of Islamic Culture (*Lajnat nashr al-thaqāfa al-Islāmiyya*). It was published in 1356–57/1937–39.[51]

In the translations from Arabic and Persian, square brackets indicate additions or explanations on my part, while texts in round brackets are clarifications that are required in the English translation in order to avoid ambiguity. In the transliteration of Arabic, I apply the standard of *The Encyclopaedia of Islam*^{THREE}. In the case of Turkish names and names from other non-Arabic and non-Persian languages, I use a less stringent system of transliteration that tries to represent the pronunciation of these names in their original language. Placenames appear the way we usually refer to them in English unless these places no longer exist. In the endnotes, I produce a short reference to the authors and the titles of publications that are listed in the bibliography.

KEY TO THE WORKS THAT ARE CITED WITH MORE THAN ONE PAGE REFERENCE:

al-Ghazālī. *al-Arbaʿīn*, edition Ṣabrī al-Kurdī 1925 / edition Jābir 1964.
———. *Fayṣal al-tafriqa*, edition Dunyā 1961 / edition Bījū 1993.
———. *Ḥimāqat-i ahl-i ibāḥat*, edition Pretzl 1933 / edition Pūrjavādī 2002.
———. *Iḥyaʾ*, edition al-Ḥalabī 1967–68 / Lajnat Nashr al-Thaqāfa edition 1937–39.
———. *Iljām al-ʿawāmm*, edition al-Ḥalabī 1891 / edition al-Baghdādī 1985.
———. *al-Imlāʾ fī ishkālāt al-Iḥyāʾ*, same as *Iḥyāʾ*.
———. *Maqāṣid al-falāsifa*, edition Ṣabrī al-Kurdī 1936 / edition Dunyā 1960.
———. *Mishkāt al-anwār*, edition ʿAfīfī 1964 / edition al-Sayrawān 1986.
———. *Mīzān al-ʿamal*, edition Ṣabrī al-Kurdī 1923 / edition Dunyā 1964.
———. *al-Mustaṣfā min ʿilm al-uṣūl*, edition Ḥamza Ḥāfiẓ 1992–93 / Būlāq edition 1904–7.
———. *Tahāfut al-falāsifa*, edition Bouyges 1927 / edition Marmura 2000.
Ibn al-Muqaffaʿ. *Kalīla wa-Dimna*, edition Cheikho 1905 / edition ʿAzzām 1941.
Ibn Sīnā. *al-Najāt*, edition Ṣabrī al-Kurdī 1938 / edition Dānishpazhūh 1985.
———. *al-Taʿlīqāt*, edition Badawī 1973 / edition al-ʿUbaydī 2002.
Rasāʾil Ikhwān al-ṣāfāʾ, edition Ziriklī 1928 / Dār Ṣādir edition, Beirut.

1

A Life between Public and Private Instruction

Al-Ghazālī's Biography

In the West, al-Ghazālī's life has frequently attracted more attention than his teachings. Every student of Islamic studies knows that at the peak of his career, al-Ghazālī left his prominent teaching position and became a Sufi. In his autobiography, *The Deliverer from Error (al-Munqidh min al-ḍalāl)*, al-Ghazālī presents this transformation in quite dramatic terms. Yet even before the seventeenth century, when this book became known in the West, European scholars were familiar with the inspiring tale of al-Ghazālī's spiritual life. In the first half of the sixteenth century, Catholic scholars at the Vatican asked the Moroccan captive al-Ḥasan ibn Muḥammad al-Wazzān (d. after 957/1550), known as Leo Africanus, to write a book on the lives of the most prominent Arabic philosophers and theologians. His biography of al-Ghazālī is the third longest of the twenty-eight biographies in that book, after those of Avicenna and Averroes—and certainly the most interesting. Al-Ghazālī's rapid rise as a scholar, his financial success, and his sudden decision to become a "hermit" (*eremita*) all figure prominently in this account.[1]

Al-Ghazālī's vocal renunciation in his autobiography of certain attitudes he held earlier in his life has always captured the imagination. At different times in his career, al-Ghazālī was considered a Sufi, a *mutakallim* who refuted *falsafa*, and, to some degree, a genuine philosopher who subscribed to philosophical teachings. This mix created numerous legends about his life. The Algerian Jewish scholar Abraham Gavison (d. 986/1578) spread one of the most curious anecdotes during the sixteenth century. He tells the story—in all earnestness—that during daytime al-Ghazālī composed his *Incoherence of the Philosophers* in response to a request by the ruler, while during the night he worked on his own accord on *The Incoherence of the Incoherence*. This book is the well-known refutation of al-Ghazālī's *Incoherence of the Philosophers* and was actually composed by Averroes (d. 595/1198) almost a century after al-Ghazālī.[2]

In the West, serious source-critical studies of al-Ghazālī's biography have made little progress in the past half-century. About forty years ago, Josef van Ess noted that of the primary sources on his life, the reports of al-Ghazālī's contemporaries and his students had not yet been fully evaluated.[3] Thirty years earlier, in 1938, Jalāl al-Dīn Humā'ī had already presented a remarkable biographic study of al-Ghazālī—written in Persian—that makes full use of the rich information in the collection of his letters.[4] In Western languages, however, the study of al-Ghazālī's life had not yet integrated these findings. The chronology of al-Ghazālī's life established by Maurice Bouyges in the 1920s and translated into Arabic by ʿAbd al-Raḥmān Badawī in 1964 is still the most comprehensive secondary literature available. This chronology—which is also the starting point of George F. Hourani's two articles on the dating of al-Ghazālī's works, published in 1959 and 1984—is more than eighty years old and is based entirely on information provided by al-Ghazālī in his autobiography or by his main biographers.[5] These sources contain substantial lacunae. For instance, considering the writings of al-Ghazālī's student Abū Bakr ibn al-ʿArabī (d. 543/1148) allows us to solve a number of problems in the chronology of al-Ghazālī's life. Abū Bakr ibn al-ʿArabī tells us when al-Ghazālī left Baghdad on his way home to Khorasan, as an example.[6] Even more important are al-Ghazālī's Persian letters, which provide us similarly with the corresponding information about when he arrived in Khorasan. Other biographic problems of concern to earlier generations of al-Ghazālī scholars involve his possible trip to Egypt and his whereabouts during the "ten years of Sufi-wandering"—a particularly deceptive verbal formulation that has caused much confusion. In all these cases, his letters as well as the testimony of his students give clear answers.

"How al-Ghazālī Created His Own Historiography" is the subtitle of ʿAbd al-Dā'im al-Baqarī's landmark study on al-Ghazālī's autobiography, and it captures well the great theologian's attitude toward his biographers.[7] Not only in his *Deliverer from Error* but also in the conversations with his biographer ʿAbd al-Ghāfir al-Fārisī (d. 529/1134) did al-Ghazāli shape the perception of his personality and effectively confuse historians for many centuries. The "ten years of Sufi wandering" are mentioned both in his autobiography as well as in ʿAbd al-Ghāfir's account of his life.[8] They create the impression that he stopped teaching and avoided all forms of public life. In particular, the authoritative nature of ʿAbd al-Ghāfir's biography, who knew al-Ghazālī personally and who based his biography on personal conversations with him, led to this misunderstanding that leaves traces even in most recent scholarship.[9] After al-Ghazālī became a professor at the Baghdad Niẓāmiyya at age thirty-five, he never stopped teaching and writing books. The circumstances under which this teaching took place and those who benefited from it became an important issue during the course of his life, as we will see.

The Main Sources for al-Ghazālī's Biography

In 1971, Dorothea Krawulsky analyzed the entries on al-Ghazālī in the major historical dictionaries of Muslim scholars and luminaries and in the chronicles of his era.[10] She concluded that only a handful of historians contributed original material, while the rest simply repeated the entries of others.[11] The main sources for the life of al-Ghazālī, these historians rely heavily on al-Ghazālī's autobiography, *Deliverer from Error*. Only in the mid-twentieth century did the value of this book as a proper reconstruction of al-Ghazālī's life become a matter of debate.[12] Observations and comments of contemporaries are the second most important source for al-Ghazālī biographers in the classical period. None of the authors of Arabic biographical dictionaries and chronicles use the collection of al-Ghazālī's Persian letters.

Among the classical biographies, the one by ʿAbd al-Ghāfir al-Fārisī stands out, as he was himself a contemporary of al-Ghazālī and integrated information he received directly from the great scholar with reports he got from others.[13] ʿAbd al-Ghāfir, a grandson of the great Sufi Abū l-Qāsim al-Qushayrī (d. 465/1072) and himself an author of works on Sufism,[14] includes an article (*tarjama*) on al-Ghazālī in his *Sequence to the History of Nishapur* (*al-Siyāq li-Taʾrīkh Nīsābūr*). This book was completed in 518/1124 and is the continuation of an earlier *History of Nishapur* by a fourth/tenth century historian. Only the second part of ʿAbd al-Ghāfir's continuation survived, and that part does not contain the entry on al-Ghazālī.[15] At the beginning of the seventh/thirteenth century, ʿAbd al-Ghāfīr al-Fārisī's book became the subject of an abridgment, which survived in full and contains an abbreviated version of his entry on al-Ghazālī.[16] The nonabbreviated version survived in the quotations of other historians, most prominently Tāj al-Dīn al-Subkī (d. 771/1370). Al-Subkī himself also lacked a copy of ʿAbd al-Ghāfir's book. He says he knew its content through Ibn ʿAsākir's history of the Ashʿarite school and through the abridged version.[17] He must have had a third source, however, since his quotations from ʿAbd al-Ghāfir's article on al-Ghazālī are more extensive than those in Ibn ʿAsākir's books.[18]

ʿAbd al-Ghāfir al-Fārisī, who was about three years younger than al-Ghazālī, knew the juvenile al-Ghazālī as a fellow student and teaching assistant (*khādim*) under al-Juwaynī (d. 478/1085). He later visited him several times and interviewed him about his life.[19] His eight-page biographical article had a huge impact on the historiography of al-Ghazālī. It is much more extensive than any other in his historical dictionary and includes personal comments on the impression al-Ghazālī made on the author. In terms of its information, however, it is not faultless. It reports that al-Ghazālī spent ten years in Syria although, in fact, he stayed there for less than two years, prompting at least one often-repeated misunderstanding.[20]

After ʿAbd al-Ghāfir, the Khorasanian historian al-Samʿānī (d. 562/1166) of Marw is the second closest biographer, both historically and geographically. He lived a generation after al-Ghazālī and studied with many scholars who knew him personally. Unfortunately, all of al-Samʿānī's documents on al-Ghazālī are lost, leaving only quotations in other historians' works.[21] There is also some evidence that al-Samʿānī's contemporary and colleague Ẓāhir al-Dīn ibn Funduq al-Bayhaqī (d. 565/1169–70) from Sabzawar in Khorasan wrote about the life of al-Ghazālī. If he did, his works on this subject are completely lost.[22]

The Damascene Ibn ʿAsākir (d. 571/1175) was the second historian after ʿAbd al-Ghāfir whose biography of al-Ghazālī is preserved. He includes a long entry in his apologetic history of the early Ashʿarite school, *The Correction of the Fabricator's Lies* (*Tabyīn kadhib al-muftarī*), and a shorter one in his history of Damascus.[23] Both entries consist of a reproduction of ʿAbd al-Ghāfir's biography, while the longer adds al-Ghazālī's brief work on the Muslim creed (*ʿaqīda*), *The Foundation on What-To-Believe* (*Qawāʿid al-ʿaqāʾid*). There is probably more original information on the life of al-Ghazālī in Ibn ʿAsākir's voluminous history of Damascus, which still needs to be fully explored.[24]

Ibn al-Jawzī's (d. 597/1201) chronicle *The Orderly Treatment in History* (*al-Muntaẓam fī l-taʾrīkh*) contains three entries on al-Ghazālī that do not always concur. Ibn al-Jawzī is the first annalist historian to include an obituary for al-Ghazālī in the year of his death. Ibn al-Jawzī reconstructs al-Ghazālī's basic life dates primarily from information given by ʿAbd al-Ghāfir al-Fārisī. Yet he also devotes significant space to his own traditionalist criticisms of and objections to al-Ghazālī's works.[25] Ibn al-Jawzī's grandson Sibṭ ibn al-Jawzī's (d. 654/1256) *The Mirror of Times* (*Mirʾāt al-zamān*) lists the available sources of information on al-Ghazālī's life. He mentions ʿAbd al-Ghāfir al-Fārisī, Ibn al-Jawzī, al-Samʿānī, and Ibn ʿAsākir.[26] Yāqūt (d. 626/1228) includes a brief sketch of al-Ghazālī's life within the entry on Ṭūs in his geographic dictionary.[27] In comparison, Ibn al-Athīr (d. 630/1233), the main chronicler of this period, writes only a very brief entry on al-Ghazālī, along with other scattered but important information.[28]

With Ibn al-Athīr ends the line of the chroniclers who were historically or locally close to al-Ghazālī and could credibly contribute original material to his biography. The major historians of Muslim luminaries such as Ibn Khallikān (d. 681/1282), al-Dhahabī (d. 748/1347), al-Ṣafadī (d. 764/1363), and Ibn Kathīr (d. 774/1373) all feature articles on al-Ghazālī in their works.[29] By the time they wrote, they had to rely on earlier works of history, some of them lost to us.[30] In the seventh/thirteenth century, Damascus became a center of Ghazālī studies, and legal scholars such as Yaḥyā al-Nawawī (d. 676/1277) wrote influential commentaries on his legal works. This activity revived the interest in al-Ghazālī's life. New information was hard to locate, however, and the dispute around al-Ghazālī's name exemplifies that it was simply too late to settle some issues of his biography. Whether the *nisba* (family name) was *al-Ghazālī* or *al-Ghazzālī* is a point disputed by various early reports. The most erudite historians of the seventh/thirteenth and eighth/fourteenth centuries gave an account of these disputes and refrained from judgment. A more plausible etymology in favor of al-Ghazzālī stood squarely against indications that the family itself—including our scholar—preferred the spelling with only one z.[31]

The new genre of monumental historical dictionaries on religious scholars, which appear in the seventh/thirteenth century and which cover not only the major luminaries but also everyone contributing to a certain field, made biographic information more readily available. Al-Ghazāli features prominently in the early examples of this genre,[32] with entries on him also integrating information that had earlier been cited only in entries on his students. Out of the interest in the Damascene Shāfiʿite circles grew the monumental compilation of earlier testimonies and comments, written by Tāj al-Dīn al-Subkī (d. 771/1370). He composed a book-length monograph on al-Ghazālī and incorporated it in his history of the Shāfiʿite scholars.[33] This is by far the most important treatment of al-Ghazālī's life and the impact he had on Muslim scholarship. Al-Subkī includes a variety of voices that have otherwise been lost.[34] He also includes a list of about forty-five of al-Ghazālī's works. One of his contemporaries, who composed an independent biography of al-Ghazālī based on similar sources, has an even more comprehensive list. Al-Wāsiṭī (d. 776/1374) lists in his history of the Shāfiʿite school almost a hundred titles written by al-Ghazālī.[35]

Much of the later contributions to al-Ghazālī's historiography still need to be discovered.[36] Writing a book on the life and the "exploits" (*manāqib*) of al-Ghazālī became a not-uncommon task of later theologians, particularly when they felt the need to defend al-Ghazālī from the rampant criticism surrounding him.[37] Most of these works are still unknown to us, although some of this material has emerged in al-Murtaḍā al-Zabīdī's (d. 1205/1791) monumental commentary on *The Revival of the Religious Sciences* (*Iḥyāʾ ʿulūm al-dīn*). He precedes his commentary with a biography of al-Ghazālī that is largely based on the one written by al-Subkī.[38]

Next to al-Ghazālī's autobiography—which was the subject of a French study as early as 1842[39]—Western scholars mostly relied on al-Subkī's and al-Murtaḍā al-Zabīdī's works when they reconstructed the life of al-Ghazālī.[40] Only during the past thirty years—after the edition of al-Ghazālī's letters published in 1955 and relevant excerpts of Abū Bakr ibn al-ʿArabī's works in 1961, 1963, and 1968—have important new sources become available in print.

Al-Ghazālī's Date of Birth: Around 448/1056

ʿAbd al-Ghāfir al-Fārisī does not mention when al-Ghazāli was born nor how old he was when he died. The year 450 AH (March 1058–February 1059), which has been accepted by most of al-Ghazālī's biographers, first appears in Ibn al-Jawzī's obituary of al-Ghazālī, composed at least sixty years after al-Ghazālī's death. Ibn al-Jawzī writes that "it is said (*dhukira*), he was born in 450."[41] Yāqūt also has this date. Ibn Khallikān repeats it, but adds that people in Ṭābarān, al-Ghazālī's birthplace, say that he was born in the year 451 AH.[42] This disagreement eventually falls prey to the times, and even al-Subkī, despite the encyclopedic character of his work, doesn't mention it anymore.[43]

The two dates of 450 or 451 AH are not without problems, however. In a letter al-Ghazālī wrote to Sanjar, who was then the vice-regent in Khorasan, he states that at the time of writing he had passed his fifty-third birthday.[44] This letter also contains a reference to al-Ghazālī's vow at the grave of Abraham in Hebron. This vow, which included the pledge never again to appear before rulers, is well known and was made in Dhū l-Qaʿda 489 / October 1096. Writing about himself in the third person, al-Ghazālī says in this letter that "he kept that vow for twelve years and the caliph as well as all the sultans considered him excused."[45] These words were written in order to convince Sanjar also to excuse al-Ghazālī from appearing before him. Thus, they allow us to date the letter and determine al-Ghazālī's year of birth.

Al-Ghazālī's words that "he kept that vow for twelve years," however, can be understood in two ways. Most straightforward would be to interpret the twelve years as the span between the vow at Hebron and the time of writing. Counting twelve lunar years after the vow at Hebron would date the letter in the final months of 501 / summer of 1108, two years after al-Ghazāli returned to teaching at the Niẓāmiyya in Nishapur. If the twelve years can be understood this way, al-Ghazālī was born in 448 AH (March 1056–March 1057), two years earlier than most of the historians report. There is the possibility to assume that he was born even a year earlier. In classical Islam, the age of persons was often counted in solar years according to the seasons.[46] If the age of fifty-three refers to solar and not lunar years, al-Ghazālī's birth would fall around 447/1055. It must be said, however, that every time al-Ghazālī refers to time spans of a certain number of years, the reference is to the Muslim lunar calendar. Since there is no evidence that he ever applied the solar calendar, the year 448/1056–57 is the most likely year of al-Ghazālī's birth.

There is, however, another way that the words, "he kept that vow for twelve years," can be understood. This alternative understanding is less likely in my opinion, but it must be mentioned and discussed. The vow at Hebron stands in connection to al-Ghazālī's decision to break his close association with the Seljuq rulers and resign from his teaching position at the Niẓāmiyya madrasa in Baghdad. Al-Ghazālī left that job and Baghdad in Dhū l-Qaʿda 488 / November 1095, almost exactly one year before the vow at Hebron was made. When in his autobiography *Deliverer from Error* (*al-Munqidh min al-ḍalāl*) al-Ghazālī writes about his return to teaching at the Niẓāmiyya school in Nishapur, he says that this happened in Dhū l-Qaʿda 499 / July–August 1106. He continues: "The period of seclusion (*ʿuzla*) amounted to eleven years."[47] The fact that he counts to his readers the number of lunar years he did not teach at state-sponsored schools is significant. In al-Ghazālī's own understanding, the date for when he began to keep the vow at Hebron may have not been the date that he made the vow. He might have understood that he began keeping the vow retroactively, so to speak, since his departure from Baghdad. Thus, he may have meant to say that he "kept that vow" since Dhū l-Qaʿda 488 / November 1095. If this was the case, the letter would have been written a year earlier in the last months of 500 / June–July 1108. Subsequently his birth would fall in 447/1055–56, if one assumes his age of fifty-three years is given in lunar years, or 446/1054–55 if one assumes solar years.

Judged from the information given in this letter to Sanjar, al-Ghazālī was born between 446/1054 and 448/1057. His most likely year of brith was 448/1056–57, two years before the date that currently appears in the literature. The period of 446/1054 to 448/1057 concurs with al-Ghazālī's own information given in his autobiography, *Deliverer from Error*. There, al-Ghazālī says that he was "over fifty" when he composed the book.[48] According to the traditional chronology of his life, which puts his birth in 450/1058–59, the *Deliverer* could not have been written before 501/1107; "over fifty" assumes that he was at least fifty-one lunar years old when he wrote the book. Yet in this book, al-Ghazālī refers vividly to the events at the end of the year 499 / summer 1106, when he returned to public teaching in Nishapur. The *Deliverer* was more likely written soon after this event, since it partly functions as an apologia for what appeared to some to be a break of his vow in Hebron.[49] In addition, the author makes the point that he should be regarded as the "renewer" (*muḥyī*) of the sixth Muslim century.[50] The beginning of the new century is identified as the turn from 499 to 500 AH, which fell on September 2, 1106. Therefore, all internal indications of the text point toward a publication soon after the beginning of the year 500 AH. According to the traditional chronology, however, that would be impossible since al-Ghazālī may have barely turned fifty and was certainly not yet "over fifty." If he was born between 446/1055 and 448/1057, however, he had by this time already passed his fifty-first, fifty-second, or fifty-third birthday—either in lunar or in solar years—and the words "over fifty" are well justified.[51]

Al-Ghazālī's birthplace Ṭābarān was one of two major towns within the district of Ṭūs, the other being Nūqān, which was situated a few miles south of Ṭābarān. During the sixth/twelfth century, Meshed (Mashhad) grew around the pilgrimage site of the Shiite Imām ʿAlī al-Riḍā (or: Riżā), who was buried in Sanābādh near Nūqān in 203/818.[52] All these places were referred to as Ṭūs, which according to Yāqūt had more than a thousand "villages" (*qarya*). Nūqān was gradually replaced by Meshed and eventually became a suburb of it. Three hundred years later, after the destruction of Ṭābarān in 791/1389 during an anti-Timurid uprising, Meshed would also replace al-Ghazālī's hometown. Ṭābarān was not rebuilt, and its water channels were redirected to Meshed.[53] It was during al-Ghazālī's lifetime that people began to refer to Nūqān, the second town of Ṭūs, as Meshed, a name al-Ghazālī, however, never used. Others among his contemporaries, however, weren't shy to use "Meshed" or even "Meshed, the holy city of Riżā."[54]

Al-Ghazālī's Early Years and His Education

Little is known about al-Ghazālī's childhood, even less about his family. In the seventh/thirteenth and eighth/fourteenth centuries, some Shāfiʿite scholars in Damascus made efforts to determine the occupation of al-Ghazālī's father. By then, however, it was already too late to get reliable information about this.

When al-Subkī claims that al-Ghazālī's father was a spinner (*ghazzāl*) of wool, he makes a leap of faith based on a spurious etymology of the family's name.[55] The *nisba* or family name "al-Ghazālī" had been in use for several generations, and its most distinguished bearer was not the first famous scholar who wore it. Another jurist by the name of al-Ghazālī lived two or three generations before him and may have been either his paternal granduncle or his great granduncle. The elder al-Ghazālī is said to have died in 435/1043–44 and was an influential teacher in Ṭūs, an author of books that have not survived.[56]

Later Muslim historians, however, gave another much humbler impression of al-Ghazālī's family. Al-Subkī tells us about the poverty of his father and how he made deathbed arrangements for his two young sons, Muḥammad and Aḥmad. The fatherless children were given up to the foster care of a Sufi friend of the family. Their small inheritance forced them to enter a madrasa for care. Thus, they entered into Muslim learning not for the sake of God, as al-Ghazālī is quoted as saying, but for the sake of food.[57] This story became a stock element of al-Ghazālī's biography, reflecting his and his younger brother's later attraction both to poverty and to Sufism. Al-Subkī gives no proper source for it. He reports it in the first person and claims that this is "just as al-Ghazālī used to tell it."[58] The story can be traced back to the lost part of Ibn al-Najjār's (d. 643/1245) *Appendix to the History of Baghdad* (*Dhayl ta'rīkh Baghdād*) which probably took it from al-Sam'ānī's lost work with the same title. Al-Dhahabī, who is our oldest extant source of this information, quotes one of al-Ghazālī's students, who heard him mentioning that when his father died he left little for his brother and him.[59] On this occasion al-Ghazālī supposedly said: "We acquired knowledge for reasons other than the sake of God; but knowledge refuses to be for anything else than for the sake of God." Although this sentence may reflect his upbringing, it is actually a well-known quote that appears both in al-Ghazālī's *Revival of the Religious Sciences* as well as in his *Scale of Action* (*Mīzān al-'amal*). There the author attributes it "to one (or: some) of those who found truth" (*ba'ḍ al-muḥaqqiqīn*).[60]

It is puzzling that al-Sam'ānī, who is most likely the first authority to report the tale, was unable to identify the unnamed Sufi who cared for the children. Al-Sam'ānī had an intimate familiarity of the intellectual life in Ṭūs during this period. Since we do not have the original text of al-Sam'ānī's version, we cannot say whether he implied it to be dubious. In my opinion, the historicity of the whole story is doubtful. Al-Subkī turns it into an emotional tale with the literary tropes of a father's deathbed remorse and two young orphans who turn toward knowledge simply to survive. Here, there is no role for al-Ghazālī's mother, who supposedly survived her husband and must have cared for her children. Yet some of these bare facts may be true; al-Ghazālī's father likely did die during his sons' childhood and left little for their education. These trappings may have given rise to further embellishments such as the Sufi friend of the family. Indeed, in this anecdote, the anonymous Sufi may stand in as a cipher for the famous Abū 'Alī al-Fāramadhī (d. 477/1084), whose youthful influence al-Ghazālī acknowledged later during his life and whose role will be explained later.

In his biography, ʿAbd al-Ghāfir al-Fārisī does not mention any of this and sticks to the bare facts of al-Ghazālī's education. There is no Sufi friend here; rather, it begins with the study of *fiqh* under a local teacher named Aḥmad al-Rādhakānī.[61] Al-Subkī says that this al-Rādhakānī had himself studied with "al-Ghazālī the elder." An Aḥmad al-Rādhakānī from Ṭābarān-Ṭūs was a member of the generation of al-Ghazālī's teachers, but it is not clear whether he was a scholar.[62] There was, however, another al-Rādhakānī in that generation who was a well-known scholar. ʿAbd al-Ghāfir mentions the scholar Abū Saʿd ʿAbd al-Malik al-Rādhakānī. He was the maternal uncle of the powerful grand vizier Niẓām al-Mulk (d. 485/1092).[63] His half-brother, Abū l-Qāsim ʿAbdallāh ibn ʿAlī (d. 499/1105–6), was a very important scholar and might have held the position of head teacher of the Niẓāmiyya madrasa in Nishapur between 493/1100 and al-Ghazālī's later appointment in 499/1106.[64] We will see that Niẓām al-Mulk was one of the most important personalities for al-Ghazālī's intellectual development. He served as grand vizier over a period of almost thirty years between 455/1063 and his violent death in 485/1092. Second in power only to the Seljuq Sultans Alp-Arslan (reg. 455–65 / 1063–72) and Malikshāh (reg. 465–485 / 1072–92), Niẓām al-Mulk formulated the religious policy for an area that stretched from Asia Minor to Afghanistan. In the intellectual centers of the Seljuq Empire, he founded religious madrasas (so-called Niẓāmiyya madrasas), which institutionalized the teaching of Sunni jurisprudence and Ashʿarite theology.[65] Niẓām al-Mulk hailed from Rādhakān, a village at the northern edge of Ṭūs.[66] His whole family became very influential among the religious scholars in Khorasan and at the Seljuq court.[67]

Their full names support the assumption that ʿAbd al-Malik al-Rādhakānī was a brother of Aḥmad. Regardless of whether Aḥmad or ʿAbd al-Malik al-Rādhakānī was al-Ghazālī's first teacher, al-Ghazālī likely made connections with the wider family of Niẓām al-Mulk. Al-Ghazālī's early teacher in Ṭābarān-Ṭūs was probably far less humble than al-Subkī assumed. He may have had family ties to the most important Shāfiʿite scholars of Khorasan during his time, perhaps even to the great vizier. Niẓām al-Mulk was a Shāfiʿite jurist educated in Ṭūs, a district small enough for all Shāfiʿite scholars to know one another well.

ʿAbd al-Ghāfir says that after al-Ghazālī's education under al-Rādhakānī, he went to study with al-Juwaynī in Nishapur, the next major city, about fifty kilometers south of Ṭūs and separated from it by a high mountain range.[68] He arrived there within a group of students from Ṭūs. Al-Subkī and other later historians say that before coming to Nishapur, al-Ghazālī went to study with someone named Abū l-Naṣr al-Ismāʿīlī in Gurgān, who is not mentioned in any other context.[69]

Al-Subkī also tells an anecdote on al-Ghazālī's early education that he traces back to Asʿad al-Mayhanī (d. 523/1129 or 527/1132–33), a prominent colleague and follower of al-Ghazālī who met with him during his later years in Ṭūs. Al-Subkī mentions a second source for the anecdote, namely the vizier Niẓām al-Mulk. This story has since gained some prominence—some scholars regard it as very significant[70]—and its origin should be looked at closely: Al-Subkī's two sources, As'ad al-Mayhanī and Niẓām al-Mulk, are probably just a single source. The historian al-Sam'ānī, whose family was close to As'ad al-Mayhanī, is the first to report the story in a *tarjama* on Niẓām al-Mulk in his lost *Appendix to the History of Baghdad*. We have his report preserved in a quotation from the historian of Aleppo Ibn al-'Adīm (d. 660/1262). There, al-Sam'ānī says that in a stack of papers left by his father he found an anecdote about how Niẓām al-Mulk taught his nephew that making notes alone is not sufficient learning. The nephew was Shihāb al-Islām 'Abd al-Razzāq (d. 525/1130), who later became a famous vizier and who during the time of this anecdote had just started studying *fiqh*:

> [Niẓām al-Mulk] told the story of how the Imām Abū Ḥāmid al-Ghazālī, the Sufi once traveled to Abū Naṣr al-Ismā'īlī in Gurgān and how he took notes from him (*'allaqa 'anhu*). When he returned to Ṭūs, he was robbed on the road and his notes (*ta'līq*) were taken away from him. He said to the captain of the highway-robbers: "Return my notes (*ta'līqa*) to me!" He asked: "What are these notes?" Al-Ghazālī answered: "A bag in which are the books of my studies." [Al-Ghazālī said:] "And I told him my story. So he asked me: 'How can it be that you have learned things that you get rid of when this bag is taken away from you? And now you remain without knowledge?' Then he returned it to me. I said: 'He was sent by God to alert me and guide me towards what is best for me. And when I entered Ṭūs, I turned my attention to this for three years until I had memorized all my notes in a way, would I have been robbed I would not have been deprived of my knowledge.'"[71]

This anecdote next appears in Ibn al-Najjār's *Appendix to the History of Baghdad*, a book whose full version is also lost.[72] It features in the *tarjama* on al-Ghazālī, and from here, it spread widely within the biographical literature on this great scholar. Al-Subkī represents just the latest stage.[73]

There are several factors that make the authenticity of this anecdote doubtful: 'Abd al-Ghāfir never mentions al-Ghazālī's studies in Gurgān, the teacher is not correctly identified, and the context of the report is anecdotal, pedagogical, and somewhat ahistorical. Most important, however, the nephew addressed by Niẓām al-Mulk is only ten years younger than al-Ghazālī and studied himself with al-Juwaynī, indicating that al-Ghazālī could not yet have been a famous Sufi when the story was allegedly told. Although the story's age does give it some credibility—it goes back almost to the days of al-Ghazālī and contains verbatim quotes—the topical nature of the story makes its historicity dubious. It is just as possible that the real experience of a less prominent scholar could have circulated among people in Ṭūs or elsewhere and become connected to the famous al-Ghazālī simply because it fit the impression that contemporaries had about his personality.

In his letter to Sanjar mentioned above, al-Ghazālī says that he started his deeper education at the age of thirteen. Using one of his favorite metaphors to compare knowledge with deep and dangerous water, al-Ghazālī writes about himself that since that age, "he had been diving into the sea of religious sciences."[74] This quotation may well refer to the beginning of his studies with al-Rādhakānī in Ṭūs, which would place it at 461/1069. A few years later he would arrive in al-Juwaynī's class in Nishapur. His famous student-colleague al-Kiyā' al-Harrāsī (d. 504/1110), who was born in 450/1158, two or three years after al-Ghazālī, entered al-Juwaynī's seminar in 468/1075–76 at the age of seventeen.

In his autobiography, al-Ghazālī briefly comments on the beginnings of his intellectual life. "The thirst for understanding the essense of things was my persistent habit from my early years and the prime of my life." This yearning, al-Ghazālī says, was not a matter of choosing but a personal instinct and a natural disposition (*gharīza wa-fiṭra*) that God had given him. This disposition allowed him to scrutinize the intellectual environment he grew up with and to thow off "the bounds of emulating others" (*rābiṭat al-taqlīd*). He broke with the convictions he inherited, he says, when he was still a boy (*ʿahd sinn al-ṣibā*).[75] Later, ʿAbd al-Ghāfir al-Fārisī would write that the young al-Ghazāli he had known had shown some "filthy strains" (*zaʿārra*) in his character. He was full of haughtiness and looked down at people with defiance. "He had a vain pride and was blinded by the ease with which God had provided him to handle words, thoughts, expressions, and the pursuit of glory."[76]

Al-Juwaynī was the most outstanding Muslim scholar of his time, an authority in both Muslim law (*fiqh*) and theology. Around 455/1063, only five years before al-Ghazālī started studying with him, he had returned from his exile at Mecca and Medina. Ten years prior, in 445/1053, he had fled from Khorasan to escape the persecution of Ashʿarites under the newly ascended Seljuqs and their sultan, Toghril-Bey (reg. 432/1040–455/1063).[77] After Toghril-Bey's death and Niẓām al-Mulk's ascension to the vizierate of the Seljuq Empire in 455/1063, this policy was reversed. Niẓām al-Mulk was sympathetic to Ashʿarism, and he actively supported this school.[78] Marw, Baghdad, Herat, and Nishapur saw the founding of Niẓāmiyya madrasas, institutions open to the theological tradition of al-Ashʿarī (d. 324/935–36). The main chair at the Niẓāmiyya madrasa in Nishapur was offered to al-Juwaynī.

Al-Juwaynī's teaching activity at the Niẓāmiyya in Nishapur proved a turning point in the history of Ashʿarite theology. Although generations of Ashʿarites—including al-Ashʿarī, the school's founder—had understood the tradition of Greek philosophy to pose a significant challenge to the epistemological edifice of Muslim theology, none of al-Juwaynī's predecessors had seriously studied the works of this school of thought. By the time of the mid-fifth/eleventh century, the philosophical tradition in Islam had evolved from its foundational texts—translations of Aristotle and their commentaries—to being dominated by the works of the Muslim philosopher Avicenna (Ibn Sīnā, d. 428/1037). Al-Juwaynī was the first Muslim theologian who seriously studied Avicenna's books. On the one hand, al-Juwaynī fully realized the methodological challenge of the Aristotelian methods of demonstration (*apodeixis / burhān*) as used by Avicenna. The Muslim philosophers (*falāsifa*) claimed, for instance, that through a chain of conclusive arguments, one can prove demonstrably that the world is pre-eternal (*qadīm*), and one can thus disprove the claim of the theologians that the world is created in time (*ḥādith*). On the other hand, al-Juwaynī also understood that the works of Avicenna and other *falāsifa* contained solutions to many theological problems the Ashʿarite school had wrestled with for centuries.

There can be little doubt that al-Ghazāli started to read philosophical literature many years before he published books about it. His preoccupation with this literature likely began in the seminary of al-Juwaynī,[79] where reading philosophical literature may have been part of the higher curriculum. The works of other scholars with a shared education reveal a detailed familiarity with the arguments of Aristotle and his Muslim followers.[80] Al-Juwaynī himself had devoted much effort to a proper study and refutation of the *falāsifa*'s arguments about the eternity of the world.[81] Despite his disagreements, he was himself influenced by Avicenna's distinction of being in (1) the being that is necessary by virtue of itself (*wājib al-wujūd*) and (2) the beings that are only contingent by themselves (*mumkin al-wujūd*). Al-Juwaynī uses both concepts in his comprehensive *summa* of Ashʿarite theology[82] as well as in his more concise directory.[83] In one of his last works, a small textbook of Ashʿarite theology written to honor his mentor Niẓām al-Mulk, al-Juwaynī expounds a proof for the existence of God that is influenced by that of Avicenna. "This is a method," al-Juwaynī writes, "that is more useful and nobler than those gathered in many volumes."[84] He starts his proof by introducing the distinction of objects of knowledge into necessary, contingent, and impossible. Nothing in the created world is necessary by virtue of itself. In fact, everything can be different, and this illustrates that everything that exists in this world is contingent (*mujawwaz*). If all things can be different from what they are, there must be a "determining agent" (*muqtaḍī*) who chooses the state of things. This "determining agent" must be continuously active and sustaining the world, which is not able to sustain itself.[85]

The Niẓāmiyya madrasa in Nishapur became the cradle of Avicenna's lasting influence on Ashʿarite theology. For the young al-Ghazālī, plunging into this sea of knowledge must have been an unforgettable moment, one he still vividly remembered forty years later in his conversation with Sanjar. In 2003, Jules Janssens suggested that there was a period in the life of the young al-Ghazālī when he was an adept of the philosophical school and a follower of Avicenna. Janssens suggested that *The Intentions of the Philosophers* (*Maqāṣid al-falāsifa*) was written during that period and that the brief preface and the short conclusion (*khātima*) of the book were added later after the appeal of *falsafa* had waned.[86] None of al-Ghazālī's biographers mentions such a period. However, in at least one passage of his works, al-Ghazālī himself seems to indicate his past attraction to philosophy. In the *Incoherence of the Philosophers*, he portrays the Muslim followers of the *falāsifa* as a group that "is convinced to be distinct from the companions and peers by virtue of a special cleverness (*fiṭna*) and quick wit (*dhakāʾ*)." He describes the followers of *falsafa* as rejecting the duties of Islam, namely, the acts of worship and ritual purity, and belittling the devotions and ordinances prescribed by the divine law.[87] They do so because they look down on religious people, al-Ghazālī claims; they see their own intelligence and methods of inquiry as making them superior to pious people who rely on revelation. In his later book, *The Jewels of the Qurʾan* (*Jawāhir al-Qurʾān*), al-Ghazālī seems to admit that he himself was once part of such a group:

> We saw among the groups of those who have a high opinion of themselves (*mutakābisūn*) some that were deceived by the literal meaning (*ẓāhir*) of revelation. They became engaged in quarrels among them, opposing each other, and pompously presenting to one another what the groups disagreed upon. Subsequently this destroyed their belief in religion and led them to the inner denial of bodily resurrection, heaven and hell, and the return (*rujūʿ*) to God the Exalted after death. They profess this in their innermost soul (*fī sarāʾirihim*). They are loose from the reins of fear of God (*taqwā*) and the bounds of piety. They are free from restraint in their pursuit of the vanities of this world. They eat what is forbidden, follow their passions, and are eager for fame, wealth, and worldly success. When they meet pious people they look down on them with pride and contempt. When they witness piety in someone whom they cannot beat intellectually because of his abundant knowledge, perfect intelligence, and penetrating mind, they bring him to a point where his goal becomes deception (*talbīs*), to win over the hearts [of these people], and to change [their] attitude towards him. When they witness piety in other people it only increases their error in the long run; while when people of religion witness piety it is one of the strongest confirmations for the convictions of the believers. (. . .) And because they do not believe in the unknown (*ghayb*) the way ordinary people believe in it, their smartness is their perdition. Ignorance is closer to salvation than the faulty cleverness and defective smartness [of these people].
>
> We were ourselves not far from this, for we had stumbled upon the tails of these errors for a while due to the calamity of bad company and our association with them until God has distanced ourselves from their errors and until He had protected us from their predicaments.[88]

Becoming a Famous Jurist and Theologian

There is a scarcity of information about the years between al-Ghazālī's entry into the Niẓāmiyya madrasa in Nishapur and his own appointment to the Niẓāmiyya in Baghdad more than twenty years later. ʿAbd al-Ghāfir al-Fārisī covers this period with a single sentence, saying that al-Ghazālī stayed with al-Juwaynī until the latter's death, that he left Nishapur afterward, and that he became part of the traveling court (*muʿaskar*) and of the assembly of scholars (*majlis*) that the vizier Niẓām al-Mulk kept around him.[89] Later historians add nothing to this description.

In earlier Turkish tradition, the court of the Seljuq sultan and his vizier would travel through the open country. The sultan's military and political strength depended on the livestock kept by his nomadic warriors, and he had to lead it through fertile pastures in order to survive. With time, however, the sultan became detached from his troops and accustomed to a more urban lifestyle. By the time Sultan Malikshāh came to power in 465/1072, the court spent much of its time in Isfahan and visited Baghdad in regular intervals.

When al-Ghazālī arrived in Baghdad in 484/1091, he came from Isfahan. Indeed, a comment in his letter to Sanjar suggests that he had spent the years after leaving Nishapur and before arriving in Baghdad exclusively in Isfahan. Talking about himself, al-Ghazālī wrote to Sanjar:

> Know that this applicant (*dāʿī*) has reached fifty-three years of age, forty years of which he has dived in the sea of religious scholarship so that he reached a point where his words are beyond the understanding of most of his contemporaries. Twenty years in the days of the martyred Sultan Malikshāh passed, while in Isfahan and Baghdad he remained in favor with the sultan. Often he was the messenger (*rasūl*) between the sultan and the caliph in their important affairs.[90]

The amount of time al-Ghazālī spent in the service of Malikshāh (see figure 1.1) is most probably exaggerated. Malikshāh reigned almost exactly twenty lunar years between Rabīʿ I 465 / January 1073 and Shawwāl 485 / November 1092, and these words suggest that al-Ghazālī served him throughout his whole period in office. With this address, al-Ghazālī aimed to impress Malikshāh's son Sanjar and to suggest that he had paid his dues of servitude to the Seljuq family. Still, these words propose that al-Ghazālī entered the court early in Malikshāh's reign, probably many years before al-Juwaynī's death in 478/1085. One of al-Ghazālī's students reports that Malikshāh commissioned one of his works in Persian; the *Proof of Truth in Responding to the Ismāʿīlites* (*Ḥujjat al-ḥaqq fī l-radd ʿalā l-bāṭiniyya*), which unfortunately is lost.[91]

During the exchange with the vice-regent Sanjar, which took place shortly after 501/1108, al-Ghazālī mentions that one of his earliest books, *The Sifted among the Notes on the Methods of Jurisprudence* (*al-Mankhūl min taʿlīqāt al-uṣūl*), was published about thirty years before.[92] That would put the publication of this book, which is an extracted version of al-Juwaynī's course curriculum (*taʿlīqa*) for Islamic law, in the years around 471/1078.[93] Ibn al-Jawzī confirms that the book was published during al-Ghazālī's teacher's lifetime; it even merited a jealous comment by al-Juwaynī.[94] Despite disagreeing with his teacher on some legal points, *The Sifted among the Notes on the Methods* was written in close cooperation with al-Juwaynī, who is honored in numerous references. Al-Ghazālī says that he "took great pain to organize the book into sections and chapters in order to facilitate the understanding when the need for consultation arises."[95] The clear and detailed organization of his material is a feature of

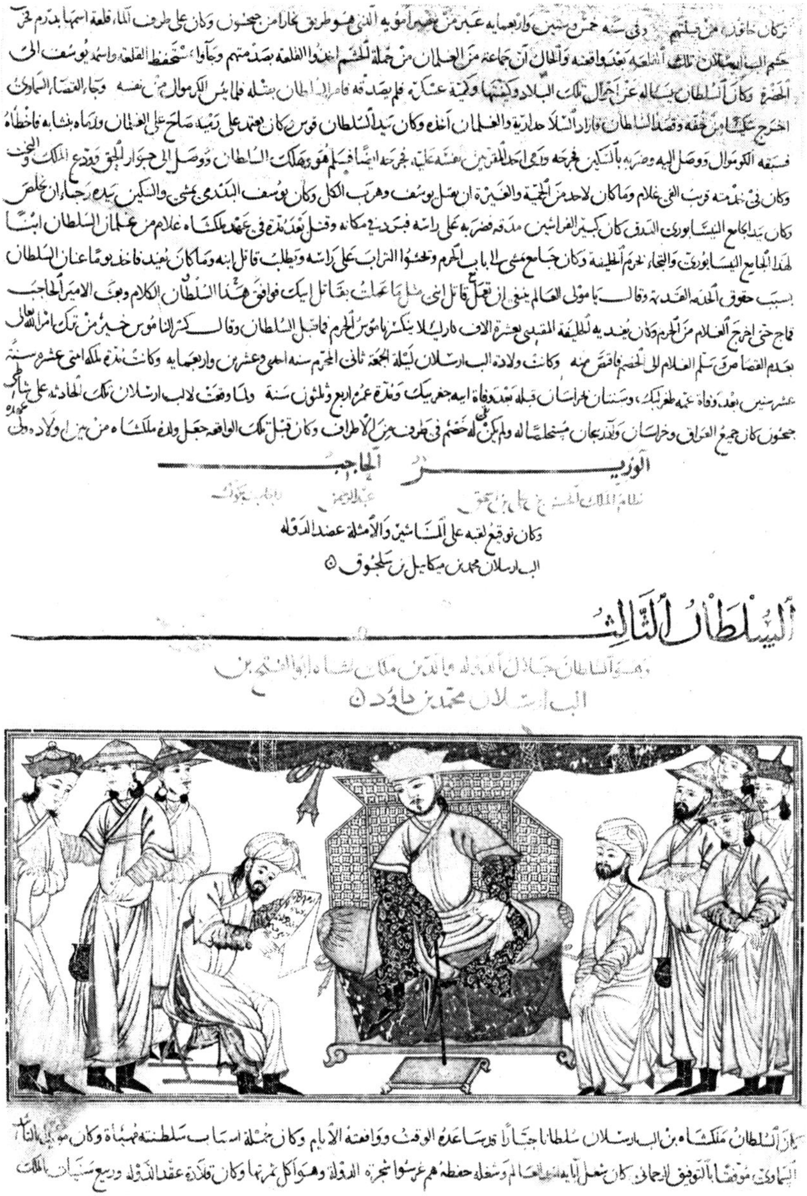

FIGURE 1.1 Sultan Malikshāh among his court. Miniature from the Arabic translation of Rashīd al-Dīn Ṭabīb's (d. 718/1318) Persian *Compendium of Chronicles* (*Jāmiʿ al-tavārīkh*), produced around 714/1314 in Tabriz, Iran (Edinburgh University Library, MS Arab 20, fol. 138a).

all of al-Ghazālī's writings, and al-Juwaynī might have understood how much his own teaching activity could benefit from it.

Ibn al-Jawzī's story of al-Juwaynī's jealousy is part of an admiring but critical account of how al-Ghazālī's intellectual brilliance was also combined with a significant amount of hubris. Ibn al-Jawzī gives the impression that the young al-Ghazālī was disrespectful toward his teacher. Given Ibn al-Jawzī's antagonism toward al-Ghazālī, one might question whether his analysis is unbiased or mere scandalmongering. Yet an earlier and more reliable source also mentions that al-Juwaynī was not entirely happy with his master student. ʿAbd al-Ghāfir al-Fārisī, who knew both and who may have witnessed what he reports, says that al-Juwaynī admired al-Ghazālī's intelligence, his eloquence, and his talent for disputations, yet "secretly (*sirr^an^*) he did not have a good opinion of al-Ghazālī."[96] Like ʿAbd al-Ghāfir himself, al-Juwaynī disliked the young al-Ghazālī's rush toward judgement and what many thought was an inborn sense of superiority. "He also was not pleased with [al-Ghazālī's] literary compositions," ʿAbd al-Ghāfir continues, "even though he had been trained by him and was associated with him." Outwardly, however, al-Juwaynī boasted the achievements of his master student and held him in high esteem.[97]

In Jumāda I 484 / July 1091, al-Ghazālī entered Baghdad as a newly appointed professor at the Niẓāmiyya madrasa. The appointment was a decision by Niẓām al-Mulk. Before he left Isfahan, Niẓām al-Mulk had bestowed upon him two honorary titles, "Brilliance of the Religion" (*zayn al-dīn*) and "Eminence among the Religious Leaders" (*sharaf al-aʾimma*).[98] Later, al-Ghazālī may have also received the title "Proof of Islam" (*hujjat al-Islām*). This latter honorific was already used during his lifetime and overshadowed all others, which might indicate that the caliph—and not a sultan or his vizier—conferred it on him.

During his court days in Isfahan, al-Ghazālī had obtainted a number of precious robes whose opulence made a significant impression in Baghdad. A contemporary noted: "When Abū Ḥāmid entered Baghdad [in 484/1091] we estimated the value of his clothing and mount to be 500 dinars. After he turned ascetic, traveled, and returned to Baghdad [in 490/1097], we valued his clothing to be worth fifteen *qirāṭ*."[99] The scholars of Baghdad must have understood the importance of al-Ghazālī's appointment to the Niẓāmiyya, since two teachers had to leave their posts to make room for him.[100] By this time, it appears that he had already published his long compendium, *The Extended One* (*al-Basīṭ*), on the individual rulings in Shāfiʿite *fiqh*, and the somewhat shorter *Middle One* (*al-Wasīṭ*) on the same subject. His fame as a brilliant scholar had most likely reached the capital.[101] When he came to Baghdad, al-Ghazālī brought with him a companion and perhaps also students from Isfahan.[102]

Al-Ghazālī's appointment to the most prestigious and most challenging teaching position of his time threw him squarely into the public light. From this point forward, there is no dearth of information about his life, and all his movements are well accounted for. Ibn al-Jawzī reports appreciatively that all major scholars of Baghdad, among them the leading Ḥanbalī jurists, sat at his feet and "were astonished by his words; they believed these teachings had great merits, and they used them in their own books."[103] Al-Ghazālī himself also attended the teaching sessions of other eminent professors at the Niẓāmiyya.[104] As his comments in the letter to Sanjar suggest, al-Ghazālī was close to the caliph's court and attended its major functions.[105] In addition to being the most prominent teacher of Muslim law and theology, al-Ghazālī was also an official of the Seljuq Empire, someone who, as he later put it critically, "consumed the riches of the ruler."[106]

Al-Ghazālī's tenure at the Niẓāmiyya in Baghdad would last only four years. The number of books he is thought to have written during this period is staggering. Al-Ghazālī himself brags about his achievements in a letter to Sanjar: before he gave up teaching in 488/1095, he writes, he had already finished seventy books.[107] In his autobiography, he claims that even while teaching three hundred students, he still found the time to study the works of the *falāsifa* and compose a refutation to them within three years.[108] Such lines should be read skeptically, as they are intended to counter the accusation that al-Ghazālī had familiarized himself with philosophical teachings even before he had learned the religious sciences. It makes little sense to assume that al-Ghazālī arrived in Baghdad in the summer of 484/1091 with empty notebooks, so to speak, without having written or drafted at least parts of the many books he would publish between his arrival at the Niẓāmiyya in Baghdad and his departure four and one-half years later. In their work on the dating of al-Ghazālī's works, Maurice Bouyges and George F. Hourani were reluctant to assume that al-Ghazālī had completed many of his works before the year 484/1091. They follow his autobiography and date the composition of *Incoherence of the Philosophers* and the many books that surround this key work in the years after 484/1091. This assumption need not be the case. The text of manuscript London, British Library Or. 3126 illustrates that al-Ghazālī studied the works of *falāsifa* such as Avicenna, al-Fārābī, and Miskawayh in an extremely close manner. Whether he composed the *Incoherence of the Philosophers* during or after this study is an interesting question that we do not have the information to answer.[109] Even if one were to assume that al-Ghazālī did not compose these works before arriving in Baghdad, there was enough time for his intense preparatory study during the twenty years between his studies with al-Juwaynī and his arrival in Baghdad. The speedy and linear process of studying and refuting, as described in his autobiography, seems overly streamlined. It is more likely that periods of philosophical study were interspersed with other activities and occupations, finally leading to the very clever response of the *Incoherence of the Philosophers*, which was published in Baghdad. Other works that came out of the study of *falsafa* such as *The Standard of Knowledge in Logics* (*Miʿyār al-ʿilm fī fann al-manṭiq*), *The Touchstone of Reasoning in Logic* (*Miḥakk al-naẓar fī l-manṭiq*), the text of manuscript London, British Library Or. 3126, and even *The Balanced Book on What-To-Believe* (*al-Iqtiṣād fī l-iʿtiqād*) may have been written or at least significantly drafted during the years before al-Ghazālī arrived in Baghdad. Similarly, al-Ghazālī's refutations of the propaganda of the Ismāʿīlite movement, which he laid down in such books as *The Scandals of the Esoterics and the Virtues of the Followers of Caliph al-Mustaẓhirī* (*Faḍāʾiḥ al-bāṭiniyya wa-faḍāʾil al-Mustaẓhiriyya*), *The Weak Positions of the Esoterics* (*Qawāṣim al-bāṭiniyya*), or *The*

Straight Balance (*al-Qisṭās al-mustaqīm*), may have been conceived or written in the period before al-Ghazālī arrived in Baghdad. One of his refutations of the Ismāʿīlite teachings was written in response to a question put to him in Hamadan, probably in the period before he came to Baghdad.[110] During the years before their takeover of the Elburz Mountains in Daylam 483/1090, Hamadan and particularly Isfahan were main centers of Ismāʿīlite activity.[111] His having developed a comprehensive response to Ismāʿīlism may have been one of the elements that qualified al-Ghazālī for his prominent position at the Baghdad Niẓāmiyya madrasa.

Many of the books written or drafted before al-Ghazālī came to Baghdad were indeed published during his tenure at the Niẓāmiyya. A manuscript of the most important book from this period, the *Incoherence of the Philosophers*, for instance, says that it was concluded on 11 Muḥarram 488 / 21 January 1095.[112] A second book that was certainly published within these years is *The Scandals of the Esoterics and the Virtues of the Followers of Caliph al-Mustaẓhirī*. The work was commissioned by the caliph's court.[113] Both the ʿAbbāsid caliph's as well as the Fāṭimid caliph's names appear in the book, and since their reigns only briefly overlapped, we know that the publication of the book fell in the year 487/1094.[114]

The years of al-Ghazālī's teaching activity at the Baghdad Niẓāmiyya were tumultuous for the city and the Seljuq Empire as a whole.[115] On 10 Ramaḍān 485 / 14 October 1092, Niẓām al-Mulk was murdered during the court's travel from Isfahan to Baghdad. A young man who appeared to be an Ismāʿīlite from Daylam assassinated him, his name recorded in the annals of the Ismāʿīlites.[116] But the murderer was immediately killed and could not be interrogated. Sultan Malikshāh, who continued on his route to Baghdad, appointed Tāj al-Mulk, a longtime rival of Niẓām al-Mulk, as his new vizier. Earlier, Malikshāh had already distanced himself from Niẓām al-Mulk and the ʿAbbāsid caliph.[117] Now, after the death of his long-serving vizier, Malikshāh demanded that the caliph move from Baghdad to another city of his preference, leaving Baghdad to the Seljuqs as their capital. The caliph al-Muqtadī asked to have at least ten days to prepare for his move, "like it is granted to any man from among the populace."[118] During this grace period, Malikshāh went hunting and returned with a fever that killed him on 16 Shawwāl / 19 November, about a month after Niẓām al-Mulk's murder.

Malikshāh's sudden death prompted his six eligible minor sons and their backers to engage in a fierce struggle for the sultan's succession. The sons were from three different mothers, and each mother attempted to build her own power base. The Seljuq generals (singl. *amīr*) and the so-called "*Niẓāmiyya*"—the family and the clients of Niẓām al-Mulk and their loyal slave-troops—followed Malikshāh's wishes and prepared for the appointment of his oldest son Berk-Yaruq, who was thirteen years old and whom they had taken to Rayy. Meanwhile in Baghdad, one of Malikshāh's widows, known as Terken Khātūn, convinced the caliph to appoint her five-year-old son, Maḥmūd, as sultan. Terken Khātūn was also the mother of one of the caliph's wives—who by this time, however, was no longer alive—and she had earlier tried to yield some influence on the caliph's own succession. After some hesitation and negotiation—in which al-Ghazālī, as we will see, played a role—the caliph responded to Terken Khātūn's demands and proclaimed Maḥmūd as the new sultan. Soon after Maḥmūd's name was called during the Friday prayers, he, his mother, and their entourage made their way to Isfahan in order to gain the support of the powerful Seljuq *amīrs*. Baghdad and the Seljuq Empire were thrown in a period of political uncertainty. Local Seljuq commanders (singl. *shaḥna*) and their garrison troops became rulers of the city.[119]

While the struggle over the sultanate was going on, the new vizier Tāj al-Mulk was murdered in Muḥarram 486 / February 1093, only three months after his predecessor was assassinated. As he had been openly accused of being responsible for the killing of Niẓām al-Mulk, the *Niẓāmiyya* avenged him and apparently killed Tāj al-Mulk. Later that year, Terken Khātūn and her son, the child sultan Maḥmūd, died of an infectious disease. Berk-Yaruq (see figure 1.2) was now free to advance to the throne; he traveled to Baghdad and was declared sultan in Muḥarram 487 / February 1094. On the following day, 15 Muḥ arram / 4 February, the Caliph al-Muqtadī died, apparently of natural causes.[120] Within sixteen months of Niẓām al-Mulk's assassination, the whole political elite of the Seljuq state was dead, including the caliph. All these deaths and upheaval led to a situation in which, according to the historian ʿAṭā-Malik Juvaynī (d. 681/1283), "the affairs of the realm were thrown into disorder and confusion; there was chaos (*harj va-marj*) in the provinces, (. . .) and turmoil and uproar in the kingdom."[121]

Erika Glassen and Carole Hillenbrand have argued that these deaths were neither coincidence nor due to the instigation of Ismāʿīlite "assassins."

FIGURE 1.2 Sultan Berk-Yaruq among his court. Miniature from Rashīd al-Dīn Ṭabīb's *Compendium of Chronicles*. The miniatures in this manuscript (same as figures 1.1 and 1.5) are the earliest extant historical illustrations in Islam (Edinburgh University Library, MS Arab 20, fol. 139b).

They were the result of a failed attempt by Terken Khātūn to bring her son Maḥmūd to power, combined with a counterintrigue instigated by the so-called *Niẓāmiyya*.[122] Al-Ghazālī's student Abū Bakr ibn al-ʿArabi gives a full account of these events that concurs with Glassen's and Hillenbrand's analysis, suggesting that Tāj al-Mulk was a clandestine Ismāʿīlite who used his contacts to arrange a contract killing.[123] Al-Ghazālī took an active part in the attempts to foil Terken Khātūn's plans. Shortly after Niẓām al-Mulk's and Malikshāh's deaths, the assumption of Terken Khātūn's son seemed all but certain. "Things went smoothly," Ibn al-Jawzī writes, "until Terken Khātūn asked the caliph for the installation of her son." This was in Shawwāl 485 / November 1092, only days after Malikshāh's death. The caliph hesitated and proposed to write three separate documents, one that would install Maḥmūd as sultan and two others that would install Maḥmūd's general as *amīr* of the army and his confidant Tāj al-Mulk as vizier and comptroller of finances. That way, the caliph would gain a chance to control the future appointment of these two vital offices, which had thus far been under the sole domain of the sultan. Terken Khātūn refused to accept this usurpation and demanded that all offices be put in the hands of her minor son. The caliph, in turn, declined, saying that religious law would not allow him to place that much power in the hands of a minor.[124]

Placed in this situation, al-Ghazālī supported the position of the caliph. The historian Ibn al-Athīr reports that when the caliph sent the letter to Terken Khātūn explaining his refusal to write a single document for Maḥmūd, she refused to receive it. To mediate between the parties, the caliph sent al-Ghazālī to Terken Khātūn. Apparently, all this happened during the week between Malikshāh's death and Maḥmūd's proclamation. Al-Ghazālī told the widowed queen in clear terms: "Your son is a minor and the religious law (*al-sharʿ*) does not allow his installation as [full] ruler." Eventually, Terken Khātūn conceded this point and accepted the caliph's conditions for the appointment of her son. When on 22 Shawwāl 485 / 25 November 1092, the *khuṭba* was read in his name, the provision for the highest military office and the vizierate was clearly spelled out.[125] Four days later, Terken Khātūn and Maḥmūd left for Isfahan, where they would both die. Al-Ghazālī was the most senior scholar who had supported the demands of the caliph; other scholars had refused this novel way of reading the *khuṭba*. Whether al-Ghazālī did this in order to boost the power of the caliph or that of the *Niẓāmiyya* is unclear. The caliph's plan was to get rid of the Seljuq overlords. Although the party of Niẓām al-Mulk would not support such a plan, the coup would fit into their plot to install Berk-Yaruq and to oust Maḥmūd. The historian Ibn Kathīr writes that the caliph initially refused to fully install Maḥmūd, "and al-Ghazālī agreed with him."[126] Al-Ghazālī's position was: "Allowed is only that what the caliph says."[127] Other scholars from the Ḥanafite school supported the claims of Terken Khātūn, but al-Ghazālī prevailed.

Eventually, Terken Khātūn, her son, and the caliph, al-Muqtadī, soon passed away, and what they had negotiated was of no value to later caliphs. The party of Niẓām al-Mulk succeeded in bringing Berk-Yaruq to power. It remains unclear whether this was what al-Ghazālī had advocated or whether he sincerely supported the advances of the caliph. In his political theory—both the early one formulated in his juvenile works on jurisprudence as well as his later ideas in *The Council for Kings* (*Naṣīḥat al-mulūk*)—the caliph plays no special role among those who bear political responsibility. If he is weak, he remains a largely ceremonial figurehead and is expected to leave the affairs of the state to officials who have real power and whom he is expected to appoint.[128] Al-Ghazālī argued in favor of strong governing bodies that could enforce the religious law effectively.[129] These strong governing individuals (*wālin*, pl. *wulāt*) could be either caliphs or sultans.[130] If the caliph is able to acquire sufficient authority and power (*shawka*), he may become himself a direct ruler and displace his appointees.[131] Al-Ghazālī's objection against the installment of a minor as a sultan may have been triggered simply by his desire for a strong executive power. Yet, he may have also supported Caliph al-Muqtadī's goal to become a direct ruler over Baghdad and Iraq. Finally, it may have also served a third interest, namely, the creation of a strong vizierate for the *Niẓāmiyya* party that could dominate a weak sultan and a weak caliph.

In a letter he wrote about ten years after these events, al-Ghazālī cites the deaths of the four viziers—Niẓām al-Mulk, Tāj al-Mulk, Majd al-Mulk, and Mu'ayyad al-Mulk—as a lesson from which to learn.[132] The letter is directed to Mujīr al-Dīn, who was then vizier to Sanjar.[133] Al-Ghazālī's elaborate prose makes no attempts to hide his opinion that the four viziers reaped what they had sowed. Niẓām al-Mulk died, the letter suggests, because he was old and could no longer control the army. "His death," al-Ghazālī writes, "was connected to treachery (*khiyānat*) and discord (*mukhālafat*)."[134] Al-Ghazālī does not mention the Ismāʿīlites.[135] Given the fact that all four viziers died violently in court intrigues, the letter's recipient is advised to take a close look at the fate of the four viziers and to draw his own conclusions. Al-Ghazālī writes that Mujīr al-Dīn's situation is worse than that of his four predecessors: "You should know that none of the four viziers had to confront what you have to confront, namely the kind of oppression (*ẓulm*) and desolation (*kharāb*) there is now."[136] Al-Ghazālī addresses Mujīr al-Dīn in blunt words, invoking fear that those who collaborate with tyrants will themselves be judged as evildoers in the hereafter. He predicts inevitable punishment if the vizier does not change his ways.

In his *Council for Kings* (*Naṣīḥat al-mulūk*), al-Ghazālī finds equally harsh words for those in power. This book was composed after 501/1108 at the request of Sanjar, when he was vice-regent of Khorasan. Governmental authority, al-Ghazālī admonishes therein, will only be firm if its holders have strong faith (*īmān*). Once the heart is deprived of faith, the talk will simply come from the tongue. Al-Ghazālī claims that true faith was rare with the government officials of the day; he wonders whether an official who squanders thousands of dinars on one of his confidants truly has anything left of his faith. On Judgment Day, this money will be demanded back from him, and he will be tormented for his waste of the community's wealth.[137]

It is hard to imagine how such a powerful state official as Mujīr al-Dīn or the members of Sanjar's courts reacted to al-Ghazālī's admonitions. In an anachronistic and probably anecdotal meeting between al-Ghazālī and the famous vizier and author Anūshirwān ibn Khālid (d. around 532/1138), the statesman rejects the scholar's reproaches as hypocrisy. Al-Ghazālī's moralistic posture is for him just another attempt to compete for worldly regard. After having listened to al-Ghazālī's warnings, Anūshirwān said: "There is no god but God! When this man started his career and sought to outdo me through merits that appeared in his honorific titles, he was dressed in gold and silk. Now, his affairs have returned to the very same state."[138] Now, Anūshirwān implied, he would try to outdo him with his moralistic posture. But even that was selfishness masquerading as virtue.

The letter to Mujīr al-Dīn was, of course, written after al-Ghazālī himself changed and refused to collaborate with rulers. In 485/1092 it appears that he was still part of the powerful political group of the *Niẓāmiyya*. He witnessed its temporary failure during the installation of Maḥmūd and later its mistake in supporting Berk-Yaruq, who as an adult was accused of sympathizing with Ismāʿīlite activities under his reign.[139] For al-Ghazālī, the events of 485/1092 and the year after must have appeared as a serious political challenge to the patrons of the Niẓāmiyya madrasa and to Sunnism as a whole. The continuing death toll among the leaders was accompanied by a civil-war-like period of religious and political subversion in Iran. Even before Niẓām al-Mulk's death, Ismāʿīlite Shiite groups, no longer loyal to the Fāṭimid caliph in Cairo, had managed to conquer and control a number of castles in Iran. In 483/1090, the stronghold of Alamūt in the northern province of Daylam had fallen into the hands of these Ismāʿīlite Shiites. By 485/1092, the Shiites, who called themselves Nizārīs and were led by Ḥasan ibn al-Ṣabbāḥ (d. 518/1124), controlled all of Daylam. A year later, the eastern province Quhistan was the place of a successful Nizārite uprising.[140] And although the Ismāʿīlites were never able to overthrow the strong Seljuq state with its numerous and powerful Turkish troops, they caused significant unrest within its cities and in some provinces.

In Baghdad and Isfahan, the Shiite insurrection led to witch hunts against suspected Ismāʿīlites, killing many.[141] The chronicler Ibn al-Jawzī refers to these events as "the days of the Esoterics."[142] Suspected agents and missionaries of the Ismāʿīlite movement were swiftly tried and executed.[143] The political crisis over Malikshāh's succession would continue until 497/1104, when Berk-Yaruq agreed to a division of power with his half-brothers Muḥammad Tapar and Sanjar. The religious confrontation between Sunnī theology and its Ismāʿīlite Shiite challengers, however, was not so easy to overcome.

Leaving Baghdad, Traveling in Syria and the Hijaz, and Returning to Khorasan

Al-Ghazālī's autobiography still offers the most detailed account of the reasons that led to his sudden departure from Baghdad in Dhū l-Qaʿda 488 / November 1095. Here he says that at some time before the month of Rajab / July of that year, he began to study the writings of such Sufis as al-Junayd, al-Shiblī, al-Ḥārith al-Muḥāsibi, Abū Yazīd al-Bisṭāmī, and Abū Ṭālib al-Makkī. In their works, he learned about epistemological paths such as "taste" (*dhawq*) and others, which had been unknown to him, according to this account. These ways of knowing are described as individual experiences of the soul, and their relationship to descriptive knowledge compares with the relationship between experiencing drunkenness and merely knowing its definition. Al-Ghazālī portrays himself during this time as being in a state in which "a strong belief in God, in prophecy, and in the Day of Judgment" had been firmly established within him.[144] After his studies and subsequent realizations, he writes, he began to understand that firm convictions about religious tenets are not relevant when it comes to the afterlife. On the Day of Judgment only an individual's actions are taken into account: "It had already become clear to me that my only hope of attaining happiness in the next world was through devoutness (*taqwā*) [towards God] and restraining the soul from the passions."[145] In his autobiography, al-Ghazālī describes his reaction after realizing this and looking at his career:

> Next, I attentively considered my circumstances, and I saw that I was immersed in attachments, which had encompassed me from all sides. I also considered my achievements—the best of them being my instructions and my teaching—and I understood that here I was applying myself to sciences that are unimportant and useless on the way to the hereafter. Then I reflected on my intentions in my instruction, and I saw that it was not directed purely to God. Rather, it was instigated and motivated by the quest for fame and widespread prestige. So I became certain that I was on the brink of a crumbling bank and already on the verge of falling into Hell unless I sat about mending my ways.[146]

These thoughts would lead to a crisis in which al-Ghazālī considered leaving his career at the Niẓāmiyya. He hesitated, however, and did not have the resolve to carry it out. In Rajab 488 / July 1095, his crisis of indecision would turn into a physical ailment: al-Ghazālī lost the ability to speak. "For God put a lock unto my tongue and I was impeded from teaching. (. . .) No word could pass my tongue and I was completely unable to say anything."[147] This also affected his eating and drinking as he became unable to swallow or even to nourish himself from broth. When a physician gave up all treatment and suggested that "this is something which had settled in the soul and from there it affects the mixture [of the four humors],"[148] it became clear to al-Ghazālī that he could find the cure nowhere else than within himself.

Now it became easy for al-Ghazālī, he wrote, to find the resolve and turn away from fame and riches (*al-jāh wa-l-māl*), from family and children, and from his colleagues (*aṣḥāb*).[149] This is one of the few passages in his autobiography in which al-Ghazālī mentions his family. Later, in a letter written around 504/1110, al-Ghazālī says that he did not yet have a family when he arrived in Baghdad in the summer of 484/1091.[150] Now, four years later, his situation has changed, and he makes provisions for them, probably sending them to Ṭūs, where they would ask him to come two years later.[151] He announced that he himself would go on a pilgrimage to Mecca, while he was in reality planning to turn his path toward Syria: "I did this as a precaution in case the caliph and all of my colleagues might learn about my plan to spend time in Damascus."[152] Escaping his obligations to the caliph and the Niẓāmiyya madrasa was an important part of al-Ghazālī's plan. On the one hand, these were professional obligations. On the other hand, they were personal, sealed by oaths (singl. *bayʿa*) toward certain individuals. While a three-month-long pilgrimage would certainly be excused, a move to Damascus would have been considered desertion and defection from the promises given to caliph, sultan, vizier, and colleagues.

In Dhū l-Qaʿda 488 / November 1095, al-Ghazālī left Baghdad and traveled to Damascus. In his autobiography, al-Ghazālī describes that he had made proper arrangements for his family and his teaching position at the Niẓāmiyya.[153] His younger brother, Aḥmad, who was then a teacher at the Tājiyya madrasa, would stand in for al-Ghazālī. Aḥmad was only his brother's substitute teacher (*nāʾib*) and not an appointed professor, and he would have to leave the Niẓāmiyya after a few months.[154] During his travels to Damascus and later to Jerusalem, Hebron, and the Hijāz, al-Ghazālī was accompanied by Abū Ṭāhir al-Shabbāk of Gurgān (d. 513/1119), who had studied with al-Juwaynī alongside al-Ghazālī and stayed close to his more brilliant classmate all through these years.[155]

There are indications that al-Ghazālī's period of retreat (*ʿuzla*), which according to his autobiography began with his well-documented departure from Baghdad in the fall of 488/1095, may have started earlier. Abū Bakr ibn al-ʿArabī, who was briefly al-Ghazālī's student, mentions that he met the great "Dānishmand"[156] in Jumāda II 490 (May–June 1097), when the theologian was on his way from Syria to Khorasan and stayed in Baghdad for about six months. In one of his books, Abū Bakr describes how al-Ghazālī gave him guidance about matters concerning the human soul. Here he writes that when he met al-Ghazālī, he had already been a practitioner of Sufism for five years. Ibn al-ʿArabī specifies that his teacher had "accepted the Sufi path (*al-ṭarīqa al-ṣūfiyya*) and made himself free for what it requires" in the year 486, which corresponds roughly to 1093. That is when al-Ghazālī had put himself in seclusion (*al-ʿuzla*), Ibn al-ʿArabī says, and when he had renounced all groups.[157]

If Abū Bakr ibn al-ʿArabī's information is correct—and we have no reason to doubt it—al-Ghazālī's turn away from fame and worldly riches and toward his "seclusion" (*ʿuzla*) would have begun at least two years before he gave up his teaching at the Niẓāmiyya and left for Syria. Ibn al-ʿArabī's report informs us that leaving Baghdad was the result of a longer process and not the five-month-long crisis that is described in al-Ghazālī's autobiography. Al-Ghazālī's presentation in his *Deliverer* may have been prompted by reports about the life of the Prophet Muḥammad and about al-Ashʿarī, who, like other figures in Islam, had a life-changing experience at the age of forty. Turning one's life around in the fortieth year is a recurring motif in Muslim biographies, and, if it applies here, it would confirm our conclusion that al-Ghazālī was born in or around 448/1056.

There has been a lot of speculation about the reasons for al-Ghazālī's turn in his lifestyle and his rapprochement with Sufism that culminated in the trip to Damascus in 488/1095.[158] In his autobiography, al-Ghazālī says such speculation had begun already during his lifetime. Those who speculated were unconvinced that the reasons for his change were purely religious.[159] There is no testimony for al-Ghazālī's motivations other than the words we quoted from his *Deliverer from Error*, and further conjecture disconnects itself from textual evidence. In the end, the reasons for al-Ghazālī's "crisis" in Baghdad are less interesting than the results. Other great minds suffered similar physical and psychological traumas, and yet such traumas do not feature as prominently in their biographies as in al-Ghazālī's.[160] Whatever he experienced in the years between 485/1092 and 488/1095, al-Ghazālī created its historiography through his highly public conduct in the aftermath of these events and their narration in his autobiography. Rather than speculating about the assumed real motives behind his decision to leave Baghdad, one should focus on the effects they have on his subsequent work.

Earlier scholarship on al-Ghazālī assumed that there was a substantial change in al-Ghazālī's thinking following the year 488/1095. Some scholars even tried to explain inconsistencies in his teachings by pointing to his "conversion." Such a hermeneutic approach is not warranted. Although the weight of certain motifs in al-Ghazālī's writing changes after 488/1095, none of his theological or philosophical positions transform from what they were before. Concurrent with the report given in the *Deliverer from Error*, evaluating the moral value of human actions gains a newfound prominence in al-Ghazālī's œuvre. The connections among an individual's "knowledge" (that is, convictions), his or her actions, and the afterlife's reward for these actions gain center stage. Al-Ghazālī saw his new understanding of the afterlifely dimension of actions in this world as a *tawba*, a "repentance" or "conversion" toward a life that cares more for happiness in the hereafter than in this world. The *tawba* is a motif in Sufi literature as well as in Muslim theological texts. It is a very public event in a Muslim's life that is often talked and written about. In all his autobiographic statements, in his *Deliverer from Error*, in his comments to ʿAbd al-Ghāfir al-Fārisī, and in his letters, al-Ghazālī approached the events of 488/1095 according to the established literary trope of a Sufi repentance (*tawba*).[161] According to this literary pattern, the experiences that led to the change and the transformation are dramatic. In reality, there might have been a more gradual development that took years to manifest itself. On one subject, however, al-Ghazālī changed his mind profoundly. From 488/1095 on, he openly declined to cooperate with rulers and tried to avoid teaching at schools they patronized.

Why did al-Ghazālī travel to Damascus? The Palestinian historian ʿAbd al-Laṭīf Ṭībāwī tried to answer that question in 1965. He suggested that al-Ghazālī was attracted by the life and teachings of Abū l-Fatḥ Naṣr ibn Ibrāhīm al-Maqdisī, a prominent Shāfiʿite and a Sufi.[162] He died during al-Ghazālī's stay in Syria in Muḥarram 490 / December 1096. Abū l-Fatḥ Naṣr enjoyed a far-reaching reputation for his austerity, asceticism, and his Sufi teachings. He taught for no payment and refused to accept gifts.[163] It was said that he lived on a loaf of bread a day that was baked from the income of a piece of land he owned in Nabulus.[164] The legitimacy of the income gained through one's teaching became an important subject for al-Ghazālī.[165] Food is illicit if it is obtained by illicit means. This includes food that is bought with money given by someone who himself has obtained it unlawfully. The property of rulers and their deputies, al-Ghazālī began to stress, should generally be regarded as unlawful.[166] The Baghdad Niẓāmiyya was funded by endowments of lands as well as direct stipends that came from the Seljuq chancellery. From its very foundation in 457/1065, pious scholars were reluctant to teach there because they could not be sure its funding was proper and licit. Was the school built from spoils of earlier buildings? Was the endowed land lawfully acquired or confiscated? Were the stipends paid with tax money that had been violently extracted from its lawful owners?[167] Fear of dealing with impure and dubious things (*wara*ʿ) is a common motif in Muslim ascetic literature, and it seems to have played an important role in al-Ghazālī's decision to leave Baghdad.[168] Abū l-Fatḥ Naṣr's ethics of unpaid instruction avoided these moral dilemmas that could easily destroy one's prospect of eternal reward for teaching rightfulness. In his *Revival*, al-Ghazālī lists the obligation of teaching one's students without payment as one of the first duties of the teachers, second only to being sympathetic to one's students and their fate in the life to come.[169]

When the local Seljuq ruler offered Abū l-Fatḥ Naṣr a sum of money that he claimed came from a lawful tax, the Sufi still refused it and sent it back.[170] From this point on, al-Ghazālī adopted a similar attitude. Ibn al-Jawzī mentions that al-Ghazālī would live from the income of his writing activity,[171] vowing on the grave of Abraham in Hebron never again "to go to any ruler, to take a ruler's money, or to engage in one of his public disputations."[172] In his *Revival*, he explained to his readers why particularly weak political leaders depend on public disputations (*munāẓarāt*) and why weak scholars are drawn to them. He warns his readers against taking part and lays down eight conditions that should be met if such disputations indeed prove necessary. The fifth condition is that these disputations should be held in small circles (*khulwa*) rather than "in presence of the grandes and the sultans."[173]

In Damascus, al-Ghazālī taught liberally, and his sessions were attended by a great number of students. The chronicler Ibn al-Athīr reports that in these sessions, he began to read from his *Revival of the Religious Sciences* (*Iḥyāʾ ʿulūm al-dīn*).[174] His teaching sessions (singl. *ḥalaqa*) took place in the Umayyad Mosque and in a school building attached to its western wall.[175] Before al-Ghazālī came to Damascus, this school was known as the *zāwiya* of Abū l-Fatḥ Naṣr. It soon became known as the Ghazāliyya-*zāwiya* and was still known by that name during the eighth/forteenth century.[176] The inhabitants of Damascus also connected al-Ghazālī's name to the southwestern minaret of the Umayyad Mosque, whose upper part has since been rebuilt in Mamlūk times. The story that al-Ghazālī lived in the minaret's highest rooms may not be too farfetched. Ibn Jubayr (d. 614/1217) reports it first, after his visit to the city in 580/1184. During his time, the spacious rooms of the minaret were a dwelling place for Sufis, and an ascetic from al-Andalus inhabited the rooms in which al-Ghazālī was said to have lived nine decades earlier.[177]

Claims that al-Ghazālī stayed in Damascus for close to ten years have become part of the local lore and were caused by ʿAbd al-Ghāfir al-Fārisī's mistaken account of al-Ghazālīs travels in Syria, which was duly copied by Ibn ʿAsākir in his book on the history of Damascus. In his own comments on the subject, Ibn ʿAsākir leaves open the question of how long al-Ghazālī resided there.[178] When in his autobiography, al-Ghazālī mentions that he "stayed for almost two years in *al-Shaʾm*," the name *al-Shaʾm* refers not only to Damascus but also to the whole of Syria, including Palestine. Even his travels from Syria to the Hijaz and back fall within these two years.[179] After no more than six months, al-Ghazālī left Damascus and traveled to Jerusalem. Al-Subkī connects al-Ghazālī's departure from Damascus with the unwelcome experience of vanity. While attending incognito the teaching session of a scholar at the Amīniyya madrasa, al-Ghazālī heard his name and his teachings being quoted. He feared that pride (*ʿujb*) might inadvertently overcome him, and he decided to leave the city.[180]

Al-Ghazālī arrived in Jerusalem during the late spring or summer of 489/1096. In his autobiography, he writes that he visited the Dome of the Rock every day and shut himself up in it.[181] Here, he published his *Letter for Jerusalem* (*al-Risāla al-Qudsiyya*), a short creed that would later be incorporated into the second book of the *Revival*. The *Letter* was considered a gift to the people of Jerusalem. It was intended to be studied "by the ordinary people" (*al-ʿawāmm*) who fear the dangers of dogmatic innovations (*bidʿa*). The popular character of this work is evident from the way al-Ghazālī introduced it within his *Revival*:

> In this book [i.e., the *Revival*] let us just present the flash-lights [of dogmatics] and let us restrict ourselves to those that we have published (*mā ḥarrarnāhu*) for the people of Jerusalem. We called it *The Letter for Jerusalem on the Foundations of What-to-believe* and it is presented here in the third chapter of the book *On the Foundations of What-to-believe* in the *Revival*.[182]

In Jerusalem, al-Ghazālī may have written or published a second book, *The Stairs of Jerusalem of the Steps Leading to Knowledge on the Soul* (*Maʿārij al-Quds fī madārij maʿrifat al-nafs*). This assumption might just be deduced from the title, however, which also allows for other interpretations.[183] Al-Ghazālī's early biographers noted that after his departure from Baghdad, he turned towards the subjects of "eliminating pride and exerting one's inner self."[184] This raised an interest in the psychological teachings of the philosophers. *The Stairs of Jerusalem* presents these psychological teachings; yet it is highly technical and not suited for popular teaching.

The local historian of Jerusalem, Mujīr al-Dīn al-ʿUlaymī (d. 928/1522), who wrote in 901/1496, provides reasonably detailed information about where al-Ghazālī lived and taught in that city. He reports that al-Ghazālī "stayed at the *zāwiya*, which is above the Gate of Mercy and was known previously as the *Nāṣiriyya*, east of the Bayt al-Maqdis. It was called the *Ghazāliyya* relating to him. Since then, it has been destroyed and fallen into oblivion."[185] The Gate of Mercy (*bāb al-raḥma*), east of the Bayt al-Maqdis, is the Golden Gate in the eastern wall of the Ḥaram al-Sharīf, which here doubles as Jerusalem's city-wall toward Gethsemane. The gate's building is either Byzantine or early Islamic. Throughout its history, it has often been closed; since Ottoman times, its two entryways have been walled shut.[186] The *Ghazāliyya* school would have been on the top of this gate, situated on a platform that is currently empty (see figure 1.3).

This account of Mujīr al-Dīn is notably similar to the one given in earlier sources about the school of Abū l-Fatḥ Naṣr in Damascus, which became known as the *Ghazāliyya*. Note that in Mujīr al-Dīn's text, the school in Jerusalem is called *al-Nāṣiriyya* (and not *al-Naṣriyya*) and that the author leaves open to whom this name initially referred.[187] Yet, in another passage of his book he writes that the *zāwiya al-Nāṣiriyya* was probably where Abū l-Fatḥ Naṣr stayed earlier "for a long time." Mujīr al-Dīn cautiously suggests that the name referred to him.[188] However, no school is known to have existed at this spot during the pre-crusader period when al-Ghazālī was there.[189] Later, an Ayyūbid school *al-Nāṣiriyya* was built above the Golden Gate in Jerusalem during the seventh/thirteenth century. Its foundation in 610/1214 was part of the refurbishment of the Ḥaram al-Sharīf by al-Malik al-Muʿaẓẓam ʿĪsā when he was governor of Damascus.[190] The name *al-Nāṣiriyya* referred to his uncle Ṣalāḥ al-Dīn (Saladin), who had reconquered Jerusalem from the crusaders in 583/1187 and whose official title was *al-Malik al-Nāṣir*—the Victorious King.[191] Ibn al-Ṣalāḥ al-Shahrazūrī (d. 643/1245), who was himself an influential commentator of al-Ghazālī's legal works, had taught at this madrasa before he settled in Damascus.[192] By the time of Mujīr al-Dīn's writing, it had long been derelict. It is most probably this school that Mujīr al-Dīn mistakenly connects

FIGURE 1.3 Jerusalem's *Gate of Mercy*. View from inside the *Noble Sanctuary* with the platform on top, the site of the Madrasa al-Nāṣiriyya during the seventh/thirteenth century.

to al-Ghazālī's time.[193] Abū l-Fatḥ Naṣr had left Jerusalem for Tyros and Damascus twenty years before al-Ghazālī arrived there, and it is unlikely that schools existed in his name in both Damascus as well as in Jerusalem when al-Ghazālī visited these places.[194]

Al-Ghazālī left Jerusalem in the fall of 489/1096 in order to take part in the annual pilgrimage at the end of that year. On his way to the Hijaz, he stopped in Hebron and visited the graves of the patriarchs, making the aforementioned vow.[195] From Hebron, al-Ghazālī continued to Mecca and Medina. His participation in the pilgrimage of 489 is a well-documented event.[196] Some Muslim

FIGURE 1.4 The *Gate of Mercy* in Jerusalem. View from outside the city wall.

historians still believed he made the pilgrimage one year earlier, emphasizing how much confusion exists about the details of al-Ghazālī's life. This confusion was, of course, created by al-Ghazālī himself when he lied about his plans before he left Baghdad in 488/1095.

Al-Ghazālī's own comment that he "stayed for almost two years in *al-Sha'm*" indicates a return to Syria after the pilgrimage. There is a report that Abū l-Fatḥ Naṣr passed away shortly before al-Ghazālī entered Damascus.[197] Since Abū l-Fatḥ died on 9 Muḥarram 490 / 27 December 1096, this report must refer to al-Ghazālī's second arrival in Damascus after his return from the Hijaz. Four months later, in Jumāda II 490 / May–June 1097, al-Ghazālī was back in Baghdad. During the year 490/1096–97, when al-Ghazālī left Syria, news of a great army of Europeans (*faranj*) reached Damascus, and the city prepared to send a contingent for the relief of the besieged Antioch.[198] The crusaders took Antioch in Jumādā I 491 / April–May 1098 when al-Ghazālī was already back in Khorasan. In his works, al-Ghazālī never refers to the arrival of the crusaders in the Levant. There is, however, a single, quite drastic reference in a Persian work that could be seen as spurious because of this atypicality. In the *Present to Kings* (*Tuḥfat al-mulūk*) attributed to al-Ghazālī, the author writes that the unbelievers (*kāfirān*) have taken over Muslim lands, removed pulpits from mosques, and turned the sanctuary of Abraham in Hebron into a pigsty (*khūk-khāne*) and Jesus' birthplace in Bethlehem into a tavern. Therefore, the author concludes, *jihād* against these unbelievers is imperative.[199]

His second sojourn in Baghdad is well documented by the reports of his student Abū Bakr ibn al-ʿArabī. Al-Ghazālī stayed at the Sufi convent named after Abū Saʿd of Nishapur right opposite to the Niẓāmiyya madrasa.[200] Every day, many people would gather there to hear him teach and read from his *Revival of the Religious Sciences*.[201] *Revival* was an unusual book for its time. It was conceived as a work on the "knowledge of the path to the afterlife" (*ʿilm ṭarīq al-ākhira*), a practical guidebook on how its readers may gain the afterlife through the actions they perform in this world.[202] In the introduction, al-Ghazālī writes that the book is on the "knowledge (or: science) of human actions" (*ʿilm al-muʿāmala*) and not on the "knowledge of the unveiling" (*ʿilm al-mukāshafa*). It wishes to provide that type of knowledge that prompts humans to act rightfully, staying clear of knoweldge that has no consequences for human actions.[203] The religious knowledge that al-Ghazālī wishes to revive is "the jurisprudence of the path to the hereafter."[204] *Revival* creates a new genre of literature by combining at least three earlier ones: the genre of *fiqh* books on the individual rulings (*furūʿ*) of Sharīʿa, the genre of philosophical tractrates on ethics and the development of character such as Miskawayh's (d. 421/1030) *Refinement of Character* (*Tahdhīb al-akhlāq*), and the genre of Sufi handbooks such as Abū Ṭālib al-Makkī's (d. 386/998) *Nourishment of the Hearts* (*Qūt al-qulūb*). About a hundred years before al-Ghazālī, authors such as al-ʿĀmirī (d. 381/992) or al-Rāghib al-Iṣfahānī (fl. c.390/1000) wrote works that combined philosophical ethics with religious literature. *Revival* stands in that tradition and borrows from it.[205]

In his autobiography, al-Ghazālī writes that "certain concerns and the pleading of the children drove me back to the native land."[206] His family, who stayed behind in Baghdad when he left in 488/1095, had already come to Ṭūs and waited for him. When he now left Baghdad, he seems to have used the same ruse he had employed two years earlier. Then, in 488/1095, al-Ghazālī had pretended to travel to the Hijaz while he had actually gone to Syria. In a letter written shortly before his departure from Baghdad in the summer of 490/1097, he claimed to be leaving for the Hijaz.[207] Another letter permits the conclusion that in Dhū l-ḥijja 490 / November 1097, he was already back in Ṭūs,[208] leaving no more than six months during the summer of 490/1097 for his second and last stay in Baghdad.

The Ideal of a Secluded Life–His Last Years in Khorasan

The Iranian historian ʿAbd al-Ḥusayn Zarrīnkūb characterized al-Ghazālī's decision to leave Baghdad in 488/1095 as an "escape from the madrasa."[209] This is correct only insofar as the madrasa was a state institution and effectively part of the Seljuq administration. His three vows at Hebron reveal that al-Ghazālī rejected the state and its officials, but not teaching in schools. Al-Ghazālī never gave up teaching, nor did he ever take time off from teaching. After 488/1095, however, his teaching largely took place at small madrasas that were not founded and financed by the Seljuq state. Such a small madrasa, or, as we would say today, a private madrasa, was often called a *zāwiya*, a "corner." In Medieval Latin, unofficial teaching that was not authorized by the church was sometimes called teaching "in corners" (*in vinculi*). Something similar might be behind the usage of the Arabic *zāwiya*. Official teaching happened in madrasas, unofficial teachings in a "corner." Abū l-Fatḥ Naṣr's school in Damascus, for instance, was a *zāwiya*. In his *Revival*, al-Ghazālī says that scholars fall into two groups: those *muftīs*, that is, scholars, who write offical *fatwās* and who are the companions of sultans, and those who "have knowledge of divine unity (*tawḥīd*) and the actions of the heart and who are the solitary and isolated inhabitants of the *zāwiyas*."[210]

In Ṭābarān, al-Ghazālī built both a *zāwiya* and a *khānqāh*.[211] As a minor madrasa that is not maintained by the state, a *zāwiya* needs the support of small endowments or donations. Al-Ghazālī was opposed to the idea that students should pay for their education. Teachers, he said, should emulate the Prophet and not require payment for teaching; knowledge should not serve its holders but rather be served by them.[212] The term *khānqāh* refers to a Sufi convent that also required the funding of donated wealth. The origin of the word *khānqāh* is Persian; and although it gains usage in Arabic during this time, some sources prefer to use the Arabic translation *ribāṭ* for ʿAbd al-Ghāfir al-Fārisī's word *khānqāh*. A *ribāṭ* originally refers to a "camp" or "convent" for those who fight in a *jihād*. In his Arabic *fatwā* on who is allowed to live in the *khānqāh*, al-Ghazālī himself uses the word *ribāṭ*, even though in a Persian text on the same subject, he employs *khānqāh*. For al-Ghazālī, the inhabitants of a *khānqāh* fight the *jihād* of the soul. Others agree: Ibn al-Jawzī, for instance, writing in Arabic, documents this synonymous usage when he reports that in Ṭūs al-Ghazālī "had in his neighborhood a madrasa and a convent (*ribāṭ*) for those who practice Sufism. He also built a nice house and planted a garden."[213]

The *khānqāh* was a relatively new institution at this time. It allowed those devoted to Sufism to stay there and pursue an ascetic lifestyle in the company of like-minded peers. Al-Ghazālī had a clear idea about who could come and stay at his *khānqāh*. He wrote a Persian *fatwā* in which he clarifies that only those who are free from such sins as adultery and homosexual intercourse and who do not adorn themselves by wearing silk and gold are allowed to live in the *khānqāh* and benefit from its facilties. He did not admit people who pursued a profession other than such things as tailoring or paper making that can be done in the *khānqāh*. The fact that there was endowed wealth (*amwāl*) on the side of the *khānqāh* should allow its attendants to withdraw from the workforce. Al-Ghazālī also excluded those who seek financial support from the sultan and who have acquired their means of living in another unlawful manner.[214]

In his autobiography written in 500 (1106–7), al-Ghazālī portrays his life back home in Ṭūs: "I chose seclusion (*ʿuzla*), desiring solitude and the purification of the heart through *dikhr*."[215] These words, together with ʿAbd al-Ghāfir al-Fārisī's report about al-Ghazālī's last years in Ṭūs, created the mistaken impression of a totally isolated scholar who had withdrawn from all public activity. ʿAbd al-Ghāfir al-Fārisī writes in a passage about al-Ghazālī's return to Khorasan that many later historians copy:

> Then he returned to his homeland where he stayed close to his family. He was preoccupied with meditation (*tafakkur*) and he was tenacious of his time. He was the precious goal and the preserveance of the hearts for those who seeked him and who came to see him.[216]

The word "seclusion" (*ʿuzla*) is used almost every time al-Ghazālī writes about his life after 488/1095. Given that he published books, taught in his *zāwiya*, and received those who came to him, this cannot mean the sort of seclusion from his contemporaries that we would describe as a hermit's retreat, fully separate from the outside world. What al-Ghazālī intended for his seclusion became clear during his written and oral exchanges with Sanjar. These conversations were collected and later edited by one of his descendents. Here, soon after 501/1108, al-Ghazālī claimed that:

> [I]n the months of the year 499, the author of these lines, Ghazālī, after having lived in seclusion (*ʿuzlat*) for twelve years and after having been devoted to the *zāwiya*, had been obliged to come to Nishapur in order to occupy himself with the spread of knowledge and of divine law (*sharīʿat*). (This was ordered), since in scholarship debility and weakness had become widespread. The hearts of those dear to him and of those who have insight (*ahl-i baṣīrat*) rushed to help him with all their good will. In sleep and in wake he was given to understand that this effort is the beginning of something good

> and the cause for a revival of scholarship and of divine law. After he agreed (to come to Nishapur) he brought splendor to the teaching position and students from all parts of the world made efforts to come to him.[217]

When al-Ghazālī writes that he lived "in seclusion (ʿ*uzla*) for twelve years devoted to the *zāwiya*," he refers to the period of eleven lunar years between his departure from Baghdad in Dhū l-Qaʿda 488 / November 1095 and the beginning of his teaching at the Niẓāmiyya in Nishapur in late 499 / summer of 1106, an event that will be discussed below. The discrepancy between twelve and eleven is either a glitch on al-Ghazālī's part or a scribal mistake.[218] The years of *zāwiya* life that al-Ghazālī mentions includes his popular teaching at Damascus and Baghdad, his writing of a letter for the people of Jerusalem, his performing of the pilgrimage, and most important, his teaching at his own *zāwiya* and *khānqāh* in his hometown Ṭābarān-Ṭūs.[219]

"Being devoted to the *zāwiya*"[220] simply means that he had dedicated himself to the teaching at private madrasas and *khānqāhs* in Damascus, Jerusalem, Baghdad, and Ṭābarān-Ṭūs. Thus "seclusion" (ʿ*uzla*) merely means not serving in a public office and not being engaged in state-sponsored teaching at one of the Niẓāmiyya schools. The key element of this seclusion is avoiding any close contact with the rulers and audiences selected by them. This principle is a Sufi topos, and it is prominent in the *Deliverer from Error*, where the two teaching engagements at Niẓāmiyya schools (separated by eleven years) are described in very similar terms. When in 504/1110, al-Ghaza¯lī is once again invited to teach at the Niẓāmiyya in Baghdad (which will also be discussed below), he declines, saying that a public office would not suit him well. In a letter to his invitor Ḍiyā al-Mulk Ah̩mad, the son of Niẓām al-Mulk and the vizier to Sultan Muh̩ammad Tapar, he excuses himself by pointing to his three vows at the grave of Abraham:

> If I fail towards these vows it will darken my heart and my life. Success won't be granted to anything that I will do in this world. In Baghdad one cannot avoid public disputations and one has to attend the palace of the caliph. During the time while I returned from Syria, I had no business in Baghdad, and since I had no official position, I was free from all responsibilities. I chose to live by my own. If I am given an office, I cannot live without burden (*musallam*). But since my innermost will yearn to give up the office and return to a free state, it will have no good effect. The most important excuse is, however, that I will be unable to earn my living, since I cannot accept money (*māl*) from a ruler (*sulṭān*) and since I have no property (*milk*) in Baghdad [to live from.] If one lives economically and in abstinence, the piece of land that I own in Ṭūs is for my humble person and the children just enough.[221]

ʿAbd al-Ghāfir al-Fārisī's personal report about this last period in al-Ghazālī's life, when he stayed in Ṭābarān-Ṭūs, centers on his conversion (*tawba*). He contrasts the al-Ghazālī of his late years with the one ʿAbd al-Ghāfir knew as a young and brilliant student-colleague under al-Juwaynī. ʿAbd al-Ghāfir's impression of the younger al-Ghazālī was far from positive: the young scholar was dominated by a feeling of superiority over others. The late al-Ghazālī had completely changed, and yet ʿAbd al-Ghāfir initially suspected his kind manners to be merely a pretense adopted to cover up his true nature as a scholar filled with hubris. By and by, however, ʿAbd al-Ghāfir became convinced of the depth of al-Ghazālī's conversion:

> I visited [al-Ghazālī] many times and it was no bare conjecture of mine that he, in spite of the maliciousness and roughness towards people that I witnessed during the times past, had become quite the opposite and was cleansed from these filthy strains. In the past he had looked at people from above and with defiance. He had a vain pride and was blinded by the ease with which God had provided him to handle words, thoughts, expressions, and the pursuit of glory. I used to think that [this new al-Ghazālī] was wrapped in the garments of false mannerism (*takalluf*) and regarding what had become of him, he was suppressing his natural disposition (*nāmūs*). But I realized after investigation that things were the opposite of what I had thought, and that the man had recovered after he had been mad.[222]

ʿAbd al-Ghāfir's report of his nightly talks with al-Ghazālī has many parallels in the autobiography *The Deliverer from Error.* Yet ʿAbd al-Ghāfir's retelling of the events are more concrete and less chronologically streamlined. The seeds of al-Ghazālī's *tawba* appear much earlier in this report than in the written autobiography. According to ʿAbd al-Ghāfir, al-Ghazālī studied the sciences and excelled in everything that had caught his interest. After these early successes, he started to meditate about the afterlife, which led him to seek the company of the influential Sufi teacher Abū ʿAlī al-Fāramadhī.[223] Al-Fāramadhī was a Shāfiʿite from Ṭūs, where he died in 477/1084 when al-Ghazālī was in his late twenties. Al-Fāramadhī was engaged in mystical practices (*tadhkīr*) and one "to whom flashes from the light of insight have been made visible."[224] ʿAbd al-Ghāfir says that the younger al-Ghazālī received from al-Fāramadhī an introduction to his Sufi method (*ṭarīqa*).

After his initiation to Sufism, al-Ghazālī experienced his first crisis of knowledge, the one he describes in the second chapter of the *Deliverer from Error*, "The Inroads of Skepticism."[225] In the autobiography, this crisis precedes al-Ghazālī's mastering of the sciences. Here, in ʿAbd al-Ghāfir's report, the epistemological crisis is a less dramatic confusion about the criteria for truth. It was prompted by the relativist impression that rational arguments seem to stand undecidedly against one another without trumping their opposites (*takāfuʾ al-adilla*). Finally, ʿAbd al-Ghāfir's conversations with al-Ghazālī illuminate the major crisis in his life, the Sufi *tawba* that led to his departure from Baghdad. The main motive from the *Deliverer*, namely, fear of the afterlife, also dominates ʿAbd al-Ghāfir's report:

> Then he related that a gate of fear had been opened for him to such an extent that he could no longer occupy himself with anything else until [his fear] got better. In this manner he remained until he was fully practiced [in matters of religion]. [Only now] the truths (*al-ḥaqāʾiq*) became apparent [to him].[226]

ʿAbd al-Ghāfir's report about al-Ghazālī's two crises calls the chronology of events in the autobiography in question. Sufism appeared much earlier in al-Ghazālī's life than he acknowledges in that book. ʿAbd al-Ghāfir also confirms the impression of some readers of the *Deliverer* that the narrative description—studying first *kalām*, then *falsafa*, then Ismāʿīlite theology, until finally reaching Sufism—stems from pedagogical conventions and does not represent the actual sequence of study in al-Ghazālī's life.[227]

In the late months of 499 / summer of 1106, shortly before the turn to a new century in the Islamic calendar, al-Ghazālī began teaching at the Niẓāmiyya madrasa in Nishapur.[228] That event prompted the writing of his autobiography, *The Deliverer from Error*, which responds to criticism from both close followers as well as hostile scholars. There, he legitimizes his return to the Niẓāmiyya schools by linking it to the needs of an epoch characterized by religious slackness (*fatra*) and the temptations of false beliefs. Al-Ghazālī says that he consulted with a group of people "who have a pure heart and religious insight (*mushāhada*)" who advised him to leave his seclusion and emerge from his *zāwiya* to lead the much-needed religious renewal at the beginning of the new century. In addition, al-Ghazālī mentions that "the sultan at that time" ordered him to come to Nishapur. From al-Ghazālī's letters, it becomes clear that Fakhr al-Mulk, a son of Niẓām al-Mulk and vizier to Sanjar (see figure 1.5), put pressure on al-Ghazālī. He wanted al-Ghazālī to return to state-sponsored

FIGURE 1.5 Sanjar as Sultan among his court. Miniature from Rashīd al-Dīn Ṭabīb's *Compendium of Chronicles* (Edinburgh University Library, MS Arab 20, fol. 142a).

teaching in Nishapur. ʿAbd al-Ghāfir reports that Fakhr al-Mulk confronted al-Ghazālī with the demand to teach at the Niẓāmiyya in Nishapur, summoned him, and listened to him—meaning he heard his excuses. ʿAbd al-Ghāfir's language suggests that the vizier did not mince words and used all means of persuasion short of brute force. Subsequently, al-Ghazālī "was taken" to Nishapur and began teaching at the Niẓāmiyya.[229] From his letters, one gets the impression that al-Ghazālī had a close relationship with Fakhr al-Mulk: the scholar addresses the vizier as his trusted intellectual mentor.

A couple of years later, when al-Ghazālī spoke about the events in 499/1106 to Sanjar, he said that he was initially afraid of returning to the Niẓāmiyya school and mentioned his fear to the vizier:

> I said to Fakhr al-Mulk that this era cannot bear my words and that during these times everybody who says the truth has walls erected right in front of him. He said to me: "This king (*scil.* Sanjar) is just and I will come to your aid."[230]

Al-Ghazālī feared the possibility that scholars who objected to his teachings might stir up the Seljuq ruler against him. This was indeed the situation in which he found himself during this conversation with Sanjar. Two years later, in late 501 / summer of 1108, a group of scholars that included all Sunni schools of jurisprudence present in Khorasan accused al-Ghazālī of

> not being a believer in Islam but rather following the beliefs of the *falāsifa* and the heretics (*mulḥidān*). All his books are infested with their words and he mixes unbelief (*kufr*) and falsehoods (*abāṭīl*) with the secrets of revelation. He calls God the "true light" [in his *Niche of Light*] and that is the teaching of the Zoroastrians (*majūs*).[231]

Abū ʿAbdallāh al-Māzarī al-Dhakī (d. 510/1116–17), a native of Mazzara in Sicily and an Ashʿarite Mālikite scholar who had come to the east from Tunisia, was particularly active in this campaign.[232] ʿAbd al-Ghāfir al-Fārisī refers to a controversy in Nishapur that began around 497/1103, when Fakhr al-Mulk became grand vizier of Sanjar and first attempted to make al-Ghazālī a teacher at the Niẓāmiyya in Nishapur.[233] Once in Nishapur, al-Ghazālī's teaching faced resistance. "His staff was struck," ʿAbd al-Ghāfir writes, "by opposition, by attacks on him, and by slanderings about what he omitted and what he committed."[234] But al-Ghazālī remained calm and did not respond to these attacks, ʿAbd al-Ghāfir says, nor did he show much ambition to correct his opponents' mistakes. Given the arrogance and the litigiousness of the younger al-Ghazālī, ʿAbd al-Ghāfir had found it hard to believe that he had changed when he returned to teaching at the Niẓāmiyya in Nishapur. Yet his calm posture in the face of numerous accusations and slandering impressed ʿAbd al-Ghāfir, and in a very personal note, he confirms that his former colleague had indeed become different. In a discussion about al-Ghazālī's bearing while teaching in Nishapur, he writes:

> What we [initially] thought was pretention (*tamarrus*) and an aqcuired mode (*takhalluq*) was, in fact, his [true] nature (*ṭab*ʿ) and the realization (*taḥaqquq*) [of what he truely was]. This was the sign of the happiness that has been ordained on him by God.[235]

The collection of al-Ghazālī's letters provides more information about the conflict. Its compiler reports that among the scholars who carried accusations to Sanjar's court were a group of Ḥanafites, who asked that al-Ghazālī be punished for a passage in one of his early legal works. In this early work, al-Ghazālī polemicizes in an aggressively partisan spirit against the founding figure of their school, Abū Ḥanīfa.[236] Since Sanjar was himself a Ḥanafite, the situation was potentially dangerous. A much later source from the tenth/sixteenth century claims that the Ḥanafite scholars had issued a *fatwā* demanding al-Ghazālī's execution.[237] Shortly after 501/1108, al-Ghazālī appeared before Sanjar. In the meantime, Ismāʿilite agents had murdered Fakhr al-Mulk. During his appearance before Sanjar, al-Ghazālī evoked the memory of the assasinated grand vizier and the promises he made to secure al-Ghazālī's safety in the midst of the accusations driven by Nishapur's notorious partisanship among the legal schools. Al-Ghazālī asked Sanjar to release him from his teaching obligation in Nishapur and in Ṭūs.[238] The name "Nishapur" refers, of course, to al-Ghazālī's teaching at the Niẓāmiyya madrasa there. It is unclear, however, what the reference to Ṭūs means. Maybe al-Ghazālī was also required to teach there at a local state-sponsored school?

Sanjar declined to release al-Ghazālī from his teaching posts. In fact, the theologian's address (*c.* 501/1108) made such a strong impression on Sanjar that he said: "We should have ordered that all scholars of Iraq and Khorasan be present to hear your words."[239] Sanjar promised to build madrasas for al-Ghazālī, "and we will order that all scholars should come to you once a year in order to learn everything what is unknown to them. If someone has a disagreement (*khilāf*) with you, he should be patient and ask you to explain the solution to his problem."[240] This version of events is the one reported by al-Ghazālī's followers and students. It does seem that he was exonerated from the accusations of his anti-Ḥanafism, and more amicable relations between him and Sanjar were established. He wrote the *Council for Kings* (*Naṣīḥat al-mulūk*) for Sanjar in response to a piece of game (*shikār*) the vice-regent sent him from one of his hunts nearby.[241]

The nature of al-Ghazālī's duties in Nishapur did not seem to keep him there long or all too often. When Sanjar summoned him to his regularly established camp near Ṭūs,[242] al-Ghazālī arrived there from Ṭābarān, although this was still during the time that he was required to teach in Nishapur. In fact, Sanjar directs all his communications with al-Ghazālī to Ṭābarān and never to Nishapur. During this time, al-Ghazālī also had students in Ṭūs; a group of them appeared before Sanjar's court to defend al-Ghazālī from enemy accusations.[243] It is not clear whether these students were those who lived at al-Ghazālī's *khānqāh* and studied with him at his *zāwiya*, or whether they were from an official teaching engagement he held in Ṭūs.

There is evidence that al-Ghazālī had temporarily left Ṭūs years before his teaching engagement in Nishapur. In one of his Persian letters to the vizier Mujīr al-Dīn, which Krawulsky has tentatively dated as shortly after 490/1097, al-Ghazālī mentions that Ṭūs had been plagued by "oppressors" (*ẓālimān*), prompting al-Ghazālī to leave that place. After a year, however, he was forced (*bi-ḥukm-i ḍarūrī*) to return to Ṭūs and saw that the oppression (*ẓulm*) was still going on.[244] If the dating of this letter and its information is correct, al-Ghazālī would have stayed in Ṭūs for very little time after he had arrived there from Baghdad in Dhū l-ḥijja 490 / November 1097. It is more likely that the dating needs to be corrected and that all this actually happened a handful of years later. About ten years after his arrival in Ṭūs, in his conversation with Sanjar, al-Ghazālī refers to the fact that the people of Ṭūs had to endure "much oppression" (*ẓulm bisyār*) and that their harvests were poor because of cold and lack of water. He implicitly accuses Sanjar of being responsible for their situation since he was the one who tolerated the people of Ṭus's being robbed.[245] Given the political situation at this time, the oppressors were most likely nomadic Turks who roamed the countryside of Ṭūs and disrupted its irrigation systems. These Turks may have been part of Sanjar's regular Seljuq army, whose choice of a camp location near Ṭūs likely led to strained area resources. The oppressors may also have been from one of the numerous groups of irregular nomadic Turks who had moved from Central Asia to Khorasan and were referred to as *ghuzz* in the sources. Sanjar had only limited power over these groups and probably little motivation to call them to order. When in 548/1153 Sanjar lauched a campaign against a group of Oğuz Turks who had failed to pay their tribute, he suffered a surprising defeat and was captured. Their real power now became evident; defenseless, the walls of most major cities of Khorasan were overrun and many of their inhabitants robbed and killed.

ʿAbd al-Ghāfir reports that al-Ghazālī quit teaching in Nishapur and returned to Ṭūs before his death. His wording suggests that the scholar handed in his resignation before the local unrest in Nishapur would lead to his dismissal.[246] Back in Ṭābarān, ʿAbd al-Ghāfir says he turned his attention to the study of *ḥadīth* in the collections of Muslim and al-Bukhārī. ʿAbd al-Ghāfir stresses that he actually studied the transmission of *ḥadīth*-material, meaning the distinction of what can and cannot be verified through chains of reliable transmitters.[247] Among the works of al-Ghazālī, there is no evidence for the kind of traditionalist *ḥadīth*-scholarship these words might invoke. Al-Ghazālī was always considered a weak transmitter of *ḥadīth*. Later critics would list this as one of his faults, and admirers filled volumes to make up for his neglect.[248] If the content of a Prophetical report fitted al-Ghazālī's purposes, he did not bother much to check whether it had a sound chain of transmitters (*isnād*). According to some historians, al-Ghazālī openly admitted this and said: "I have little expertise in the *ḥadīth*-science."[249] In fact, in the first book of *Revival*, he criticizes those who wrote the earliest collections of *ḥadīth*.[250] ʿAbd al-Ghāfir's report reproduces a literary trope in classical Islamic literature: a rationalist scholar who neglects the outward meaning of revelation and the *sunna* of the Prophet for much of his life, finally repenting shortly before his death and returning to these sources. There is little evidence for al-Ghazālī becoming a traditionalist *ḥadīth*-scholar late in his life, and perhaps behind this report is a different kind of *ḥadīth*-study than the verification of reports through the study of their chains of transmission. In his two late books, *The Criterion of Distinction between Islam and Clandestine Apostasy* (*Fayṣal al-tafriqa bayna l-Islām wa-l-zandaqa*) and *Restraining the Ordinary People from the Science of Kalām* (*Iljām al-ʿawāmm ʿan ʿilm al-kalām*), al-Ghazālī is deeply concerned with the anthropomorphic descriptions of God that appear in the *ḥadīth*-corpus. Both books teach an appropriate attitude toward those reports and the correct interpretation (*taʾwīl*) of them, and maybe this is what ʿAbd al-Ghāfir tried to turn apologetically into a more traditionalist understanding of *ḥadīth*-scholarship.

Whether al-Ghazālī was ever officially released from his teaching position in Nishapur is unclear. Neither do we know when his teaching engagement ended nor who succeeded him as the head teacher of the Niẓāmiyya madrasa in Nishapur. An obvious candidate is Abū l-Qāsim al-Anṣārī (d. 512/1118), one of the most prominent theologians of his time in Nishapur and, like al-Ghazālī, a student of al-Juwaynī. He seems to have been younger than al-Ghazālī. He is the author of two important works that stand much deeper in the teaching tradition of al-Juwaynī than al-Ghazālī's œuvre.[251] Al-Anṣārī was initially a teacher at the Bayhaqī madrasa, the second most important institution for Shāfiʿites in Nishapur.[252] If he had ever become the head teacher at the Niẓāmiyya in Nishapur, he did so after al-Ghazālī left that position.[253]

According to the *Deliverer from Error*, al-Ghazālī seems to have accepted that the return to a Niẓāmiyya was necessary for reasons other than just the pressure of Sanjar and Fakhr al-Mulk. The letters clearly reveal that al-Ghazālī never liked this assignment.[254] There are at least two reasons why he would detest teaching at the Niẓāmiyya madrasa. First was his decision not to work for state authorities. Second, al-Ghazālī may not have liked the fact that he had to teach in a public space where whomever wanted could join the teaching circle. In his conversations with Sanjar, it becomes clear that he feared eavesdroppers on his lectures and potential spies for other scholars or for the Seljuq authorities. This is why he starts his apologetic address to Sanjar by saying that he is intellectually so remote from other scholars that they are unable to understand the real meaning of his words. In his own *zāwiya* in Ṭābarān, where he apparently taught all through these years, he could handpick those who would become his students and expel those he did not trust.

In 504/1110, Ḍiyāʾ al-Mulk Aḥmad ibn Niẓām al-Mulk, the vizier to the Supreme Sultan Muḥammad Tapar, who was Sanjar's older brother, invited al-Ghazālī to return to the Niẓāmiyya madrasa in Baghdad and take up the chair he once held. Its recent holder, al-Kiyāʾ al-Harrāsī, who had been teaching on this position since 493/1100, had just died.[255] The exchange of letters on this occasion is preserved. Al-Ghazālī responded in a letter that later became widely known.[256] He declines Ḍiyāʾ al-Mulk's offer and excuses himself by saying that "pursuing the increase of worldly goods" (*ṭalab bi-ziyādat-i dunyā*) has been removed from his heart. He mentions his madrasa in Ṭūs and says that he has a family and 150 students to care for.[257] It seems that at this time, al-Ghazālī no longer taught at the Niẓāmiyya in Nishapur.

On 14 Jumāda II 505 / 18 December 1111, al-Ghazālī died in Ṭābarān, at approximately fifty-five years old. His death came only a few days after he had finished work on his last book, *Restraining the Ordinary People from the Science of Kalām* (*Iljām al-ʿawāmm ʿan ʿilm al-kalām*). His brother, Aḥmad, was probably present during his death, since he left us a description of al-Ghazālī's last day.[258] When the news of his death reached Baghdad, the court poet al-Abīwardī (d. 507/1113) eulogized al-Ghazālī in a short poem.[259] Al-Ghazālī was

FIGURE 1.6 The Hārūniyya mausoleum in Ṭābarān-Ṭūs at the beginning of the twentieth century. Watercolor by André Sevruguin (from Diez, *Die Kunst der islamischen Völker*).

buried in a mausoleum right outside the walls of Ṭābarān's citadel (*qaṣaba*).[260] After Ṭābarān's destruction in 791/1389, al-Ghazālī's mausoleum fell into decay and could at one point barely be identified. It is most likely a heavily reconstructed building that is today erroneously named *al-Hārūniyya*, that is, the mausoleum of Hārūn al-Rashīd (see figure 1.6).[261] Significant funds went into the contruction of this impressive building. It bears some architectural resemblance to Sanjar's mausoleum in Marw, which suggests that he or some high dignitary at the Seljuq court commissioned al-Ghazālī's mausoleum.

There is no information as to what became of al-Ghazālī's children. ʿAbd al-Ghāfir al-Fārisī provides the information that he had only girls.[262] There was, in fact, no prominent male descendent of al-Ghazālī, at least not someone who merited mention in the biographical dictionaries. A manuscript of one of his legal works copied two years after al-Ghazālī's death in 507/1113 contains an *ijāza* issued by a Muḥammad al-Ghazālī who, if he existed, may have been the author's son.[263] Of course, the note may simply be a forgery, intended to increase the manuscript's market value. We do not hear of his descendents until some time later, when the unknown collector of al-Ghazālī's letters claims to be related to the author.[264]

A direct descendent of al-Ghazālī is mentioned during the Īl-Khānid period in Baghdad. The Egyptian lexicographer al-Fayyūmī reports that in 710/1310–11, he met a sheikh in Baghdad who was an eighth-generation descendant of al-Ghazālī.[265] According to his lineage, which is fully recorded by al-Fayyūmī, one of al-Ghazālī's daughters was the great-grandmother of Ṭāhir ibn Abī l-Faḍāʾil Fakhrāwir, who appears in this chain as a Shirwānshāh, that is, a king of the independent region of Shirwān in northern Azerbaijan. It might be a coincidence that around the time that Ṭāhir lived, a member of the family of the Shirwānshāhs was a student of Fakhr al-Dīn al-Rāzī, who commissioned one of his books.[266] Later references to the family of al-Ghazālī are much more vague. The historian Ibn al-ʿImād (d. 1089/1679) mentions a direct decendent of al-Ghazālī, a Ḥanbalī scholar who died in Aleppo in 830/1427; but this information seems unreliable.[267] In the twelfth/eighteenth century, al-Murtaḍā al-Zabīdī reports that Aḥmad al-Ṭahṭāʾī (d. 1186/1772), one of the Egyptian Shādhilī Sufis, claimed that he once met descendants (*awlād*) of al-Ghazālī in Abnūd in Upper Egypt.[268]

2

Al-Ghazālī's Most Influential Students and Early Followers

Al-Ghazālī was the most influential teacher of Islamic law and theology during the fifth/eleventh and the sixth/twelfth centuries. He had a particularly monumental impact on the intellectual life of the century after his death. Indeed, his writings on the relationship between the philosophical sciences and Muslim theology profoundly affected all Muslim thinkers until the early twentieth century and still carry weight in the Muslim discourse on reason and revelation today. The biographical dictionaries of the Shāfiʿite school of law feature numerous articles on the many scholars who studied with al-Ghazālī. In 1972, Henri Laoust made a cautious attempt to view this material.[1] The writings of his students are an important source for our understanding of al-Ghazālī's theology. In particular, his early followers offer contextualized insight into his teachings that later literature cannot offer. Although the reactions to al-Ghazālī by authors from the Muslim West (al-Andalus, specifically) have been studied since Ernest Renan's *Averroès et l'averroïsme* of 1852, comparatively little is known about the intellectual history of the sixth/twelfth century in the Muslim East. Key figures of the reception of al-Ghazālī's thought during this century remain largely unknown today. Sharaf al-Dīn al-Masʿūdī (d. after 582/1186), for instance, lived around the middle of the sixth/twelfth century in Transoxania and wrote what is probably the very first commentary on Avicenna's *Pointers and Reminders*. In this work, *Doubts and Uncertainties on the Pointers*, al-Masʿūdī takes a critical stand toward Avicenna's most theological work.[2] Al-Masʿūdī's student Ibn Ghaylān al-Balkhī, who lived close to end of the century, composed a harsh criticism of Avicenna's arguments in favor of the world's pre-eternity. In it, he praises his teacher al-Masʿūdī as someone who had developed an understanding of the philosophical sciences similar only to that of al-Ghazālī.[3] We will hear more about the typical Ghazalian approach of al-Masʿūdī and Ibn Ghaylān al-Balkhī at the end of the next chapter.

One of the most important early followers of al-Ghazālī was his brother, Abū-l Futūḥ Aḥmad al-Ghazālī. He outlived his older sibling Muḥammad by either eighteen or twenty-one years and was an influential scholar in his own right.[4] He became famous for his preaching activity in the cities of Iraq and Iran. His brother Muḥammad confessed that he had no talent for preaching and would rather leave that to others.[5] He saw the role of highly educated religious scholars (*ʿulamāʾ bi-Llāh*) as addressing the intellectual elite, while preachers (*al-wuʿʿāẓ*) would speak to the masses.[6] Muḥammad clearly saw himself in the first category; his brother Aḥmad likely understood himself as also belonging to the latter class.

The most widespread epitome of al-Ghazālī's *Revival* is a book named *The Kernels of the Revival* (*al-Lubāb min al-Iḥyāʾ*), which is sometimes attributed to Aḥmad, although most manuscripts, including the one(s) on which the printed version is based, clearly identify it as a work of Muḥammad's.[7] In his own œuvre, Aḥmad was concerned with the same subjects that his brother discussed in his *Revival*. In one of Aḥmad's short epistles, for instance, he explains what the confession of monotheism (*tawḥīd*) truly entails.[8] This is a prominent subject in the thirty-fifth book of his brother's *Revival* and in his *Niche of Lights*. ʿAyn al-Quḍāt al-Hamadhānī, who will be discussed below, became deeply acquainted with the works of Muḥmmad al-Ghazālī through his personal contact with Aḥmad.[9] Yet, unlike ʿAyn al-Quḍāt, for instance, Aḥmad was not so much attracted to the philosophical Sufism that al-Ghazālī taught, and he pursued in his own works a less rationalist mysticism that focused around the *leitmotif* of love for God. It is interesting to note that in his *Revival*, Muḥammad shows little patience with some Sufis' "long and pleonastic invocations on the love of God," since they distract one's attention from outward human actions.[10] Richard Gramlich judged that the particular appeal of Aḥmad's collection of aphorisms on Sufi love is neither the result of his intellectual depth or penetration, nor is it due to some strength in poetic creativity. Rather, his sometimes strange and baroque technique of interweaving thoughts is what creates the beauty of Aḥmad's writing.[11] Further studies are necessary to explicate the relationship between the theological teachings of the two brothers.

In the following pages, I will introduce those students and followers of al-Ghazālī who may contribute significantly to the reconstruction of his teachings. From what is available to us, their texts are particularly important, since al-Ghazālī seems to have been more outspoken with those students with whom he had a close relationship than with those who were more on the periphery. Considering the views of the close students and well-informed followers should significantly enhance our understanding of his theology.

Abū Bakr ibn al-ʿArabī (d. 543/1148)

Among his contemporaries, Abū Bakr ibn al-ʿArabī (468/1076–543/1148) is the most important source of information about al-Ghazālī's life and his teachings.

A native of Seville in al-Andalus, he and his father went on a long trip to the Muslim East. The purpose of this travel was partly political: Abū Bakr's father, Abū Muḥammad ibn al-ʿArabī had been an administrator in the local Sevillian government of the ʿAbbādids. When in 484/1091 the Almoravids conquered Seville, he felt that it would be prudent to leave al-Andalus.[12] He knew that the ruler (*amīr*) of the Almoravids, Yūsuf ibn Tāshifīn (d. 500/1107), longed for an official recognition from the ʿAbbāsid caliph in Baghdad. For Abū Muḥammad, this was a welcome opportunity to flee al-Andalus and await the outcome of the conflict between the Almoravids and the Taifa-Kings. Abū Muḥammad ibn al-ʿArabī offered Yūsuf ibn Tāshifīn the opportunity to perform a political mission on his behalf and achieve official recognition from the caliph. Caution made him take his son with him.[13] In any case, he and his son were in no haste to return with the desired documents, and they spent much time among the scholars of Jerusalem, Damascus, and Baghdad before they even started to lobby on behalf of Yūsuf ibn Tāshifīn four years after their departure.[14]

The two Ibn al-ʿArabīs left al-Andalus in the spring of 485/1092 when Abū Bakr was just sixteen years old.[15] They traveled on ships, which took them—not without an incident of shipwreck—to Bougie, Mahdiyya, and finally Egypt. From there they turned toward Jerusalem, where they spent most of their time between the years 486/1093 and 489/1096. In Jerusalem, the Ibn al-ʿArabīs met their fellow Andalusian al-Ṭurṭūshī (d. 520/1126), who was a staunch supporter of Yūsuf ibn Tāshifīn and the Almoravids, and the young Abū Bakr studied with him.[16] Ibn al-ʿArabī reports on his travels in an autobiographical book with the title *Book on the Arrangement of the Travel That Raised My Interests in Religion.*[17] This book has not come down to us.[18] There is, however, a second book by Abū Bakr ibn al-ʿArabī, in which he briefly reports on his travels and meetings with eminent scholars. This work, *Experiences of the Great Authorities and Eminent People by the Observer of Islam and the Various Lands,* presents detailed information about the two Ibn al-ʿArabīs' travels in the service of the Berber king Yūsuf ibn Tāshifīn.[19] According to this text and to information in Ibn al-ʿArabī's book *The Rule of Interpretation,* the two Ibn al-ʿArabīs traveled from Jerusalem via Ascalon, Acre, and Damascus to Baghdad, where they arrived in the early days of Ramaḍān 489 / August 1096.[20] Al-Ghazālī arrived in Jerusalem during the summer of 489/1096, almost a year after the Ibn al-ʿArabīs had left the city. During the four or five months al-Ghazālī stayed in Jerusalem, Ibn al-ʿArabī's erstwhile teacher al-Ṭurṭūshī tried unsuccessfully to meet with al-Ghazālī.[21] In the meantime, in Baghdad, the two Ibn al-ʿArabīs joined the pilgrimage caravan that would leave Iraq in the fall of 489/1096. This was the pilgrimage in which al-Ghazālī also took part, although he joined the caravan that started in Syria.

The father, Abū Muḥammad ibn al-ʿArabī used the gathering of scholars during the pilgrimage to propagate the virtues of the Almoravids and of Yūsuf ibn Tāshifīn. Although this did not have an immediate effect, his son, Abū Bakr, claimed that the good tidings about the Almoravids reached al-Ghazālī and prepared him to respond positively to a later request by the Ibn al-ʿArabīs. Indeed, a year later al-Ghazālī wrote a letter and a legal opinion in support of

Yūsuf ibn Tāshifīn.[22] During the pilgrimage, however, the three did not meet: Al-Ghazālī was on the Syrian caravan of the pilgrims, while the two Ibn al-ʿArabīs were on the Iraqi one. The two Andalusians only glimpsed the great scholar from afar.[23] They returned to Baghdad in early 490/1097, and al-Ghazālī returned to Damascus.

In his book *Protective Guards Against Strong Objections*, Abū Bakr ibn al-ʿArabī says that he finally met al-Ghazālī in Baghdad in Jumāda II 490 / May–June 1097.[24] This was right after al-Ghazālī arrived in Baghdad from Syria. The personal acquaintance with al-Ghazālī was an important event for the young Ibn al-ʿArabī. By now, he was twenty-one years old, and al-Ghazālī was the great "Dānishmand"[25] who had left his posts in Baghdad less than two years earlier and was now on his way back to his hometown, Ṭūs. Abū Bakr ibn al-ʿArabī studied closely with al-Ghazālī, and the latter devoted some considerable interest to his disciple. Abū Bakr nowhere mentions that he accompanied al-Ghazālī during his travels to Khorasan; the two likely just spent a couple of months together in the summer of 490/1097, when al-Ghazālī stayed at the "Ribāṭ of Abū Saʿd right across from the Niẓāmiyya madrasa."[26]

Abū Bakr and his father remained in Baghdad after al-Ghazālī's departure in the late summer or early fall of the same year they met. The two Ibn al-ʿArabīs had their audience with the twenty-two-year-old caliph al-Mustaẓhir and his vizier ʿAmīd al-Dawla ibn Jahīr, a son-in-law of Niẓām al-Mulk, in Rajab 491 / June 1098. They achieved their goal and secured a caliphal document supporting Yūsuf ibn Tāshifīn.[27] After this success, they traveled back via Syria and Egypt. In Alexandria, Abū Bakr ibn al-ʿArabī studied a second time with al-Ṭurṭūshī.[28] At this point, al-Ṭurṭūshī had already become a fierce opponent of al-Ghazālī's teachings. In 503/1109 or later, he wrote a response to a yet-unidentified Ibn Muẓaffar who had asked him about al-Ghazālī's works. In his answer, al-Ṭurṭūshī claims to have met al-Ghazālī and shows appreciation for his "understanding and intelligence" (*al-fahm wa-l-ʿaql*). Yet the letter mostly expresses al-Ṭurṭūshī's serious critiques of what he regarded as contradictions in al-Ghazālī's œuvre and his adaptation of philosophical doctrines, particularly in his *Revival of the Religious Sciences*.[29] This epistle was quoted later by influential biographers of al-Ghazālī such as al-Dhahabī, al-Subkī, and al-Murtaḍā al-Zabīdī.[30]

In Muḥarram 493 / November–December 1099, Abū Bakr's father died at age fifty-seven. That same month, Abū Bakr left Alexandria, where he had spent about a year. In one of his works, he lists all the books he took home from the Muslim East. This list offers a helpful clue to the dating of some of al-Ghazālī's books, as it verifies that certain of his books were indeed published before 493/1099.[31]

Via Tunis, Tlemcen, and Fès, Abū Bakr ibn al-ʿArabī made his way back to Seville where he arrived in 495/1102. By now, he was twenty-six years old, having spent ten of his years in the Muslim East. Back in Seville, he became a venerated scholar and teacher and the main source for the spread of al-Ghazālī's works and doctrines in the Muslim West.

Ibn al-ʿArabī's First Report of His Meeting with al-Ghazālī

In his extant works, Abū Bakr ibn al-ʿArabī describes his first meeting with al-Ghazālī at least twice. The most vivid picture of al-Ghazālī is given in *The Rule of Interpretation* (*Qānūn al-taʾwīl*), a book that he wrote in 533/1138–39 in Seville,[32] forty-eight years after the reported event took place in Baghdad. Ibn al-ʿArabī describes the intellectual climate in Baghdad:[33]

> [In Baghdad,] I engaged in exchanges with the scholars and I regularly went to their teaching sessions. In particular I went to Fakhr al-Islām Abū Bakr al-Shāshī[34] the *faqīh* and the *imām* of the times. Here, suns of insight rose for me and I said to myself: "God is great! This is the goal that I always wanted to achieve and the kind of time that I always wanted to spend and that I longed for." [In Baghdad,] I studied, I restricted myself [to study], and I quenched my thirst [for knowledge]. I listened [to the scholars] and retained [their teachings] in my memory, until the *Dānishmand* [al-Ghazālī] came across us [*scil.* Abū Bakr and his father]. He stayed in the Ribāṭ of Abū Saʿd right opposite the Niẓāmiyya Madrasa.[35] He had turned away from this world and had turned towards God the Exalted. We walked towards him, presented our credentials, and I said to him: "You are the guide that we were looking for and the *imām* that will give us right guidance." We met with him and our meeting was by way of *maʿrifa*. We took from him what is above the ledge (*al-ṣuffa*); and we realized that whatever has come down to us in terms of information about the unknown is beyond theoretical inisight (*fawqa l-mushāhada*) and is not for the ordinary people (*al-ʿumūm*). And had the poet Ibn al-Rūmī known [al-Ghazālī], he would not have said:
>
> If you praise a man who is absent,
> do not exaggerate in his glory and be to the point!
> Because, when you exaggerate,
> you go to the utmost extreme with him.
> So he falls short where you glorify him,
> because of the advantage of the absent over him who is there.[36]
>
> [Al-Ghazālī] was a man, who when you saw him with your own eyes, you saw an outward beauty (*jamāl*), and when you experienced his knowledge you found that it was a swelling sea. The more you learned from him, the greater your delight would be.
>
> I developed strong ties with him and I became inseparable from his carpet. I seized his isolation and his agility, and every time he attended to me, I exhausted him with my expectations. He allowed me [to share] his place and I was with him in the morning, the afternoon, at lunchtime, and at dinner, whether he was in casual clothes or in his formal attire. During these times, I could ask him without

restraint, like a scholar at a place where the shackles of enquiry are entrusted [to him]. I found him to be welcoming towards me regarding instruction and I found him true to his word.

One of God's friendly deeds towards me and His granting of success to me was that, when He let me stay in Syria, He did so at a blessed spot among scholars. This would become a stepping-stone for my meeting with those who had found the truth (*al-muḥaqqiqūn*), who could correct what I had understood, who could comment on what I had assembled, who could clarify what I had made obscure, and who could complete what I had left diminished. Whatever I had understood from these preliminaries, it made me ready to receive the real truths (*al-ḥaqā'iq*) hidden within them, and it limited the risk that their meaning would evade me. It was as if someone enters the Garden of Eden and gathers the gold together with the sand, and then carries it to the foundry for his later use.[37]

Ibn al-ʿArabī's Second Report of His Meeting with al-Ghazālī

In a second book, *Protective Guards Against Strong Objections* (*al-ʿAwāṣim min al-qawāṣim*), Ibn al-ʿArabī gives another account of his meeting with al-Ghazālī. The context is different from the one in *The Rule of Interpretation*, as this second work is much more concerned with al-Ghazālī's doctrine than the first book. It is less enthusiastic about al-Ghazālī and more critical of his teachings. Ibn al-ʿArabī understood well that al-Ghazālī theology was heavily influenced by his reading of *falsafa*, and indeed, he criticizes this theology in more than one passage of his œuvre. As Ibn Taymiyya quotes Abū Bakr ibn al-ʿArabī, "Our Sheikh Abū Ḥāmid entered deeply into the bellies of the *falāsifa* and when he wanted to get out, he couldn't."[38] Here, Ibn al-ʿArabī's critique falls in line with some of the criticism voiced in al-Ṭurṭūshī's *Letter to Ibn Muẓaffar*. But although al-Ṭurṭūshī went as far as recommending the burning of al-Ghazālī's books,[39] Ibn al-ʿArabī always respected al-Ghazālī, despite their differences regarding the teachings of *falsafa*.

The following passage from *Protective Guards Against Strong Objections* expresses Ibn al-ʿArabī's reservations about al-Ghazālī's teachings on the soul. This book is essentially a popular reworking of some of al-Ghazālī's own objections against the arguments of the *falāsifa* and the Ismāʿīlites.[40] It quotes "strong objections" (*qawāṣim*) presented by the *falāsifa* as well as by the Ismāʿīlites and counters them with "protective guards" (*ʿawāṣim*), that is, counterarguments. On the one hand, Ibn al-ʿArabī's book relies heavily on several of al-Ghazālī's works: *The Intentions of the Philosophers*, *The Incoherence of the Philosophers*, and *Infamies of the Esoterics*.[41] On the other hand, al-Ghazālī appears sometimes on the side of those who bring forward "strong objections" (*qawāṣim*) that need to be refuted, particularly when he restates philosophical teachings without what Ibn al-ʿArabī considered the appropriate measure of criticism.

Ibn al-ʿArabī begins his book with a discussion of epistemological questions, leading him to reflections on the nature of the soul. He reports the position of some Sufis, in particular al-Ḥārith al-Muḥāsibī (d. 243/857) and the Ashʿarite Abū l-Qāsim al-Qushayrī (d. 465/1071). They said "that knowledge will only be achieved through purity (*ṭahāra*) of the soul, chastening (*tazkiya*) of the heart, the untying of the relationship between the heart and the body, and the disenfranchising from material motives such as fame and riches."[42] This is an extreme position (*ghulūw*), Ibn al-ʿArabī says, because there is no connection between the knowledge that a person acquires and any pious deeds that his heart—meaning his soul—has performed. Similarly, there is no connection between certain practices in one's worship and the unveiling of some kind of hidden knowledge. The subject of whether Sufi practice or the asceticism of the "friends of God" (*awliyāʾ*) leads to superior religious insight seems to be the focal point of the dispute about al-Ghazālī's work in the Muslim West. Before Ibn al-ʿArabī wrote this book, the grandfather of the philosopher Ibn Rushd, Ibn Rushd al-Jadd (d. 520/1126), had issued a *fatwā* dismissing the position of al-Ghazālī and other Sufis on this subject.[43] Although Ibn Rushd al-Jadd exempted "moderate" Sufis such as al-Qushayrī from his criticism, Ibn al-ʿArabī specifically names him as one who presented the problematic position that pious deeds—such as the Sufi practice of invoking the names of God—may lead to superior religious knowledge. Ibn al-ʿArabī vigorously denounces this: It is simply not true, he writes, that the practitioner of Sufi *dhikr* "will see the angels and hear what they say; until he will reach to the spirits of the prophets and hear their words."[44] Using the book's method to discuss the pros (*ʿawāṣim*) and the cons (*qawāṣim*) of a certain position, Ibn al-ʿArabī cites an objection (a *qāṣima*) to his position. This objection was presented by al-Ghazālī during the months that they studied together in Baghdad:

> I conferred about this with Abū Ḥāmid when I met him in Baghdad in the month of Jumāda II 490 (May–June 1097). Earlier, namely in the year eighty-six (1093), which was at this time about five years ago, he had accepted the Sufi path (*al-ṭarīqa al-ṣūfiyya*) and made himself free for what it requires. He had put himself in seclusion (*al-ʿuzla*) and renounced all groups. Due to reasons that we have explained in the *Book on the Arrangement of the Travel*[45] he devoted himself exclusively to me and I read all of his books under his instruction and heard the book that he named *The Revival for the Religious Sciences (al-Iḥyāʾ li-ʿulūm al-dīn)*.[46] I asked him for guidance in order to reach his convictions (*ʿaqīda*). I also asked for an explanation of his method (*ṭarīqa*) so that I could reach complete insight (*tāmm al-maʿrifa*) into the secret of those hints and indications that he had put into his books. And yes, he answered me. His response opened the right way for the postulant to reach the loftiness of his level and the heights of his station.[47]

Al-Ghazālī gave a long response, heavily influenced by Avicenna's explanation of how prophets reach their superior level of insight and why they may have an almost supernatural influence on the world around them. Avicenna had taught that prophets spontaneously receive their insights in either their imaginative faculty or their intellect.[48] Prophets lack the impeding forces of ordinary people that suppress visions while they are awake and receive sense data. Therefore, prophets receive in their waking hours visions that ordinary people receive in their sleep.[49] Prophets also benefit from the power of intuition (*quwwat al-ḥads*) and have the capacity of immediately finding the middle term of a syllogism. This capacity gives a prophet perfect theoretical knowledge without instruction, solely through intellectual intuition (*ḥads*).[50] Finally, prophets also have a strong practical faculty of the soul (*quwwa nafsiyya ʿamaliyya*) that can affect other beings and worldly processes. All souls have the capacity to effect physical changes in their own bodies; the extraordinary powers of a prophet's soul have also the capacity to bring about changes in natural objects outside their own bodies. Prophets have the capacity, for instance, to cause storms, rain, and earthquakes or even to cause people to sink into the ground.[51]

Al-Ghazālī's answer to his student draws on these teachings. Here, al-Ghazālī applies these teachings, not only to prophets, but also to everyone who has purified his soul from the bodily passions:

> If the heart purifies itself (*ṭahhara*) from the relationship with the sensibly perceived body and devotes itself to the intelligibles (*al-maʿqūl*), the truths (*al-ḥaqāʾiq*) are revealed to it. You will understand these things only through personal experience (*tajriba*) and keeping company with those who have already mastered it. Being in their presence and rubbing shoulders with them will help you understand these things.[52]

A certain school of thought (*ṭarīq min al-naẓar*), al-Ghazālī says, claims that the heart is a refined substance or a polished gem—a *jawhar ṣaqīl* (meaning both)—and that it reflects knowledge like a mirror reflects. A mirror can be used to present to us things that cannot be seen without a reflection, such as things in the next room or around a corner. If the mirror is not constantly polished, al-Ghazālī says, it becomes tarnished. Likewise, the heart suffers from "certain harms that accumulate and befall it." If the heart is purified, however, it is "like a mirror from which the tarnish has disappeared and that now reflects those things perfectly." Sometimes these truths (*ḥaqāʾiq*) that are received by the purified heart appear as clear insights; sometimes they appear as symbols or representations (*mithāl*).

Ibn al-ʿArabī proceeds in his report of al-Ghazālī's response by saying that the soul (*nafs*) gets stronger when the heart is purified and becomes cleaner. Every soul has "an influencing faculty" (*quwwa taʾthīriyya*), giving it an influence over its own body as well as over the bodies of other people:

> An example is given by the man who walks on a line on the ground that is as wide as the span of a hand. Would he walk on such a line up on a highly elevated wall that is as wide as a forearm, he would be unable to hold on to it since he imagines himself falling from the

> wall. When the soul realizes these circumstances and it becomes set upon it, the body becomes affected by it and it quickly falls.[53]

When the purification of the heart makes the soul stronger, the soul develops the capacity to influence and affect bodies other than its own. Al-Ghazālī here gives the example of a person's strong love or desire for another person. Once the other person knows that he or she is loved and desired, this person often also develops a strong love and desire for the one who loves. Thus can one soul affect the feelings of another soul. This happens whenever the affection of a soul is strong. The soul is particularly strongly affected when it is purified:

> The soul's influencing faculty and its readiness to receive insights increases with its purity (*bi-ṣafāʾihā*). Now, you believe in the sending down of abundant rain showers and the spontaneous growth of plants and similar things that are miracles violating the habitual courses. And what I have spoken about is similar to this. These are the souls of the prophets and their influence on other bodies are the signs that give evidence of the prophets' conditions.[54]

In his answer to Ibn al-ʿArabī, al-Ghazālī mentions all elements of Avicenna's prophetical psychology: imaginative revelation, intellectual revelation, and the prophets' strong practical and motive faculty, which is here called "the influencing faculty." The letter is, in fact, heavily influenced by Avicenna's presentation of the particular properties (*khawāṣṣ*) of prophets and of "friends of God" (*awliyāʾ Allāh*) in his *Pointers and Reminders*.[55] Like Avicenna, al-Ghazālī claims that every human has a small portion of these faculties, not only prophets; prophets are only the most distinguished examples of purified souls. These faculties become stronger if one purifies one's soul by cleansing it (*ṭahhara*) from worldly desires. The miracles that the prophets perform—which earlier Ashʿarites regarded, as we will see, as a break on God's habit—are simply the causal effects of the strong influencing power of the prophet's soul.

We should note that al-Ghazālī remained somewhat uncommitted to the teachings presented in this letter, introducing them as "a school of thought" to teach the existence of a close link between the purity of a person's heart and his or her level of knowledge and insight. Ibn al-ʿArabī, however, understood that al-Ghazālī was a member of this "school of thought"—a reference to none other than Avicenna—which teaches that purified souls are able to achieve higher insights than those hearts that remain tarnished.

Ibn al-ʿArabī's book *Protective Guards Against Strong Objections* (*al-ʿAwāṣim min al-qawāṣim*) is a rich source for comments al-Ghazālī made to his students. At one point, Ibn al-ʿArabī reports that al-Ghazālī inclined toward the position of some *falāsifa* that rationality (*ʿaql*) offers a path toward knowledge about the afterlife. In his published works, al-Ghazālī exhibits no such leanings, always maintaining that the revealed information about the afterlife is so detailed and so clear that it overrules all rational speculation and allows no figurative interpretation of the revealed text.[56] Ibn al-ʿArabī, however, says that al-Ghazālī leaned toward the opposite position and held it in high esteem.[57]

In another passage of this book, Ibn al-ʿArabī reports al-Ghazālī's opinion on those who claim to see the Prophet in their dreams.[58] Ibn al-ʿArabī's other books, such as his voluminous commentary on al-Tirmidhī's *ḥadīth* collection, may yield more relevant information on al-Ghazālī's teachings,[59] as may some other works that still lie in manuscripts. In his commentary on the noble divine names, for instance, Ibn al-ʿArabī seems to be taking issue with al-Ghazālī's rationalist teachings on that subject. Ibn al-ʿArabī's dictum that his teacher entered so deep into the bellies of the *falāsifa* that he could not get out may be taken from this work or from his equally unedited *Lamp of the Novices* (*Sirāj al-murīdīn*), in which Ibn al-ʿArabī argues against al-Ghazālī's view of the best of all possible worlds.[60]

Al-Ghazālī wrote a few texts in response to Ibn al-ʿArabī's questions. One manuscript of the book *Breathing of the Spirit and the Shaping* (*Nafkh al-rūḥ wa-l-taswiya*) says that work is a response to Ibn al-ʿArabī's questions.[61] This text, whose abbreviated form is known as *The Short Text to Be Withheld* (*al-Maḍnūn al-ṣaghīr*) or possibly also as *Ghazalian Answers to Questions about the Afterlife* (*al-Ajwiba al-Ghazāliyya fī l-masāʾil al-ukhrawiyya*), discusses the nature of the human soul and the human spirit (*rūḥ*) and the latter's relation to God's act of shaping the body and breathing His life force into it (Q 32:9, 15:29, 38:72). Judging from the considerable number of manuscripts and modern prints, the book was and is very popular among al-Ghazālī's readers. Its authenticity, however, is not fully established. Another Ghazalian text connected to Abū Bakr ibn al-ʿArabī also deals with the dispute between *mutakallimūn* and *falāsifa* on the nature of the human soul.[62] In a letter al-Ghazālī addressed to Abū Bakr, he answers three questions on various subjects, among them, whether the soul is a self-subsisting substance (*jawhar*) or just an accident that inheres in a body. The existence of this brief text was noted by Iḥsān ʿAbbās in 1968;[63] the text is still unedited and was not available to me.[64] The first question Ibn al-ʿArabī asks is: "Is the spirit (*al-rūḥ*) lightened particles (. . .) or is it a spiritual substance (*jawhar*) that each body encounters in the form of rays like one encounters the sun (. . .)?" The second question enquires about "the difference between a bird['s flight] and a good omen." The third questions is: "What is the meaning of the Prophet's saying: 'The devil runs with one of you in his veins?' "[65]

Al-Ghazālī answers these questions cautiously, first reminding Ibn al-ʿArabī and other students that they should not strive to answer each and every question that they find raised within themselves. Second, they should not assume that the results of a demonstration (*burhān*) could ever be false. The intellect—if properly applied—does not lead to false results. Third, they should keep in mind that when it comes to figurative interpretation (*taʾwīl*) of revelation, it is insufficient to specify an interpretation that is merely probable. It is dangerous to judge what God might have intended in his revelation and what the Prophet might have intended in his sayings by assumptions and guesses.

The text of al-Ghazālī's letter appears to be at least partly identical to al-Ghazālī's *The Universal Rule in Interpreting Revelation* (*al-Qānūn al-kullī fī l-taʾwīl*), a short work of a dozen pages discovered in a Cairo manuscript and published in 1940.[66] There al-Ghazālī discusses several suggested interpretations of the *ḥadīth* about the devil running in the veins of some of the Prophet's companions. He begins his own explanations with three recommendations, namely, (1) that one should not aspire to know everything, (2) that one should not assume a valid demonstration could result in a falsehood, and (3) that one should not engage in interpretation (*taʾwīl*) if one is uncertain about the meaning of the revealed text.[67] It appears that *The Universal Rule in Interpreting Revelation*, which is mentioned in the work lists of al-Subkī and al-Wāsiṭī, was generated from a letter al-Ghazālī wrote in response to Abū Bakr ibn al-ʿArabi.[68]

Despite their brief period of personal contact, Abū Bakr ibn al-ʿArabī was probably the master student of al-Ghazālī's—at least when it comes to his theology. Abū Bakr was particularly interested in all questions dealing with the human soul and with epistemology. By the time he met al-Ghazālī, the great *Dānishmand* (al-Ghazālī) had adopted an Avicennan psychology regarding the human soul as a self-subsisting substance, able to continue existence after the body's death. Yet in some books of the *Revival*—most evidently in the *Letter for Jerusalem* in the Second Book—he expresses the relationship between the human soul and the human body in the language of the *mutakallimūn* as dependent accidents (meaning the soul) that inhere in the atoms of the body.[69] The *Letter to Jerusalem* was written only a few months before Abū Bakr met al-Ghazālī. In other writings, I tried to resolve this apparent contradiction.[70] Although al-Ghazālī personally preferred the theory of the human soul as a self-subsisting substance that he ascribed to the *falāsifa* and the Sufis, neither through reason nor through revelation are humans able to decide whether this theory is true or the alternative explanation held by the *mutakallimūn*. Neither of the two competing views can be demonstrably proven, and both are viable explanations of the text of revelation. It was important for al-Ghazālī that all Muslim scholars become convinced of the corporeal character of resurrection in the afterlife. One should find a way to teach this essential element of the Muslim creed without needing to change the views of one's readership on the nature of the soul and thus confuse their convictions. We will see that this strategy is a result of what I will call al-Ghazālī's nominalist approach to human knowledge.

Asʿad al-Mayhanī (d. 523/1130 or 527/1132–33)

Abū l-Fatḥ Asʿad ibn Muḥammad al-Mayhanī was probably the most influential immediate follower of al-Ghazālī in the Muslim East. Whether he was a student of the great theologian is not entirely clear; the entries on him in chronicles and biographical dictionaries do not mention such a relationship. In fact, there is something enigmatic about his education that challenges the currently prevailing understanding of the educational patterns of this period. According to the historical reports written by religious authorities such as al-Subkī or Ibn al-Jawzī, Asʿad al-Mayhanī was a successful and highly regarded teacher of Islamic law; nothing would suggest that he ever taught disputed positions associated with *falsafa*. However, al-Bayhaqī's biographical dictionary of scholars connected to the philosophical movement features a short article on As'ad al-Mayhanī. There he writes that As'ad had studied with al-Lawkarī, who was a student of one of Avicenna's students. Al-Lawkarī was the most important figure for the introduction of Avicennism in Khorasan.[71] As'ad al-Mayhanī was the first Muslim scholar with a dual intellectual pedigree: he was a reputable religious scholar who taught at theological madrasas, while still participating in the philosophical teaching tradition established by Avicenna.

As'ad al-Mayhanī was born 461/1068–69 in Mayhana, a town in northern Khorasan that is less than 100 km northeast of Ṭūs.[72] He studied *fiqh* with Abū l-Muẓaffar al-Sam'ānī at the Niẓāmiyya madrasa in Merw,[73] where he later became a teacher. He then moved to Ghazna, where his fame grew. In 507/1113–14, the youthful Maḥmūd ibn Muḥammad Tapar ibn Malikshāh, who ruled as governor over Baghdad, invited him to teach at the local Niẓāmiyya. Like al-Ghazālī twenty-two years earlier, As'ad was a Seljuq appointee and close to the caliph's court. Al-Bayhaqī says that everybody who witnessed him at the caliphal court was highly impressed. Other historians add that the caliph, the sultan, and all the other dignitaries held him in high esteem and mention that As'ad soon acquired significant riches. In 510/1117, As'ad gave his friend, the famous author al-Shahrastānī (d. 548/1153), a teaching post at the Baghdad Niẓāmiyya.[74] In 513/1119–20, both al-Shahrastānī and As'ad ceased teaching at the Niẓāmiyya—perhaps because their Seljuq patron temporarily lost authority over Baghdad.[75] As'ad taught at the Baghdad Niẓāmiyya for a second period of six months in 517/1123. He died either in 523/1129 or 527/1132–33 in Hamadan.[76]

As'ad al-Mayhanī composed a curriculum of studies or a textbook that was adopted by the Niẓāmiyya in Baghdad and by other schools. In Baghdad, it remained in use many decades after his death. The work is referred to as "The Notes" (*al-Ta'līqa*), and it is credited for its masterful treatments of the techniques used in disputations (*khilāf*).[77] It seems to have followed al-Ghazālī's approach and included the study of formal logics in the area of jurisprudence (*fiqh*). The philosopher 'Abd al-Laṭīf al-Baghdādī (d. 629/1231) says that his father studied at the Baghdad Niẓāmiyya "the sciences of law, Shāfi'ite *fiqh*, and the disputations between the schools (*khilāf*) with the 'Notes' of As'ad al-Mayhanī, who was famous during that time."[78] This was in the middle of the sixth/twelfth century when As'ad was no longer alive. Twenty years later, the rationalist theologian Sayf al-Dīn al-Āmidī (d. 631/1233) studied As'ad's "Notes" diligently and considered himself a follower of al-Mayhanī.[79] At the end of the century, the Ḥanbalite jurist Ibn al-Jawzī (d. 597/1201) wrote that many students of his school use As'ad's "Notes" even though it teaches primarily Shāfi'ite and not Ḥanbalite law.[80] In the fourteenth century, the conservative Ibn Kathīr (d. 774/1373) confirms that the work was still famous despite the fact that he considered it of little value.[81]

As'ad was only slightly more than ten years younger than al-Ghazālī, so a proper teacher-student relationship must be ruled out. The Muslim historians report details of al-Ghazālī's life on As'ad's authority. The two were thus known or plausibly thought to be in contact. Al-Subkī quotes a very appreciative comment of As'ad that aims to defend al-Ghazālī against the criticism of lesser-accomplished theologians: "Nobody will arrive at al-Ghazālī's level of insights and his virtue unless he reaches—or at least almost reaches—intellectual perfection."[82]

The historians, however, do not report that the two ever met. Their meeting can only be deduced from two separate narratives about an episode in al-Ghazālī's late life. Each of the two narratives is incomplete, and at least one must be partly erroneous. The first report is from the collection of al-Ghazālī's letters. The anonymous collector tells of a group of scholars at al-Ghazālī's *khānqāh* in Ṭūs who asked him, "which school do you belong to?"[83] The story immediately follows al-Ghazālī's exchange with Sanjar, which took place soon after 501/1108. As already explained, al-Ghazālī was asked to appear before Sanjar and defend himself against the accusation brought forward by Ḥanafite scholars that al-Ghazālī had shunned their Imam Abū Ḥanīfa in one of his earlier books. As Sanjar and many within the Seljuq court were Ḥanafites,[84] thirty-year-old derogatory comments on the Ḥanafite school's founder could still harm al-Ghazālī. The accusations and how al-Ghazālī successfully parried them are reported in the collection of his letters.[85] Although the group of scholars that visited al-Ghazālī and asked this question is not identified, they are brought in connection with "his enemies" (*muta'annitān-i way*) from the court of Sanjar. In al-Ghazālī's answer to their question, he gives a short version of his epistemological approach to Muslim theology and ethics, mirroring his "law of figurative interpretation" (*qānūn al-ta'wīl*). He says:

> Regarding the subjects that are settled by reason (*ma'qūlāt*) my school (*madhhab*) is that of demonstration, following what a rational argument (*dalīl 'aqlī*) mandates. Regarding the subjects that are settled by revelation (*shar'iyyāt*) my school is the Qur'an and I do not follow one of the Imams by way of emulation (*taqlīd*). Neither al-Shāfi'ī nor Abū Ḥanīfa may take a line of writing away from me and claim it.[86]

The second narrative of this incident appears in Dawlatshāh Samarqandī's (d. ca. 900/1494) history of Persian poets. We must assume that Dawlatshāh wrote after the collection of letters; in fact, it seems likely that he took most of his information from there. He writes:

> The scholar As'ad of Mayhana, a chronicler who was at the court of Sultan Muḥammad Tapar, engaged in a public disputation (*munāẓara*) with Abū Ḥāmid al-Ghazālī. The scholars of Khorasan supported As'ad and during a session at Sultan Muḥammad's court he asked al-Ghazālī the first question: "Are you of the legal school of Abū Ḥanīfa or of al-Shāfi'ī?"[87]

Al-Ghazālī responded with the same answer that is noted in the collection of letters. It is striking that in Dawlatshāh's report, As'ad al-Mayhanī appears as al-Ghazālī's enemy. There were, however, two As'ad al-Mayhanīs who were contemporaries. One studied Shāfi'ite *fiqh*, theology, and *falsafa* and became a teacher at the Niẓāmiyya in Baghdad after al-Ghazālī's death. The second was about seven years older and was a Sufi and *ḥadīth* scholar who died shortly after al-Ghazālī.[88] The two were apparently not related. It is thus likely that Dawlatshāh had the second more conservative As'ad from Mayhana in mind, who may have belonged to the Ḥanafite school of law. We can thus assume that Dawlatshāh constructed this encounter based on his knowledge of al-Ghazālī's letters.

Not so easily solved is the fact that in Dawlatshāh's story, the exchange between As'ad and al-Ghazālī happens at the court of Sultan Muḥammad Tapar, rather than at the court of Sanjar. The name Muḥammad Tapar cannot simply be an erroneous substitution for Sanjar, since in Dawlatshāh's book, al-Ghazālī refers to an earlier exchange with Sanjar.[89] There are, however, no reliable reports of a confrontation between Sultan Muḥammad Tapar and al-Ghazālī. Sultan Muḥammad Tapar resided in Isfahan and had left the affairs of Khorasan in the hands of his brother Sanjar, who would succeed him as supreme sultan of the Seljuq Empire after his death in 511/1118. Regarding this piece of information, Dawlatshāh's story is probably wrong; it may be again based on an erroneous reading of the collection of al-Ghazālī's letters.[90]

What then is the grain of truth in all this? The collector of al-Ghazālī's letters vaguely suggests that those who put the question to him were also hostile. That, however, need not be the case. The story of al-Ghazālī's memorable comment on his *madhhab*—a word that may also mean his "method"—might have been mixed up with earlier accusations about his *fiqh* brought forward by Ḥanafite scholars. This confusion might have already existed when the collection of letters was put together. Initially, the two might have been different episodes and unconnected reports. Al-Ghazālī's answer to the question of his *madhhab* reads very much like one that he would have given to close students or to followers rather than to hostile accusers. Al-Ghazālī is known to have been very careful about what he conveyed to whom.[91] His blunt answer would certainly make him vulnerable to the accusation of being too rationalist even to follow al-Shāfiʿī. Putting himself in such a position was unnecessary, as the question—if put by adversaries—simply asks about his formal allegiance in *fiqh*.

One way to reconcile the discrepancies is to accept the historical accuracy of the answer and the name of the questioner. The question was probably put forth by Asʿad al-Mayhanī, just as Dawlatshāh reports—but not by the conservative Sufi but rather by the As'ad al-Mayhanī who was the Shāfiʿite theologian sympathetic to al-Ghazālī. It has already been said that this As'ad conveyed information on al-Ghazālī's life. We may assume that the Shāfiʿite As'ad al-Mayanī was a follower of al-Ghazālī who visited him in his *khānqāh* in Ṭūs. Later, As'ad's report was used by the collector of the letters as well as by Dawlatshāh, both of whom somewhat misrepresent its original context.

Muḥammad ibn Yaḥyā al-Janzī (d. 549/1154)

If Asʿad al-Mayhanī represents the continuation of the Ghazalian teaching tradition at the Niẓāmiyya madrasa in Baghdad, Muḥammad ibn Yaḥyā represents it in Nishapur. He was born 476/1083–84 in Ṭuraythīth, a village in the vicinity of Nishapur. His family came from Janza in Arran, a town that was also known as Ganja and today is known as Kirovabad in Azerbaijan. Two generations later, Janza would become known as the home of the famous Persian poet Niẓāmī (d. *c.* 604/1207). The historian al-Samʿānī, who studied with Muḥammad ibn Yaḥyā, says that his father came to Nishapur for the famous Ashʿarite Sufi al-Qushayrī. He became one of his disciples, and after having performed the pilgrimage, he settled in Ṭuraythīth. His son, Abū Saʿd Muḥammad ibn Yahyā, studied with Aḥmad al-Khawāfī (d. 500/1106–7) and Abū Ḥāmid al-Ghazālī. Al-Khawāfī was a student of al-Juwaynī and became the judge (*qāḍī*) of Ṭūs shortly before 478/1085. The historians describe him as a companion (*rafīq*) of al-Ghazālī, renowned for his expertise in the techniques of disputation (*jadal* and *munāẓara*) and in the "silencing of one's opponent" (*ifḥām al-khuṣūm*).[92] Since al-Khawāfī is associated with Ṭūs rather than with Nishapur, it is most likely that Muḥammad ibn Yaḥyā studied with al-Ghazālī at his *zāwiya* there and not exclusively during al-Ghazālī's tenure at the Niẓāmiyya madrasa in Nishapur in the years after 499/506.

Muḥammad ibn Yaḥyā himself became an influential teacher of Islamic law who attracted students from far away. He was appointed head teacher at the Niẓāmiyya in Nishapur.[93] His name is associated with a great number of students, and he figures in countless intellectual lineages. Two of his students are credited with the introduction of Ashʿarite theology in Ayyūbid Syria, for instance.[94] When the famous theologian, philosopher, and jurist Fakhr al-Dīn al-Rāzī (d. 606/1210) came to Nishapur in his youth, he studied with al-Kamāl al-Simnānī (d. 575/1179–80), who was a student of Muḥammad ibn Yaḥyā. Fakhr al-Dīn al-Rāzī's biographers stress that, through al-Simnānī and Muḥammad ibn Yaḥyā, he is linked to al-Ghazālī's teaching activity.[95]

Muḥammad ibn Yaḥyā is particularly connected to the spread of al-Ghazālī's work in Shāfiʿite law. He was called the "Renewer of Religion" (*muḥyī l-dīn*), a title that al-Ghazālī earlier had claimed for himself in his autobiography;[96] perhaps his student acquired it in his place. Muḥammad wrote the first commentary on one of al-Ghazālī's books on Shāfiʿite law, *The Middle One* (*al-Wasīṭ fī l-madhhab*). Al-Ghazālī wrote at least three books on the individual rulings or the substantive law (*furūʿ*) of the Shāfiʿite school, the most voluminous being a book with the title *The Extended*.[97] This large work and the less extensive *Middle One*, which became the subject of Muḥammad ibn Yaḥyā's commentary, were written early in al-Ghazālī's life and are mentioned in books that he composed soon after 488/1095.[98] The shortest of al-Ghazālī's books on applied law, *The Succinct One* (*al-Wajīz*), was completed in the year 495/1101 while he was teaching at his *zāwiya* in Ṭūs.[99] The titles of these three works are inspired by three works of Qurʾan commentary (*tafsīr*) by the Nishapurian commentator al-Wāḥidī (d. 468/1076), who lived two generations before al-Ghazālī.[100] As in al-Wāḥidī's three works, these books represent three set levels of depth (*miqdār makhṣūṣ*) in which the subject is treated,[101] and they do not imply, for instance, that the book *The Middle One* was composed after the longer and the shorter one. Muḥammad ibn Yaḥyā's commentary, *The Comprehensive Book about the Commentary on The Middle One* (*al-Muḥīṭ fī sharḥ al-Wasīṭ*), is unfortunately lost.[102]

Muḥammad ibn Yaḥyā's *Comprehensive Book* was the first of many commentaries on al-Ghazālī's two shorter works on the substantive law (*furūʿ*) of the Shāfiʿites, *The Middle One* and *The Succinct One*. Some of these commentaries are among the most successful works in Islamic law. Three generations after Muḥammad ibn Yaḥyā, the Shāfiʿite Abū l-Qāsim al-Rāfiʿī (d. 623/1226), of Qazvin in northern Iran, wrote a commentary on al-Ghazālī's *The Succinct One*.[103] As a commentator on al-Ghazālī's legal works, al-Rāfiʿī has been overshadowed only by Yaḥyā al-Nawawī (d. 676/1277), who composed a super-commentary on his work. Al-Nawawī was a student of Ibn al-Ṣalāḥ al-Shahrazūrī (d. 643/1245), who wrote himself a commentary on al-Ghazālī's *The Middle One*. Al-Shahrazūrī, who had studied in Nishapur, moved to Damascus and founded a prominent tradition of al-Ghazālī studies. His student al-Nawawī also composed a commentary on *The Middle One*.[104] Yet much more successful was his book, *The Plentiful Garden for the Students and the Support of the Muftīs* (*Rawḍat al-ṭālibīn wa-ʿumdat al-muftiyīn*), the super commentary on Abū l-Qāsim al-Rāfiʿī's commentary on al-Ghazālī's *The Succinct One* mentioned earlier. Al-Nawawī's *Plentiful Garden* is the fruit of a productive period of Ghazālī reception among the Damascene Shāfiʿites in the seventh/thirteenth and eighth/fourteenth centuries. Both al-Nawawī's and al-Rāfiʿī's commentaries are still used among jurists of Shāfiʿīte law today, and they are doubtless among the most influential references in that field.

Muḥammad ibn Yaḥyā, who had a significant part in securing al-Ghazālī's influential position among Shāfiʿite jurists, died at the age of seventy during the tragic sacking of Nishapur by the Oğuz nomads. In 548/1153, Sanjar's Seljuq-Turk army suffered a surprise defeat by one of the larger groups of Oğuz Turks that had newly entered into Khorasan. The nomads took Sanjar prisoner and pillaged the cities in his realm. When they arrived in Nishapur in Ramaḍan / November of that year, they sacked the outer city and killed many of its inhabitants in search for hidden treasures. Soon afterward, they returned and overran Nishapur's inner city. Muḥammad ibn Yaḥyā was killed either in Ramaḍān 548 / November–December 1153 or—which is more likely—on 11 Shawwāl 549 / 19 December 1154 in the New Mosque of Nishapur. It is said that the Oğuz forced dirt down his throat until he died.[105]

The destruction of Nishapur in 548/1153 was only one step in the steady decline of that city as a center of Muslim scholarship. In 553/1158, the long-standing differences between the Ḥanafites and the Shāfiʿites erupted in a civil war that lasted until 557/1162 and caused more destruction than the two sackings by the Oğuz nomads. Merw, Isfaraʾin, Ṭābarān-Ṭūs, and other cities in Khorasan also suffered from the breakdown of the Seljuq military force in 548/1153. In addition, the region was hit by a number of devastating earthquakes, so that during the second half of the seventh/twelfth century, urban life in Khorasan went through a severe crisis. With it suffered the cities' institutions of learning such as the Niẓāmiyya madrasas in Nishapur, Merw, and Herat. Fakhr al-Dīn al-Rāzī was among the last generation of scholars who could connect themselves to al-Ghazālī's teaching tradition in Nishapur. The very last head teacher at the Nishapurian Niẓāmiyya mentioned in the sources was Abū Bakr ibn al-Ṣaffār, a member of the rich and influential Ṣaffār family of Nishapur. He had taught a course on al-Ghazālī's *Middle One* forty times before he was killed in 618/1221 at age eighty-two, when the Mongol armies under Chingiz Khān's son Toluy captured Nishapur and systematically slaughtered its inhabitants. One of his many students was Ibn al-Ṣalāḥ al-Shahrazūrī from the region of Irbil in Iraq. Ibn al-Ṣalāḥ al-Shahrazūrī and others would carry the teaching tradition of al-Ghazālī's legal works from Nishapur to its new center in Damascus.[106]

Ibn Tūmart (d. 524/1130)

Ibn Tūmart, the founder of the Almohad Empire in North Africa and al-Andalus, never met al-Ghazālī. He traveled from Morocco to Baghdad and studied at the Niẓāmiyya madrasa at a time when al-Ghazāli was no longer there. He became an accomplished and quite innovative theologian, developing a number of positions in theology and *fiqh* that can be connected to al-Ghazālī's teachings.

Ibn Tūmart was born in the Sūs Valley of southern Morocco some time between 470/1077 and 480/1088. He was a contemporary of Abū Bakr ibn al-ʿArabī, whom he also never met. At some time before 500/1106, Ibn Tūmart left Morocco in pursuit of religious knowledge. He first traveled to al-Andalus but soon turned his attention to the east and made his way to Baghdad. There he studied at the Niẓāmiyya for an undetermined period between the years 500/1106 and 511/1117. Ibn Tūmart's biographers mention a number of scholars as his teachers at the Niẓāmiyya, including Abū Bakr al-Shāshī (d. 507/1114),[107] Abū l-Ḥasan al-Ṣayrafī (d. 500/1107),[108] and al-Kiyāʾ al-Harrāsī, all venerated scholars of their time. Ibn Tūmart might have also studied with Asʿad al-Mayhanī and al-Shahrastānī, both of whom taught at the Niẓāmiyya during this period. Some historians also claim that Ibn Tūmart was a student of al-Ghazālī, and that the two had a memorable encounter in which the great theologian entrusted Ibn Tūmart with his theological legacy. That, however, is a myth spread by Ibn Tūmart's political heirs after his death. By the time Ibn Tūmart arrived in Baghdad, al-Ghazālī was already in Khorasan.[109]

There are no reliable reports about Ibn Tūmart's life before 510/1116 or 511/1117 when he returned to the Maghrib from the Muslim East. It follows that there is no reliable information that he did definitely study at the Niẓāmiyya in Baghdad. In an article published in 2005, I compare some of Ibn Tūmart's theological teachings—particularly his proof of God's existence—with those of al-Juwaynī and al-Ghazālī, concluding that he was indeed at the Niẓāmiyya in Baghdad. The argument supporting the historians' claim is based on the continuity of ideas rather than evidence of his whereabouts. His teachings are distinctly Juwaynian and to some degree Ghazalian. Just as al-Juwaynī and al-Ghazālī were influenced by philosophical arguments, so was Ibn Tūmart. The philosophical influence need not be direct and has most probably been mediated through theological ideas taught at the Niẓamiyya during this time. Even after al-Ghazālī's departure, the Baghdad Niẓāmiyya remained a hotbed of Nishapurian Ashʿarite theology and its adaptation of philosophical teachings.

Ibn Tūmart's career as a religious leader began soon after 510/1116, when he appeared in Tunis. In the Maghrib, he made a name for himself by preaching strict morality of the sort al-Ghazālī taught in his *Revival of the Religious Sciences*. In particular, on the duty of "commanding good and forbidding wrong," Ibn Tūmart followed al-Ghazālī's moralistic approach.[110] On his way back to Morocco, he gathered more and more followers, a zealous group that accompanied him and tried to enforce his high moral standards. By the time Ibn Tūmart arrived at Marrakesh in 515/1121, his followers had emerged into the avant-garde of a religious and political movement, primarily of Maṣmūda-Berbers, that would soon conquer North Africa and Muslim Spain.

Ibn Tūmart did not witness the full success of the movement that he started. His followers called themselves "those who profess divine unity" (*al-muwaḥḥidūn*), becoming known as Almohads in Western literature.[111] Ibn Tūmart died in 524/1130, during the early years of the military campaign that led to the conquest of almost all of the Maghrib, including al-Andalus. His successor (his "caliph") ʿAbd al-Muʾmin ibn ʿAlī (d. 558/1163) was one of those who joined the preacher on his way from Tunis to Marrakesh, and he became the real political founder of the Almohad movement. Under his rule, so it is said, the works of Ibn Tūmart were collected and written down. The writings that he supposedly edited were collected in *The Book That Contains All the Notes on the Infallible Imam and Acknowledged Mahdi . . . According to How the Caliph ʿAbd al-Muʾmin Dictated It*. It is preserved in two manuscript copies from this time and has since been edited.[112]

Most of the works contained in this book are quite complex in their language and written with great care. These texts claim to represent the oral teachings of Ibn Tūmart, edited more than twenty years after his death by the "caliph" ʿAbd al-Muʾmin from notes (*taʿālīq*) taken by Ibn Tūmart's companions. This is conspicuously similar to what is known about the collection of the Qur'an by Caliph ʿUthman ibn ʿAffān, and probably is not true. It is hard to imagine that Ibn Tūmart himself did not compose these works. For our purposes in understanding Ibn Tūmart's theology and his intellectual connection to al-Ghazālī, three texts will prove to be most important. These texts are his *Creed on the Creator's Divine Unity* (*Tawḥīd al-Bāriʾ*) and two short texts of about one page each, referred to as his *Guides No. I* and *II* (*Murshida* I and II).

Ibn Tūmart's proof for the existence of God follows in its outward structure the traditional *kalām* proof for God's existence: we know from observation that all things either change or, if they do not change, have the potential to change. Things change their place, their position, sometimes their color, and so forth. All these changes happen in time, that is, they appear from one moment to the next. A substance in which temporal change occurs must be generated in time and cannot be eternal. If the temporal changes in a thing are caused by another thing that is subject to temporal change, then the series of things that are subject to such changes cannot regress indefinitely. Thus, these changes must be introduced by something that is itself not subject to temporal change,and this must be eternal and not generated in time. This is God.[113] Al-Juwaynī has this proof fully worked out in his late work *The Creed for Niẓām al-Mulk*.[114] Al-Ghazālī gives a version of this proof at the beginning of his *Balanced Book on What-To-Believe* (*al-Iqtiṣād fī l-iʿtiqād*). He devotes much space to the proof that all created things are subject to change.[115] This is the key premise of the *kalām* proof for God's existence, and it is challenged by an objection of the *falāsifa*, namely, that the celestial bodies and their spheres are not—and have never been—subject to change. Ibn Tūmart's *Creed on the Creator's Divine Unity* shows that he was well familiar with this problem. He develops an innovative argument that aims to extend judgments about objects of our experience to things that we cannot experience. Ibn Tūmart's wishes to establish a valid analogy that extends to all created beings. If such an analogy is possible, judgments about things that we experience directly can be extended to things that we experience only indirectly or from a distance, such as celestial objects.

Ibn Tūmart's analogy is inspired by the division of judgments into necessary, contingent, and impossible. These divisions are a prominent feature of Avicenna's philosophy, who introduced them to the philosophical genre of proofs for God's existence. In Avicenna, however, necessary, contingent, and impossible are not predicates of judgments but rather of things in the outside world. We will see that al-Ghazālī criticized Avicenna by saying that these divisions are not ontological, meaning they cannot be found within the world; but they are rather epistemological, meaning the three predicates of necessary, contingent, and impossible apply only to our judgments and not to objects in the world. This contention can already be found in the works of al-Juwaynī. Ibn Tūmart applies the threefold division of necessary, possible, and impossible as an epistemological distinction about human judgments. In our mind, we find that the truth of some judgments is necessary, the truth of others is contingent, and again others cannot at all be true. An example of the first kind of judgment is: "Everything has a maker (*fāʿil*)." This judgment is always true, says Ibn Tūmart, and this leads us to know that there cannot be anything in this world that doesn't have a maker, or, in the parlance of the philosophers, an efficient cause (*fāʿil*). For al-Ghazālī, the principle that all that comes to be must have a cause that brings it about is an axiom of reason and a necessary truth.[116] Ibn Tūmart understands that humans are given this truth *a priori*, and through it, God has given us a way to prove His existence.[117] The necessary truth of the principle that everything has a maker leads humans to realize that everything is created, even the stars in heaven. Ibn Tūmart's detailed inquiry into who could be the maker of such complicated objects as a human body leads to the realization that only God can create such complex things as a human. Other beings wouldn't even be able to create a single limb. If only God can be the creator of a human limb, He is *a fortiori* the creator of the stars and of everything else in the world.

Ibn Tūmart's argument for God's existence shares many of the notions and ideas important in the theology of al-Ghazālī. He uses Avicenna's ontological distinction of necessary, contingent, and impossible in an epistemological way. Already al-Juwaynī employed it thus in his *Creed for Niẓām al-Mulk*.[118] Like al-Ghazālī, Ibn Tūmart is impressed by the ingenuity of God's creation and by the well-fitted function and place of all individual elements in an overall plan. Ibn

Tūmart and al-Ghazālī both saw a most skillful plan at work in God's creation. This conviction made them introduce arguments for God's existence from design and teleological motifs in their respective proofs.[119]

Yet the clearest indicator of al-Ghazālī's profound influence on Ibn Tūmart is their common teachings about God's determination of every event in the created world. Both taught that the plan for God's creation existed even before the first creature came into being. Every event is predetermined by God's decree. God is viewed as the omniscient engineer of an ingenious network of what appears to people to be causes and effects. Once that network runs, however, God does not change it. After its initial creation, the world follows the plan that God made in His eternity:

> The Omniscient determines [all] this in His eternity (*fi azalihi*) and the things become manifest through His wisdom in accordance with what He has determined. Then, they take place according to His determination, which follows an undisturbed calculus (*ḥisāb lā yukhtalla*) and an unbroken order.[120]

These words recall al-Ghazālī's comparison of the created world with a water clock (*ṣandūq al-sāʿāt*), which will be discussed below.[121] Yet, unlike al-Ghazālī, Ibn Tūmart uses language that is—as far as I can see—unambiguously occasionalist.[122] We will see that occasionalism is very much within the range of what might be called Ghazalian theology. We will also see that the great theologian from Khorasan was reluctant to express his ideas about God's predetermination of future events all too candidly. The works of Ibn Tūmart, who shows fewer scruples in this respect, are therefore a welcome and helpful interpretation of what was taught at the Niẓāmiyya in Baghdad during the first years of the sixth/twelfth century.

Ibn Tūmart's view of divine creation and predetermination reflects much of what al-Ghazālī has written on this subject. At the beginning of his *Creed*, for instance, Ibn Tūmart says that a Muslim's belief (*īmān*) and piety (*ikhlāṣ*) are the result of chains of events that eventually go back to the miracle that confirms the mission of the Prophet.[123] We will see that these chains of events (al-Ghazālī says: "the chaining of causes," *tasalsul al-asbāb*) play a very important role in al-Ghazālī's theology. For instance, al-Ghazālī mentions a very similar chain in the thirty-first book of his *Revival of the Religious Sciences* on the subject of patience and thankfulness (*Kitāb al-Ṣabr wa-l-shukr*).[124] These chains of events are a novel concept and cannot be found in the works of earlier Ashʿarite thinkers.

After Ibn Tūmart's death in 524/1131 and the Almohads' conquest of Morocco and al-Andalus, Ghazalism became firmly established in a region where thus far the political leaders had been openly hostile toward it. In 503/1109, the Almoravids, the predecessors of the Almohads as rulers over the Maghrib, burned al-Ghazālī's *Revival of the Religious Science* in the courtyard of the mosque in Cordoba. The Almoravids were conservative Mālikites who rejected al-Ghazālī's critique of their legal method as well as his rationalist and Sufi tendencies.[125] After their demise at the hands of the Almohads, al-Ghazālī's position within the theological climate in the Maghrib changed dramatically. While under the Almoravids, al-Ghazālī's teachings were regarded as unbelief; they flourished under the Almohads, who actively promoted them.[126] The philosophical and theological teachings of such important Almohad thinkers as Ibn Ṭufayl (d. 456/1061) and Averroes are part of the Ghazalian tradition, despite the fact that both made a point of criticizing al-Ghazālī.[127] Almohad theology and philosophy is said to have disappeared after the defeat of the Almohad Empire by the Christian Reconquista in the first half of the seventh/thirteenth century. Yet the rationalist attitude of Almohadism and Ghazalism continued to have a long-lasting effect on intellectuals of the Maghrib. The Mālikite jurist al-Shāṭibī (d. 790/1388), who was active in Granada during the Naṣrid era, is a good example of the application of Ghazalian principles in jurisprudence (*fiqh*). His stress on public benefit (*maṣlaḥa*) as a source of Islamic law is a development of al-Ghazālī's earlier rationalist teachings along these lines.[128] In theology and law, scholars in the Maghrib became more open to accepting the view that these disciplines must be accompanied by the study of philosophical logic. In the Muslim East, influential interpreters of al-Ghazālī, such as the two Damascenes Ibn al-Ṣalāḥ al-Shahrazūrī and Yaḥyā al-Nawawī, rejected this element of his teachings. Like Jalāl al-Dīn al-Suyūṭī (d. 911/1505), they regarded Aristotelian logic as a dangerous innovation that would lead students to become receptive to the heterodox thought of the *falāsifa*.[129] In the Maghrib, however, the study of Aristotelian logic flourished and produced a great number of works written throughout the eighth/fourteenth to the twelfth/eighteenth centuries.[130] At the end of the twelfth/eighteenth century, the Egyptian-based scholar and Ghazālī commentator al-Murtaḍā al-Zābidī observed that Maghribī scholars had reintroduced the study of philosophical logic into Egypt two generations before.[131] By this time, the Mālikite Maghrib, where al-Ghazālī's books were burned during his lifetime, had become more Ghazalian than the Muslim East.

ʿAyn al-Quḍāt al-Hamadhānī (d. 525/1131)

Like Ibn Tūmart, ʿAyn al-Quḍāt al-Hamadhānī (or: ʿAyn al-Qużāt-i Hamadānī) was not a direct student of al-Ghazālī's, never having even met the great scholar. ʿAyn al-Quḍāt was born in 492/1098 in Hamadan in central Iran to a family of scholars. The historian al-Bayhaqī characterizes ʿAyn al-Quḍāt as an author who "mixed the teachings of the Sufis with those of the philosophers."[132] As a young adult, ʿAyn al-Quḍāt had met al-Ghazālī's brother Aḥmad and was so impressed by him that, despite his age, ʿAyn al-Quḍāt became his student (figure 2.1).

Although ʿAyn al-Quḍāt had studied al-Ghazālī's *Revival* before, his close contact with Aḥmad caused him to immerse himself again in the works of Muḥmmad al-Ghazālī and to appreciate them greatly. In one of his books, ʿAyn al-Quḍāt writes how he had come to the conclusion that Muḥammad al-Ghazālī belongs, like his brother Aḥmad and himself, to a select group of ten scholars firmly rooted (*rāsikh*) in knowledge and knowing the outer as well as the inner

FIGURE 2.1 ʿAyn al-Quḍāt al-Ḥamadhānī meets Aḥmad al-Ghazālī in a garden. Miniature from a manuscript of Kamāl al-Dīn Gāzurgāhī's (d. after 909/1503–4) *Assemblies of God-Lovers* (*Majālis al-ʿushshāq*), produced c. 967/1560 in India. (MS London, British Library, Or. 11837, fol. 57b).

meanings of the Qur'an (cf. Q 3:7).[133] Like Asʿad al-Mayhanī, ʿAyn al-Quḍāt believed that al-Ghazālī's intelligence (*ʿaql*) reached a stage that few other humans can match.[134] He considered himself a disciple of al-Ghazālī's books.[135]

A close reading of ʿAyn al-Quḍāt's works shows that he was well acquainted with the most important motifs in al-Ghazālī's theology, frequently adopting them as his own. He criticizes, for instance, the *falāsifa*'s concept of efficient causality with arguments that are inspired by al-Ghazālī's seventeenth chapter in his *Incoherence*.[136] Like Ibn Tūmart, ʿAyn al-Quḍāt was influenced by Avicenna's proof of God's existence. Unlike his contemporary in the Muslim West, however, he was aware of this philosophical influence and discussed it openly. ʿAyn al-Quḍāt begins his most theological work, *The Essence of Truths* (*Zubdat al-ḥaqāʾiq*), with a brief autobiography in which the rediscovery of al-Ghazālī's books shortly before 512/1118 takes center stage. Once he had been pointed to these books, he studied them for four years, and it was only by reading them that he began to understand the religious sciences.[137] The discussion of theology in this book starts with a comparison between the merits of the *kalām* proof for God's existence and the one developed by Avicenna. ʿAyn al-Quḍāt clearly prefers the latter and excuses al-Ghazālī for having produced a version of the *kalām* proof in his *Balanced Book on What-To-Believe*.[138] Al-Ghazālī's attitude to Avicenna's so-called "Proof of the Reliable Ones" (*burhān al-ṣiddīqīn*) was ambiguous. In his *Scandals of the Esoterics*, he produces a version of this proof and says its conclusion is necessary.[139] In the fourth and fifth discussion of his *Incoherence*, however, he criticizes several elements of the Avicennan proof and suggests that it is demonstrative only after adding the additional premise that the world was created in time.[140] The original Avicennan proof from contingency has indeed some implications that are undesirable for al-Ghazālī. It proves the existence of God as the origin of all being (*wujūd*) and as the only being that is necessary by virtue of itself (*wājib al-wujūd bi-dhātihi*). This implies that all aspects of God's being are necessary, including His will and His actions. It also implies that God creates necessarily, meaning continuously from pre-eternity. We will see that al-Ghazālī harshly criticizes Avicenna for teaching that God's will and His actions are necessary. Some aspects of the Avicennan proof, however, were quite appealing to al-Ghazālī, primarily the fact that it enables humans "to give evidence to the created things by way of their Creator, rather than giving evidence to Him by way of the created things."[141] Thus, Avicenna's proof avoids the ascent from the low to the high and allows one to prove God's existence solely by contemplating on the nature of existence. This is more reliable and nobler (*awthaq wa-ashraf*) than any other argument for God's existence.[142]

Al-Ghazālī accepted a version of Avicenna's proof that avoids the implication of eternal creation, and he seems to have regarded it as equivalent—or maybe even preferable—to the traditional *kalām* proof.[143] The fact that all existence is either by itself possible or by itself necessary opens a way to proof that God's existence is the origin for the existence of all other things. In such works as *The Niche of Lights*, al-Ghazālī expresses approval for this aspect of the Avicennan proof. In that book, al-Ghazālī explains that the sun's light is the best metaphor to show how everything in this world emerges from God's existence. ʿAyn al-Quḍāt enthusiastically follows him in this approach. In his collection *Preludes* (*Tamhīdāt*), ʿAyn al-Quḍāt explains how the simile of light works, and he expands upon it with a much more complicated notion of lightness and darkness, personified by the pre-Islamic dualistic figures of Yazdān and Ahriman.[144]

ʿAyn al-Quḍāt's theology is influenced by the Ghazalian notion that God bestows existence onto the created world. God is the only real existence, while all other things have their existence borrowed for a limited time from Him. Everything is, by itself, sheer nothing: "Every contingent being (*mumkin*), in so far as it is looked at in itself and not considered sustained by the Necessary, is by itself non-existent (*maʿdūm*)."[145] Things only come into existence when the conditions (*shurūṭ*) are fulfilled for a particular possible existent to receive existence from God. This idea of the conditions for future contingencies had already been put forward by al-Ghazālī in his *Revival of the Religious Sciences* as an attempt to reconcile the limitless world of an occasionalist cosmology with the necessary restrictions to which any future moment is subject. What can possibly be created in the next moment depends on what already exists in this one.[146] God's plan of creation responds to these limitations. He determines necessarily what has been created in the past and what will be created in the future. There is no arbitrariness in God's plan; it exists in a timeless sphere and was already there when creation began. Thus, whatever will exist in the future is already determined in God's timeless knowledge.[147]

ʿAyn al-Quḍāt was particularly attracted to al-Ghazālī's ontology. He quotes and explains, for instance, al-Ghazālī's ideas on semantics in his account of the relationship between a name and what it names (*ism wa-musammā*) from the introduction to al-Ghazālī's *Highest Goal in Explaining the Beautiful Names of God*.[148] There is no evidence that ʿAyn al-Quḍāt was aware of the philosophical background of these particular teachings, although he clearly did understand the intellectual connection between Avicenna and al-Ghazālī. ʿAyn al-Quḍāt criticizes the *falāsifa* together with the *mutakallimūn* because their negative theology cannot lead to an adequate understanding of the Divine,[149] yet he also expresses a fondness toward Avicenna. One of ʿAyn al-Quḍāt's original teachings is that the true seeker after God should be acquainted with a certain kind of unbelief (*kufr*) in order to reach a higher degree of belief. This position, ʿAyn al-Quḍāt claims, had already been expressed by Avicenna in his *Epistle on the Occasion of the Feast of Sacrifice* (*al-Risāla al-Aḍḥawiyya*). According to ʿAyn al-Quḍāt's account, when a Sufi asked Avicenna to provide a proof—for what exactly remains obscure—he simply said:

> [The proof is] to enter true unbelief (*al-kufr al-ḥaqīqī*) and to leave what is (only) metaphorical Islam (*al-Islām al-majāzī*) and to pay attention only to what is beyond the three [types of] people until you are a believing Muslim *and* an unbeliever. If you are beyond this [level] you are neither believer nor unbeliever. If you remain below this, then you are a polytheist Muslim. If you are ignorant of this, then you will know that there will be no resurrection for you, nor will you return as one of the existing beings.[150]

None of this can be found in Avicenna's *Epistle on the Occasion of the Feast of Sacrifice* or elsewhere in his writings. It is, in fact, a Ghazalian notion inspired by his explanation of four levels of believe in divine unity (*tawḥīd*) at the beginning of the thirty-fifth book of his *Revival*. There, al-Ghazālī says that the true seeker of God should aim for the fourth and highest level of insight as to what belief in one single God (*tawḥīd*) really means. On this level, he understands that all being is God. The three lower levels represent lesser insights, insufficient for the true seeker.[151] In ʿAyn al-Quḍāt's pseudo-Avicennan quote, this notion is combined with the idea that the true believer is one who cannot be defined by categories such as "Muslim" or "unbeliever." A portion of unbelief is required to reach the highest level of understanding divine unity (*tawḥīd*).

What is meant by requiring such a portion of unbelief is illuminated in another passage in ʿAyn al-Quḍāt's *Preludes*. Here, he defends Avicenna's position of the world's pre-eternity. When Avicenna said that the four prime elements are pre-eternal (*qadīm*), he did not mean to say, ʿAyn al-Quḍāt explains, that anything in the sublunar sphere and the world of coming-to-be and passing-away is pre-eternal. Only the building materials of the earthly world are pre-eternal, and these are the "real elements" (*ʿanāṣir-i ḥaqīqī*). This teaching is correct, says ʿAyn al-Quḍāt, and "Avicenna should be excused for saying this."[152] Yet, al-Ghazālī had branded this position as unbelief and apostasy from Islam. It seems that ʿAyn al-Quḍāt aimed to turn his condemnation into something positive that the Sufi should embrace.

ʿAyn al-Quḍāt also incorporates many of the major ideas of al-Ghazālī's moral teachings. He follows al-Ghazālī closely in his critique of *kalām*.[153] Like al-Ghazālī, he criticizes the political elite for their corruption and calls them in one of his letters "a Satan among the Satans of humanity and an enemy among the enemies of God and His messenger."[154] Those scholars who seek the rulers' patronage and who do not use their knowledge to earn the afterlife are condemned. He advises his students to "serve the sandals" (*khidmat-i kafsh*) rather than to serve the sultan,[155] using the Ghazalian expression "serving of the sandals" coined in his *Niche of Lights*. It means that one should follow the example of Moses, whom God had asked in the valley of Ṭuwā to "take off the two sandals" (Q 20:12). Al-Ghazālī interprets this verse as meaning that Moses was asked to leave the worldly affairs (*al-dunyā*) behind him and concentrate fully on the afterlife. Several mystics in the generation after al-Ghazālī picked up this metaphor. ʿAyn al-Quḍāt's usage is joined, for instance, by his contemporary Ibn Qasī (d. 546/1151) from al-Andalus. In 539/1144, he was the leader of a Sufi revolt against the antimystical Almoravids. Ibn Qasī's movement had its center in what is today the Algarve in southern Portugal.[156] Ibn Qasī's main work is *The Book on Taking Off the Two Sandals* (*Kitāb Khalʿ al-naʿlayn*), and here he pursues the same Ghazalian motif as ʿAyn al-Quḍāt persued, "to throw off the two worlds (*kawnān*)." Moses, al-Ghazālī says, obeyed God's imperative outwardly by taking off his sandals and inwardly by throwing off the two worlds.[157]

Western scholarship on ʿAyn al-Quḍāt has mostly focused on his political significance. In 525/1131, at age thirty-three, he was crucified in Hamadan along with other officials with whom he had close ties. This happened during the reign of the Seljuq sultan Maḥmūd ibn Muḥammad Tapar (reg. 511–25 / 1118–31) and during the vizierate of Qawwām al-Dīn al-Dargazīnī (d. 527/1133). This is the same Sultan Maḥmūd who, as a child, when he held the governorship of Baghdad, had invited Asʿad al-Mayhanī to teach at the local Niẓāmiyya. He was not known for antirationalist or antiphilosophical tendencies. The Seljuq ruling family, particular Maḥmūd's uncle, the Supreme Sultan Sanjar, had formally embraced the teachings of al-Ghazālī after the accusations against him were dismissed.[158] The sources do not allow us to determine fully why ʿAyn al-Quḍāt was executed and whether this was a reaction to his teachings. Most historians have tried to explain his execution as the outcome of a court intrigue in which al-Dargazīnī is usually assigned the role of the villain.[159]

The scholarly community during ʿAyn al-Quḍāt's days did not share these misgivings at the Seljuq court. His contemporary al-Samʿānī has high praise for ʿAyn al-Quḍāt's virtue and his Sufi scholarship.[160] Another early historian wrote: "He was one of the great imams and friends of God (*awliyāʾ*) who was noble-hearted and who followed in his works Abū Ḥāmid al-Ghazālī."[161] From his prison cell in Baghdad, ʿAyn al-Quḍāt wrote a treatise in his defense addressed to the scholars of Islam. It reveals that he was formally charged with heretical teachings, some of them regarded as apostasy from Islam.[162] Among the accusations were: (1) adhering to the Ismāʿīlite doctrine of slavishly learning from a teacher (*taʿlīm*) and (2) teaching two heterodox philosophical positions, namely that the world is pre-eternal and that God does not know individuals.[163] Al-Ghazālī had condemned these two teachings as apostasy from Islam, punishable by death. ʿAyn al-Quḍāt admits that he used philosophical language that may lead to weak minds getting the impression that he believed in these two condemned doctrines.[164] Yet he maintains that these weak minds misunderstand his words, that he never accepted these teachings, and that he, in fact, refutes them in his writings:

> Those of my words that they hold against me are all also in the books of al-Ghazālī—the same expressions in the same meanings. For example our words regarding the Creator of the world, namely that He is the source of being (*yanbūʿ al-wujūd*) and the origin of being (*maṣdar al-wujūd*), that He is the universe (*al-kull*) and that He is the true being (*al-wujūd al-ḥaqīqī*) and that everything that is not He is with regard to its essence empty, fading, annihilating, and non-existent. And only that exists whose existence the Eternal Power (*al-qudra al-azaliyya*) sustains. These are well-known words that appear in many passages in the *Revival of the Religious Sciences*, in the *Niche of Lights*, and in the *Deliverer from Error*, and all these books were written by al-Ghazālī.[165]

In the case of ʿAyn al-Quḍāt al-Hamadhānī, the persecuting spirit that al-Ghazālī created by adding a legal judgment to his epistemological discussion in his *Incoherence of the Philosophers* came to haunt one of his own close followers. A careful study of ʿAyn al-Quḍāt's teaching on theology and Sufism is still a desideratum. In the 1970s, Toshihiko Izutsu and Hermann Landolt made valuable contributions that still require further study.[166] ʿAyn al-Quḍāt's personal acquaintance with the brother Aḥmad and his philosophically inspired Sufism make him one of the most significant early followers of al-Ghazālī.

The Anonymous Author of *The Lion and the Diver* (*al-Asad wa-l-ghawwāṣ*)

In 1978, Riḍwān al-Sayyid edited an Arabic animal fable that is extant in at least four manuscripts. The colophon of one of those manuscripts notes that the source from which the copy was made (*al-umm al-mansūkh minhā*) was completed in Ramaḍān 530 / June 1136.[167] The novel tells the story of a wise and learned jackal who seeks to become a member of the lion-king's court in order to counsel him and help him benefit from his insight. In its overall composition as well as in the style of its dialogues and its shorter fables and parables, the novel owes much to *Kalīla and Dimna* (*Kalīla wa-Dimna*), a collection of animal fables that Ibn al-Muqaffaʿ (d. *c.*137/755) translated from Pahlevi into Arabic during the mid-second/eighth century. Most plot elements in *The Lion and the Diver*—with the notable exception of its ending—are taken from the tenth chapter, "The Lion and the Jackal" (*al-Asad wa-bn Āwā*) in *Kalīla and Dimna* (see figure 2.2).[168]

The anonymous author of *The Lion and the Diver* was a highly accomplished literate who had studied the genre of Arabic animal fables well. *Kalīla and Dimna* is formally written as a *fürstenspiegel*, a book addressed to a prince, aiming to entertain the ruler while at the same time educating him and giving him council. The book is thus a guidebook in ethics, in politics, and in theology. These descriptors are also true for *The Lion and the Diver*. In addition to being a highly talented writer, its author was educated in medicine, and he knew some history of Sasanid Persia and the legends of the pre-Islamic Arabia (*ayyām al-ʿarab*). He was a Sunni Muslim with a highly rationalist mind-set. Most important, he was a Ghazalian, meaning that he expressed many of the motifs, maxims, and insights that appear prominently in the works of al-Ghazālī. In fact, the wise jackal, known as "the diver" (*al-ghawwāṣ*), has so much in common with al-Ghazālī, in opinions and in biography, that the original readers may have perceived him as a literary personification of the famous scholar. *The Lion and the Diver* may well be a *roman à clef* of al-Ghazālī's dealings with those in power.

The novel's scholar-jackal protagonist is a virtuous, highly reflective, and immensely educated soldier in the lion's army. Later in the novel, it is revealed that his goal in life is to earn the afterlife rather than succeed in this world.[169] At the beginning, he is presented as well aware of the potential dangers of approaching the lion's court. In a dialogue with his best friend, the two remind each other that scholarship and political power do not go well with each other. The scholar may easily offend the ruler with all-too-candid advice. He may become the victim of the ruler's anger or of his whims. Rulers tend to surround themselves with courtiers who satisfy their vanity rather than those who give honest and sometimes tactless counsel. When a violent water buffalo threatens the jackal's community, however, his sense of duty makes him overcome his reservations, and he decides that he must approach the king and give him

FIGURE 2.2 The lion and the jackal. Miniature from a manuscript of *Kalīla and Dimna*, dated 755/1354 (Bodleian Library, University of Oxford, MS Pococke 400, fol. 138b).

wise counsel. In order to avoid the well-known hazards, he aims to become a loyal member of the court, to please the king and win his confidence, and only then will he give honest council. This strategy is successful, until the other courtiers become jealous of the jackal's success and start to plot against him. They employ underhand tactics to cast suspicion on the jackal's sincerity. The king throws the jackal in jail and has him surveyed by his agents, who tell him

FIGURE 2.3 The king and the philosopher, the king wearing a Seljuq crown and sitting on a Seljuq throne (cf. figs. 1.1, 1.2, and 1.5). Miniature from a manuscript of *Kalīla and Dimna*, dated 755/1354 (Bodleian Library, University of Oxford, MS Pococke 400, fol. 136b).

that the jackal is innocent. The lion learns about the tricks played on the jackal by his enemies and rehabilitates him. The jackal, however, rejects the king's invitation to become his close advisor. He leaves the king with a "testament" (*waṣiyya*) and chooses to withdraw himself to one of the "houses of worship" (*buyūt al-ʿibāda*) in the mountains. The jackal purifies himself from his harmful experience by admonishing his soul and preaching to it in an inner dialogue.[170] Al-Ghazālī uses the same literary technique of talking to one's soul and admonishing its desires in one of his letters.[171] The novel ends with this sentence: "the lion used to visit the jackal from time to time until fate (*al-dahr*) parted them."

An attentive reader of al-Ghazālī finds numerous explicit hints to his biography. During his introduction, the king asks the jackal why he is called "the diver"; the answer is: "Because I dive deep for the subtle meanings and because I bring out the hidden secrets of the sciences."[172] One of al-Ghazālī's favorite metaphors for the dangers of scholarship was that of a deep sea. While the trained scholar plunges into the deep sea of scholarship and swims through it, others, who lack a sufficient education, are drawn to these depths but often drown.[173] The metaphor appears so frequently in al-Ghazālī's books that Ibn Ṭufayl almost mockingly alludes to it in the introduction to his *Ḥayy ibn Yaqẓān*.[174] If the well-trained scholar can swim in the sea of knowledge, then the most accomplished scholar is a diver who picks up secrets from the dark depths of that sea like a pearl diver (*ghawwāṣ*) collecting precious pearls. In one of his letters to the "king of Khorasan," Sanjar, al-Ghazālī depicted himself as having spent forty years of his life "diving into the sea of religious sciences."[175]

The collection of al-Ghazālī's letters was compiled at some time during the sixth/twelfth century. Al-Ghazālī's letter to Sanjar was likely written before the composition of *The Lion and the Diver*, in fact the two may have been published at roughly the same time. The relationship between the scholar-jackal and the lion-king develops very much along the lines of an idealized and even exaggerated picture that a follower of al-Ghazālī might have painted of the relationship between him and members of the Seljuq dynasty. Throughout the novel, the virtues of the jackal are unquestioned, and his temporary downfall is solely the result of other people's jealousy. That mirrors al-Ghazālī's own perception as to why some people have accused him at Sanjar's court. In his written response to these accusations and in his *Decisive Criterion*, al-Ghazālī quotes an anonymous one-liner that he may have picked up from al-Qushayrī's *Epistle* (*al-Risāla*).[176] The quoted poem is meant to explain why al-Ghazālī's work triggered so much enmity: while truly virtuous scholars are impressed by his scholarship and often convinced by the force of his arguments, some are jealous of al-Ghazālī's natural gifts. These jealous colleagues are blind to his achievements, and their enmity cannot be resolved. The quoted poem goes:

> One can overcome all kinds of hostility,
> except for that which is due to jealousy.[177]

There are more general parallels between the story and al-Ghazālī's biography. During his early life, the jackal educates himself. His education is driven by a universal curiosity and an independent mind. He says that he did not benefit from teachers. A similar picture is painted in al-Ghazālī's autobiography in which teachers are not given any credit. Where the jackal grew up, his "love of wisdom" (*ḥubb al-ḥikma*)—a literary translation of the Greek *philosophía*—was discouraged. When the jackal enters the king's court, he says about his education:

> O King, I grew up among people who regard the pursuit of knowledge as a mistake and love of wisdom as a blemish (*ʿayb*). Therefore, I first concealed everything of this kind that I had within myself because I was ashamed, and I tricked the others until this became a habit, and the habit became a natural impulse (*gharīza*) that I followed. (. . .)
>
> I took it upon myself to think and I often refrained from speaking. I never quarreled with others and searched knowledge for my own sake so that I spend my life as a prisoner of books and as a companion of thoughts. The tongue needs incitement in order to become fluent, and exercise in order to make it agile and sharp.[178]

Initially, the wise jackal endeavors to be of service to the king, and he becomes a member of his court. The two have intelligent conversations in which the jackal reminds him that the wise man acts decisively and shows no neglect or fatalism, although he also knows that God predetermines all events. This allows him, for instance, to benefit from astrological predictions.[179] Appearing prominently in the thirty-second as well as the thirty-fifth books of the *Revival,* the maxim of accepting predetermination yet not falling into fatalism is just one of many Ghazalian motifs that appears throughout the anonymous novel.[180] The king is reminded of his duty to maintain order (*niẓām*) and justice in the world and to defend the *sunna* of the Prophet and the *sharīʿa* of Islam. In fact, the order in the empire depends directly on the power of its ruler, who is the guardian of the *sunna*.[181] The king's reign over his realm resembles the reign of the human's soul over his limbs.[182] Like al-Ghazālī, the jackal-scholar in the book is strongly opposed to emulating higher authorities (*taqlīd*). Truth (*al-ḥaqq*) is known by itself and not by the authority of those who testify to it.[183] Other Ghazalian notions include the ideas that people—including kings—should be addressed according to their intellectual capacities[184] and that there is no good in this world that is not also accompanied by some harm.[185] Yet, the whole world has been designed "with utmost wisdom and good craftsmanship" (*ʿalā ghāyat al-ḥikma wa-ḥusn al-ṣināʿa*), and careful attention has been paid to even the tiniest of its details.[186]

One of the most forcefully presented messages of the book is the Ghazalian idea that knowledge by itself is useless when it does not lead to right action.[187] People who are particularly knowledgeable do not act virtuously simply as a consequence of their knowledge. Only the correct kind of knowledge leads to inner virtue, which in turn leads to right action. Such virtue is gained through the training of the soul. At the end of the novel, when the jackal takes his leave, the lion asks him why he cannot discipline his soul and also remain an advisor at his court. The jackal's answer makes productive use of the novel's technique to illustrates Ghazalian principles with parables and fables:

> Then the lion asked the jackal: What prevents you from worshipping God where you are [right now]?
>
> The jackal answered: In order for the animal soul, dear King, to gather itself and be trained it must be separated from the things it loves. I have been damaged by the exposure to things that are naturally deemed nice and pleasant. I fear that this situation will become a habit to me and prevent me from removing these things once they have been firmly installed [in my life]. Then, it would happen to me what happened to the owner of the stallion.
>
> The king asked: What happened to him?
>
> The jackal: The story goes that a courageous man had a foal that grew up in his possession. The animal was of utmost grace and beauty, had straight limbs and a strong body. The old man was infatuated with it and it occupied all his concerns. He ceaselessly indulged it and provided an abundance of fodder. The man was too old to train the animal himself, yet he was also too anxious to have someone else ride it, train it, and break it in. So the animal was not trained by anyone. Its character traits became spoiled and its temperament bad. Next to the young stallion stood a mare, whose scent aroused his passion. The stallion's owner had great difficulty any time he wanted to ride it. The days went by and as the old man became more and more frail, the coat became stronger and stronger. The time came that he needed to ride on the stallion and engage in an attack against his enemies. But any time the horse was not bound—at its feet, for instance—it did not obey its owner's instructions. Once the old man mounted the stallion, it broke with him through the lines of the enemy to reach a mare it had scented. The enemies struck down the horse and killed the old man.
>
> This is similar to a man and his soul. A man is like the owner of the horse. Had he trained it regularly, he would have had a tame riding animal that would have gotten him wherever he wanted to go. But if he doesn't break it in and teach it good manners, it acquires these repulsive habits—and maybe it will gain the upper hand over its rider and destroy him together with itself.[188]

In Book 22 of his *Revival,* "Disciplining the Self, Refinement of Character, and Treating the Diseases of the Heart," al-Ghazālī writes that "an allegory for desire (*shahwa*) is the horse which one rides during a chase. It is sometimes well-disciplined and well-behaved and sometimes it is defiant."[189] Desire is a character trait (*khulq*) that needs to be trained like any other. Earlier generations of scholars that had worked on Muslim ethics, al-Ghazālī complains, had hardly ever dealt with the human character, but were merely concerned with its fruits, the human actions.[190] Muslim jurists are mostly concerned with the bare compliance to the rules of Sharīʿa and thus cannot give council on matters of good character. They are mere "scholars of this world" (*ʿulamāʾ al-dunyā*) who cannot guide Muslims on the best way to gain the afterlife.[191] The substance of the human (*gawhar-i ādamī*), al-Ghazālī says in his Persian *Alchemy of Happiness* (*Kīmyā-yi saʿādat*), is initially deficient and ignoble (*nāqiṣ wa-khasīs*); only strict efforts and patient treatment can lead the soul from its deficient state to its perfection. The human soul's temperament becomes imbalanced through the influence of other people and needs to undergo disciplining (*riyāḍa*) and training (*tarbiya*) in order to keep the character traits (*akhlāq*) at equilibrium.[192] Al-Ghazālī rejected the notion that one should try to give up potentially harmful affections such as anger or sexual desire. These character traits are part of human nature, he teaches, and they cannot be given up. Rather, disciplining the soul allows control over these potentially harmful traits through one's rationality (*ʿaql*). Al-Ghazālī compares the human pursuit of redemption in the afterlife with the hunter's pursuit of game. Sexual desire and anger are not always negative. Anger, for instance, is a positive character trait in the war against infidels. Sexual desire and anger are to the human rational faculty what the horse and the dog are to the hunter. The hunter trains his horse and dog in order to benefit from their service. In the hunt for the afterlife's reward, anger and sexual desire are just as useful to the human, yet rationality must train them and control them like the hunter trains and controls his horse and dog.[193]

None of these notions and ideas, which I identify as "Ghazalian," is particularly unique to al-Ghazālī and could not also have been picked up from other Muslim rationalist literature of this or earlier times. Many of the theological motifs and the moral teachings in *The Lion and the Diver*, such as the imperative to develop one's inner virtuous character rather than to focus on the fulfillment of Sharīʿa's prescriptions, come from philosophical and from Sufi literature. These philosophical and mystical motifs became more widespread during the sixth/twelfth century, particularly in mainstream religious literature, in which earlier obedience to the rules of Sharīʿa had dominated the debate on morality. Al-Ghazālī's work played a significant part in this development. The accumulation of teachings in this novel that appear prominently in al-Ghazālī's work is significant. Most difficult to determine is, however, whether the author has been directly inspired by elements in al-Ghazālī's biography. The novel is clearly modeled after the story of the lion and the jackal in *Kalīla and Dimna*; in that book, the jackal stays with the lion-king and again becomes his trusted advisor after his suffering from the ruses of the courtiers and subsequent rehabilitation. By staying at court, the jackal fulfills the wishes of the king. In the *Lion and the Diver*, the rift between the scholar-jackal and the ruler has become too deep for him to stay, and here it is the jackal who eventually determines the terms of their relationship. He leaves royal service and becomes an ascetic.

The appearance of the Seljuq warrior-kings during the mid-fifth/eleventh century brought a new aspect to the age-old conflict in Islamic civilization between those who use the sword as their weapon and those who use the pen. The Turk nomads' military hegemony—today we would say their military dictatorship—triggered much reflection about the mores and the conduct of this new type of rulers who, unlike earlier sovereigns, were most often not participants in the literary, moral, or theological discourses of their time. Such reflections appear frequently in the Islamic literature of the sixth/twelfth century. The Persian poet Niẓāmī, for instance, who wrote at the end of that century, includes in the first of the five epics of his highly popular *Quintet* (*Khamsah*) a story of how an old woman admonished the powerful sultan Sanjar. The poor woman who had been wronged by one of Sanjar's troops approaches his entourage and grabs the sultan's garment (figure 2.4).

FIGURE 2.4 Sultan Sanjar admonished by an old woman. Miniature illustrating a story from Niẓāmī's *Quintet* (*Khamsah*), attributed to the famous painter Bihzād (d. 942/1535–36) or his workshop in Herat (Afghanistan) and dated 901/1495–96. (MS London, British Library, Or. 6810, fol. 16a).

She complains and accuses Sanjar of neglecting justice (*āzarm*), and she makes dire predictions about his future. Sanjar sets her admonishments at naught, and, according to Niẓāmī's implied message, would regret to have done so once his own fate had turned and he has fallen into the hands of the oppressive Oğuz Turks. The story ends in Niẓāmī's lament that "in our time, justice can no longer be found."[194]

The arrival of the warrior-kings carried with it a new kind of relationship between Muslim scholars and political power. *The Lion and the Diver* explores these new types of relationships. With regard to this subject, al-Ghazālī was an almost unavoidable focus point. He began his career as an influential and highly visible supporter and advisor of the Seljuq dynasty, yet he ended it in the seclusion of his private madrasa and *khānqāh* in Ṭūs after a very vocal disillusionment with those who hold power. The striking parallels between the jackal's biography and the way al-Ghazālī wrote about his own life may indeed simply be because of a similar analysis of the historical situation.

The widespread appearance of Ghazalian notions in books of the early sixth/twelfth century should not be surprising. Reading the works of Ibn Bājja, Ibn Ghaylān al-Balkhī, Ibn Ṭufayl, Averroes, Shihāb al-Dīn Yaḥyā al-Suhrawardī, or Ibn al-Jawzī reveals that there is hardly any religious writer of this century who does not grapple in one way or another with al-Ghazālī's legacy, and probably none who does not refer to him. Al-Ghazạlī was by far the most influential religious figure during the sixth/twelfth century, and he left his traces in all kinds of religious writing of this period.

3

Al-Ghazālī on the Role of *falsafa* in Islam

Al-Ghazālī's *Incoherence of the Philosophers* (*Tahāfut al-falāsifa*) marks the start of a significant development in medieval philosophy. With its publication, the particular Neoplatonic understanding of Aristotle that developed in late antiquity and dominated the Middle Ages until the fourteenth century began to be challenged by what later became known as nominalism. Nominalism is the position that abstract concepts and universals have no independent existence on their own. As we will see, many of the arguments used by al-Ghazālī are nominalist. The move toward a nominalist critique of Neoplatonist Aristotelianism occurred not only in Arabic and Islamic philosophy but also in the Hebrew and, most of all, Latin traditions. Al-Ghazālī stands at the beginning of this development.

In his *Incoherence*, al-Ghazālī critiques twenty teachings of the *falāsifa*, sixteen from their metaphysics and four from their natural sciences. He writes in his autobiography that during his time at the Baghdad Niẓāmiyya, he studied the works of the *falāsifa* for two years before writing his *Incoherence of the Philosophers* in the third year.[1] Most likely apologetic, this account is designed to reject the claim of some of al-Ghazālī's critics that he had learned *falsafa* before completing his own religious education.[2] *The Incoherence of the Philosophers* is a masterwork of philosophical literature, perhaps decades in the making. Several other texts exist in which al-Ghazālī provides faithful reports of the philosophers' teachings. At least two of those reports are available to us. The first is an untitled and almost complete fragment of a long book in which al-Ghazālī copies or paraphrases passages from the works of philosophers and produces a comprehensive account about their teachings in metaphysics. In an earlier publication, I described this text and showed that it is, in fact, written by al-Ghazālī.[3] The second report of the philosophers' teachings is the *Intentions of the Philosophers* (*Maqāsid al-falāsifa*), an adapted Arabic translation of the parts on logics, metaphysics, and the natural sciences in Avicenna's Persian work *Philosophy for 'Alā' al-Dawla* (*Dānishnamah-yi 'Alā'ī*).[4] Earlier scholars assumed that the *Intentions of the Philosophers* was written as a preparatory study to his major work, the *Incoherence*.[5] This contention no longer seems viable. *The Intentions of the Philosophers* bears only a very loose connection to the text of the *Incoherence*. For example, the *Incoherence* and the *Intentions* use different terminologies, and the latter presents its material in ways that do not support the criticism in the *Incoherence*.[6] *The Intentions of the Philosophers* may have been a text that was initially unconnected to the *Incoherence* or was one that was generated after the composition of the latter. Only its introduction and its brief *explicit* at the end of the book create a connection to the refutation in the *Incoherence*.[7] These parts were almost certainly written (or added) after the publication of the *Incoherence*.[8]

The Refutation of the *falāsifa* in the *Incoherence* (*Tahāfut*)

Al-Ghazālī describes *The Incoherence of the Philosophers* as a "refutation" (*radd*) of the philosophical movement.[9] This professed stance has contributed to the scholarly misconception that he opposed Aristotelianism and rejected its teachings. In reality, his response to *falsafa* was far more complex, even allowing him to adopt many of its teachings. By "refutation," he does not mean the plain rejection of the philosophical teachings discussed in that book. It is clear that in his *Incoherence*, al-Ghazālī does not set out to prove the falsehood of all of—or even of most of—the philosophical teachings discussed there. The great majority of its twenty chapters focus on the *falāsifa*'s inability to demonstrate given elements of their teachings. In a 1924 article, David Z. Baneth reminded his readers that al-Ghazālī's criticism of the *falāsifa*'s teachings had often been overestimated. Al-Ghazālī's goal is to show that the metaphysics of al-Fārābī and Avicenna are "unscientific," as Baneth put it, meaning they are not backed by demonstrative proofs. Even unproven positions can still be correct. Whether or not these teachings are wrong depends upon a second criterion: only if these unproven teachings are incompatible with the literal wording of revelation must their truth be rejected. In the fifth and the ninth chapters, for instance, al-Ghazālī attacks the *falāsifa*'s proofs for their view that God is one and that He cannot have a body. Despite his critiques, al-Ghazālī shares these positions; what he attacks are the *falāsifa*'s arguments and not their results. He claims that these arguments are not demonstrative and do not establish certain knowledge about God's unity or His incorporeity. Humans do have knowledge about these two facts, al-Ghazālī says, yet not the kind of knowledge that the philosophers claim. Al-Ghazālī also attacks the *falāsifa*'s arguments for the existence of souls in the heavens and for the incorruptibility of the human soul in the afterlife. Other of his works show, however, that al-Ghazālī taught these same things. According to Baneth, al-Ghazālī's explicit goal was "to remove these questions from the realm of pure rational knowledge and assign their answer to another source of truth, namely revelation."[10] In doing so, the *Incoherence* follows the technique of *kalām* disputations. Any reader of the *Incoherence* is struck by its careful composition and the economy of its language. Al-Ghazālī's reports of philosophical teachings are short and precise. His counterarguments make productive use of the *kalām* technique of "exhaustive investigation and disjunction" (*al-sabr wa-l-taqsīm*), where the consequences or implications of an adversary's position are fully investigated and individually discussed and, in this case, dismissed and refuted one by one. The book's twenty discussions are interspersed with objections and with further rejections, with secondary discussions, and with parallel attempts to convince the reader that alternative explanations to those put forward by the *falāsifa* are just as plausible and tenable.

In the twenty detailed and intricate philosophical discussions of the *Incoherence,* al-Ghazālī aims to show that none of the arguments supporting the twenty convictions fulfills the high epistemological standard of demonstration (*burhān*) that the *falāsifa* have set for themselves. Rather, the arguments that the *falāsifa* bring to support these teachings rely upon unproven premises that are accepted only among the *falāsifa*, not established by reason.[11] The twenty discussions of the *Incoherence* are one element in a larger case about the authority of revelation. In the thirteenth discussion, for instance, al-Ghazālī maintains that when Avicenna argues that God does not know individuals and has knowledge only of the classes of beings, none of the arguments he uses is a demonstration. The truth of the opposite position—that God knows *everything* in this world—is established in countless passages in the Qur'an and in the prophetical *ḥadīth*.

By criticizing a selected number of teachings in the *falāsifa*'s metaphysics and the natural sciences, al-Ghazālī aims to make room for the epistemological claims of revelation. At the beginning of the *Incoherence,* al-Ghazālī complains that a group among the *falāsifa* flatly denies the claims of revelation because it believes its way of arguing to be superior to that of the religious scholars who accept revelation.[12] The claim that their teachings are based on demonstrative arguments has been repeated from generation to generation of philosophers, leading them to accept this claim as a fact that has passed from teacher to student. However, al-Ghazālī maintains that if someone who is not tainted by their blind acceptance (*taqlīd*) of the authorities of Aristotle and Plato thoroughly investigates the teachings of the *falāsifa,* he will find that the *falāsifa*'s arguments do not fulfill their own standard for apodictic proofs (singl. *burhān*). This standard is set in their own books of logic, following the *Organon* of Aristotle. The demonstrative method is most clearly explained in those books of the *falāsifa*'s works on logics that are equivalent to Aristotle's *Second Analytics.* Demonstration relies on the method of syllogistics, which is explained in the *First Analytics.* In Avicenna's *Healing* (*al-Shifāʾ*), for instance, the books on logic follow Aristotle's curriculum of studies and have the same titles as those of the Stagirite. Al-Ghazālī claims that Avicenna's arguments in his metaphysics do not comply with the standard set out in his logical writings. In the introduction of the *Incoherence,* he writes:

> We will make it plain that in their metaphysical sciences they have not been able to fulfill the claims laid out in the different parts of the [textbook on] logics and in the introduction to it, i.e. what they have set down in the *Second Analytics* (*Kitāb al-Burhān*) on the conditions for the truth of the premise of a syllogism, and what they have set down in the *First Analytics* (*Kitāb al-Qiyās*) on the conditions of the syllogism's figures, and the various things they posit in the *Isagoge* and the *Categories.*[13]

In his autobiography, al-Ghazālī repeats this charge without referring to the individual books of the *Organon*, the standard textbook on logics:

> The majority of their errors (*aghālīṭ*) are in metaphysics. [Here,] they are unable to fulfill demonstration (*burhān*) as they have set it out as a condition in logics. This is why most of the disagreements amongst them is in (the field of) metaphysics.[14]

If the metaphysics of the *falāsifa* cannot maintain the standards for demonstrative arguments made by them in their textbooks for logics, their teachings cannot stand up against the competing authority of revelation. This is an important element of what al-Ghazālī will later call his "rule of interpretation" (*qānūn al-taʾwīl*). We will be dealing with this rule in the next chapter.

Many of the twenty discussions in the *Incoherence*, however, discuss questions that do not contradict the literal wording of revelation. We learn from many of his later works that al-Ghazālī did not object to the position discussed in the fifteenth discussion, namely, that the heavens are moved by souls. Like the *falāsifa*, he thought that the heavens are indeed moved by souls, referred to as angels in the Qur'an. In these and in other cases, al-Ghazālī accepts the truth of the *falāsifa*'s teaching but rejects their claim to knowing it through demonstration. These things are known from revelation, he objects, and the *falāsifa*'s so-called demonstrations are merely attempts of proving this knowledge *post factum* with arguments that do not fully convince. Al-Ghazālī held that many philosophical teachings come from sources that are not acknowledged by the *falāsifa*, most important from the revelations sent to Abraham and Moses that were available to the nations before Jesus and Muḥammad. Through making use of arguments, these revelations teach syllogistic logics to humankind. The philosophers simply extracted (*istakhraja*) this method from there.[15] Humanity learned all the sciences, including the "method of reasoning" (*ṭarīq al-naẓar*), from prophets who were given this knowledge in revelations.[16] Once the rational sciences (*al-ʿulūm al-ʿaqliyya al-naẓariyya*) such as logics and mathematics were made available to humans, each individual had the ability to learn them from a good teacher (*fāḍil*), without resorting to a prophet or someone who claims to have been given divine insight.[17]

The initial argument of the *Incoherence* focuses on *apodeixis* and the demonstrative character of the philosophical teachings that it refutes. While the book does touch on the truth of many of these teachings, it clearly "refutes" numerous positions whose truths al-Ghazālī acknowledges or to which he subscribed in his later works. In these cases, al-Ghazālī wishes to show that although these particular philosophical teachings may be sound and true, they are not demonstrated by proofs. If anything, the details of God's arrangements in the heavenly spheres are made known to prophets by way of inspiration (*ilhām*) and have not been made known by way of rational arguments.[18] The ultimate source of the *falāsifa*'s knowledge about God's nature, the human soul, or the heavenly spheres is the revelations given to early prophets such as Abraham and Moses. Their information made it into the books of the ancient philosophers who falsely claimed that they gained these insights by reason alone.

Al-Ghazālī's *fatwā* against Three Teachings of the *falāsifa*

In his *Incoherence*, al-Ghazālī does more than simply make room for the epistemological claims of revelation. One of the first things students of Islamic history or of the history of philosophy learn is that al-Ghazālī condemned the tradition of Aristotelian philosophy in Islam. That condemnation is first expressed at the end of the *Incoherence of the Philosophers* (*Tahāfut al-falāsifa*), published in 487/1095, and later repeated in his *Decisive Criterion for Distinguishing Islam from Clandestine Apostasy* (*Fayṣal al-tafriqa bayna l-Islām wa-l-zandaqa*) and in his widely read autobiography *The Deliverer from Error* (*al-Munqidh min al-ḍalāl*), both works written around 500/1106.[19] Earlier intellectual historians of Islam claimed that this condemnation destroyed the philosophical tradition in Islam,[20] while today we know that this is not true.

Al-Ghazālī's legal verdict in the *Incoherence* extends to no more than a single page at the end of the book. It is, in effect, a *fatwā*, a legal response to a question posed by a real or fictitious inquirer. In its original version on the last page of the *Incoherence*, it reads:

> If someone asks: "Now that you have discussed in detail the teachings of these [philosophers], do you [also] say decisively that they hold unbelief (*kufr*) and that the killing of someone who upholds their convictions is obligatory?"
>
> We answer: Pronouncing them unbelievers must be done in three questions. One of them is the question of the world's pre-eternity and their saying that the substances are all pre-eternal. The second is their statement that God's knowledge does not encompass the temporally created particulars among individual [existents]. The third is their denial of the resurrection of bodies and assembly of bodies [on Judgment Day].
>
> These three teachings do not agree with Islam in any way. Whoever holds them [also] holds that prophets utter falsehoods and that they said whatever they have said in order to promote the public benefit, [meaning that the prophets] use symbols for the multitude of people in order to make them understand. Such [a position] is manifest unbelief (*kufr ṣirāḥ*) which none of the [various] groups of Muslims [ever] held.[21]

In his verdict against the *falāsifa*, al-Ghazālī singles out a limited number of theological or philosophical positions as unbelief. Here in the *Incoherence*, he lists three teachings: (1) that the word has no beginning in the past and is not created in time; (2) that God's knowledge includes only classes of beings (universals) and does not extend to individual beings and their circumstances (particulars); and (3) that the rewards and punishments in the next life are only spiritual in character and not also bodily. In his *Scandals of the Esoterics* (*Faḍāʾiḥ al-bāṭiniyya*), he adds (4) instances of blatant violations to the monotheism of Islam as well as the position (5) that although the teachings of the prophets provide some benefit (*maṣlaḥa*) to both the individual and to society, they are not actually true.[22] In al-Ghazālī's usual formulations of the verdict, however, this last position is listed as a mere result of the earlier three (or four) points in the list.

With the exception of the world's pre-eternity, all positions that al-Ghazālī condemns as unbelief are connected to the political authority of the religious law.[23] Al-Ghazālī feared that the teachings of the *falāsifa* and the Ismāʿīlites—effectively the only two Muslim groups that he brands as unbelievers—undermine the moral and legal authority of revelation. In his *Balanced Book on What-to-Believe* (*al-Iqtiṣād fī l-iʿtiqād*), he implies why one may not interpret figuratively passages in revelation that speak of a bodily reward in the hereafter. The *falāsifa* read these passages as symbols that stand for purely spiritual and immaterial bliss. Such interpretation is unbelief, he says, since it leads to a situation in which people no longer take their guidance from the Qur'an or from the teachings of the prophets.[24] For most people, spiritual bliss has no meaning; only the belief in the bodily character of the afterlife will enable the system of rewards and punishment to function sufficiently drastic as an effective incitement and deterrent in this world. In his *Revival of the Religious Sciences*, al-Ghazālī clarifies that, in addition to the two elements of the Muslim creed expressed in the Muslim profession of faith (*shahāda*)—monotheism and Muḥammad's prophecy—the most important belief for Muslims to hold is the belief in an afterlife with paradise and hell, with the rewards and punishments for this world's actions affecting both body and soul.[25]

Al-Ghazālī's *fatwā* is appended to a work where the specific legal status of the twenty teachings discussed in that book never comes up.[26] Indeed, there is a certain argumentative gap between the philosophical discussions of the twenty teachings of the *falāsifa* in the main part of the *Incoherence* and the brief *fatwā* at the end.[27] In the several introductory chapters of the *Incoherence*, al-Ghazālī explains his motivation for writing the work. Providing the basis for a legal condemnation of some of the *falāsifa*'s teachings is neither mentioned there nor anywhere else in the discussions of the twenty teachings. The long chapter on the world's pre-eternity, for instance—a discussion that amounts to almost a third of the whole work—focuses on the question of whether the philosophers' claim of a pre-eternal world is tenable and correct. Although al-Ghazālī denies this, he never engages in a justification why this position cannot be tolerated.

Yet there is an additional aspect to his *fatwā* that is rarely ever mentioned. The above quoted passage continues with a declaration that all teachings other than the listed three must be tolerated and should not become the subject of a legal condemnation. Al-Ghazālī says that the *falāsifa*'s views of God's unity and His attributes are similar to the ones held by the Muʿtazilites. If someone thinks the Muʿtazilites are unbelievers, he might also believe that the *falāsifa* are unbelievers with regard to these teachings. Al-Ghazālī recommends, however, that one should not regard the Muʿtazilites as unbelievers, and consequently also not regard the other teachings of the *falāsifa* as unbelief. Regarding teachings other than the three listed, the *falāsifa* may be accused of undue innovation (*bidʿa*). Accusing them of introducing such innovations (*tabdīʿ*), however, is a mere moral or dogmatic judgment that bears no legal consequences:

> As far as we are concerned, we do not prefer to plunge into the [question] of pronouncing those who uphold innovations (*bidʿa*) as unbelievers and what is or is not adequate for them lest the discourse could stray from the objective of this book.[28]

What is here only cautiously expressed will become a much more forceful motif in al-Ghazālī's subsequent works: Muslims who hold teachings that are not singled out as unbelief (*kufr*) may be accused of being unorthodox, but that accusation cannot lead to legal sanctions. It may, of course, trigger certain social stigmas by those who consider themselves orthodox, stigmas such as refusing to pray behind such a person, refusing to attend his funeral, or even avoiding social contact. Accusing someone of innovation (*tabdīʿ*), of error (*takhtiʾa*), or of deviation (*taḍlīl*), however, should not lead to punishments imposed by the judicial and political authorities. Although the state authorities should use coercion to prevent the teaching of unbelief (*kufr*), they should not interfere in the teaching of innovation (*bidʿa*), error (*khaṭāʾ*), or deviation (*ḍalal*),

Unbelief and Apostasy

In terms of its legal contents, there is something strikingly new in al-Ghazālī's *fatwā* against three teachings of the *falāsifa*. Al-Ghazālī responds to the question of whether those who teach these three positions can be killed. The real or imagined questioner asks: "Do you say decisively that they hold unbelief (*kufr*) and that the killing of someone who upholds their convictions is obligatory?" Al-Ghazālī answers yes. Whoever publicly supports or teaches the three named positions indeed deserves to be killed. In his legal works, some of them written before the *Incoherence*, al-Ghazālī explains that those who claim to be Muslims while teaching and propagating opinions established as unbelief can be killed without further delay.[29] This judgment is based on the law of apostasy. If Muslims fall into unbelief after having been believers (*kufr baʿda imānihi*), this constitutes apostasy from Islam, which carries the death penalty, according to all four schools of Muslim jurisprudence.

In early Islam, the apostasy of Muslims was established when they openly renounced Islam to follow a different religion. Only a public renouncement of Islam would constitute apostasy, or a public act such as giving up the Muslim prayer and attending the Christian mass. Any legal condemnation of apostasy required an unambiguous declaration from the side of the accused apostate. Accusing someone of secretly renouncing Islam and clandestinely practicing a different religion always led to a public interrogation. In the early centuries of Islam, a public declaration of belief in Islam by repeating the Muslim profession of faith (*shahāda*) was always accepted and would end the legal proceedings.[30] Apostasy could only be punished if the accused openly renounced the Muslim faith and was unwilling to pay public lip service to Islam. Hence, apostasy and unbelief were two very different things in early Islam. Muslims might be accused of being unbelievers without bringing them anywhere close to the accusation of apostasy.

In contrast to these legal formulations, al-Ghazālī equates the unbelief of Muslims with their apostasy from Islam. This required a significant change in the legal meaning of the word "unbelief" (*kufr*). Elsewhere, I give a detailed account of how "unbelief" (*kufr*) was understood in early Islam and how its understanding changed roughly two generations before al-Ghazālī.[31] Al-Ghazālī uses "unbelief" (*kufr*) as a legal term, meaning that the legal and political institutions—the jurists, the rulers, and their military—must act whenever unbelief is detected within the Muslim community.[32] Such an understanding reveals a major development away from the earlier meaning of the term. According to the majority opinion of Muslim legal scholars before the mid-fifth/eleventh century, unbelief (*kufr*) was a matter that God will punish in the afterlife, while in this world it would warrant no more than social sanctions for those associated with it. Consequently, accusing one's theological opponent as an unbeliever was quite widespread. "Declaring someone an unbeliever" (*takfīr*) was often used to brand and slander one's theological opponent; it very rarely implied legal sanctions, and certainly not the death penalty. Abū Mūsā al-Murdār, for instance, a Muʿtazilite of the third/ninth century, was known to have accused all people of unbelief who did not share Muʿtazilite positions on the most controversial theological issues of his days. When a fellow theologian pointed out that this would apply to almost all people, al-Murdār shrugged his shoulders. His colleague wondered in astonishment why the Qur'an says that paradise is as wide as heaven and earth (Q 3:133), when according to al-Murdār's view only he and the three people who agreed with him will enter.[33] The remark illustrates that in the third/ninth century, "unbelief" simply meant that the persons accused of it will—in the opinion of their accusers—suffer in hell.

Only in the mid-fifth/eleventh century did the jurists in the Shāfiʿite legal tradition begin to connect the unbelief of Muslims with what they called "clandestine apostasy" (*zandaqa*).[34] Apostasy (*irtitād*) from Islam had always been a punishable offence in Islam: a prophetical *ḥadīth* says that whoever changes his religion shall be killed.[35] Although this judgment established the death penalty for apostasy from Islam, it was limited to those who made an explicit and clear statement that he or she was renouncing Islam. A philosopher who teaches the pre-eternity of the world did not usually regard himself as a renegade or apostate from Islam. Avicenna, for instance, considered himself not only a faithful Muslim but also among the religious elite in Islam.[36] We must assume the same was true for his followers—al-Ghazālī's target readership in his *Incoherence*. They certainly considered themselves full members of the Muslim community. Al-Ghazālī and his colleagues in Islamic law effectively created a new legal status, that of "clandestine apostasy" (*zandaqa*). The accused no longer needed to declare or acknowledge his apostasy from Islam: he could be found guilty of clandestine apostasy when he violated certain principles of Islam or refused to subscribe to core elements of the Muslim creed.

Along with this judgment of clandestine apostasy comes a systematic effort to disentangle the question of what constitutes unbelief and apostasy from the criteria for religious orthodoxy. Al-Ghazālī understands that orthodoxy is in the eye of the beholder; from the viewpoint of an Ashʿarite, other Muslim groups such as the Muʿtazilites or moderate Shiites are certainly not orthodox. Such heterodox groups, however, were not considered clandestine apostates from Islam, and they continued to enjoy legal status as Muslims. The Ashʿarites regarded them as tolerated groups within Islam. Distinguishing the criteria for apostasy from simple heterodoxy is one of al-Ghazālī's most important contributions to the legal discourse about unbelief and apostasy in Islam. He firmly establishes the legal status of tolerated heterodoxy, a category containing Muʿtazilites and most Shiites, for instance. According to this qualification, philosophers who avoid the three condemned teachings fall under this category of tolerated nonconformists or dissenters. Al-Ghazālī's distinction between taxing someone with unbelief (*takfīr*) and taxing someone with error (*takhtiʾa*), deviation (*taḍlīl*), or innovation (*tabdīʿ*) creates two different categories of deviators. The three latter judgments are mere pronouncements that the adversaries hold positions that are not correct and that will, in the opinion of al-Ghazālī, lead them toward punishment in the afterlife. Taxing someone with error, deviation, or innovation has no legal implication; in fact, it amounts to the declaration that the Muslim community tolerates such theological positions.[37]

The Decisive Criterion (*Fayṣal al-tafriqa*)

In the earlier centuries of Islam, someone who accused a Muslim of unbelief (*takfīr*) would assume that his adversary would burn in hell but should not burn at the stake. Now that the parameters of unbelief as a legal judgment had changed, an attitude of frequent accusations could lead to an atmosphere of legal persecution and to a wave of capital punishments. Al-Ghazālī was quite concerned with this dangerous situation. In response, he wrote *The Decisive Criterion for Distinguishing Islam from Clandestine Apostasy* (*Fayṣal al-tafriqa bayna l-Islām wa-l-zandaqa*). It is a book primarily about who should *not* be accused of unbelief and clandestine apostasy. As such, it establishes a legal and theological place for religious tolerance in Islam. The book also clarifies the background of al-Ghazālī's earlier judgments about apostasy from Islam.

Richard M. Frank remarked that al-Ghazālī wrote *Distinctive Criterion* in response to accusations that he himself was an unbeliever because he deviated from some early Ashʿarite teachings in his *Revival*.[38] This might well have been the case, since at the beginning of the book, al-Ghazālī speaks of "a group of envious people who discredit one of our books about the truths concerning religious practices"—a clear hint to his *Revival of the Religious Sciences*—"and who claim that there are things in it which are contrary to the teachings of the earlier colleagues of the school," meaning the Ash'arite school.[39] In this case, the goal to defend himself against accusations of unbelief coincides with the objective to limit the practice of accusing one's theological opponent and also with the desire to clarify the criteria for unbelief and apostasy from Islam. The last point was still a desideratum from the days when he wrote his *fatwā* at the end of *Incoherence*. *The Distinctive Criterion* is a systematic work on the boundaries of Islam, and it explains al-Ghazālī's reasoning for condemning the *falāsifa* and the Ismāʿīlite Shiites.

At the beginning of this short book of thirty pages, al-Ghazālī approaches his readers to use an "indicator" (*ʿalāma*), or a rule of thumb, any time they feel the urge to accuse someone of unbelief:

> Based on this indicator you should refrain from accusing any group of unbelief and from spreading rumors about the people of Islam—even if they differ in their ways—as long as they firmly confess that there is no god but God and that Muḥammad is His messenger, and as long as they hold this true and do not contradict it. [The indicator is:] Unbelief (*kufr*) is the accusation that something that comes from the Prophet—peace and prayers be upon him—is wrong. Belief (*īmān*) is to consider him true and truthful (*ṣidq*) in everything that comes from him.[40]

The full implications of this rule of thumb are too manifold to explore here. Compared to earlier Ash'arite views, it no longer assumes that a Muslim's faith and belief (*īmān*) consist in accepting the truthfulness of God (*taṣdīq Allāh*), rather al-Ghazālī teaches that Muslim faith means accepting the truthfulness of the Prophet Muḥammad (*taṣdīq al-rasūl*) in everything that is reliably reported of him.[41] This blurs the line between the Qur'an and the *ḥadīth* corpus—al-Ghazālī regards both as revelation—and it shifts the burden of proof from the realm of the divine to the truthfulness of a person. This change results from al-Ghazālī's adaptation of Avicenna's prophetology. Avicenna offers a comprehensive theory of how revelation comes about in the mind of prophets. Accepting this explanation allowed al-Ghazālī to propose ways of verifying a person's belief in Muḥammad's truthfulness (*ṣidq*).

Al-Ghazālī presents his readers with a demanding theory of how to verify that a certain position implies the accusation that the Prophet Muḥammad has uttered an untruth (*kidhb*). It requires the reader to accept a quite difficult theory of language signification: any given statement from the mouth of the Prophet—no matter whether it has become part of the Qur'an or the *ḥadīth* corpus—constitutes a sign that refers to a "being" (*wujūd*). The prophetic statement—one can also say the word that comes from the mouth of the Prophet—is a linguistic marker that stands in for a certain entity ("being"). In most cases, these entities are objects in the outside world: places, animals, people, actions, and so on. For example, in a usual statement of the Qur'an, such as God's imperative to Moses, "Go to Pharaoh since he does wrong!" (Q 20:24 and 20:43), all the "beings" to which the text refers, such as Pharaoh or wrongdoing, are well-known entities or actions in the outside world. Al-Ghazālī discusses this example in the first book of his *Revival* and points out that the outward meaning of this sentence must not be denied. "Pharaoh is an individual that can be perceived by the senses and [historical] reports reliably confirm his existence."[42]

This appeal to outward meaning differs in the case of the following prophetical *ḥadīth*: "Paradise appeared to me on the width of this wall."[43] A demonstrative argument (*burhān*) establishes, al-Ghazālī argues, that paradise does not fit on the surface of whatever wall. This argument justifies that in this context, the original meaning of the word "paradise" must be abandoned. Here the word does not refer to a real being, meaning the true paradise, a material entity that exists somewhere in the outside world, but only to the Prophet's sense perception. To be more precise, the entity that this *ḥadīth* describes exists only within the Prophet's faculty of sense perception (*ḥiss*). Therefore, the word "paradise" does not correspond to a "real being" (*wujūd dhātī*) but merely to a "sensible being" (*wujūd ḥissī*). In this case, the reader must apply "interpretation" (*ta'wīl*), meaning he or she must understand the sentence, not in the original sense of its wording, but in an interpreted sense, and acknowledge that Muḥammad had the sensory impression of paradise on the surface of a wall while the real, material paradise was not at all involved.

In the Arabic tradition, "to interpret" (*awwala*) etymologically means "to bring something to its origin." Those who practice interpretation (*ta'wīl*) apply the meaning originally intended by the author of the text.[44] Al-Ghazālī assumes that the author of revelation sometimes chose to express himself in metaphors and symbols (singl. *majāz*). The interpreter traces these metaphors back to their *ḥaqīqa*, that to which they truly refer. In the case just discussed, the *ḥaqīqa* that corresponds to the text is a "sensible being," a mere impression or perception in the Prophet's faculty of sight. In other cases, it may be an "imaginative being" (*wujūd khayālī*), a *ḥaqīqa* that exists only in the Prophet's faculty of imagination (*khayāl*). In others, it may be "conceptual" (*'aqlī*), meaning the phrase in question refers to a universal concept in the mind of the Prophet. Finally, the *ḥaqīqa* may be just a "similar being" (*wujūd shibhī*), an idea with an analogous relationship to the concept originally intended in the text.

The two latter levels of existence need to be further explained. According to al-Ghazālī, when the Qur'an says that God has a hand, there exists no entity in the outside world called "God's hand" to which it could refer. There is also no possible sense perception of such a hand, nor can we imagine such a hand in our faculty of imagination, al-Ghazālī says. In this case, the entity that the word "God's hand" refers to can only be a concept in the mind of the Prophet. When we use the word "hand" in ordinary language, we mean "that with which one seizes something and makes something, with which one gives and takes." This is, according to al-Ghazālī, the *ḥaqīqa* of the word "hand," meaning that what it essentially refers to.[45] God does not have a hand, but He has a capacity to give and take. Since there is a correspondence of essential qualities between God's capacity to give and take and that of a human "hand," the latter word substitutes for the concept of "giving and taking." Thus, when the Qur'an mentions God's hand, it intends to refer to God's capacity to give and to take.

Roughly the same applies in the case of the "similar being" (*wujūd shibhī*), only that the correspondence between an attribute on God's side and a word in human language is in the field of nonessential qualities. When the Qur'an says that "God is angry" (*ghaḍiba Allāh*, see Q 4:93, 48.6, or 58.14), for instance, it cannot mean something that is in its essence similar to human anger. An essential part of human anger is the desire to seek satisfaction: God's transcendence and His exaltation over any deficiencies clearly preclude His having such a desire. What human anger has in common with God's anger is that both seek to punish. That, al-Ghazālī says, is a correspondence in the field of nonessential qualities, which is why the word "anger," when applied to God, corresponds to an entity similar to the known meaning of "anger" as it applies to humans.[46]

All propositions in the Qur'an and in the *ḥadīth* corpus refer to one of these five levels of beings: real, sensible, imaginative, conceptual, or similar. This theory has numerous implications that scholars have analyzed and explained.[47] Several key elements of this theory are based on Avicenna's theory of prophecy and the "inner senses" (*ḥawāss bāṭina*), meaning the human inner faculties of sense perception and of thinking.[48] For our purpose, it is important to realize that only the first of these five levels of being represents the literal meaning of a word; the other four represent a level of figurative interpretation (*ta'wīl*) that minimizes or denies the validity of the literal meaning.

As long as a Muslim acknowledges that the words in revelation refer to one of these five levels of existence, al-Ghazālī teaches, he or she cannot be considered an unbeliever or a clandestine apostate: "You should know that everybody who reduces a statement of the lawgiver to one of these degrees is one of those who believe."[49] A Muslim must acknowledge that all the words in revelation refer to *something*—be it either a being in the outward world or a being in the Prophet's sense perception, imagination, or intellect. Unbelief and apostasy is the denial that a word in revelation refers to anything of this kind. Such a denial assumes that the statements of the Prophet are not sincere. As al-Ghazālī writes, unbelief occurs:

> (. . .) when all these meanings are denied and when it is said that the statements (of the lawgiver) have no meaning and are only pure falsehood (*kidhb*), that the only goal behind (such a false statement) is to present things as they are not (*talbīs*), or to improve the conditions in the present world (*maṣlaḥat al-dunyā*). This is pure unbelief and clandestine apostasy.[50]

Unbelief is the failure to acknowledge that there are beings that correspond to the reports of revelation. "Corresponds" in this regard means a correspondence of words not only to objects of the outside world but also to the Prophet's sensible perceptions, to his imaginations, and to his metaphors either as metaphors based on similarities of essential or of accidental attributes. Unbelief is the case only when all these possibilities are denied, when it is said that some reports of the books of revelation do not correspond to *anything*. In all other cases, however, jurists should not concern themselves with the alleged apostasy of people who interpret revelation figuratively.

Al-Ghazālī probably realized that his colleagues in Islamic law would hardly follow him through this quite complicated text and apply a criterion that does seem rather hard to verify. Later in his book, he presents a much more easily applicable criterion of distinction between Islam and apostasy. It is based on the identification of three core elements of the Muslim faith that are differentiated from less important elements, which al-Ghazālī calls branches (*furūʿ*). Only teachings that violate certain "fundamental doctrines" (*uṣūl al-ʿaqāʾid*) should be deemed unbelief and apostasy. These doctrines are limited to three: (1) monotheism, (2) Muḥammad's prophecy, and (3) the descriptions of life after death in the Muslim revelation.[51] If a Muslim explicitly as well as implicitly acknowledges these three dogmas, he or she should not be accused of unbelief and apostasy. Only the open or the implicit denial of one of these three dogmas can lead to the accusation of clandestine apostasy. Once a violation of one of the three elements becomes evident, however, the state authorities have a right and a responsibility to persecute the violators.

4

The Reconciliation of Reason and Revelation through the "Rule of Interpretation" (*Qānūn al-taʾwīl*)

After establishing his criteria for unbelief and clandestine apostasy from Islam, al-Ghazālī addresses another distinction, that of a correct and a false understanding of revelation. As we discussed earlier, al-Ghazālī uses two sets of criteria for classifying heterodox beliefs, one that determines the boundary between Islam and apostasy and one that distinguishes orthodoxy from heterodoxy. The "rule of interpretation" is concerned with the latter distinction between correct and false beliefs. This rule of al-Ghazālī classifies how particular understandings of revelation represent the meanings intended by their author and how others do not—in other words, which readings are considered orthodox and which are heterodox.

Al-Ghazālī approaches the distinction between what he sees as a correct belief (Greek *ortho-doxus*) and what as an incorrect one from the perspective of Qur'an interpretation. Which verses, he asks, should be interpreted in a way that deviates from the literal meaning, and which verses must be understood in their literal sense? In order to establish a correct balance between the authority of the literal text of revelation and other competing sources of knowledge—most important the human capacity of reason—al-Ghazālī presents a "rule of interpretation" (*qānūn al-taʾwīl*). After introducing the five levels of being (*marātib al-wujūd*), he continues in his *Decisive Criterion*:

> Hear now the rule of interpretation: You learned that with regard to interpretation (*taʾwīl*) the different groups [of Islam] agree upon these five levels of being (. . .). They also agree that allowing [a reading that deviates from the literal

> meaning] depends on the production of a demonstration (*burhān*) that the literal meaning (*al-ẓāhir*) is impossible.[1]

This passage boldly assumes that all Muslim scholars agree on the five levels of existence introduced earlier in this book. Invoking this kind of agreement among all Muslim scholars is more than just a rhetorical device. Al-Ghazālī is convinced that disputes about the meaning of revelation go back to disagreements about what must be considered certain knowledge. Even the most literalist groups among the Muslims must sometimes understand a passage in revelation in deviation from its strictly literal wording, al-Ghazālī says.[2] The criterion for applying a figurative reading depends on the "production of a demonstration" (*qiyām al-burhān*) that proves the impossibility of the outward meaning (*istiḥālat al-ẓāhir*). If an argument can be produced saying that the words in the passage in question cannot be valid in their usual meaning, and if this argument reaches the high standard of a demonstration, then these words must be understood as symbols or metaphors. In this case, the demonstration invalidates the reading of the passage on the level of "real being" (*wujūd dhātī*), allowing one to consider the reading on the next level of being, the "sensible being" (*wujūd ḥissī*):

> The literal meaning (*ẓāhir*), which is the first, is the real being (*al-wujūd al-dhātī*). If it is affirmed it includes all [the other beings]. If it is invalidated, the sensible being applies (*al-wujūd al-ḥissī*). If it is affirmed it includes what comes after it. If it is invalidated the imaginative being (*al-wujūd al-khayālī*) applies, or the conceptual (*ʿaqlī*). If it is invalidated, the similar being (*al-wujūd al-shibhī*) applies, which is metaphorical.[3]

The principle is clear: The scholar must first try to understand a word or a passage in revelation according to its literal meaning. If, as a result of a demonstration, that is impossible, he must read it on the level of the sensible being and assume the word refers to a sensible perception of the Prophet. Again, if a demonstration proves that this is impossible, he applies the imaginative being and tries to understand the word as a reference to something in the Prophet's imagination. Eventually he will reach a point at which no demonstration establishes the invalidity of one of the five levels. This is the level on which the word or passage must be understood.

Dismissing a higher level of being and advancing to a lower one is only justified if a demonstrative argument invalidates (lit. "excuses," *ʿadhara*) the higher level: "There is no foregoing one level for a level that does not include the earlier one without the necessity of a demonstration."[4] The many disagreements about how passages in the Qur'an should be read, al-Ghazālī maintains, are merely disagreements about what can be proven demonstratively. A Ḥanbalite, for instance, will not accept a demonstration proving that God cannot be "above" (*fawq*). Thus he accepts that the word "above" (e.g. Q 12:76 or 6:18) refers to a "real being," meaning a spatial relationship, and does not allow interpreting this word in a way that deviates from its literal meaning.

Al-Ghazālī, however, assumes that such a demonstrative argument can be produced. He therefore concludes that "above," when used as a description of God, cannot have a spatial meaning. Rather, it is meant metaphorically to indicate a superior rank.[5]

Michael E. Marmura stresses in many of his publications that for al-Ghazālī, the literal sense of revelation can only become subject to "interpretation" (*taʾwīl*) if a demonstration (*burhān*) shows that it is impossible.[6] "Interpretation" is for al-Ghazālī the abandoning of the outward or literal sense, using a reading of the word or the passage as a symbol or metaphor. To what the metaphor refers is again determined by a demonstrative argument. It should be stressed that for al-Ghazālī, the text of revelation can have more than one meaning. The "rule of interpretation" establishes the most authoritative reading of the text, the one referring to the highest possible of the five levels of beings. It determines what kind of descriptive information the passage conveys. Once this reading is established, it allows all lower levels. These levels establish additional meanings of the text.

This point should be briefly explained. In his *Niche of Lights* (*Mishkāt al-anwār*), al-Ghazālī discusses the meaning of the Qur'anic passage about Moses and the burning bush. Sura 20 reports how Moses saw the burning bush and approached it, and when he came to it, a voice spoke to him and asked him to take off his two sandals. The voice identified itself as that of God and engaged in a dialogue with Moses (Q 20:9–36). The Ismāʿīlites and some Sufis claimed that God did not truly speak to Moses and that the imperative to Moses to remove his sandals (Q 20:12) carries purely metaphorical meaning. Al-Ghazālī, however, insists that there is no demonstrative argument that invalidates the narrative of the Qur'an. It is not impossible that God spoke to Moses from the direction of a burning bush. In his *Balanced Book on What-to-Believe*, al-Ghazālī clarifies that God's speech is different from that of humans and does not consist of words (singl. *ḥarf*) or sound (*ṣawt*) but is more akin to an inner human speech.[7] In any case, no argument can invalidate the imperative to remove one's sandals. The passage informs us that Moses wore sandals and that he was asked to remove them, which we assume he did. This reading is on the level of the real being (*wujūd dhātī*) and refers to real historical events. In addition, however, God's imperative had a symbolic meaning. The two sandals also refer to two parts of the world, the "world of sovereignty" (*ʿālam al-malakūt*) and the "world of sense perception" (*ʿālam al-shahāda*). God asked Moses to leave these two realms, which may also mean this world and the hereafter, behind and turn fully toward God. Thus, the passage also has an inner meaning. Moses did two things: he took off his sandals, and he threw off the two worlds. He acted outwardly and inwardly.[8] Acknowledging an inner meaning of this passage by no means invalidates its outward historical narrative.

The strategy of reading an additional inner meaning in a verse whose literal meaning has already been acknowledged is covered by al-Ghazālī's "rule of interpretation." In his *Decisive Criterion*, he says that "the literal meaning (*al-ẓāhir*) (. . .) is the first (. . .), and if it is affirmed it includes (*taḍammana*) all [the beings]."[9] Before one engages in the exploration of the inner meaning of these verses, one must make sure to acknowledge that their literal sense is true and that Moses did actually take his sandals off. In his *Revival of the Religious Sciences,* al-Ghazālī says that searching for an inner meaning in passages of the revelation is in some ways similar to dream interpretation.[10] In the interpretation of dreams, however, we do not acknowledge that there is truth in the literal narrative (*ẓāhir*) of the dream. The Arabic word for dream interpretation—*taʿbīr*—means to go from one thing to another, al-Ghazālī points out. This term illustrates that in a dream interpretation, we abandon one thing for the next. Where a bridge appears in a dream, for instance, we may regard it as a symbol for some kind of transition. The visual image of the bridge is abandoned once its symbolic character is understood. This technique of dream interpretation, however, is not the right way to approach the text of revelation. Al-Ghazālī claims that this is what the Ismāʿīlites do when they unduly dismiss the outward wording and jump toward an inner meaning (*bāṭin*). Indeed, one must acknowledge the outward sense and seek the meaning that the text also "calls attention to." The text contains words that "call attention to the inner meanings" (*al-tanbīh li-l-bawāṭin*). The difference between the technique of dream interpretation (*taʿbīr*) and that of "detecting the suggestions" (*tanbīh*) of the text is that the latter acknowledges the outward wording while the former ignores it.[11]

In the books of revelation, the prophets give expression to visions or to pictures that appear similar to visions that other people receive during their sleep. Unlike ordinary dreams, however, the prophets' visions are truthful (*ṣādiq*) and relate events that have either happened in the past or will happen in the future.[12] Like ordinary dreams, the prophets' visions also carry an inner meaning. This inner meaning is expressed in symbols (*mithāl*)—which are not allegories—that require deciphering by the attentive reader of revelation.[13]

In his *Revival,* al-Ghazālī says that when the prophets convey their message to humanity (*khalq*), they must make use of symbols and parables (*ḍarb al-amthāl*). God commissioned the prophets to speak with those to whom they have been sent according to their understanding (*ʿalā qadr ʿuqūlihim*). Ignorant people will only understand the outward meaning of these symbols and parables; only the educated (*al-ʿālimūn*) understand the inner meaning (*al-maʿnā*).[14] In another passage from the introduction of the *Revival,* al-Ghazālī clarifies that prophets can only convey information about the *general method* of how humans can achieve theoretical knowledge (*ʿilm al-mukāshafa*) about God, His attributes, and the fundamental constitution of the universe. The prophets can guide (*arshada*) their followers toward this knowledge, but they cannot convey it openly. The limited understanding of the people would spoil their efforts and lead to dangerous confusions. When the prophets talk about this kind of theoretical knowledge, they must use allusions (*rumūz*), the method of symbolization (*sabīl al-tamthīl*), and summarization (*ijmāl*). The scholars are the heirs of the prophets in this regard. They must take the prophets as their models and refrain from conveying the highest theoretical knowledge openly to the people.[15]

Whereas the prophets cannot talk openly about theoretical knowledge, they do bring a clear message with regard to human actions (*muʿāmalāt*). The prophets have a distinctly political function. Politics for al-Ghazālī aims at reforming people and guiding them toward achieving salvation in the afterlife. Prophets hold the highest political authority. They stand above the other three groups of holders of political authority, namely, (1) the caliphs, kings, and sultans, (2) the religious scholars, and (3) the popular preachers. The superiority of the prophets rests on the fact that they alone have the means to reach the inner convictions of the educated as well as those of the ordinary people. Caliphs, kings, and sultans use compulsion to achieve outward compliance with the law, and the scholars and preachers only reach either the educated people or the masses.[16]

Three Different Types of Passages in Revelation

When seen in the light of the larger conflict between reason and the literal meaning of revelation, al-Ghazālī's rule of interpretation effectively divides the verses and passages of the Qur'an and the prophetical *ḥadīth* into three different groups.

(1) In the first group fall those passages that are contradicted by a demonstrative argument. Here reason abrogates the literal wording of revelation. It should be stressed, however, that from al-Ghazālī's perspective, reason in no way abrogates or even overrules revelation. It simply determines how this text is meant to be understood by its author. Reason and revelation come from the same source and must teach the same deeper truths. In these cases, rationality establishes the inner meaning (*bāṭin*) of the text, with the outer meaning (*ẓāhir*) becoming invalid. This group of passages in revelation consists largely of those verses in which God is described in anthropomorphic language. Valid demonstrations establish that God is not like a human, which requires us to interpret these verses in a way that deviates from the literal meaning.

The large majority of passages in revelation are those that are not contradicted by a demonstrative argument. These fall into two subgroups, which together with the first group of passages add up to three groups overall. Among those that are not contradicted by reason are a group of passages (2) in which the results of demonstrative proofs either agree with or do not affect the text of revelation. These are the great majority of verses, in which the literal wording of the text remains untouched. Most historical narratives fall into this group.

The last group (3) are those passages in which no possible demonstrative proof exists to contradict the information conveyed in revelation. This is the most vital category in al-Ghazālī's theory of interpreting revelation. Al-Ghazālī is convinced, for instance, that rationality cannot possibly convey certain knowledge about what will happen in the afterlife. "No rational argument averts that what has been promised with regard to paradise and hell in the afterlife will happen."[17] In the twentieth discussion of the *Incoherence*, al-Ghazālī attempts to show that no demonstration proves that the bodily character of the afterlife is impossible.[18] Since humans have no certain way of knowing about the afterlife other than through revelation, they must accept the information that revelation conveys in this field: "The indications in the Qur'an and the prophetical *ḥadīth* that explain [what will happen] on the Day of Judgment and those that explain the connection of God's knowledge with detailed events reach a limit that is not receptive to interpretation."[19]

The reader of revelation must first determine into which of the three groups a verse or a passage falls. If a passage of revelation gives information that cannot be challenged by a demonstration, because no rational argument can establish knowledge about this aspect of revelation, the reader must then accept the outward meaning of the text. This analysis requires a thorough exploration of the fields of knowledge in which demonstration cannot be accomplished. If, however, the information given in a certain verse or passage falls into a field of knowledge where demonstration is possible, then the results of demonstration determine whether the literal meaning is the intended one or whether it must be read as a symbol or metaphor.

According to al-Ghazālī, our understanding of revelation depends on a thorough determination of what can be established demonstratively and what cannot. From the discussion in the *Incoherence,* it is clear that when al-Ghazālī uses the word "demonstration" (*burhān*), he has the high standard of Aristotelian apodeixis in mind.[20] Demonstration produces "necessity" (*ḍarūra*), and only this can justify foregoing the authority of revelation's literal wording.[21] In order to learn what "demonstration" truly entails, al-Ghazālī refers his readers to his *Touchstone of Reasoning in Logic* (*Miḥakk al-naẓar*), a work in which he introduces Aristotelian logics to a readership within the religious sciences.[22] Al-Ghazālī also recommends his short book, *The Correct Balance* (*al-Qisṭās al-mustaqīm*), initially composed as a *streitschrift* against the Ismāʿīlites, who in their propaganda rejected logics and the necessity of rational arguments.[23] There al-Ghazālī teaches five different types of "balances"—a cipher for syllogisms. Once these five different syllogisms are fully understood, he believes that they can solve a myriad of disagreements about what constitutes "certain knowledge" (*yaqīn*). Following this path—meaning applying Aristotelian logic—would likely settle most disagreements about what can be demonstrably proven and what cannot be proven.[24]

Al-Ghazālī follows Aristotle and the *falāsifa* in their opinion that reason (*ʿaql*) is executed most purely and precisely by formulating demonstrative arguments, which reach a level at which their conclusions are beyond doubt. He remains true to the rationalist approach, which was shared by both Ashʿarites as well as *falāsifa,* that our understanding of revelation is determined by what can and cannot be established through reason. Unlike the *falāsifa,* however, al-Ghazālī assumes that there is a surplus of information on the side of revelation that rationality cannot match. His rule of interpretation responds to this situation and makes room for the epistemological authority of Qur'an and *sunna.*

A Dispute about al-Ghazālī's Approach: Ibn Ghaylān versus Fakhr al-Dīn al-Rāzī

Around the year 580/1185, more than seventy years after al-Ghazālī's death, the Muslim theologian Fakhr al-Dīn al-Rāzī (d. 606/1210) passed through the town of Ṭūs-Ṭābarān and visited the small madrasa where al-Ghazālī had taught. Al-Rāzī, who was in his early forties at this time, had already published books on Islamic theology and a commentary on Avicenna's *Pointers and Reminders* (*al-Ishārāt wa-l-tanbīhāt*). He must have had a significant reputation, as he reports that the local scholars of Ṭūs put him in al-Ghazālī's chamber (*ṣawmaʿa*) and disputed with him. Al-Rāzī provoked the scholars of the place, who may have considered themselves the custodians of al-Ghazālī's heritage, by offering one hundred dinars—a very significant sum—to anyone who could successfully defend any of al-Ghazālī's teachings from the logical part of his *Choice Essentials* (*al-Mustaṣfā*).[25] In his own account of this wager, al-Rāzī departs as the uncontested victor, his adversaries readily admitting their inability to defend their teacher and conceding the weakness of al-Ghazālī's teachings.[26]

Later, during his travels in Khorasan and Transoxania, al-Rāzī came to Samarkand and visited its most famous scholar, Farīd al-Dīn ibn Ghaylān al-Balkhī (d. *c.* 590/1195).[27] Like many scholars whom al-Rāzī met on his travels, he was a Ghazalian. Ibn Ghaylān, who was now in his seventies, had studied at the Niẓāmiyya madrasa in Merw and in Nishapur and later became associated with Sharaf al-Dīn al-Masʿūdī, a scholar with whom al-Rāzī also disputed in 582/1186.[28] Al-Masʿūdī had written what is probably the earliest commentary to Avicenna's *Pointers and Reminders*.[29] His student Ibn Ghaylān held al-Masʿūdī's scholarship in high regard. He mentions him in the same breath as al-Ghazālī, describing both as prime examples of *kalām* scholars who had mastered logics and metaphysics and who were able to distinguish the correct teachings of the *falāsifa* from the incorrect ones.[30] Ibn Ghaylān ventures to do the same when soon after the violent uprising of the Turk nomads in 548–49 / 1153–54, he wrote a book of refutation against Avicenna's teaching on the pre-eternity of the world.[31] In this work, *The Creation of the World in Time* (*Ḥudūth al-ʿālam*), he refutes a short epistle by Avicenna in which the philosopher collects the arguments in favor of the world's pre-eternity. In the first part of his book, Ibn Ghaylān engages in a detailed refutation of these arguments, thus aiming to establish the world's creation in time.[32]

Fakhr al-Dīn al-Rāzī took issue with this approach. In his autobiographic report of the disputes he had with scholars in Transoxania, he writes that he was eager to talk with Ibn Ghaylān. When he reached Samarkand, he rushed to Ibn Ghaylān's house, but his host treated him with indifference. When he finally took time for his guest, al-Rāzī asked him in a curt manner about his book on the creation of the world in time. "Avicenna wrote an epistle," Ibn Ghaylān answered, "as a response to well-known arguments refuting [the position that] temporary created things can have no beginning. I responded to that epistle and showed that his arguments are weak."[33] At this point al-Rāzī apparently lost his temper and confronted Ibn Ghaylān with the objection that nothing is gained from refuting the position of a single scholar. Different philosophers held different opinions about the pre-eternity of the world, and whereas Aristotle's teachings on this subject agree with Avicenna's teachings, other philosophers produced different arguments. Refuting Avicenna's arguments leaves the arguments of Abū Bakr al-Rāzī (d. *c.* 323/935), for instance, intact. "If one follows this method," Fakhr al-Dīn al-Rāzī continued, "one will not arrive at rational and scholarly research (*baḥth*); rather this is merely a sort of disputation (*mujādala*) with a certain person on a certain subject."[34]

Fakhr al-Dīn al-Rāzī's criticism is unjustified, at least in its vigor. Ibn Ghaylān's method is not simply a rhetorical disputation that lacks scholarly rigor. Rather, he follows al-Ghazālī's method to establish the authority of revelation on the question of whether the world was created in time or exists from past eternity. Ibn Ghaylān's goal is to show that the arguments that support the position that he opposes are baseless. Behind this strategy stands the conviction that the point he wishes to prove is already established in revelation. Like al-Ghazālī, Ibn Ghaylān believes that creation in time is an established tenet of Islam. As long as there is no convincing rational argument that proves the world's pre-eternity, the temporal origination of all bodies in this world remains established. Since Avicenna enjoyed the reputation of having produced the most convincing rational arguments in favor of the world's pre-eternity, disproving him establishes—according to Ghazalian principles—the world's creation in time. "With regard to this question," Ibn Ghaylān responded to Fakhr al-Dīn, "I only respond to Avicenna. Once I had refuted his teachings on eternal motion, this was sufficient for me to uphold [the position that] bodies are created in time."[35]

There is, however, a problem with al-Ghazālī's and Ibn Ghaylān's approach to this particular question that Fakhr al-Dīn seems to have recognized. Although there is ample evidence in revelation for the positions that God has detailed knowledge of His creations and that reward and punishment in the afterlife take the forms of bodily pains and pleasures, there is no Qur'anic passage that clearly expresses that the world was created from nothing at one point in time. Averroes closely examined al-Ghazālī's argument as to why the *falāsifa* should be condemned for the three teachings mentioned. He agreed that God's detailed knowledge of his creation and the bodily character of the afterlife are elements of the Muslim creeds and that even a philosopher must acknowledge these two points.[36] No such acknowledgment needs to be made in the case of the eternity of the world: revelation is silent on this issue. In his *Decisive Treatise* (*Faṣl al-maqāl*), Averroes writes:

> If the outward meaning of revelation (*ẓāhir al-sharʿ*) is scrutinized it will become evident that the verses that provide information about the bringing into existence of the world [say] that its [current] form is really created in time but that existence itself and time extend continuously in both directions, [past and future], I mean without interruption.[37]

Averroes discusses a few verses from the Qur'an that actually point toward this interpretation. Verse 11:7, on the one hand, mentions the creation of the heavens and earth in six days, while also assuming that God's throne (*al-ʿarsh*) existed before. Verse 14:48, on the other hand, implies that after the end of this world, another world will come into existence. "It is not [said] in revelation,"

Averroes concludes, "that God existed together with pure nothingness (ʿ*adam*). There is simply no [conclusive] text (*naṣṣ*) that says so."[38] Fakhr al-Dīn al-Rāzī, who most probably had no knowledge of Averroes's *Decisive Treatise*, comes to a similar conclusion. In his theological summa, *The Elevated Issues in the Divine Science* (*al-Maṭālib al-ʿāliya min al-ʿilm al-ilāhī*), in a chapter that stretches over five pages, he reviews the textual evidence in revelation for either creation in time or the pre-eternity of the world. He finds no clear statement in favor of either position and concludes that revelation leaves this issue open.[39]

Al-Ghazālī was evidently aware of this problem. In his *Decisive Criterion for Distinguishing Islam from Clandestine Apostasy*, in which he argues that the legal judgment of unbelief (*kufr*) requires that the accused has violated a core tenet of Islam, he nowhere mentions the philosophical position of the world's pre-eternity. *The Decisive Criterion* lists three doctrines of Islam whose violation must be considered unbelief and apostasy: monotheism, prophecy, and revelations' descriptions of life after death.[40] The world's creation in time is not brought up in this book. Here the *falāsifa* are only condemned because they negate God's knowledge of the individuals and the bodily character of the afterlife.[41] In neither the *Incoherence* nor other works, in which the position of the world's pre-eternity is condemned as unbelief and apostasy, does al-Ghazālī succeed in providing a convincing legal justification for his ruling.[42] Averroes, who generally accepted al-Ghazālī's rules for determining unbelief and apostasy, assumed quite correctly that al-Ghazālī had condemned the position of the world's pre-eternity because it violates the consensus of the Muslims (*ijmāʿ al-muslimīn*). Averroes, however, protested that no such consensus exists. As long as the Muslim *falāsifa* uphold the world's pre-eternity with demonstrative arguments, a consensus may exist only among the *mutakallimūn*. Such a limited accord, however, cannot justify a judgment on the unbelief and apostasy of the Muslim *falāsifa*.[43]

For Ibn Ghaylān, who approaches the issue of the world's eternity from a distinctly Ghazalian perspective, the world's pre-eternity can be seen as a false position simply because it cannot be demonstratively proven. The opposing position—that the world was created in time—takes its truth from a higher authority than reason. Ibn Ghaylān says that Avicenna's teachings oppose Islam (*mukhālafa li-l-Islām*), particularly those on the pre-eternal character of the world.[44] Ibn Ghaylān was well aware that he was walking in the shoes of al-Ghazālī's *Incoherence*, and he gives him ample credit.[45] Unlike his adversary Fakhr al-Dīn al-Rāzī, however, Ibn Ghaylān does not discuss the Qur'anic evidence before he engages in a refutation of Avicenna's arguments. Also unlike al-Ghazālī in his *Incoherence*, Ibn Ghaylān presents and discusses the *kalām* arguments in favor of the world's creation in time.[46] Yet like al-Ghazālī, he silently assumes that the world's temporal creation is established on an authority that transcends reason and that refuting all claims of demonstrating otherwise establishes this doctrine on religious grounds. Showing that there are no demonstrative arguments that prove the world's pre-eternity is, of course, easier than establishing creation in time with one's own demonstrative arguments. Given that there is a certain balance of arguments, of which none truly reaches the threshold of apodeixis, al-Ghazālī and many religious scholars after him assumed that, in this question, a religious authority—meaning revelation or the consensus of Muslims—tips the scale.

Demonstrative Knowledge (*burhān*) and Its Opposite—Emulation of Authorities (*taqlīd*)

The dispute between Fakhr al-Dīn al-Rāzī and Ibn Ghaylān reveals both the strengths and the weaknesses of al-Ghazālī's approach toward conflicts between reason and the revealed text. First of all, al-Ghazālī was a distinctly rationalist theologian who generally accepted the Aristotelian notion of apodeixis (*burhān*) and the *falāsifa*'s claim that they can resolve certain scientific problems beyond doubt. On numerous instances, al-Ghazālī argued against other notions of rationality that are, in his view, more vague and less verifiable than demonstrations. In his *Straight Balance* (*al-Qisṭās al-mustaqīm*), for instance, he polemicizes against the use of "opinion" (*ra'i*) and "legal analogy" (*qiyās*) as criteria for truth, and he censures other Muslim scholars for deviating from the outward text of revelation on such feeble grounds.[47] "Let there be a rule for what counts as a demonstrative proof (*burhān*) among [the scholars of Islam] that they all agree upon and acknoweldge," al-Ghazālī demands in his *Decisive Criterion*.[48] This rule (*qānūn*) is available in the form of the standard for necessary (*ḍarūrī*), certain (*yaqīn*), and decisive (*qāṭiʿ*) knowledge, he says. While the five outer senses such as eyesight, smell, and so forth commit many sorts of errors when they perceive the world, the human faculty of rationality (*ʿaql*) is—if pursued in a correct way as demonstration—immune (*munazzih*) to error.[49] Once all Muslim scholars accept demonstration, the doctrinal disputes will likely near an end.[50] In fact, the errors of unbelievers, innovators, and deviators (*gumrāhān*) are a direct result of their faults in the method of rational arguments (*ṭariq-i ḥujjat*). Knowledge is the cure of all error.[51]

In his autobiography, al-Ghazālī asks his readers to take a sober look at the teachings of the *falāsifa*. Those that are unbelief must be condemned, and those that are heretical innovations (singl. *bidʿa*) should be rejected. However, other teachings of the *falāsifa* may be correct, al-Ghazālī adds; and despite their philosophical background, they should be accepted by the Muslim community. Each teaching must be judged by itself, and if found sound and in accordance with revelation, it should be adopted.[52] This attitude leads to a widespread application of Aristotelian teachings in al-Ghazālī's works on Muslim theology and ethics. When in his autobiography he defends himself against the accusation of having reproduced a philosophical position in his own works, he explains that no group has a monopoly on truth. It is false to assume that these positions can only be found in the books of the *falāsifa*:

> If these teachings are by themselves based on reason (*maʿqūl*), [if they are] corroborated by demonstration (*burhān*), and are not contrary to the Qur'an and the *sunna*, why should they be shunned and abandoned?[53]

If all truth must be abandoned, if it comes from a person who previously had voiced some false ideas, one would have to forgo much of what is commonly considered true. The greatest mistake people make, al-Ghazālī continues, is that they assess the truth of a statement by the standard of who says it. Truth is never known by means of an authority; rather, authorities are known by the fact that they speak truth.[54]

Although demonstration is for al-Ghazālī a God-given standard of rationality—it is the "touchstone of reasoning" taught in his book with that title—he sees a human tendency to deviate from this measure and to accept as true those teachings that are familiar from youth. This tendency to fall into an uncritical acceptance (*taqlīd*) of what is familiar is the enemy of the inborn faculty (*fiṭra*) of accepting demonstrative arguments. For al-Ghazālī, uncritical acceptance (*taqlīd*) is the root of all falsehood. The above described tendency of judging a teaching by its teacher is just one of the many varieties in which *taqlīd* manifests itself.

It must be stressed that al-Ghazālī held two teachings with regard to *taqlīd*. In the case of the ordinary people (*ʿawāmm*), who are not scholars and therefore unfamiliar with Muslim theology, reliance on *taqlīd* is recommended and indeed necessary. "The firmly-grounded belief (*al-īmān al-rāsikh*) is the belief of the ordinary people that attains in their hearts during [their] youth through the repeated appearance of what is heard."[55] For al-Ghazālī, the belief (*īmān*) of the masses is a naive religious assent to something one hasn't understood. It is not firm enough to count as knowledge (*ʿilm*).[56] The scholars of Islam must base their opinions and judgments on knowledge.[57] In his *Revival*, he defines *taqlīd* as "relying upon something one has heard from someone else (. . .) or upon books and texts."[58] Other than in the case of prophets, scholars should never rely on other people's opinions. Such reliance is "unsatisfactory" (*ghayr murḍin*) and cannot be justified. Unjustified *taqlīd* and demonstration are for al-Ghazālī opposites; and while the partisans of truth are those who apply demonstration, all those who oppose al-Ghazālī and his teachings are guilty of some kind of *taqlīd*. His conservative adversaries among the Sunni groups cannot disentangle the truth of a statement from the reputation of whoever says it. The Ismāʿīlites' greatest fault is that they slavishly follow the teachings of their Imam, who is infallible in their opinion.[59]

In the case of the *falāsifa*, uncritical acceptance has taken a curious form. Because of the development of the demonstrative method by philosophers such as Socrates, Plato, and Aristotle, the Muslim *falāsifa* uncritically repeat the view that philosophy is superior to revelation and that they are superior to the Muslim theologians.[60] The Muslim *falāsifa* have developed a hubris that leads them to uncritically accept the arguments for the pre-eternity of the world, for instance, or to favor the view that God cannot know particulars. The philosophers claim to "be distinct from their companions and peers [in the other sciences] by virtue of a special clever talent and intelligence."[61] This sense of superiority stems from the claims they make in their logic. The demonstrative method claims indubitability and the sense of possessing an infallible scientific method.[62] For the *falāsifa,* this sense shapes a conviction that they have knowledge and intelligence superior to their peers in the religious sciences. Because of their belief in demonstration, some have lost all respect for revelation and no longer perform the ritual duties prescribed therein. In his *Incoherence,* al-Ghazālī sets out to prove that many of the *falāsifa*'s arguments cannot be considered demonstrations. For generations, the *falāsifa* deluded themselves by uncritically repeating that they could answer these particular questions demonstratively. Al-Ghazālī accepts *taqlīd* only in the case of the prophets: they are the only humans whose teachings should be uncritically accepted. Following any other person uncritically inevitably leads into error.[63]

Fakhr al-Dīn al-Rāzī's harsh accusations against Ibn Ghaylān illustrate, however, that the Ghazalian method can fail to produce clear-cut positions to those questions to which neither demonstration nor revelation can offer a conclusive answer. Fakhr al-Dīn realized that revelation does not settle the dispute over the world's pre-eternity. Attacking the arguments of the *falāsifa* has little effect in this situation. If one accepts that there are no demonstrative arguments in favor of the world's pre-eternity—as some Aristotelians in the generations after al-Ghazālī were indeed willing to do—the situation requires careful consideration and weighing arguments that may not be demonstrative and that carry different convincing forces.[64] Al-Ghazālī's epistemology was unprepared for this situation.

Cosmology is precisely one of those questions in which al-Ghazālī believed that neither revelation nor demonstration provides a conclusive answer as to how God acts upon His creation. We will see that the position that al-Ghazālī developed for cosmology is sincere and true to his principles. Once he realized that neither of the two principal sources in his own epistemology—reason and revelation—could settle the matter, al-Ghazālī simply lost interest in cosmology as a scientific question. Additionally, al-Ghazālī deliberately aimed to avoid ambiguities in his writings. Because he had no clear position to posit, he never explained his stance on the conflict between occasionalism and secondary causality. The failure to clarify his position on cosmology, however, did lead to profound confusions among many of his interpreters.

5

Cosmology in Early Islam

Developments That Led to al-Ghazālī's Incoherence of the Philosophers

According to the German philosopher Christian Wolff (1679–1754), who first used the word, the term "cosmology" refers to the most general knowledge of the world and the universe, of the composite and modifiable nature of its being. Cosmology, however, existed long before the eighteenth century in the form of theories about the general structure and composition of the world. Often it has been connected to cosmogony, which refers to the explanation of how this world came about. For instance, the first chapter of the Bible, the book of Genesis, offers a report about how God created the heavens and the earth, light and darkness, water and land, and all the plants and creatures of this world. The Qur'an refers at several points to the creation of the heavens and the earth in six days (e.g. in Q 7:54); yet in the Muslim revelation, there is no single passage that is as central to its cosmogony as the Genesis report is to the Bible. The Qur'an doesn't introduce its readers to how God created the world; rather, it assumes that the readers or listeners already have some basic knowledge about this process and clarifies certain details.

Short accounts of creation are sprinkled all over the Qur'an. They mention that the seven heavens were created from smoke, forming layers, one above the other (Q 41:11–12, 67:3). These heavens are spheres (singl. *falak*), in each of which swims a celestial body such as the sun or the moon (Q 21:33, 36:40). In the seventh heaven, in which the angels praise God and seek forgiveness for the believers, sits the divine throne (*ʿarsh*), carried by angels who move in rows (Q 40:7, 89:22). This throne "extends over" (*wasiʿa*) the heavens and the earth (Q 2:255), with God as the Lord of this throne (Q: 9:129). The lowest heaven is adorned with lights (Q 41:12), which are the sun and the moon (Q 71:16, 78:13), the stars, and the constellations of the zodiac (Q 37:6, 15:16). The earth was created within two days (Q 41:9) from an integrated disk-shaped mass (Q 21:30). Paralleling the seven heavens, there are seven layers of "earths" (Q 65:12). The whole edifice of heavens and earth is surrounded by two waters, separated by a barrier (*barzakh,* Q 55:19–20). God created the first humans from dust or from various kinds of clay (Q 3:52, 23:12, 55:14, 15:26). While creating humans, God also created the demons (singl. *jinn*) from smokeless fire (Q 55:15).

Like the two different strains of narrative that have been collated to the Genesis report of the Old Testament, the creation narratives in the Qur'an are not always compatible with one another.[1] Yet they do convey a sense of purpose for each element of God's creation. God creates effortlessly but deliberately, and He chooses between alternatives (Q 4:133, 5:17, 14:19–20, 35:16–17). God has merely to say, "'Be!' And it is" (*kun fa-yakūnu,* Q 3:47, 3:59, 6:73, etc.); He has power over all things (*ʿalā kull shayʾ qadīr,* Q 64.1, 65.12, 66.8, 67.1).

Ashʿarite Occasionalism in the Generations before al-Ghazālī

It is the task of theologians to make sense of revelation and develop cohesive explanations to clarify these verses and make them consistent with what we know about the world from other sources, including our daily experience. Disputes about cosmology are prompted by concerns that have little to do with the creation reports in revelation. In Islamic theology, comprehensive cosmological theories developed in the context of an early theological debate on the nature of human actions. If God has power over all things, how can we explain that humans are also under the impression that they have power over their own actions? Do humans have the power (*qudra*) to carry out their own actions, or is God the force actualizing this power? And if God solely possesses this power, why does the human earn God's blame for bad actions and His reward for good ones?

In the first/seventh century, the theological conflict between a human's responsibility for his or her actions and God's omnipotence initiated discussions that subsequently led to the development of comprehensive theological systems. During the second/eighth century, a group of theologians who defended the view that humans—and not God—decide and execute their own actions developed systematic positions about the nature of God and the effects of His obligations on His human creations. This group, the Muʿtazilites, argued that humans have a free choice (*ikhtiyār*) whether to obey or disobey God's commands. On the other side of the argument, the opponents of the Muʿtazilites pointed to verses in the Qur'an in which God claims responsibility for people becoming unbelievers since He "seals their heart" and thus prevents them from obeying His command to believe (Q 2:7). People become unbelievers, not because they choose, but rather because God makes them become unbelievers. This notion was unacceptable to the Muʿtazilites, who held that God is supremely just and would never commit an act of injustice. Preventing someone from becoming a believer and later punishing the same person for his or her unbelief would be unjust.

At the beginning of the fourth/tenth century, al-Ashʿarī (d. 324/935–36), a renegade Muʿtazilite theologian, pointed to what he saw as a fundamental incompatibility in the Muʿtazilite system: God cannot both be just and also leave humans a free choice over their actions. Assuming that God knows whether people will act good or bad during their lives and that it is God who decides about their time of death, how do the Muʿtazilites explain why an infant, who dies without doing either good or bad deeds, lacks the chance to earn rewards in the afterlife, even though numerous wretched people are allowed to live long lives in which they thoughtlessly waste their chances to obey God—chances that the infant craved in vain. If we apply to God the same principles of justice that we apply to human actions, it is unjust that He would let the wrongdoers continue to do wrong when He knows they will end up in hell. It would be more just to let them die as infants, as He allows with many of His creatures.[2]

Al-Ashʿarī and his students developed a radical critique of Muʿtazilite theology. Among the central motifs of early Ashʿarite theology was the preservation of God's complete control over His creation. In their desire to safeguard the Creator's omnipotence, Ashʿarites developed a truly original cosmology that came to be known as occasionalism. One key element of Ashʿarite occasionalism is atomism. Earlier, Muʿtazilites had argued that all physical objects consist of smaller parts, which at one point can no longer be divided (*lā yatajazzaʾu*). All bodies consist of such parts—atoms—which are the indivisible substances (singl. *jawhar*) of the bodies. Atoms are the smallest units of matter and are by themselves bare of all color, structure, smell, or taste. Atoms gain these sensory attributes only after they are assembled into bodies. Their attributes are viewed as "accidents" (singl. *ʿaraḍ*) that inhere in the substances, that is, the atoms of bodies. Accidents exist only when they subsist in the atoms of a body. And while they cannot exist without bodies, bodies also need accidents in order to exist because the atoms are by themselves without any attributes. All accidents together constitute the content of the present reality of any given particular thing.[3]

The atomist theory developed in *kalām* literature is different from modern ideas about the atom, for instance, because it assumes that atoms are by themselves completely powerless and have no predetermined way of reacting to other atoms or to accidents. Every nonmaterial being—such as an odor, an impression, or an idea—is understood as an accident of a material being. The *mutakallimūn* taught that when a human believes in God's existence, the atoms of his heart carry the accident of "belief in God." When an architect has a plan for a building, the atoms of her brain carry the accident of that plan. Both the atoms and the accidents are by themselves devoid of all power and need to be combined in order to create bodies, be they animated or lifeless. Atoms are empty building blocks, so to speak, and they only constitute the shape of a body. All other characteristics are formed by the accidents that inhere in the body. This kind of atomism appealed to al-Ashʿarī because it does not assume that potentialities in things limit how these things will develop in the future. Such potentialities would limit God's action. Al-Ashʿarī insisted upon the nonexistence of any true potentiality outside of God.[4] In principle, any atom can adopt any kind of accident as long as God has created the association of this particular atom with that particular accident.

Ashʿarites adopted their understanding of physical processes from earlier theories developed in Muʿtazilite *kalām*. The Muʿtazilite movement was particularly rich in attempts to explain physical processes. Some Muʿtazilites speculated that movements are not continuous processes but consist of smaller leaps (singl. *ṭafra*) that our senses cannot detect and whose sum we perceive as a continuously flowing movement. This theory, in turn, led other Muʿtazilite thinkers to assume that time itself is not a continuous flow but is rather a fast procession of "moments" (singl. *waqt*), which again is concealed from our senses.[5]

Muʿtazilite thinkers had already discussed these ideas when al-Ashʿarī adopted them, combined them, and formulated what became known as an occasionalist cosmology. Its main components are the atomism of the earlier *mutakallimūn* plus the idea that time is a leaped sequence of moments. The latter idea is sometimes called an "atomism of time."[6] Muʿtazilites had already developed the idea that accidents cannot subsist from one moment to another. They need to be created every moment anew. And since bodies cannot exist without accidents, bodies exist from one moment to the next only because God creates their accidents anew in every moment. In order for an atom to exist from one moment to another, God has to create the accident of "subsistence" (*baqāʾ*) every moment He wants the atom to persist. This leads to a cosmology in which in each moment, God must assign the accidents to the atoms and to the bodies they form. When one moment ends, He creates new accidents in the next moments; and through these new accidents, He ensures that the atoms persist. None of the accidents created in the second moment has any causal relation to the accidents in the earlier moment. If a body has a certain attribute from one moment to another, then God created two identical accidents inhering in that body. Movement and development occur when God decides to deviate from the arrangement of the moment before. A ball is moved, for example, when in the second moment of two, the atoms of the ball are created at a specific distance from the locus of the first moment. The distance determines the speed of the movement. The ball thus jumps in leaps over the playing field, as do the players' limbs and their whole bodies. This also applies to the atoms of the air if there is some wind. In every moment, God rearranges all the atoms of this world and creates their accidents anew—thus creating a new world every moment.[7]

Occasionalism was conceived out of a strong desire to grant God control over each and every single element of His creation at every point in time. This desire is connected to the Ashʿarites' dispute with the Muʿtazilites over the character of human actions. Al-Ashʿarī taught that something that is created has neither influence on nor power over itself or any other being:[8] "Everything that is created in time is created spontaneously and new by God exalted, without a reason (*sabab*) that makes it necessary or a cause (*ʿilla*) that generates it."[9] Al-Ashʿarī denied that things could be caused by anything other than God. There is no causal efficacy among God's creation: a ball on a playing field appears to be moved by a player, but in fact it is moved by God. There is only one single cause for all events in the universe, which is God. He has the most immediate effect on all His creatures and no being other than He has any effect on others:

> The fact that the stone moves when it is pushed is not an act of him who pushes, but a direct act of God (*ikhtirāʾ min Allāh*). It would be perfectly possible that one of us pushed it whithout it being moved because God did not produce its movement, or that there is none who pushes it and it still moves because God directly produces its movement.[10]

Al-Ashʿarī's line of argument was directed against the Muʿtazilite way of saying that humans "create" (*khalaqa*) their actions and "generate" (*tawallada*) the subsequent effects. The Muʿtazilites taught that human voluntary actions are neither created by God nor known to Him before they happen; rather, they are the autonomous creation (*khalq*) of the human agents. According to the Muʿtazilites, God does not will the wrongful actions of men, and He does not create their consequences. These consequences are causally "generated" by human wrongdoing.[11] Al-Ashʿarī argued that the idea of human "generation" assumes that God controls neither human actions nor their effects, and thus it must be wrong.

At the heart of al-Ashʿarī's ontology lies the denial of any unrealized potentialities in the created world. Al-Ashʿarī rejected the idea that created beings are compelled to act according to their nature (*ṭabʿ*). We usually assume that if a date stone, for instance, is planted and fed, it can only develop into a date palm and not into an apple tree. Although this is true for all practical purposes, in theology, this assumption unduly limits God's freedom to act. After discussing where such natures would be located in his cosmology, al-Ashʿarī determined that they can be classified neither as atoms nor as accidents. Thus, he concluded, the word "nature" (*ṭabʿ*) is empty of any comprehensible meaning. Those who use it wish to indicate that there is some regularity in the production of accidents in certain bodies, nothing more.[12]

These regularities in God's actions are what lead some humans to assume the existence of "causal laws" or "laws of nature." Yet in reality, al-Ashʿarī argued, God doesn't create according to such laws, which would only limit His omnipotence and His free choice. God deliberately chooses to create satiety after having eaten food and hunger in the absence of it. If He wished to do it the other way round, He certainly could: "But God follows a habit (*ajrā al-ʿāda*) in the temporal order in which He brings these events about, and doing it the other way would be a violation of His habit."[13] For al-Ashʿarī, there is neither causality nor laws of nature. Observing God's habits brings some humans to the false conclusion that such laws exist. But an omnipotent God is not bound to laws of nature. It is easy for Him to break His habits; indeed, He does so when one of His prophets calls upon Him to bring about a miracle and confirm the prophet's mission. The prophetical miracle consists of "events that are produced in violation of the previous habit."[14]

Secondary Causes in Ashʿarite Theology

The term "occasionalism" defines the cosmology of what has become known as the early Ashʿarite school, as we refer to the Ashʿarites up to the generations of al-Juwaynī and al-Ghazālī. A brief look at the teachings of al-Bāqillānī (d. 403/1013), Abū Isḥāq al-Isfarāʾīnī (d. 418/1027), and ʿAbd al-Qāhir al-Baghdādī (d. 429/1037) reveals that all of them denied the existence of "natures" (*ṭabāʾiʿ*).[15] It has often been said that in their denial of natures, Ashʿarite occasionalists implied the denial of *any* causal relation between created beings.[16] Richard M. Frank, however, has argued that this is not the case. According to Frank, the Ashʿarite rejection of the existence of natures results from their denial of potentialities that could limit God's creative activity. At the core of Ashʿarite occasionalism stood the denial of potentialities in the created world.[17] The question of whether a created being may have efficacy on another created being was only secondary to that concern. Al-Ashʿarī taught, for instance, that when humans act, their actions are the causal effects of a power-to-act that God creates on behalf of the humans. This power, Frank argues, is a "power of causation" that is created by God.[18] For al-Ashʿarī, a human is the "agent" (*fāʿil*) of his or her own actions and thus the true cause of them. God still remains the creator of man's causation. At the moment of the realization of the human voluntary act, God creates a "temporarily created power-to-act" (*quwwa muḥdatha* or *qudra muḥdatha*), through (*bi-*) which the act is realized. Frank describes the relationship between the created power-to-act and the human act in terms of secondary causality. The created power is a secondary cause that is employed by God in order to achieve its effect.[19] God creates the human action through (*bi-*) a temporarily created power that is created on behalf of the human.[20]

In their theory of human actions, Ashʿarites were torn between their denial of efficacy (*taʾthīr*) on the side of created beings and their desire to express that humans truly perform the actions for which they bear responsibility on Judgment Day. This latter notion led to the acknowledgment of some kind of secondary causality in the performance of the human act. More detailed studies are needed to see whether there was a development between these two poles of thinking particular among the Nishapurian Ashʿarites. With Ibn Fūrak (d. 406/1015), al-Isfarāʾīnī, and al-Baghdādī, the intellectual center of the Ashʿarite school moved from Baghdad to Nishapur. In regards to the question that prompted this issue, namely whether humans "cause" their own actions, al-Bāqillānī, Ibn Fūrak, and al-Isfarāʾīnī followed the general theory of al-Ashʿarī that humans are the agents of their own actions. Daniel Gimaret describes this position as a concession to the Muʿtazilite position that otherwise humans would be punished for something over which they had no agency.[21]

Al-Ghazālī and al-Juwaynī, pinnacles of the Ashʿarite tradition in Nishapur, were both quite ambiguous regarding secondary causality. Al-Juwaynī emphasized different motifs of Ashʿarite thinking in different works. In his influential textbook of Ashʿarite theology, *The Book of Guidance* (*Kitāb al-Irshād*), al-Juwaynī stresses the notion that created beings have no causal efficacy. A comment by one of his students reveals that al-Juwaynī believed that this was al-Ashʿarī's original position.[22] When humans act voluntarily, al-Juwaynī teaches, they have a temporarily created power-to-act (*qudra ḥāditha*), which is one of the accidents (sing. *ʿaraḍ*) of their bodies. God creates this temporary power for the sole purpose of allowing a human the performance of a single act. The temporary power is an accident and thus cannot subsist from one moment to another; it exists only in the moment when the human acts. In his *Book of Guidance*, al-Juwaynī denied categorically (*aṣlan*) that the temporarily created power has any efficacy (*taʾthīr*) on the human action (*al-maqdūr*).[23] The temporarily created power does not cause the existence of the human act. Only God can cause the act. The temporarily created power applies to the act like a human's knowledge applies to what is known to him or her. The knowledge corresponds to what is known, but it does not cause it, nor is it caused by it. Similarly the human volition to perform a certain act corresponds to the act, but it does not cause it.[24] God creates the human act independently from the human volition yet still in correspondence to it.

In a short work on the Muslim creed that al-Juwaynī wrote late in his life and that he dedicated to his benefactor, Niẓām al-Mulk, he emphasizes the second notion that humans truly perform their action. Here, al-Juwaynī points to the well-known fact that God has given humans certain obligations (*taklīf*). God promises reward if they are fulfilled and threatens punishment if violated. The text of the Qur'an clearly assumes, al-Juwaynī argues, that God has given humans power to fulfill what He asks them to do, and that He sets them in a position (*makkana*) to be obedient. In light of all this, it makes no sense "to doubt that the actions of humans happen according to the humans' efficacy (*īthār*), their choice (*ikhtiyār*), and their capacity to act (*iqtidār*)." In fact, to deny the human power-to-act and its efficacy to perform actions would void the obligations of the Sharīʿa.[25]

Still, al-Juwaynī nowhere says that humans have efficacy on objects that exist outside of themselves, such as having the ability to move a stone, for instance. He focuses on the generation of human acts and acknowledges that there must be a causal connection between the human's decision and the human act. He does not seem to be arguing against Muʿtazilites here but rather against more radical occasionalists who claim that no event in the world can be caused by anything other than God. This cannot be true, al-Juwaynī objects, since the human's action must be caused by the human's choice. Otherwise, the whole idea of God imposing obligations upon humans would be meaningless:

> He who claims that the temporarily created power has no effect (*athar*) on the human action (*ilā maqdūrihā*) like [as if] knowledge had no effect on what the human knows, holds that God's demand towards humans to perform certain acts is as if God would demand from humans to produce by themselves colors and [other] perceptions.[26] That would be beyond the limits of equitability and an imposition of something vain and impossible. It implies the negation of the Sharīʿa and the rejection of the prophets' message.[27]

A more radical occasionalist would assume that the temporarily created power and the human act itself are two accidents, which are—like all accidents—created independently by God. Al-Juwaynī's student al-Anṣārī associates the school founder al-Ash'arī with such a view. Al-Ash'arī taught, al-Anṣārī reports, "that the temporarily created power has no effect on its corresponding action; nor has it any part on the production of the act or on one of its attributes."[28] For al-Ash'arī, the coherence between the human's decision and his or her act would result from God's habit to create a human act in accord with its corresponding temporarily created power. Such an accord, al-Juwaynī objects, cannot be the basis of God's later judgment about the human's choice. The action would not be prompted by a human choice. In fact, in al-Ash'arī's theory, it is not clear whether there is a human choice after all, since all al-Ash'arī discusses is the power to act (*qudra*) and its object (*maqdūr*), which is the human action. For al-Juwaynī, the human decision in favor of a certain action and its corresponding temporary power to perform it are the sufficient causes of the action. Only this position takes into account that God obliges humans to acts according to His commands and prohibitions.

Al-Juwaynī consciously departs from what he believes was al-Ash'arī's strict principle that no created being can have any influence upon another. Some created beings do have efficacy, he says, namely, the human decisions about our actions. Still, this does not mean that the human creates his acts independent from God.[29] Rather, when humans decide about an action, God gives them a temporarily created power, and like the human decision, that power is among the necessary causes for the performance of the action:

> The human's power is created by God (. . .) and the act, which is possible through (*bi-*) the temporarily created power, is definitely produced through (*bi-*) that power. Yet it is related to God in terms of it being determined and being created [by Him]. It is produced through God's action, i.e. through the power-to-act (*al-qudra*). The power-to-act is not an action performed by a human. It is simply one of God's attributes. (. . .) God has given the human a free choice (*ikhtiyār*). By means of this choice, the human disposes freely (*ṣarrafa*) over the power-to-act. Whenever he produces something by means of the power-to-act, that what is produced is attributable to God with regard to it being produced by God's action.[30]

When humans freely decide to perform an action, God cedes control over His power-to-act (*qudra*) to the human. God creates a temporary power for the human's usage. As the human decides whether to perform the action, it is God's power that performs it. Yet therein lies a causal determination: the human decision to perform the act leads to the act's performance. The human's free decision in favor of a certain act becomes a means of God's execution of His power over His creation. Only when the human's decision to act and the temporarily created power-to-act coincide will the action occur. These two together are the sufficient cause for the human action.

For al-Juwaynī, the human is not the creator of his or her actions; such an idea would violate the opinions of the forefathers (*salaf*).[31] Humans cannot be the creator of their actions, because they are ignorant of the true essence of the acts and of the full implications (*ḥawādith*) these acts have. For al-Juwaynī, the creator of an act must have a detailed knowledge about all aspects of it.[32] God, however, withholds such knowledge from humans.[33]

One might ask whether for al-Juwaynī, God's knowledge of His creation is in any way affected by the human's free choice? After all, if the human's decision is truly free, it cannot be predicted, and God would not know how the human uses the divine creative power. Such a limitation of God's knowledge and His omnipotence, however, is unacceptable to al-Juwaynī. All things that come into being are willed by God;[34] including those that are created by means of the human's temporarily created power. Everything is subject to God's determination (*taqdīr*):

> God wills that the human acts and He creates (*aḥdatha*) in him motives (*dawāʿī*), a will (*irāda*), and knowledge (*ʿilm*) that the actions will be produced to the extent the human knows of it. The actions are produced through (*bi-*) the power-to-act, whose creation for the human is in accord with what he knows and wants. Humans have a free choice (*ikhtiyār*) and are distinguished by a capacity to act (*iqtidār*). (. . .)[35]
>
> The human is a free actor (*fāʿil mukhtār*) who receives commands and prohibitions. [Yet at the same time] his actions are determined by God, willed by Him, created by Him, and determined by Him.[36]

The human is like a servant, al-Juwaynī says, who is not permitted free control over the money of his master. If the servant would act on his own accord and buy or sell, the master would not execute his transactions. Once the servant is given a power of attorney for certain transactions and once he decides to make such a transaction, his master will honor the arrangements and execute them. In all these cases, the true buyer or seller is not the servant but the master, and only he can empower the servant to perform a transaction. Without the master's will and his permission, there would be no transaction.[37] For al-Juwaynī, the human is a trustee of God's power, able to use it freely within the limits that God creates for him. Within these limits, however, the human causes his own actions. This comparison with the servant can also illustrate a major problem with al-Juwaynī's theory of human actions. Someone who issues a power of attorney cannot expect his agent to negotiate within certain limits *and* also determine all details of the transaction. The agent's freedom is hard to reconcile with a complete predetermination of his actions.

One and a half centuries later, Fakhr al-Dīn al-Rāzī saw in al-Juwaynī's teachings an early version of his own position about the determination of human actions through "motives."[38] According to al-Rāzī, al-Juwaynī taught that the human motive (*dāʿin*) together with the divine power (*qudra*) causes the human act. God is still the creator (*khāliq*) of the human act, in the sense that he "lays down" (*waḍaʿa*) the causes that necessitate the act. Fakhr al-Dīn al-Rāzī, however, realized that there can be no free choice for humans as long as God has a preknowledge of their actions. For him, there was only an illusion of freedom on the human's side: God uses causes to determine the motives, which then determine the human's actions: "The human is a compelled actor in the guise of a free agent."[39]

Often occasionalism is so closely connected to early Ashʿarism that it is almost regarded as a necessary constituent of that theology. That, however, is not the case. Daniel Gimaret and Richard M. Frank have shown that at no point in Ashʿarite history did they defend a radical occasionalist position that completely denies efficacy to created beings.[40] Most early Ashʿarites acknowledged that human decisions trigger their actions even if they are not the only sufficient cause. When al-Juwaynī says, for instance, that the human is a *fāʿil mukhtār*, meaning a free agent or a freely choosing efficient cause, he accepted efficient causation in the case of human actions.[41]

According to al-Shahrastānī, who wrote two generations after him, al-Juwaynī went much further and departed more radically from the cosmological axioms of early Ashʿarism. Following his report of al-Juwaynī's view that the existence of the human act relies on a power-to-act (*qudra*) on the side of the human, al-Shahrastānī continues that according to al-Juwaynī,

> (. . .) the [human] power-to-act relies for its existence on another cause (*sabab*). The relationship between the power-to-act to and that cause is like the relationship between the act and the power-to-act. Similarly, a cause relies on [another] cause until it ends with the one who arranges the causes (*musabbib al-asbāb*) and that is the Creator of the causes and of their effects (*musababāt*), who is the Self-sufficient (*al-mustaghnī*) in the true sense [of that word]. For every cause is self-sufficient in a certain way and it is dependent (*muḥtāj*) in another way.[42]

According to al-Shahrastānī, al-Juwaynī taught that causal efficacy is not limited to the connection between the human's choice and the performance of the act. Rather, the human decision is itself determined by certain causes—here he may have the motives in mind that Fakhr al-Dīn al-Rāzī also mentioned. These motives are, in turn, the effects of other causes. All these causes and effects are elements in long causal chains that have their starting point in God. Human acts are prompted by a consecutive succession (*tasalsul*) of secondary causes, which go back to their first cause in God. This, al-Shahrastānī adds, was clearly not a position previously known in the field of *kalām*; rather, it was newly introduced by al-Juwaynī. He took it from the teachings of the philosophical metaphysicians, al-Shahrastāni remarks, "who hold that causal dependency is not restricted to [the relation between] the human act and the power-to-act, but rather between everything that comes into being."[43]

None of this is expressed in those of al-Juwaynī's works that have come down to us. Yet even in these works, there are clear indications of a change of direction in Ashʿarite theology. In his *Creed for Niẓām al-Mulk*, al-Juwaynī mentions the existence of "motives" (*dawāʿī*) that determine human actions.[44] Already in his *Book of Guidance*, al-Juwaynī had acknowledged that God creates right-guidance (*hudā*) and error (*ḍalāl*) either directly in His creatures or by confronting them in the form of a "summons" or "call" (*daʿwa*) that He communicates to them in His revelation.[45] This latter teaching is again more developed in his *Creed for Niẓām al-Mulk*:

> If God wills good for a human, He makes his intelligence perfect, completes his insight, and removes from him obstacles, adverse incentives and hindrances. He brings him together with beneficial companions, and makes His path easy for him (. . .)[46]

In other words, if God wants a human to become a believer, He does not do so by creating the accident of "belief" in his heart, but rather He creates conditions that make it highly likely—or maybe even necessary—for the human to become a believer. Ash'arite theology is no longer expressing itself in a purely occasionalist cosmology, but rather in one where—at least in the case of human actions—God achieves his desired effect by means of secondary causes.

The *falāsifa*'s View of Creation by Means of Secondary Causality

"We live at a time," al-Juwaynī writes in his *Creed for Niẓām al-Mulk*, "where people draw from a sea of principles (*uṣūl*), and that sea cannot all be emptied with ladles."[47] These many principles derive from the often drastically different epistemologies of the major intellectual currents of al-Juwaynī's time. There were, of course, the Ash'arites and their traditional adversaries, the Mu'tazilites, whose prime concerns in theology were starkly different. Yet in his time, al-Juwaynī also saw the increasing success of a group with which earlier Ash'arites had been only marginally concerned: the Arab philosophers (*falāsifa*). The contacts and influences between *kalām* and *falsafa* during the fourth/tenth and fifth/eleventh centuries need to be studied more closely than it can be done in this book. The traditional account, which is significantly influenced by a report in Ibn Khaldūn's *Introduction* (*al-Muqaddima*), assumes that up to the end of the fifth/eleventh century, there were few links between scholars of these two disciplines. Al-Juwaynī was the first Ash'arite theologian who was affected by the works of the *falāsifa*. His student al-Ghazālī began a new theological approach (*ṭarīqat al-muta'akhkhirīn*) that took full account of philosophical logics, and in doing Muslim theology, says Ibn Khaldūn, it meddled with (*khālaṭa*) philosophical works.[48]

Ibn Khaldūn, however, is not entirely correct. Recently, Robert Wisnovsky argued that the beginning of the blending of *kalām* and *falsafa* should be predated to Avicenna's activity at the turn of the fifth/eleventh century. As a philosopher, Avicenna was well aware of developments in Mu'tazilite *kalām*. He responded in his works to concerns posed by their theology and tried to give thorough philosophical explanations to religious phenomena such as revelation and prophetical miracles. According to Wisnovsky, Avicenna's works mark the beginning of a synthesis between the Neoplatonist peripatetic tradition in Arabic and the tradition of Muslim *kalām*.[49] But even if one maintains Ibn Khaldūn's perspective and looks at developments only from the side of Sunni *kalām*, it was—as far as we know—al-Juwaynī and not al-Ghazālī who first gave detailed and correct reports of the philosophers' teachings and who addressed their theories.[50] Whether al-Juwaynī's late work *The Creed for Niẓām al-Mulk* is influenced more by Muʿtazilites such as Abū l-Ḥusayn al-Baṣrī (d. 436/1044)—another figure neglected in Ibn Khaldūn's report—or by al-Juwaynī's knowledge of Avicenna's philosophy is difficult to establish at this point.[51] It is quite evident, though, that within the context of Ashʿarite theology, there is something distinctly innovative in al-Juwaynī's short *Creed for Niẓām al-Mulk*. It ushers in the new theological approach discussed by Ibn Khaldūn.[52] The works of al-Juwaynī's students al-Kiyāʾ al-Harrāsī, al-Anṣārī, and most of all al-Ghazālī show a deep familiarity with the *falāsifa*'s teachings and the challenges they put forward.

Causality is at the very heart of every Aristotelian approach to physics and metaphysics. "For every corruptible thing," Avicenna says in his *Physics*, "and for everything occurring in motion, or everything composed of matter and form, there are existing causes."[53] Causality, he adds, is a principle (*mabdaʾ*) of the natural sciences that is proven in metaphysics. Causality in Avicenna's metaphysics is in some ways even more important than in the metaphysics of Aristotle, the starting point of many of Avicenna's ideas.[54] Robert Wisnovsky has shown that Avicenna's understanding of causality had been influenced and in many ways determined by the commentary tradition of Aristotle's works. These commentaries—written in both Greek and Arabic—were not all available to Avicenna. He did not read Greek and had no access to many of the early commentaries of the Alexandrian tradition. Yet, what Avicenna gleaned from those books available to him helped him develop a certain perspective on Aristotle's teachings that reflected developments in earlier commentaries. Greek Neoplatonist thinkers such as Ammonius Hermiae (fl. *c.* 500) of the school of Alexandria had the most profound influence on Avicenna's understanding of causality. His distinctly Neoplatonist interpretation of Aristotle's ideas on causality came to Avicenna not by way of Neoplatonic treatises that were translated from Greek to Arabic. By the time Avicenna crafted his philosophy, Neoplatonism had become part of the overall tradition of Aristotelianism. To Arabic philosophers such as Avicenna, Neoplatonism did not come through a funnel, as Wisnovsky put it, but through a sieve.[55]

Aristotle had taught that when we ask about the "why" of a certain thing or event, our different and sometimes ambiguous answers confirm to one of four aspects. In the writings of the Aristotelians, the word "cause" can be understood in one of two ways: either as something that effects or produces the item, or as an explanation of the need for or function of the thing. When we explain, for instance, why the chiseling tool known as an adze (*qādūm*) chisels wood, we provide answers that refer either (1) to the specific shape of the tool, or its form, or (2) to the material of which it is made, in this case, iron; or we explain the "why" (3) by referring to the goal that we would like to achieve by using the tool, namely, chiseling, or, last, (4) by referring to the agent, that is, the craftsman who has produced the adze.[56] Aristotle said that the word "cause" refers to a (1) a formal cause (*ṣūra*), (2) a material cause (*ʿunṣur*), (3) a final cause (*ghāya*), and (4) an efficient cause (*fāʿil*).[57]

The Neoplatonist commentary literature on Aristotle focused mainly on the two latter causes, the final and the efficient ones. Both are external causes, as, unlike matter and form, they are not constituents of the thing itself. In his *Metaphysics*, Aristotle had explained what he saw as a principle of being: things are disposed to realize the possibilities with which they have come to exist.[58] Like an apple seed, which strives to become an apple tree, all beings endeavor to realize their inherent potentials. Humans, for instance, make great efforts to acquire knowledge and to perfect their intellect. Neoplatonist philosophers came to understand this Aristotelian principle of *energeia* or *entelekheia* as meaning that everything strives toward its perfection (*teleiotes*). They combined this idea with the notion of final causality and created a cosmology in which things are ranked according to how close their perfect state reaches toward the final cause of all being, which is God. The heavenly intellects, for instance, exist in a state of perfect rationality. Subsequently, their being is ranked higher than that of humans who just strive to perfect their rational intellects. The celestial intellects are regarded as more perfect than humans. A more perfect being is also regarded as more perfect in terms of its existence. A more perfect being passes the existence it receives from what is above it in the cosmic hierarchy down to what is below it.

For Aristotelians, every effect is necessary in relation to its efficient cause. Existence is viewed as downwardly progressing; a higher efficient cause passes it to a lesser one. The higher efficient cause is thus responsible for the existence of a lower object[59] This does not mean, however, that an efficient cause must exist before its effect. Cause and effect coexist in time. The effect cannot be delayed once its sufficient cause exists. The cause necessitates the effect and precedes it only "with respect to its attaining existence," but not necessarily in time. Since God is the only sufficient cause of the world, the world must have existed for as long as God has existed.[60] God and the world exist for Avicenna from eternity.

God causes the world by emanation of the first creation, the intellect of the highest sphere. From the One, from God, Avicenna proclaims, only one creation proceeds. Creation proceeds in successive steps during which an efficient cause gives existence to an effect, which itself becomes the efficient cause for the next effect.[61] Again, there is no temporal priority on the side of the cause but only an ontological priority. Viewed as a whole, God can be seen as both the world's agent and its efficient cause (*fāʿil*). By "agent" or "efficient cause," Avicenna means "a cause that bestows existence which differs from itself."[62] The relationship of God to the world is one that Avicenna calls "essential causality." An essential cause (*ʿilla dhātiyya*) is a sufficient efficient cause, meaning that its existence alone necessitates the existence of its effect.[63] For Avicenna, the relation between an essential cause and its effect is necessary; meaning every moment the essential cause exists, its effect *must* also exist.

Avicenna presents in his works two different arguments that aim to prove the necessity of causal relations. The first is invoked more often than the second. Closely connected with Avicenna's argument for God's existence, it starts by arguing that in every existent thing, the existence can be distinguished from the essence of the thing. The fact that a particular thing—a horse, for instance—exists in actuality implies that the freestanding idea of "a horse" is a possible existence. Being possible, however, does not also mean that "a horse" must exist in actuality. Something that is by itself possible may or may not exist in any given moment. In order for the possible to be actualized, there must be something that gives it existence. With regard to a given object that we witness around us, this something cannot be the object itself; it must be something other than the object. Whenever a particular thing that is by itself possible exists, its existence must be caused by its efficient cause (*ʿilla* or *fāʿil*).[64]

Jon McGinnis has argued that in his response to the philosophical theory of efficient causality, al-Ghazālī is less concerned with this first argument but he is very concerned with a second one that appears in a brief passage in Avicenna's *Rescue* (*al-Najāt*). Avicenna refers to the example of fire burning a piece of cotton. According to Aristotle's theory of power or faculty (*dynamis*) in the ninth book of his *Metaphysics*, fire has the active power (*quwwa fāʿiliyya*) to burn, and cotton has the passive power (*quwwa munfaʿila*) to be burned.[65] Once the two come together, their powers, which are a part of their natures, are necessarily actualized. The fire becomes the "agent" (*fāʿil*) that burns the cotton or—in a different translation of the Arabic—the "efficient cause" of the cotton's combustion. It is impossible that the fire would not cause the combustion, because postulating the opposite would lead to one of two contradictions: either fire does not have the active power to burn, or cotton does not have the passive power to be burned. Either of these assumptions would contradict the accepted premise of the argument, which means the argument is necessary.[66] One can also say that accepting the existence of natures that have passive and active powers implies that causal relations are necessary.

Avicenna's views about how everything that exists receives its being (*wujūd*) from a higher efficient cause are in many ways identical to those of al-Fārābī. As a writer, however, al-Fārābī was much more explicit than Avicenna about how the chains of being work and about how the higher efficient causes in the heavens determine the existence of lower beings. Based on earlier philosophical and astronomical models of cosmology, al-Fārābī taught that there are ten spheres, with the lowest being the sublunar sphere of generation and corruption in which humans, animals, and plants live. The nine other spheres are in the heavens, wrapped around one another like layers of an onion. Al-Fārābī's cosmology relies on Ptolemy's (d. *c.* 165) geocentric model of the planetary system, although it disregards movements within the planetary spheres, the so-called epicycles. For al-Fārābī, each of the five planets known before the invention of the telescope as well as the sun and the moon move with their own celestial sphere. The sphere of the earth—the sublunar sphere—is a true globe at the center of this system enveloped by the nine celestial spheres. At the upper end of the visible universe, above the spheres of the sun, the moon, and the five planets, sits the ninth sphere of the fixed stars. In order to account for the extremely slow rotation of the earth's axis around the celestial pole—a rotation completed only every 25,700 years and causing the precession of the equinoxes—Ptolemy added a tenth sphere at the outermost end, right above the sphere of the fixed stars. The celestial spheres move in circles with different speeds, the higher spheres always faster than the ones below them as they drag the lower ones with their movement. The outermost sphere moves exactly at the speed of one rotation per day.[67] It contains neither a planet nor any fixed stars nor any other visible object. To the Arabs, it was known as the "supreme sphere" (*falak al-aflāk*), or the "sphere of Atlas." Since it is the highest-ranking moving object, the Latin interpreters of this planetary system referred to it as the *primum mobile*, or, the highest moving object.

Each of the ten spheres in al-Fārābī's model of the universe consists of a material body and a soul. The soul is dominated by an intellect that governs the sphere and causes its movement. The intellect that governs the *primum mobile* is the highest created being. Beyond it is only the being that causes all this, that is, the First Principle, of which al-Fārābī says, "one should believe this is God."[68] In thinking itself, al-Fārābī's God emanates a single being, the intellect that governs the *primum mobile*. God directly acts only upon one being, which is this particular intellect. God's oneness prevents Him from acting upon anything else. What is truly single in all its aspects is unchanging and can only have one effect, the highest created being. This is the first intellect that causes, in turn, the existence of its sphere, and it also causes the intellect of the sphere right below it, that is, that of the fixed stars. Every celestial intellect—with the exception of the lowest one, the active intellect—is the cause of two things: its own sphere and the intellect directly below it. In contrast to the "First Cause," which is God, al-Fārābī calls the celestial intellects "secondary causes" (*asbāb thawānī*).[69] God mediates His creative activity through these secondary causes to the lowest celestial intellect, the tenth one. This is the active intellect (*al-ʿaql al-faʿʿāl*), and it has more than just two effects. It causes the existence of all the beings in the sublunar sphere, all beings on earth.[70] Of these ten celestial intellects, al-Fārābī says, "one should believe they are the angels."[71]

Avicenna parted ways with al-Fārābī's cosmology on such minor issues as the number of spheres and intellects in the lower celestial orbs or whether the celestial souls are purely rational or also have imagination.[72] Yet, with regard to the principle of secondary causality—that is, the fact that God creates the world and controls it by passing existence along a line of secondary causes,—there was no disagreement between any of the Arabic philosophers in the peripatetic tradition. God creates through the mediation of efficient secondary causes. These causes cannot stand by themselves but depend on higher causes for their being, which eventually receive their existence from God. In terms of any specific causal connection, the higher efficient cause establishes the existence of its effect in a predetermined and necessary way. If all conditions are fulfilled for a certain cause to have its effect, the connection between the cause and effect must occur and cannot be suspended. If fire reaches a cotton ball, to use the most prominent example in Arabic literature on causality, the cotton ball will necessarily start burning. Nothing, not even God himself, can suspend this connection. The cause is both the necessary and the necessitating condition of the effect's existence, even if ultimately God is the one who creates this necessary connection through the mediation of many multiple steps of secondary causes.

In his *Letter on the Secret of Predestination* (*Risāla Fī sirr al-qadar*), Avicenna writes that

> (. . .) in the world as a whole and in its parts, both upper and earthly, there is nothing which forms an exception to the fact that God is the cause (*sabab*) of its existence and origination and that God has knowledge of it, governs it, and wills its coming into being; it is all subject to His government (*tadbīr*), determination (*taqdīr*), knowledge, and will.[73]

Avicenna adds that this is "a general and superficial statement" (*ʿalā l-jumla wa-l-ẓāhir*), and attentive readers of his works understand that here he lumps together "the upper as well as the earthly" parts of God's creation, which are to be treated differently with respect to God's government, determination, knowledge, and will. The upper, celestial part of creation consists of the celestial spheres, which are governed by intellects. They exist from past eternity, function in the most orderly way, and move in complete and permanent circles, the most perfect kind of movement. Each sphere is its own class of being, of which it is the only individual. The active intellect (*al-ʿaql al-faʿʿāl*) that governs the lowest sphere contains all classes of beings that exist within the lowest sphere below the moon. In the lowest sphere, however, things become less regulated and less perfect than in the upper world. Beings in the sublunar sphere come to be and pass away, meaning they are corruptible and not pre-eternal. Once the causal chains have traversed the celestial realm and enter the lowest sphere, they create multiple individuals of each class of being. These individuals have individual traits, which are the result of the contact between the immaterial forms of the active intellect with physical matter.

When the philosophers say that God is the principle or the "starting-point" (*mabdaʾ*) of the world, they mean that both matter as well as all the rules that govern this world are a result of His nature. This is not that different from a modern deist or rationalist view of God as the sum of all laws that govern physical and psychological processes, human behavior, language, rational thinking, and all the other domains that are determined by rules. This is, of course, a very impersonal view of God. For Avicenna, this view implied that only the rules that govern God's creation are contained in the divine knowledge. In an Aristotelian understanding of nature, the classes of beings—meaning the nine celestial spheres and all the sublunar species contained in the active intellect—are the substrates where these rules are conserved. How cotton reacts when it is touched by fire is part of the cotton's nature, that is, the rules that are enshrined in the universal species "cotton." God has foreseen that once the classes of beings, which are universal and purely intellectual entities, mix with matter, they form individuals; but according to Avicenna, God has no awareness of these individuals. He does not know the individuals; He only "knows" the immaterial and universal classes of beings because they are the ones that are determined directly by His nature. The individuals are also determined by His nature, since the interplay between the universal forms and the individuating matter takes place according to the rules enshrined in the universals. But what happens in the sublunar sphere of generation and corruption is too mediated a result of God's nature and is therefore not "known" to Him.[74]

Avicenna teaches that the divine knowledge cannot contain events in the sublunar sphere. There seemed to have been a tension in Avicenna's thought regarding the second question of whether God also determines all events in the sublunar world, or, alternatively, whether some events in the sublunar world are related to chance and the haphazard influence from matter. In some of his works at least, Avicenna stresses that there are no arbitrary effects and that the events in the sublunar sphere are fully determined by God's creative activity. There are no causeless events or substances in this world. The effects of the celestial causes reach into the sublunar sphere and determine everything that happens there.[75] But how, one might ask, can such a fully determined world be squared with our impression that some future events are contingent on what precedes them, particularly those events that are the effects of human actions? Do humans not have a free will whose effect cannot be determined fully by the existing causes?

Al-Fārābī was the first Arabic philosopher to address this problem in his *Commentary on Aristotle's De interpretatione*. In that book's ninth chapter—the *locus classicus* for the discussion of the predetermination of future contingencies—Aristotle analyzes the meaning of the sentence: "There will be a sea battle tomorrow." This is not a statement that can be true and at the same time false. It must be either true or false, even if we cannot say which it is.[76] In al-Fārābī's discussion of this passage, he stresses that humans inherently understand that such an event is the effect of human free will: "We know right from the beginning, from our primordial nature that many things have a possibility of occurring and of not occurring, above all, those we know to be left to our choice and will."[77] A few pages later, al-Fārābī brings a well-known argument from Muʿtazilite theology that aims to prove the existence of human free will: if all future events were predetermined, human free will and deliberation would be void, and thus whatever punishment were to befall humans for their actions would be unjust. This denial of free will not only is absurd, al-Fārābī argues, but also it damages severely the social and political purpose of revealed religion.[78] It seems that here al-Fārābī adopts the Muʿtazilite position, denying a fully determined future and the possibility of divine foreknowledge of future events. Now, however, he raises another theological concern that also results from his position about the social and political function of revealed religion. The moral order in a state is upheld by the people's belief that God knows their actions and that He will reward them for right ones and punish them for wrong. Saying, however, that the future existence of a certain event is unknown to God denies divine omniscience. The indefiniteness (*ʿadam al-taḥṣīl*) of a future possibility, al-Fārābī says, exists only in our human knowledge because of our minds' deficiencies. Attributing similar deficiencies to God would be detrimental to the public benefit of religion.[79] Once humans no longer assume that God is omniscient, al-Fārābī implies, they loose respect for the moral injunctions and the legal impositions that are derived from revelation and no longer fear God's punishment for violating these rules.

The dilemma al-Fārābī finds himself in is the same as that of al-Juwaynī in his *Creed for Niẓām al-Mulk*. How can we say that humans decide their actions freely while God has a foreknowledge of all future events? Al-Fārābī's solution will become very important for al-Ghazālī, and we must examine it closely. For al-Fārābī, some future contingencies are the result of human free will, but they are also foreknown by God. Al-Fārābī tries to reconcile this apparent contradiction by distinguishing between two types of necessities, namely, "necessity in itself" (*ḍarūra fī nafsihi*) and "necessity from something else" (*ḍarūrat al-shayʾ ʿan al-shayʾ*). Future contingencies are not necessary by themselves, yet if they become existent, they are necessary from something else, meaning they are necessary by virtue of their causes. If God knows that Zayd will set out on a journey tomorrow, to use one of al-Fārābī's examples, then Zayd will necessarily travel tomorrow. The event is necessary due to something else, in this case, God's creative activity that manifests itself in God's foreknowledge. If the event is looked at solely by itself, however, Zayd's decision to travel is not necessary but merely possible, as it is still within Zayd's power (*qudra*) not to travel. Divine foreknowledge does not remove human free will or the ability to act differently from what is foreknown. Although God knows that Zayd will travel before he does so, His knowledge does not exclude the possibility of Zayd staying at home. It just excludes that this possibility will be realized. By distinguishing between these two types of necessity, al-Fārābī tries to maintain that (1) humans have the capacity (*qudra*) to perform or not to perform their acts and to choose between these options while (2) God also has a detailed foreknowledge of the future. God judges over human acts not according to His foreknowledge, al-Fārābī says, but in terms of the choices that humans make. God's foreknowledge, therefore, does not deprive humans from their freedom of choice and is not contrary to justice.[80]

Al-Fārābī's distinction between these two types of necessity initiated an important development in Arabic philosophy as well as in Muslim theology.[81] Avicenna was one of the first to adopt the distinction that all created events are "possible by virtue of themselves" (*mumkin bi-nafsihi*) and "necessary by virtue of something else" (*wājib bi-ghayrihi*), meaning necessary by virtue of their causes. This distinction is a cornerstone of Avicennan metaphysics on which the whole edifice of how God relates to His creation is built.[82] Avicenna, however, did not follow al-Fārābī in taking up the cudgel on behalf of human free will. Like al-Fārābī, he opted for a fully determined universe in which all events, including human actions, are fully predetermined by God.[83] Unlike al-Fārābī, however, Avicenna did not assume that God knows such events as Zayd's journey. The impact that the universal celestial causes have on matter in the sublunar sphere of generation and corruption are not all part of the divine knowledge. For Avicenna, God is an intellect and has no body. He thus lacks the epistemological faculty to grasp individual objects. In humans, these faculties, such as sense perception or the faculty of imagination, are closely connected to the body. Being pure intellect, God's knowledge contains only universals. Thus, the universal concept of a human is part of God's knowledge, as is the fact of Zayd having all the essential attributes of a human, such as a soul and rationality. God knows these things because they are the effect of His knowledge. The accidental attributes of Zayd, however, cannot be part of God's knowledge on account of the fact that He is pure intellect.[84] Whether Zayd travels tomorrow is therefore not part of the divine knowledge. God also lacks the knowledge of whether Zayd ever committed a sin.

Avicenna was not particularly forthcoming about this element of his teaching, and there is a certain degree of obfuscation in his writings about God's ignorance of the accidents. Avicenna rarely speaks of the "collisions" (*muṣādamāt*) in the sublunar sphere, and he tries to give the impression that a detailed knowledge of events in this sphere is, in fact, possible.[85] Humans, for instance, would be able to know the future if they knew all the temporal events on earth and in heaven, including the natures of the things that are involved.[86] Once one knows *all* the causes in one moment, one would be able to deduce the effects of the next moment and predict the future. The souls of the heavenly bodies have such perfect knowledge, and they can reveal it, for instance, to the prophets.[87] Humans and celestial spheres are composed of intellects as well as bodies and therefore have in their souls the faculties to know accidents. The divine knowledge, in contrast, is pure intellect and contains only universal principles. God's knowledge is a single one (*wāḥid*); it is changeless and outside of time. It does not consist of individual cognitions (*ʿulūm*) that refer to multiple objects. Individual events are part of God's knowledge only insofar as they result directly from principles, such as the celestial rotations, for instance, or the eclipse of one celestial body by another.[88] Avicenna admits indirectly that God cannot know the accidents in the sublunar sphere: he says that both the celestial souls as well as "that which is above them" (*mā fawqahā*) have knowledge of the particulars (*al-juzʾiyyāt*). However, that which is above the celestial souls—meaning God—he adds, "knows the particulars only in a universal way."[89]

The *falāsifa*'s View That This World Is Necessary

According to al-Fārābī and Avicenna, everything in this world is, first of all, determined by its proximate efficient cause, which is a created being within this world. This proximate efficient cause—or these causes, as in the case of the birth of a human at which more than one proximate efficient cause is required—is itself determined by other efficient causes and so on, until the causal chains are eventually traced back to their divine origin. The secondary causes have active and passive powers only because they receive these powers from God, who is the absolute efficient cause of everything other than Him. All created things depend necessarily on God for their existence, for their active and passive powers, and for the specific way how they are created.

In the teachings of Avicenna, there lies a second aspect of God's necessity, one much more problematic from a theological point of view. Avicenna taught that the creation of the world has its starting point in God's knowledge, which may be viewed as the blueprint of His creation. God's knowledge is, according to Avicenna, an aspect of the divine essence, and as such it does not change. God's essence is total unity, and it is not possible for there to be division or change within something that is totally unified in its nature. This view challenges the position that God's creative activity involves free choice. Although Avicenna maintained that God has *ikhtiyār*, a term usually understood as referring to a free choice between alternatives, he never explained what he meant by it, and a critical reader may surmise that he simply wished to say that God's actions are not determined by anything outside of His essence, such as in the case of human actions that are caused by motives, for instance.[90] From reading Avicenna—and particularly from reading the reactions to Avicenna—it becomes clear that his God cannot choose between creating a blue heaven, for instance, and the alternative of creating a yellow one. The blue heaven is necessary since that is what is part of God's knowledge. God's knowledge is unchangeable, but it is also perfect.

These elements come together in the philosopher's teaching on divine providence (*ʿināya ilāhiyya*). In his book *Pointers and Reminders* (*al-Ishārāt wa-l-tanbīhāt*), Avicenna explains that divine providence is the combination of three aspects that are included in God's knowledge. The first aspect is that God's knowledge accounts for everything there is. The second is that God's knowledge arranges everything in a necessary way so that it follows the best order (*aḥsan al-niẓām*). The third aspect is that this necessity of creation comes from God Himself, since the necessity of the world's order is itself included in God's knowledge. This means that God's knowledge itself is necessary and cannot be any different from what it is. In Avicenna, the combination of these three aspects, that (1) God's knowledge is the creator of everything, (2) everything is in a necessary order, and (3) God's knowledge itself is necessary, leads to a concept of creation in which nothing can be different from the way it is:[91]

> The existing things correspond to the objects of God's knowledge according to the best order (*ʿalā aḥsan al-niẓām*)—without a motivating intention on the side of the First Being (. . .) and without Him desiring something. Thus, the First Being's knowledge of how to best arrange the existence of everything is the source of the emanation of the good and of everything.[92]

According to the *falāsifa*, God has no goal (*qaṣd*), pursuit (*ṭalab*), desire (*ārzū*), or intention (*gharaḍ*) present when He creates.[93] If God's actions followed any intention to produce things, He would act for something that is not Himself, which would introduce multiplicity to the divine essence. God is the perfect good, and the perfect good creates because it has to do so. One underlying principle in the *falāsifa*'s cosmology is that being is always better than nonbeing. The perfect good therefore has to create; it does not create according to what it chooses but rather according to what is necessary as the best creation. The implication of the *falāsifa*'s view that everything follows necessarily from God's knowledge and that God's knowledge itself is necessary is that God does not have the sort of will that enables Him to choose between alternative creations. Nevertheless, the philosophers claimed that there is a will on God's part. In his Persian introductory work on philosophy, Avicenna claims that we must ascribe a will to God. God, he argues, has knowledge (*dānish*) of the fact that everything emanates from His nature. If one has knowledge of one's actions, Avicenna argues, one cannot say that these actions are only the result of one's nature. The existence of such a knowledge on God's part leads Avicenna to conclude that God does not solely act out of His nature and has indeed some kind of will (*khʷāst*).[94] In his doxographic report of philosophical teachings, *The Intentions of the Philosophers* (*Maqāṣid al-falāsifa*), al-Ghazālī distinguishes these two ways of creation: creation through one's nature and creation by one's will. Here he reports the position of the philosophers that wherever there is knowledge of the action, there is will:

> One can be an agent in two ways, either by pure nature or by a will. An action is out of pure nature if it is without knowledge of either what is done or of the doing itself. All actions that involve a knowledge of the act of doing involve a will.[95]

The *falāsifa* therefore maintain that there is some kind of a will on the part of God, even if there is no decision about the action. These they implicitly admit: the God of the *falāsifa* has no free choice in what to create, and in His creation He does not choose between alternatives. For the *falāsifa*, God creates out of the necessity of His being. God is the one being that is necessary by virtue of Himself (*wājib al-wujūd bi-dhātihi*), and everything about Him is necessary. Avicenna writes that the First Principle is necessary in all its aspects (*min jamīʿ jihātihi*).[96] This entails that God's actions follow from Him with necessity. God is the source of the necessity that turns everything that exists in itself as a sheer contingency into actuality. As such, God cannot himself be contingent, and His actions cannot have an element of possibility within them. In a letter to one of his contemporaries, Avicenna sums up his teachings on the predetermination of all events, on God creating without pursuing a goal or a desire, and on this world being the necessary result of God's essence:

> Pre-determination (*al-qadar*) is the existence of reasons (*ʿilal*) and causes (*asbāb*) and their harmonization (*ittisāq*) in accordance with their arrangement (*tadbīr*) and their order (*niẓām*), leading to the results (*maʿlūlāt*) and effects (*musabbabāt*). This is what is necessitated (*mūjab*) by the decree (*al-qaḍāʾ*) and what follows from it. There is no "why" (*limiyya*) for the action of the Creator because His action is due to (*li-*) His essence and not due to a motive (*dāʿin*) that would motivate Him to do something. (. . .)
>
> "The Decree" (*al-qaḍāʾ*) is God's foreknowledge (*sābiq ʿilm Allāh*) from which that which is determined (*al-muqaddar*) derives (*inbaʿashat*). Every existent whose existence comes about through a smaller number of intermediaries (*bi-wasāʾiṭ aqall*) is of an existence that is stronger (*aqwā*) [than the one that comes about through a greater number of intermediaries].[97]

Al-Ghazālī's Treatment of Causality in MS London, Or. 3126

The Incoherence of the Philosophers is the first work in which al-Ghazālī presents his own ideas about fundamental cosmological issues. We will see that his treatment of causality in the seventeenth discussion of that book is—despite its brevity—so comprehensive that he hardly needed to add anything during his later writings. We will also find that in his later writings, al-Ghazālī stressed certain aspects of what he postulates in this chapter over others. These aspects are not always the same, and in different works he stresses different aspects. Almost everything that he will teach later in his life on the subject of causality, however, has already been put down in the seventeenth chapter of the *Incoherence*. There is no notable development of his views on causality.

An earlier level of al-Ghazālī's occupation with causality is preserved in the text of a London manuscript. This text, whose title is lost, represents al-Ghazālī's efforts to report the teachings of the philosophers rather than to refute them. Unlike his much better known *Intentions of the Philosophers*, here, al-Ghazālī almost exclusively quotes from philosophical works rather than paraphrasing their teachings in his own words. The book was written in the same period that al-Ghazālī worked on the *Incoherence*, or at least shortly after its publication. The text of the London manuscript allows us to reconstruct which philosophical subjects and which works attracted his interest during this period.

The text of the London manuscript contains a very thorough report of the *falāsifa*'s teachings on causality. In his autobiography, al-Ghazālī says that developing a meticulous understanding of the adversary's teachings is an important prerequisite to properly responding to false teachings. A proper refutation is not achieved by simply answering the adversaries' accusations with numerous unsystematic counterarguments. Rather, one must give a thorough report (*ḥikāya*) of the adversaries' teachings,[98] identify the key element in one's own teaching that the adversaries deny, and turn this element against them (*qalb* or *inqilāb*) by showing that they cannot uphold their own teachings without it.[99] The London manuscript devotes almost one-fifth of its text to the subject of causality.[100] The material al-Ghazālī presents on these pages is proportionally more than what Avicenna wrote on this subject in the section on metaphysics of his *Healing* (*al-Shifāʾ*). Al-Ghazālī uses all these passages from Avicenna's metaphysics in the *Healing*, either copying them into his book or paraphrasing them.[101] In these passages, Avicenna introduces the four Aristotelian types of causes. The final and the efficient cause are singled out for more thorough treatment.

Avicenna presents the argument that no causal series, from any of the four types of causes, can regress indefinitely.[102] Every series of causes and effects must have three components: a first element, a middle element, and a last element. The last element is solely an effect and not a cause. The first element of any causal chain is solely a cause and not an effect and causes everything that follows after it. The middle element is the cause for the last one and also the effect of the first. The first element is the absolute cause (*ʿilla muṭlaqa*) of both the middle element and the last. It causes these two either "through an intermediary" (*bi-mutawassaṭ*[in])—namely another middle element of the chain—or without it.[103] Looking at a chain of efficient causes, the "finiteness of the causes" (*tanāhī l-ʿilal*) serves for Avicenna as the basis of a proof of God's existence. Tracing back all efficient causes in the universe will lead to a first efficient cause, which is itself uncaused. When the First Cause is also shown to be incorporeal and one in number, we have achieved a proof of the deity.[104]

While paraphrasing or copying these teachings verbatim from the metaphysics of Avicenna's *Healing*, al-Ghazālī adds material from other non-Avicennan sources, as well as occasionally adding his own original comments.[105] These passages are not meant to criticize Avicenna's approach but rather to explain the philosopher's teachings and make them more accessible to readers not trained in philosophy. In the following passage, for instance, he encourages his readers to reflect on the *falāsifa*'s understanding of causes and to compare them with the way we use words such as "cause" in ordinary language:

> It may appear to some weak minds (*awhām*) that the connection between the thing that we call "an efficient cause," (*fāʿil*) with the thing that we call "caused by it" (*munfaʿil*) or "an efficient effect" (*mafʿūl*) is of the same kind of meaning when the ordinary people (*al-ʿāmma*) name it "that what is made" (*al-mafʿūl*) and "the maker" (*al-fāʿil*). The former kind [of meaning] is that the [efficient cause] generates, and produces, and makes, while the [efficient effect] is generated, is produced, and is made. All this goes back to the fact that one thing attains (*ḥasala*) existence from another thing.[106]

When the *falāsifa* use the word "efficient cause" (*fāʿil*), they mean something different from what we in our ordinary language mean when we use the word "maker" (*fāʿil*). In many instances this meaning is the same, as in the case of the adze, for instance, in which case its maker, the workman, is also one of its efficient causes. Al-Ghazālī explains, however, that sometimes we use words such as "he makes" (*faʿala*), "he produces" (*ṣanaʿa*), or "he generates" (*awjada*) in order to express aspects that belong to the final cause (*gharaḍ*) and not the efficient one. Al-Ghazālī neglects to discuss this in more detail, but what he seems to have in mind is when we say something like, "The doctor makes the patient take the medicine," or "The teacher generates knowledge in his students." These sentences are ambiguous as to the efficient causes of the actions, and both doctor and teacher are more part of the final cause than the efficient one. Al-Ghazālī wishes to stress that the philosophical usage of the Arabic word *fāʿil* knows no such ambiguities. It means "that one thing comes into being after non-being by means of a cause."

In addition to such clarifications, al-Ghazālī stresses in his report the secondary nature of causality more than Avicenna did. He chooses two passages from the works of al-Fārābī that are explicit about the way causes proceed from God. The effects are mediated through the intermediary causes in the heavens and arrive at the sublunar sphere of coming-to-be and passing-away through the mediation of the active intellect. Al-Ghazālī reproduces al-Fārābī's explanation of how "the First, which is God, is the proximate cause of the existence of the secondary causes and of the active intellect."[107] Avicenna avoided giving such a detailed account about the celestial causes because unlike al-Fārābī, he was unsure about their precise number and other matters of detail. In his report, al-Ghazālī prefers outspokenness over precision. He adds another account from the works of al-Fārābī on how the second cause, which is the first intellect, emanates from the First Cause. This chapter also explains how through a procession of secondary causes—each of them an intellect residing in the spheres of Atlas, of the zodiac, of Saturn, Jupiter, Mars, the sun, Venus, Mercury, and the moon—the active intellect is reached. At this point, al-Ghazālī returns to the Avicennan perspective and identifies the active intellect as the "giver of forms" (*wāhib al-ṣuwar*) of the sublunar sphere. An interesting detail in this report is a seemingly minor change of terminology. In the original, al-Fārābī refers to the spheres with the Arabic word *kura*. Al-Ghazālī replaces it throughout the whole passage with the word *falak*, which has the same technical meaning.[108] Unlike *kura*, however, *falak* appears in two verses of the Qur'an (21:33, 36:40), where it refers to the spheres in which the celestial objects swim. Readers in the religious sciences are familiar with *falak*, and using this word might make al-Fārābī's explanation of the heavens more acceptable to them.

Overall, al-Ghazālī tried to make philosophical cosmology more approachable to the religiously trained reader. Later, in his *Revival of the Religious Sciences*, al-Ghazālī writes that it is not contrary to the religious law for a Muslim to believe that the celestial objects are compelled by God's command to act as causes (*asbāb*) in accord with His wisdom. It is forbidden, however, to assume that the stars would be by themselves the efficient causes (*fāʿila*) of their effects, and that there would not be a being that governs (*yudabbir*) over all of them. This assumption would be considered unbelief (*kufr*).[109] Here, in his report on the philosophical teachings of metaphysics, al-Ghazālī makes sure that the readers understand the *secondary* nature of philosophical causality. None of the intellects that reside in the ten celestial spheres is an ultimate efficient cause. Each one of them is a secondary cause and an intermediary employed by God. Al-Ghazālī reproduces a distinctly Avicennan position of causality and adds some of the more detailed accounts of the secondary causes (*asbāb thawānī*) from al-Fārābī's works.

6

The Seventeenth Discussion of *The Incoherence of the Philosophers*

The seventeenth discussion of al-Ghazālī's *Incoherence of the Philosophers* has become famous for its criticism of causality. When Solomon Munk, the first Western analyst of the *Incoherence*, read the seventeenth discussion, he understood al-Ghazālī as saying that "the philosophers' theory of causality is false, and that they are not right when they deny that things can happen contrary to what they call the law of nature and contrary to what happens *habitually*."[1] For Munk, this was an expression of al-Ghazālī's skepticism, which simply denied the existence of causality in the outside world. For students of philosophy and theology, the seventeenth discussion of the *Incoherence* has become a *locus classicus* for pious and yet intelligent criticism of the existence of causal connection. The mistaken understanding that here al-Ghazālī denies the existence of causal connections still persists today. Michael E. Marmura, for instance, goes as far as saying that for al-Ghazālī, "the Aristotelian theory of natural efficient causation is false."[2]

A close reading of the seventeenth discussion shows, however, that on its two dozen or so pages, al-Ghazālī does not deny the existence of causal connections—and thus of causality—and he certainly does not argue that efficient causality as an explanation of physical change is false. Among the many things he does in this discussion is open ways to uphold causality as an epistemological principle of the natural sciences, while remaining uncommitted whether those things in this world that we regard as causes truly have efficacy on their assumed effects. More important, however, the seventeenth discussion is a criticism of Avicenna's necessarianism, that is, the position that events in this world are necessarily determined and could not be any different from what they are.

Al-Ghazālī begins his analysis of the seventeenth discussion by stating a much more limited goal. In its preceding introduction, he says that he aims to convince the followers of the philosophical movement and those who are attracted to its teachings that the things they deem impossible—namely, some prophetical miracles like the changing of a staff into a serpent,[3] the revivication of the dead,[4] or the splitting of the moon (Q 54.1)—should be considered possible events. If they are possible, the Qur'anic accounts of these events are literally true and do not need to be interpreted as metaphors.[5] In our earlier discussion of al-Ghazālī's interpretation of the Qur'an, we saw that according to his rule of interpretation, one's understanding of the text of revelation depends on what one considers possible or impossible. This premise determines al-Ghazālī's perspective in this discussion of the *Incoherence*. It is less a discussion about whether causality is a fact than it is a dispute about modalities and the way we know them. In the seventeenth discussion, al-Ghazālī argues with the Muslim philosophers about what is possible for God to create.[6]

Al-Ghazālī presents the subject of causality as a problem of Qur'an interpretation. Although the *falāsifa* acknowledge that prophets are capable of performing extraordinary feats and can influence their surroundings through the practical faculty (*al-quwwa al-ʿamaliyya*) of their souls by creating rains, storms, and earthquakes, they did not accept that the prophets could change an inanimate being such as a piece of wood or a corpse into a living being such as a serpent or a human or that they could transform celestial objects such as the moon.[7] In their theories, a substance (*jawhar*)—here understood in the Aristotelian sense of a clearly defined object with a number of essential and unchanging characteristics—such as a piece of wood cannot change into another substance such as a living serpent. Celestial bodies are uncomposed in the *falāsifa*'s opinion and thus are not divisible. Yet the Qur'an and the *ḥadīth* describe miracles such as these as confirming the prophecies of Moses and Muḥammad. "For this reason," al-Ghazālī says at the end of the introduction to the seventeenth discussion, "it becomes necessary to plunge into the question [of causality] in order to affirm the existence of miracles." This all happens, he adds, in the interest of upholding the Muslim religious tenet that God is omnipotent (*qādir ʿalā kull shayʾ*).[8]

In the seventeenth discussion itself, the claim of upholding God's omnipotence is nowhere mentioned. Indeed, only a very limited part of that chapter can be seen as responding to this concern. Al-Ghazālī's goal in this discussion is rather limited. In the opening sentence, he formulates the position of which he wishes to convince his readers: the connection between the generally accepted ideas of "the cause" and "the effect" is not a necessary one. If the readers accept this position, so goes the implicit assumption, their acceptance of the reported miracles will follow. Behind this understanding lies the principle that one must fully accept the authority of revelation in places where its literal wording is deemed possible. If the readers acknowledge that God's reports of prophetical miracles in the Qur'an are possible in their outward sense (*ẓāhir*), they must accept the reports' truth.

In accordance with the general strategy of the *Incoherence* to alert the followers of the philosophical movements to mistakes their teachers make in their reasoning, al-Ghazālī first presents an argument that aims to shake the reader's conviction as to the necessity of causal connections and then presents an alternative model for explaining these connections. Al-Ghazālī briefly introduces the counterargument as well as the alternative explanation in an opening statement that is a masterwork of philosophical literature:

> The connection (*iqtirān*) between what is habitually believed to be a cause and what is habitually believed to be an effect is not necessary (*ḍarūriyan*) according to us. But [with] any two things that are not identical and which do not imply one another[9] it is not necessary that the existence or the nonexistence of one follows necessarily (*min ḍarūra*) out of the existence or the nonexistence of the other. (. . .) Their connection is due to the prior decree (*taqdīr*) of God who creates them side by side (*'alā l-tasāwuq*), not to its being necessary by itself, incapable of separation.[10]

Here, al-Ghazālī lays out four conditions for explaining physical processes. The requirements are: (1) that the connection between a cause and its effect is not necessary; (2) that the effect can exist without the cause ("they are not incapable of separation"); (3) that God creates two events concomitantly, side by side; and (4) that God's creation follows a prior decree. Earlier in the introduction to the discussion, al-Ghazālī had said that from a Muslim's point of view, a physical theory is acceptable only if it leaves space for unusual creations "that disrupt the habitual course [of events]."[11] This condition is no longer part of the four in this initial statement of the discussion. This omission is an important indicator. Additionally, upholding divine omnipotence, which is mentioned as a motive for this debate at the end of the introductory statement, does not appear in the seventeenth discussion itself. In the discussion, al-Ghazālī focuses purely on the *possibility* of the reported miracles, and he does not claim that we should consider God capable of doing all those things the philosophers deny that He can do. It is important to understand that al-Ghazālī does not deny the existence of a connection between a cause and its effect; rather he denies the *necessary character* of this connection.[12]

On first sight, it seems that only a consequent occasionalist explanation of physical processes would fulfill these four conditions. Ulrich Rudolph, however, pointed out that not only occasionalism but also other types of explanations fulfill these four criteria. Most misleading is the third requirement that God would need to create events "side by side." These words seem to point exclusively to an occasionalist understanding of creation. One should keep in mind, however, that this formula leaves open *how* God creates events. Even an Avicennan philosopher holds that God creates the cause concomitant to its effect through secondary causality. Rudolph convincingly argues that although the seventeenth discussion of the *Incoherence* points toward occasionalism as a possible solution, it also allows for other solutions.[13] Al-Ghazālī chooses language that can be easily associated with occasionalist theories, which has led many interpreters of this discussion to believe that here he argues exclusively in favor of it. On at least two occasions, however, al-Ghazālī alerts his occasionalist readers to some very undesired consequences of their position. He implicitly cautions his readers against subscribing to consequent occasionalist explanations of physical processes.[14] Simultaneously, al-Ghazālī alerts his target readership—Muslim scholars attracted to philosophical explanations—to a fundamental mistake they make when they talk about necessity and possibility. From that place, he develops several alternative explanations likely to satisfy the requirements for physical explanations as described by Aristotelian natural sciences. These alternative explanations accept the possibility of the reported prophetical miracles.

Prior analyses of the seventeenth chapter of the *Incoherence* do not always note its division into three different "positions" (singl. *maqām*).[15] Each "position" cites a claim within the teachings of a group of *falāsifa* and points out why this claim is either untenable or must be modified. These different claims come from different groups among the *falāsifa*. The "position" (*maqām*) is that of an opponent, which is rebuffed by al-Ghazālī's objections to it.[16] In one case, this rebuff is divided into two "approaches" (singl. *maslak*). It should be noted that a "position" within this text consists of the citation of a philosophical position *plus* al-Ghazālī's answer to it.[17] The character of the *Incoherence* allows al-Ghazālī to cite all sorts of objections in his answers, whether he subscribes to them or not. In order to make his point most effectively, al-Ghazālī puts forward more than just one explanation as to how the reported miracles are possible. In the Second and the Third Positions, he presents in total three different interpretations of the relationship between what is called cause and effect. These explanations are *different* theories; each is consistent only within itself. The seventeenth discussion leaves open whether al-Ghazālī subscribes to any one of them. Although the first of his alternative explanations denies the existence of natures, meaning the unchanging character of the relation between cause and effect, the second alternative accepts that natures do exist.[18] Al-Ghazālī presents various theories that shake the convictions of his opponents on different levels, sometimes more and sometimes less radically.

The First Position: Observation Does Not Establish Causal Connections

The First Position (*al-maqām al-awwal*) cites the claim that in a given example in which fire comes into contact with a cotton ball, "the efficient cause of the [cotton's] combustion is the fire alone."[19] The fire is the agent or the efficient cause (*fāʿil*) igniting the cotton in accord with its nature (*fāʿil bi-ṭabʿ*), and it has no choice over its actions. According to this position, fire is the *only* efficient cause of the ignition; it is the only sufficient cause that by itself makes ignition necessary. This is not the position of Avicenna: he taught that in any given chain of efficient causes, only the first element is the cause in the real sense of that word. That first element is the absolute cause (*ʿilla muṭlaqa*) of all that follows after it. Thus, with regard to efficient causality, there is only one absolute cause, and that is God. For Avicenna, who believed in secondary causality, the fire would only be a middle element in a causal chain. The fire would be both a cause and an effect, and it could not be called the *only* efficient cause of the ignition. At other places in his writing, al-Ghazālī ascribed this First Position somehow vaguely to a group of people he calls "eternalist" (*dahriyyūn*) for their belief in an eternal world without a cause or a maker. These people, he adds, are clandestine apostates (*zanādiqa*), meaning they could not be counted among the various groups of Muslims.[20] Later in this book, al-Ghazālī adds that this position is closely akin to the one held by Muʿtazilites with regard to the generation (*tawallud*) of human actions and their effects.[21]

From his later comment in the *Revival*, we know that al-Ghazālī condemned as unbelief (*kufr*) the view that stars would be by themselves efficient causes that are not governed by higher ones. The First Position in this discussion presents this view. It is not surprising that al-Ghazālī responds vigorously in response to this theory: this position must be denied. Rather, the efficient cause for the burning of the cotton, and it being reduced to ashes, is God. Again, these words seem to suggest that al-Ghazālī refers exclusively to occasionalism as the only acceptable alternative explanation. An Avicennan, however, could easily agree with the statement that God is the ultimate or absolute efficient cause of the cotton's combustion. This alternate explanation is taken into account in the statement in which al-Ghazālī rejects the initial position:

> This [position] is one of those that we deny. Rather we say that the efficient cause (*fāʿil*) of the combustion through the creation of blackness in the cotton and through causing the separation of its parts and turning it into coal or ashes is God, either through the mediation of the angels or without mediation.[22]

The angels here are the celestial intellects. The correct position is either an occasionalist explanation *or* Avicenna's view of creation by means of secondary causality. In both theories, not the fire but God is the absolute efficient cause of the burning.

In this First Position, al-Ghazālī implies agreement with Avicenna and the Aristotelian philosophers when he says that events such as the birth of a baby are not simply caused by the parents but rather by "the First" (*al-awwal*), meaning God, "either without mediation or through the mediation of the angels who are entrusted with these temporal things."[23] Here again, the word "angels" (*malāʾika*) refers to the celestial intellects, who in Avicenna's cosmology are causal intermediaries between God and the sublunar sphere. For events in the sublunar sphere, al-Ghazālī names the active intellect as one of their causes. The intellect is named as the "giver of forms" (*wāhib al-ṣuwar*) in the sphere of generation and corruption. Here in the First Position, al-Ghazālī accepts that the "giver of forms" is the angel (*malak*) from which the "events that occur when contacts between bodies take place" have their source (or emanate).[24]

This is the position of those who search diligently for truth among the philosophers (*muḥaqqiqūhum*), al-Ghazālī says.

After finding common ground with the Avicennans, al-Ghazālī attacks the adversary's position that fire can be the only efficient cause. His objection is based on epistemology: the simple observation of one thing following another does not justify denying the involvement of causes that are not visible. Earlier Ashʿarites such as al-Bāqillānī had used the same line of reasoning with a more radical scope, arguing that sense perception does not establish any connection between cause and effect.[25] According to al-Bāqillānī, all we can know without doubt is that these two things usually follow each other in our observation or our sense perception (*mushāhada*). Such perceptions, however, are unable to inform us about a causal connection between these two events. Like earlier Ashʿarites, al-Ghazālī uses this argument in a radical sense. The fact that we experience cotton as burning every time fire touches it informs us neither (1) about *any* causal connection between the fire and the burning of the cotton nor (2) whether fire is the only cause:

> Observation (*mushāhada*) points towards a concomitant occurrence (*al-ḥuṣūl ʿindahu*) but not to a combined occurrence (*al-ḥuṣūl bihi*) and that there is no other cause (*ʿilla*) for it.[26]

In the context of the First Position, al-Ghazālī focuses on the latter point; we have no means to know whether fire is the only efficient cause, as these people claim. Nobody would say, for instance, that the parents (al-Ghazālī says elliptically: the father) are the only efficient causes of a child. There may be hidden causes everywhere, and it is next to impossible to say that any given cause is the only sufficient one for the effect it appears to trigger.

Al-Ghazālī's denial of the claim that an event may have a single immanent efficient cause is based on the wider-ranging epistemological objection that sense perception creates no knowledge of causal dependencies. When a thing exists together with (*ʿinda*) another, it does not mean that it exists through (*bi-*) it.[27] Concurrent events need not be connected with one another; and even if they are, the connection may be much more complex than what we witness.

By using this argument, al-Ghazālī introduces some confusion into this First Position. Apparently, al-Ghazālī intends to argue against the position that fire is the *absolute* efficient cause of the cotton's burning, a point at which he rightfully claims agreement with the Avicennan *falāsifa*. But by referring to the epistemological objection that observation can prove concomitance of two events but no connection between them, he has justifiably been understood as being more radical. He seems to object not only to those who teach there are (absolute) efficient causes other than God, but also to those who teach that causes have efficacy on their effects.

This is not where the confusion ends. While arguing that fire cannot be the only efficient cause for the cotton's combustion, al-Ghazālī brings a very brief side argument: "As for the fire, it is an inanimate being (*jamād*) and it has no action (*fiʿl*)."[28] Here al-Ghazālī refers back to an objection he made in the third discussion in the *Incoherence* about what can be called a *fāʿil*, or, an agent or an efficient cause. Motivated by considerations that will become clear later during this study, al-Ghazālī simply rejects the terminology of the *falāsifa*—the Avicennans as well as any other group. For Avicenna, for instance, the word *fāʿil* merely describes the efficient cause: it is the thing that gives existence to another thing.[29] In the third discussion of the *Incoherence*, al-Ghazālī rejects that usage on the grounds that according to common understanding, the word *fāʿil* describes the originator of an act—al-Ghazālī uses a pronoun that refers to a person and not a thing—who has a will, has chosen the act freely, and has knowledge of what is willed.[30] This sense of *fāʿil* is totally alien to Avicenna, and al-Ghazālī's statement here shows a fundamental disagreement between him and Avicenna about the meaning of the word *fāʿil*. For al-Ghazālī, it means "voluntary agent"; for Avicenna, simply "efficient cause." In the seventeenth discussion, al-Ghazālī throws in this earlier argument without further pursuing the point. Although primarily directed against a nonsecondary understanding of causality, the sentence is ultimately also directed against Avicenna's particular understanding of secondary causality. In the context of the First Position here, which does not represent Avicenna's view on causality, the sentence is somewhat misleading and has, in fact, led to misunderstandings among al-Ghazālī's modern interpreters.[31]

The First Approach of the Second Position: How the Natural Sciences Are Possible Even in an Occasionalist Universe

The Second Position (*al-maqām al-thānī*) solves some of the confusion that remains from the First. It begins with the claim of a philosophical opponent who concedes that fire is not the true efficient cause of the cotton's ignition. This philosopher admits that events emanate from "the principles of temporary events" (*mabādīʾ al-ḥawādith*). He maintains that the connection between the cause and the effect is inseparable and necessary. Causal processes proceed with necessity and in accord with the natures of things, not by means of deliberation and choice by the efficient cause. The philosophical adversary argues that all things have a certain predisposition (*istiʿdād*) that determines how they react to other things. This predisposition is part of the thing's nature (*ṭabʿ*).[32] Because these natures cannot change, the things react necessarily to given circumstances. Cotton, for instance, necessarily burns when it comes in contact with fire. Here, al-Ghazālī paraphrases the position of Avicenna and other Aristotelians. The philosopher of the Second Position teaches secondary causality; he believes in the necessity of causal connection and in the existence of natures (*ṭabāʾiʿ*).

Al-Ghazālī divides his response to this position into two "approaches" (singl. *maslak*). The First Approach counters this philosophical position with that of a consistent occasionalist. Al-Ghazālī asks his philosopher-opponent to consider that nothing in this world follows its given natures. Everything can be changed if so willed by God.[33] Pointing to God's omnipotence prompts the opponent to bring his most forceful objection against al-Ghazālī's criticism of causality. If there are no natures and no given predispositions, the philosopher-opponent says, how are we to know anything about the world? If we do not take our judgments from the nature of things, we may well take them from any random source, and then they simply become arbitrary:

> If one denies that the effects follow necessarily from their causes and relates them to the will of the Creator, the will having no specific designated course but [a course that] can vary and change in kind, then let each of us allow the possibility of there being in front of someone ferocious beasts, raging fires, high mountains, or enemies ready with their weapons [to kill him], but [also the possibility] that he does not see them because God does not create [vision of them] for him. And if someone leaves a book in the house, let him allow as possible its change on his returning home into a beardless slave boy (. . .) or into an animal (. . .).[34]

Al-Ghazālī admits that this is a strong objection by saying that it brings up the vilifying or hideous impossibilities (*muḥālāt shanīʿa*) of a consequent occasionalist position, impossibilities that one might not want to be associated with.[35] Much of what follows in the seventeenth discussion may be understood as al-Ghazālī's response to what he evidently considered a quite compelling point.

In his most immediate answer, al-Ghazālī brings two arguments that defend the occasionalist's position. In the first, he introduces a difference between two types of possibilities. This passage in the seventeenth discussion is very similar to one in al-Ghazālī's *Balanced Book on What-to-Believe*, yet here in the *Incoherence*, the language he uses is surprisingly untechnical. Al-Ghazālī says that although all of the possibilities the adversary mentions are possible, there is a difference between possibility and actuality. Admitting that something is possible involves no commitment that it is true. If God had created this world in such a way that we would make no distinction between what is possible and what exists in actuality, we would indeed be confused about the possibility of a book transforming into a horse. However, God created human knowledge in such a way that we *do* distinguish what is merely possible from what occurs in actuality. Granted that it is possible—and thus within God's power—to change books into horses at any moment, we know that in our world such an event never occurs, whether in our presence or in our absence. God's past habits have given us some guidance about what we consider possible or impossible: "The continuous habit of their occurrence repeatedly, one time after another, fixes unshakably in our minds the belief in their occurrence according to past habit."[36] Al-Ghazālī makes his point again in an opaque passage with an example that he explicates fully in the *Balanced Book on What-to-Believe*. The philosophers agree, al-Ghazālī says, that prophets have been given the ability to look into the future. When they do, they have certain knowledge about which future contingencies will become actual and which will not be realized. The clairvoyance of the prophets shows that the distinctions between what possibilities will and will not occur in the future already exist today. In the *Balanced Book*, al-Ghazālī says that those future contingencies, which will remain unrealized, are possible with regard to themselves but impossible with regard to something else.[37] In other words, an event such as a book changing into a horse is possible with regard to itself, but with regard to the "something else" of God's habit, such an event will not occur.

William Courtenay, who was unaware of the discussion in the *Balanced Book*, understood that here al-Ghazālī applies a distinction between God's absolute power-to-act and the exercised or ordained power of God.[38] This distinction can be also understood as analogous to al-Fārābī's distinction between what is possible or necessary "in itself" and "from something else." Regarded purely in itself, it is within God's power to change books into horses. But God operates consistently and does not alter his operations by whim or caprice. Regarded from the perspective of God's preknowledge and the consistency of His action, we do not think it possible for books to turn into animals. God will not interrupt the habitual operations of what appears to be cause and effect without good reason. The only reason why God would suspend the habitual relationship between causes and effects—so it seems in the seventeenth discussion—is the confirmation of one of His prophets. If God's preknowledge includes the enactment of a miracle, He suspends His habit.

Al-Ghazālī brings a second argument in defense of the occasionalist's position, one that focuses on the relationship between events in the created world and our knowledge of them. Usually we say their relationship is causal: outside events cause our knowledge of them. For the occasionalist, this translates into saying that this connection is not by itself determined. Given that there are no causes among creatures, the outside events cannot cause our knowledge, the occasionalist claims. Rather, God both creates the event in the outside world and creates our knowledge independently to accord with the event.[39] Here again, the relationship is habitual but not necessary. Although we have reason to trust in God and assume that our knowledge of the world corresponds to its actual function, there is no direct connection between the events and our knowledge of them.[40]

Michael E. Marmura and Ulrich Rudolph suggest that al-Ghazālī tried to rebuff the objection that occasionalism leads to ignorance by augmenting an occasionalist view of causality in the outside world with an occasionalist understanding of human knowledge. Since God has direct control over our knowledge as well as over our imaginations, and since we witness that nobody is seriously concerned about books changing into an animal zoo, God evidently prevents us from being confused by not creating in us absurd thoughts such as these.[41] The force of this line of argument seems to rest on the common observations (1) that nobody experiences the transformations of books into animals and also (2) that humans with a sound intellect do not draw false conclusions about what is likely to happen. The second experience is just as important as the first. God creates human knowledge to be neither discontinuous nor capricious. Agreeing with his philosopher-opponent, al-Ghazālī believes that true knowledge corresponds with its objects in the outside world. Here he aims to strengthen the notion that humans do have true knowledge. He argues that God creates our knowledge of the world habitually in accord with it; truth is therefore a result of God's habit and not of causal connections between objects and their perception.

The philosopher-opponent suggests that an omnipotent God may act arbitrarily. As in the first point, al-Ghazālī's rebuff is based on the strictly habitual character of God's actions. He responds that God's habit is manifest in two ways. First, books habitually do not change into animals. Second, our knowledge of the actual (and not possible) transformation of books habitually corresponds to what actually happens in the outside world. Stressing the strictly habitual character of God's operations aims at rejecting the ideas that this world could be chaotic or that we do not have true knowledge of it. It is indeed possible in principle for books to turn into horses while still giving us the impression that they had remained books. If God were to will that sort of thing, He could prevent us from ever finding out what had really happened to our books. Neither of these incidents would ever happen, al-Ghazālī says, because past experience shows that God habitually does not act this way. Humans are therefore not confused about books turning into horses, because it is part of God's habit to prevent our confusion. When God made His plan of creation, He chose not to enact these possibilities that the philosophers evoke, and He created human knowledge accordingly. God already knows in His divine foreknowledge that He would not do a certain act and thus break His habit.[42]

Miracles are naturally part of God's foreknowledge. When they occur, God adjusts the knowledge of those humans who witness it. The witnesses' habitual foregone conclusions about the expected course of events will be suspended in order for them to realize that they are, in fact, witnessing a miracle:

> If, then, God disrupts (*kharaqa*) the habitual [course of events] by making [the miracle] occur at a time when a disruption of the habitual events takes place, these cognitions [about the habitual course of events] have slipped away from people's minds since God didn't create them.[43]

The two points al-Ghazālī makes in the First Approach of the Second Position are those of a fully consistent occasionalist who stresses the reliability of God's habit. God directly creates all events in His creation, including the knowledge of humans. Yet the strictly habitual character of God's actions avoids epistemological solipsism and creates the possibility of natural science. Humans successfully master the world by knowing, for instance, that books will remain books. This fact is a clear indication about the strictly habitual character of God's actions.

The Second Approach of the Second Position: An Immanent Explanation of Miracles

Al-Ghazālī presents to his readers a second consistent theory to explain miracles. This theory promises "deliverance from these vilifications," meaning the absurdities of having to reckon with books changing into horses and similar things.[44] This Second Approach (*al-maslak al-thānī*) lacks the radical spirit of the first. In fact, it has often been regarded as a wide-ranging concession to al-Ghazālī's philosophical opponents that subscribe to the necessary character of the connection between cause and effect.[45] Al-Ghazālī proposes that physical processes, which are simply unknown to us, explain those prophetical miracles that the *falāsifa* deny. We are unaware of these processes because they occur so rarely that we may not have witnessed them. The Qu'ran depicts Abraham's being thrown into a blazing fire (Q 21:68, 29:24, 37:97) and surviving unharmed; his survival can be seen as similar to people who coat themselves with talc and sit in fiery furnaces, unaffected by the heat. Similarly, Moses' stick changing into a serpent can be seen as the rapid version of the natural recycling of a stick's wood into fertile earth, into new plants, into the flesh of herbivores, and from there into the flesh of carnivores such as snakes. There is no limitation to how fast these processes can unfold.[46] Miracles are sometimes hard to distinguish from what may be called magic or sorcery. Talismanic art, for instance, has at times repelled snakes, scorpions, or bedbugs from towns and villages.[47]

The likely confusion of sorcery and prophetic miracles is an important motif in al-Ghazālī's later works, most prominently in his autobiography, *Deliverer from Error.* These later passages will be discussed further on. This explanation of prophetical "miracles" provided in the Second Approach is certainly the one most conducive to a philosophical reader. We also note that this approach does not uphold the initial stipulation of the discussion's introduction that physical theories must leave God space for "disrupting (*kharaqa*) the habitual course [of events]."[48] Indeed, at the beginning of the seventeenth discussion, this condition fails to be mentioned. In any case, the kinds of explanations proposed in this Second Approach are not disruptions of the physical course of events. Here prophetical "miracles" are merely understood as marvels, seemingly wondrous events that, if all factors are taken into consideration, can be explained as effects of natural causes. They are effects and permutations that may be witnessed rarely or may not have been witnessed at all. Still, al-Ghazālī says, the serious natural philosopher should consider them possible. He must acknowledge that the natural sciences cannot explain all phenomena that humans have witnessed in the past: "Among the objects lying within God's power there are strange and wondrous things, not all which we have seen. Why, then, should we deny their possibility and judge them impossible?"[49] Such a denial of the reported "miracles" would be because of a lack of understanding the ways of God's creation: "Whoever studies the wonders of the sciences will not regard whatever has been reported of the prophetical miracles in any way remote from the power of God."[50]

Overcoming Occasionalism: The Third Position

Al-Ghazālī quotes another claim of an opposing philosopher in what we find as the third and last position (*maqām*).[51] This third philosopher-adversary proposes a seemingly simple understanding: both parties must agree upon the fact that God can only create what is possible and that He cannot create what is impossible. This leads the philosopher to ask al-Ghazālī: what does he believe is impossible?[52] If he would say that impossibility is just the negation of two contradictory things existing together, he would simply render himself ridiculous, since according to the opponent, it is obvious that many other things are also impossible for God to create. God cannot move a dead man's hand, and He cannot create a will in a creature that has no knowledge. There can also be no knowledge in creatures that have no life.

The imaginary opponent puts his finger on a significant discrepancy between the two parties that explains much of their differences. The Aristotelian philosophers regard creation as a necessary process that flows from God's unchanging knowledge. God's knowledge and His power to create are together sufficient causes for the world to be as it is. God's knowledge is the determining factor that necessitates the world in its current state, and His knowledge is itself determined by His unchanging and eternal nature. Presuming that God's knowledge is eternal and unchanging makes the world's history determined and necessary. This necessity does not permit the creation of anything other than what actually is. Any actual creation is necessitated by the combination of long chains of causes that all have its starting point in God's nature. God cannot change the continuous realization of these chains of causes and effects, just as He cannot make water flow uphill. For the *falāsifa*, everything that does not exist in actuality is therefore impossible to be created. It is impossible for the world to be anything other than it is.

Modern Western interpreters of al-Ghazālī disagree about his answer to this challenge. The majority holds that al-Ghazālī's response makes a significant concession to the position of the *falāsifa*: he acknowledges that there are certain limits to God's creative power, boundaries much narrower than that which is logically impossible. Al-Ghazālī concedes that some assumptions imply others. A stone, for instance, can have no knowledge. The assumption of knowledge in a thing implies that this thing has life. The same is true for will and knowledge, as the former implies the latter. We cannot say that something has a will without also assuming that it has prior knowledge about the object of its will. In his interpretation of the Third Position, Ulrich Rudolph points to the fact that from the very beginning of the seventeenth discussion, relationships of identification and implication were exempt from al-Ghazālī's critique of causality. The initial statement of this discussion says that, "[with] any two things that are *not identical* and which do *not imply one another*, it is not necessary that the existence or the nonexistence of one follows necessarily out of the existence or the nonexistence of the other."[53] Here at the end of the discussion, al-Ghazālī clarifies what he meant when he had said that two things are identical or imply each other.

At the start of this Third Position, in his response to the philosopher's challenge al-Ghazali postulates three principles that God's creative power is subject to. In his creation, God is bound by three norms: First of all, God cannot violate the rule of excluded contradiction. He thus cannot affirm (meaning create) and also deny (meaning not create) a specific thing at a given time. Second, God must accept relationships of implications. This is closely connected to the principle just mentioned: God cannot "affirm the special and at the same time deny the more general [when it includes the special]" (*ithbāt al-akhaṣṣ ma'anafī l-a'amm*). Third, God cannot "affirm two things and at the same time deny one of them" (*ithbāt al-ithayn ma'a nafī l-wāḥid*). These three rules define what is impossible. Everything that is not limited by these three rules is, according to al-Ghazālī, possible for God to create.[54]

In the next step, al-Ghazālī explains how these three norms are to be applied. He gives some examples: God cannot create black and white in the same substrate or locus (*maḥall*), and he cannot create a person in two places at once since this would violate the principle of excluded contradiction. The second rule on the binding character of implications says that God can neither create a will without knowledge nor create knowledge without life.[55] Lenn E. Goodman suggests that acknowledging this principle introduces the Aristotelian schema of genera and differentia and of essences and accidental properties. Identifying a thing as X carries with it all further specification of X's definition.[56] If God wishes to create an animal, for instance, He must create it animated and cannot leave it lifeless.

The third rule brings with it an equally wide-raging consequence, since it disallows, in al-Ghazālī's view, "the changing of genera" (*qalb al-ajnās*). Goodman probably goes too far when he argues that with this principle, al-Ghazālī accepts the whole apparatus of Aristotelian hylemorphism.[57] More likely, al-Ghazālī means that transformations can only happen within the "genera" and not across their lines. Blood can change into sperm, and water can change into steam, but a color cannot be changed into a material object, for instance. In the permitted cases, the matter (*mādda*) of the initial substance assumes a different form (*ṣūra*). For al-Ghazālī, matter is generally receptive to change and may be transformed into another material being. A stick may therefore be transformed into a serpent, since the two share a "common matter" (*mādda mushtarika*). It is impossible, however, that an attribute such as "blackness" could change into a material being such as a cooking pot.[58] Thus the word "genera" (*ajnās*) describes for al-Ghazālī not the Aristotelian classes of beings but the two traditional classes of beings in the ontology of *kalām*: bodies that consist of atoms (*jawāhir*) and attributes, that is, accidents (*a'rāḍ*) that subsist in bodies.[59] This is indeed how the word "genera" (*ajnās*) has been used by earlier Ash'arites.[60] Transformation between bodies and accidents is impossible. All changes within the genera are possible, says al-Ghazālī, and it is, for instance, easy for God to move the body of a dead man. This would not require the creation of life in a corpse, for God could just move the limbs of the corpse without putting life into it. Not the man but God would be the mover.

Lenn E. Goodman's and Ulrich Rudolph's readings of the Third Position represent the majority opinion of modern interpreters.[61] They understand that in the concluding part of the seventeenth discussion, al-Ghazālī makes significant concessions to his philosophical opponents. He acknowledges that God is bound not only by certain rules of logic, such as the principle of excluded contradiction, but also to a limited number of natural laws that we know to be true and binding from experience.[62] The impossibility of "changing the genera" (*qalb al-ajnās*) would be part of this second group of limitations on God's power.

Julian Obermann's "Subjectivist" Interpretation of the Seventeenth Discussion

There is also a minority interpretation whose understanding of the Third Position is probably just as consistent with the text as the one we have just discussed. In its scope, however, it is much more radical. Julian Obermann, who was the first Western scholar to critically analyze the seventeenth discussion of the *Incoherence*, presented the results of his 1915 dissertation in a long article and a considerably expanded book, both published in Vienna shortly before and after the First World War.[63] His interpretation, however, did not have much impact on later scholarship.[64]

Obermann connects al-Ghazālī's denial that anything in this world could be an absolute efficient cause to arguments presented in earlier discussions of the *Incoherence*. In the first discussion on the subject of the eternity of the world, al-Ghazālī argues that "will" (*irāda*) is something that is not determined by the things we find in this world. If a thirsty man is given two glasses of water that are identical to each other and equal in their position to him, the man is not at all paralyzed by the choice between these two identically beneficial options. His choice between the two glasses is not determined by his experience of the outside world. For al-Ghazālī, will is the capacity to distinguish one thing from another that is exactly similar to it.[65] The lack of difference between the two glasses has no effect on the thirsty man's choice to pick one. It is the human's will that distinguishes the two glasses and not the human's knowledge of them. This shows al-Ghazālī that the *falāsifa*'s causal determinism cannot explain why the thirsty man picks a glass. For them, his choice should be determined by the differences he perceives. Since there are no differences, a deterministic explanation of this situation would have the man die of thirst, unable to pick either of the two glasses.[66]

Obermann argued more generally that for al-Ghazālī, humans distinguish things by means of their will and not by what the things really are or by how they interact with our epistemological apparatus. The criteria of the human will are often random and arbitrary. They are certainly not determined by the outside world. The lack of distinction between the two glasses is not in any way causally connected to the choice of the man. More generally, our position toward causal connections in the outside world is independent of what we perceive there. Our senses do not perceive the agency of a cause on its effect: causality is the result of a choice within us. It is "solely due to the continuity of a habitual action that our memory and our imagination are imprinted with the validity of an action according to its repeated observation."[67]

For Obermann, who wrote his analysis of al-Ghazālī's critique during the late 1910s, this is the position of "philosophical subjectivism." Obermann interpreted al-Ghazālī's criticism of causality from the point of view of the post-Kantian debate about "subjectivism" and "psychologism" in early twentieth-century Vienna.[68] Al-Ghazālī's thought, however, even if it is understood along Obermann's lines, can hardly be compared with modern subjectivism. There is not enough evidence that the Muslim theologian argued in favor of a relativist view of human knowledge, one in which knowledge is dependent on epistemological decisions by the perceiving subject. In fact, in the face of philosophical accusations of epistemological relativism, al-Ghazālī maintains that truth is the correspondence of human knowledge with the outside reality. He believes that humans do have true knowledge in this sense. Therefore, Hans Heinrich Schaeler, who criticized Obermann's choice of "subjectivism," suggested that if Obermann's interpretation is correct, al-Ghazālī's approach should rather be called "anthropocentric." It is not occupied with subjectivist concern but aims to gain further insight into the way God created humanity.[69]

Obermann welcomed al-Ghazālī's critique of Avicenna's epistemological realism and considered it a major philosophical achievement.[70] His analysis places al-Ghazālī as a predecessor of Immanuel Kant and proposes that, whereas for the Muslim theologian empirical observation stands on shaky grounds, human judgments remain the solid foundation of certain and firm knowledge. Obermann understood that in the Third Position of the seventeenth discussion, al-Ghazālī reconsiders his earlier suggestion that our knowledge is not necessarily connected to the world. But although there may not be a necessary connection between the world and our knowledge of it, just as there is no necessary connection between any two events within the world, our knowledge is bound to certain conditions of our judgments. The most important judgments are those about what is possible, what is impossible, and what is necessary.

Thus, according to Obermann, al-Ghazālī objects to what he believes is a naive empiricism of the *falāsifa* by saying that possibility and impossibility are not contained within the things themselves. They are predicates of human judgments:

> Science only accepts necessary connections where they have to be thought of as necessary and impossibilities where they have to be thought of as impossibilities. The standard for the value of scientific knowledge, for its dignity, its right, and its claims is created only within our minds.[71]

According to our mutual judgments, it is impossible that one object is at two places at the same time. This impossibility we know not from observation—as we cannot inspect all places of the world simultaneously—but rather we hold it as a principle of our judgment. When we say that an individual is within the house, as al-Ghazālī writes in the Third Position, it implies that we deny that he or she is outside of the house.[72] We deny the existence of the individual outside the house, not because we cannot find him or her outside, but because we cannot think of a person as being at the same time in- and outside of the house.[73] The same applies to the other implications discussed above. When we say that we know that things without life cannot possess knowledge, we refer to a principle of our judgment, rather than the world as such. It is inconceivable that inanimate beings are knowledgeable, and thus it is impossible for us to assume the existence of a knowledgeable stone.[74]

All this leads to the acknowledgment of certain conditions for human knowledge, according to Obermann. If we talk about something having a will (*irāda*), we implicitly assume that this something also has knowledge because we cannot imagine will without knowledge.[75] The necessary connection between will and knowledge is not something that we find in the objects of the world; rather, it is generated by our judgments. In the outside world, there may or may not be a connection between will and knowledge.

In the First Position of the seventeenth discussion, al-Ghazālī had disputed that our sense perception (*mushāhada*) can detect necessity in the outside world. Thus, Obermann's implicit question: would he give up this position during the later course of the discussion in the Second and Third Positions? In the Third Position, which is for Obermann something like a summary conclusion to the whole seventeenth discussion, al-Ghazālī proposes that the principle of causality is valid not in an absolute sense but in a logical-intellectual one. It is valid as a law within the sciences, although its empirical verification transcends the boundaries of human knowledge and leads into the field of religion.

Al-Ghazālī's Critique of Avicenna's Conception of the Modalities

Obermann's use of the category "subjectivism" may not have been an auspicious one. It seems evident today that al-Ghazālī's approach has nothing to do with modern subjectivism. He does not say that human knowledge of what is possible is merely an impressed belief that has no connection to reality. It is true, says al-Ghazālī in the First Approach of the Second Position, that God could, in principle, disconnect our knowledge from the outside world. But that is only a thought experiment, similar to the possibility that books could change into animals, another possibility that God does not enact. We will see that trust in God (*tawakkul*) is a major condition for investigating the natural sciences. Such trust requires the certainty to know that God will not change books into horses or disconnect our knowledge from reality. Given that God habitually creates our knowledge to accord with reality, we can rely on our senses and our judgment and confidently pursue the natural sciences.

Yet there is a more moderate way to understand Obermann's interpretation of al-Ghazālī. Certain words and formulas used by al-Ghazālī support Obermann's suggestion that in the Third Position, al-Ghazālī is talking not about what God might possibly enact but rather what is possible for a human's judgments. The opponent in the Third Position starts the discussion by assuming that the modalities exist both within the power of God as well as in our knowledge.[76] Al-Ghazālī quotes the position of his Avicennan opponent who says that the outside world is divided into two basic modalities, meaning it is divided into two categories of beings: (1) those that are necessary by themselves and (2) those that are by themselves possible (but not necessary).[77] The opponent implies that the mental existence of the modalities—meaning our judgments that something is necessary, possible, or impossible—is derived from their existence in reality. We will see that al-Ghazālī rejects such an understanding of the modalities. In his response, he does concede that God cannot *enact* the impossible. Yet he then immediately shifts the whole debate away from what God can do to what can be *affirmed* or *denied*, that is, to the level of human judgments.[78] Throughout the Third Position, al-Ghazālī combines language that refers to God's power to act—using such words as "power" (*qudra*) and "object of power" (*maqdūr*), words that refer to the outside world—with language that refers exclusively to human judgments, such as "affirmation" (*ithbāt*) and "negation" (*nafī*). The "impossible" is defined as the combination of an affirmation with its negation (*al-muḥāl ithbāt. . . ma'a nafī. . .*).[79] Impossibility seems to exist only in human judgments. If the interpreter of al-Ghazālī follows the hermeneutic strategy to replace the word "impossible" with its given definition, al-Ghazālī is saying: "God cannot enact an affirmation that is combined with its negation." This sentences, if it makes any sense at all, points to a nominalist interpretation of God's power to create and says: God cannot create judgments in our minds that combine an affirmation with its mutual negation.

Avicenna's position stands in opposition to this. He teaches that the mental existence of modalities derives from their existence in reality.[80] Avicenna taught that human knowledge is determined by the way God creates the world. Like most thinkers of his tradition, Avicenna was an epistemological realist; and like Plato and Aristotle, he believed in an eternal and invariant formal level of being that makes individual objects what they are and that makes the human soul a conscious copy of the formal basic structure of reality. Aristotle teaches that actual knowledge is identical with its object.[81] In being thought of, the formal basis of reality—the forms and ideas that are the backbones of reality—is actualized in the human mind. The human mind is thus directly acquainted with the formal underpinnings of reality. The knowledge it contains is "an inside view into the ultimate foundations of being and sees the visible world as its imitation or explication."[82] When we see a horse, for instance, we connect our sensual perception to the formal concept of "horseness," which is the universal essence or quiddity (*māhiyya*) of every individual horse. In Avicenna's opinion, knowledge can be achieved only by identifying a given individual object as a member of a class of being, a universal. Understanding means reducing any given multitude of sensual perceptions to a combination of universals. The horse may be white, male, and strong. Whiteness, maleness, and strength are universals that exist not only as categories of descriptions in our mind but also as entities that exist *in realiter* in the active intellect, from which humans receive them. The same applies to the modalities.

Al-Ghazālī questions the assumption of an ontological coherence between this world and our knowledge of it. Certain predications—which, for Avicenna, apply to things in the real world—apply, for al-Ghazālī, only to human judgments. Al-Ghazālī's position can be clarified from the final sentences of the Third Position of the seventeenth discussion. Here al-Ghazālī makes the point that when we see a person acting orderly without a tremor or other freak movements, we cannot help assuming that the person has control over his or her movements. The orderly movements of a person lead to (*ḥaṣala*) the knowledge about his or her control. This connection, however, cannot be made solely from sensory perceptions. According to al-Ghazālī, our judgment that "the person is in control of the movements" is already understood from our observation of the orderly movements. This implication follows from how God has created the human mind:

> These are cognitions (*ʿulūm*) that God creates according to the habitual course [of events], by which we know the existence of one of the two alternatives [namely the person's control or non-control over his or her movements] but by which the impossibility of the other alternative is not shown (. . .).[83]

Neither the sheer fact of the orderly movement nor our perception of it can create our judgment that the person is in control of his or her body. Even the fact that there are only two mutually exclusive alternatives ("in control" and "not in control") can be inferred neither from the world nor from our visual perception of the orderly movement. These predicates do not exist in the outside world; rather, they are names that we connect to certain sensual perceptions. Reality itself does not guaranty its own intelligibility.[84] Our understanding of the world relies on parameters that are not part of the world's formal structure. Saying that these parameters are—like all human cognitions (*ʿulūm*)—God's creations and that God produces our knowledge about the person's control by creating such categories in our mind only means that we cannot expect to understand the world by simply looking at it and studying its ontological structure.

Al-Ghazālī was particularly unsatisfied with the *falāsifa*'s use of the modalities, as he makes clear in the first discussion of the *Incoherence* on pre-eternity of the world. Here al-Ghazālī rebuffs two arguments that stem from the implications of saying that something is possible. In the third argument of the first discussion, the philosophical opponent claims that the existence of the world is and has always been possible because the world cannot change from a state of impossibility into a state of possibility. Since the world's possibility has no beginning, it is eternally possible.[85] In other parts of *Incoherence*, al-Ghazālī denies that the world *can* be eternal. Based on arguments first proposed by John Philoponus (d. *c.* 570 CE), he says elsewhere in this book that it is impossible for the world to be pre-eternal because an action (*fiʿl*) must have a temporal beginning.[86] What did the opponent mean, however, when he said that the world's existence has always been possible? Al-Ghazālī does not object to this particular statement. Considered just by itself, he says at the end of the discussion, the statement that the creation of the world was possible at any time before or after its actual creation is true. In that sense, the world is eternally possible.[87]

However, that is not how the opponent understands the sentence: "The world is always possible to exist" (*lam yazal al-ʿālam mumkin*^{an} *wujūduhu*). The difference between the two readings of this sentence can be explained by using what became known in the Latin West as the *de re* and *de dicto* distinctions of modality. Later Arab logicians would refer to this distinction as the *dhātī* and the *waṣfī* readings of modal sentences. The distinction goes back to Aristotle's *Sophistic Refutations*.[88] When we say it is possible for the world to always exist, one way to understand the sentence is to attribute possible truth to the proposition "the world exists always" (*lam yazal wujūd al-ʿālam*).[89] This seems to be what the *falāsifa* are doing when they make their point that the existence of the world has always been possible. Here, a predication or proposition (*dictum/waṣf*) is considered possibly true. For al-Ghazālī, this *de dicto/waṣfī* interpretation of possibility is unacceptable in this context because, for him, that sentence can never be true. If it can never be true, the sentence cannot be seen as possibly true. However, we may mean to attribute to the world the possibility of having always existed, that is, at any given time before or after its actual creation. Here the predicate "exist" is attached in a possible predication to the thing (*res/dhāt*), that is, the world. This proposition does not require the world to be eternal; it is true as long as the world could have come into existence at any time other than it actually did. This is what al-Ghazālī stresses in his objection to the *falāsifa*'s third proof:

> The world is such that it is eternally possible for it to be temporally originated. No doubt then that there is no [single] moment of time but wherein its creation could not but be conceived. But if it is supposed to exist eternally, then it would not be temporally originated. The factual then would not be in conformity with possibility, but contrary to it.[90]

Regarded by itself, al-Ghazālī considers the statement "The world is always possible to exist" as true. Yet he reads it *de re* or *dhātī* and rejects the competing *de dicto/waṣfī* interpretation of the statement. The distinction of modal statements into these two readings is not prominently represented in Avicenna's logical works.[91] Some interpreters believe that Avicenna did not apply the distinction at all. The third argument that al-Ghazālī objects to in the first discussion about the world's pre-eternity is thus probably not from the works of Avicenna.[92] From a discussion in a later work, it becomes clear that al-Ghazālī understood the difference between the *de re/dhātī* and *de dicto/waṣfī* meaning of modal statements. In that later work, such as in this example, he was willing to understand modal statements *de re/dhātī* rather than *de dicto/waṣfī*.[93]

Al-Ghazālī's irritation with the *falāsifa*'s treatment of modalities becomes clearer in the next passage of the *Incoherence* in which al-Ghazālī's criticism is more radical. In two articles published in 2000 and 2001, Taneli Kukkonen and Blake D. Dutton examine al-Ghazālī's interpretation of modal terms in the *Incoherence*.[94] Both focus on al-Ghazālī's response to the philosophers' fourth proof for the eternity of the world, which is also debated in the first discussion of the *Incoherence*. Again, the *falāsifa* try to prove the pre-eternity of the world from the fact that it has always been possible. This time the argument that al-Ghazālī addresses comes from Avicenna. It is based on the premise that possibility cannot be self-subsistent but requires a substrate (*maḥall*) in which to inhere.[95] Following Aristotle's argument, Avicenna says that this substrate is the *hylé*, the prime matter that exists eternally. Its receptivity to the forms makes it the substrate of the world's possibility. Thus, the fact that the world is eternally possible proves that the substrate of this possibility, which is prime matter, must exist eternally.[96]

In his response, al-Ghazālī denies the premise that possibility needs a substrate. Possibility does not exist in the outside world; rather, it is merely a judgment of the mind:

> The possibility which they mention reverts to a judgment of the mind (*qaḍā l-ʿaql*). Anything whose existence the mind supposes, [nothing] preventing its supposing it possible, we call "possible," and if it is prevented we call it "impossible." If [the mind] is unable to suppose its nonexistence, we name it "necessary." For these are rational propositions (*qaḍāyā ʿaqliyya*) that do not require an existent so as to be rendered a description thereof.[97]

Al-Ghazālī repeats this argument in the nineteenth discussion, in which Avicenna claims that the possibility of perishing (*imkān al-ʿadam*) can only subsist in matter and that purely immaterial beings such as human souls are incorruptible. If that were true, al-Ghazālī says, it would imply that a thing could be simultaneously potential and actual with regard to a certain predicate. Affirming both the potentiality and the actuality of a given predicate is a contradiction, al-Ghazālī objects. As long as a thing is potentially something, it cannot be the same thing in actuality. At the root of the problem, al-Ghazālī says, is Avicenna's view that possibility (*imkān*) requires a material substrate in which to subsist. This substrate is not required, al-Ghazālī maintains, since when we talk about possibility we make no distinction whether it were to apply to a material substance or to an immaterial one such as the human soul.[98]

As Kukkonen puts it, al-Ghazālī shifts the locus of the presumption of a thing's actual existence from the plane of the actualized reality to the plane of mental conceivability.[99] The domain of possibility is not part of what actually exists in the outside world, al-Ghazālī argues. These modalities are like universal concepts, and like the universals such as color or like the judgment that all animals have a soul, for instance, their existence is in the mind only. The outside world consists of individual objects, and these individuals cannot be the objects of our universal knowledge. The universals are abstracted from the individual objects that we perceive. "What exists in the outside world (*fī l-aʿyān*) are individual particulars that are perceptible in our senses (*maḥsūsa*) and not in our mind (*maʿqūla*)."[100] Like the universal concept of "being a color" (*lawniyya*) that we cannot find anywhere in the outside world, the predicates "possible," "impossible," and "necessary" do not apply to objects outside of our mind. Al-Ghazālī takes a nominalist position with regard to the modalities and argues that modal judgments are abstract notions that our minds develop on the basis of sense perception.[101]

In his objection to Avicenna's conception of the modalities, al-Ghazālī makes innovative use of Ash'arite ontological principles.[102] When the Ash'arites denied the existence of natures, they rejected the limitations that come with the Aristotelian theory of entelechy. Viewing things as the carriers of possibilities that are bound to be actualized restricts the way these things may exist in the future. These restrictions unduly limit God's omnipotence, the Ash'arites say; and as long as things are regarded by themselves, the possibilities of how they exist are limited only by our mental conceivability. Additionally, when Ash'arites talk about something that exists, they mean something that can be affirmed (*athbata*).[103] To claim that there presently exists in a thing an inactive capacity to be different from how it presently is—meaning that there exists such a possibility in that thing—is really to say that there presently exists something that does not exist.[104] This is a contradiction, and Ash'arites subsequently denied the existence of nonactive capacities: existence is always actual existence.[105] This is why Ash'arites refused to acknowledge the existence of natures that determine how things react to given situations. Natures are, in essence, such nonactive capacities. In the course of this study, it will become clear that the status of modalities marks an important crossroads between Avicenna and al-Ghazālī that determines their positions on ontology. Al-Ghazālī's philosophical shift stems from a background in *kalām* literature, a change that merits closer look.

The Different Conceptions of the Modalities in *falsafa* and *kalām*

Ancient Greek philosophy used and distinguished several different modal paradigms, but none included the view of synchronic alternatives. Our modern view of modalities is that of synchronic alternative states of affairs. In that model, "[t]he notion of logical necessity refers to what obtains in all alternatives, the notion of possibility refers to what obtains at least in one alternative, and that which is logically impossible does not obtain in any conceivable state of affairs."[106] In contrast, Aristotle's modal theory has been described as a statistical interpretation of modal concepts as applied to temporal indefinite sentences. To explain a temporally unqualified sentence of the form "S is P" contains an implicit or explicit reference to the time of utterance as part of its meaning. If this sentence is true whenever uttered, it is necessarily true. If its truth-value can change in the course of time, it is possible. If such a sentence is false whenever uttered, it is impossible.[107] Simo Knuuttila clarifies that in ancient Greek, modal terms were understood to refer to the one and only historical world of ours, and "it was commonly thought that all generic types of possibility had to prove their mettle through actualization."[108]

Avicenna's view of the modalities is not significantly different from the statistical model of Aristotle that connects the possibility of a thing to its temporal actuality.[109] Here he followed al-Fārābī, who teaches that the word "possible" or, to be more precise, "contingent" (*mumkin*)[110] is best applied to what is in a state of nonexistence in the present and stands ready either to exist or not to exist (*yatahayyi'u an yūjada wa an lā yūjada*) at any moment in the future.[111]

Avicenna shares this temporal attitude toward the modalities: the necessary is what holds always, and the contingent is what neither holds always nor holds never.[112] This position, which represents mainstream Aristotelianism, seems to imply that something has to exist at one point in time in order to be possible. For Avicenna, however, "what neither holds always nor holds never" refers to predications about things in the outside world as well as those that exist only in the mind. The "heptagonal house" (*al-bayt al-musabbaʿ*), for instance, may never exist in the outside word but will at one point in time exist in a human mind and is therefore a possible being.[113] For Avicenna, the principle of plentitude is valid for existence in the mind (*fī l-dhihn*) but not for existence *in re* (*fī l-aʿyān*), that is, in the outside world. It is contingent that some houses, or all houses, are heptagonal, since the combination of "house" and "heptagonal" is neither necessary nor impossible. Here Avicenna clearly divorces modality from time. The possibility of a thing is not understood in terms of its actual existence in the future but in terms of its mental conceivability.[114] By acknowledging that some beings such as the chiliagon—a polygon with so many sides that it cannot be distinguished from a circle—exist in the mind but will probably never exist in the outside world, Avicenna recognizes possibilities that are never actualized *in re*.[115] To say that "all animals are humans" is a contingent proposition because we can imagine a time in which there is no animal but man, in spite of the fact that such a time probably never existed *in re*.[116] The contingency of the proposition is not verified by the future or past existence of a certain state of affairs *in re* but rather through a mental process, namely, whether such a state can be imagined to exist without contradictions.[117] The phrase, "all white things," may have two different meanings according to the context in which it is uttered. It may refer to all beings that are white at the particular time when the statement is made or to those possible beings that are always described as being white every time they appear in the mind (*ʿinda l-ʿaql*).[118]

In principle, Avicenna does not part with the Aristotelian statistical understanding of the modalities. In order to be possible, something must exist for at least one moment in the past or future. Mental existence (*al-wujūd fī-l-dhihn*), however, is one of the two modes of existence in Avicenna's ontology. Whether something exists in our minds depends upon whether it is the subject of a predication. There is no ontological difference between whether a thing exists in reality or merely in the human mind.[119]

Avicenna's understanding of existence is significantly different from that of his predecessors. Al-Fārābī, for instance, followed Aristotle and taught that predication itself includes no statement of existence. When one states that "Socrates is just," it need not follow that Socrates is existent. Avicenna disagreed because the nonexistent cannot be the subject of a predication; any predication gives mental existence to Socrates.[120] Allowing two modes of existence and accepting mental existence as equal to existence *in re* leads Avicenna to develop an understanding of possibility as that which is actually conceived in the mind (*maʿqūl bi-l-fiʿl*).[121] Any possible subject of a true predication is a possible being. This dovetails with Avicenna's view that what is possible by itself (*mumkin bi-dhātihi*) is determined on the level of the quiddities (*māhiyyāt*). The quiddities have three modes: in themselves, in individuals (*fī aʿyān al-ashyāʾ*), and as singular objects of thought (*fī l-taṣawwur*).[122] In themselves, the quiddities are in a state prior to existence and are pure possibility by themselves; the moment a quiddity is conceived in the human mind, it is given existence. When the mind proceeds to another thought, the thing just pondered or imagined falls from existence. This example highlights that for Avicenna, the concepts of possibility and existence are closely connected. Possibility is what can be existent at any moment in our mind, and existence is actualized possibility either *in re* or in the mind. The modalities can, therefore, also be expressed as simple modes of existence: necessary is what cannot but exist; possible (or rather: contingent) is what can exist but must not exist; impossible is what cannot exist. In each of the three modes, existence is understood as being either *in re* or in the mind, although in most contexts it is both. For Avicenna, the division between necessary and contingent is one of the prime divisions of being that is known as *a priori*.[123] Although strictly speaking, this is still a temporal understanding of the modalities, it puts the modalities on the plane of mental conceivability. For all practical matters, the modalities are not connected to existence in time but to existence in the mind (*fī l-dhihn*).[124]

Avicenna took an important step toward understanding possibility as a synchronic alternative state of affairs. He himself never achieved such an understanding, however, because in his ontology, there can be no alternatives to what actually exists. We have already said that Avicenna's metaphysics was necessitarian, meaning that whatever exists either in the outside world or in the human mind is the necessary result of God's essence.[125] In chapter nine of *De interpretatione,* Aristotle had already argued that what presently exists can be defined as necessary: what is, is by necessity. Avicenna applies the distinction—known to us from al-Fārābī's commentary on this section of *De interpretatione*—between the modal status a being has by itself and its modal status as coexisting with other things. By itself, there is only one being that is necessary by virtue of itself (*wājib al-wujūd bi-dhātihi*), and that is God. This being cannot but exist. Considered by themselves, all other beings are merely possible (*mumkin al-wujūd bi-dhātihi*); God's creative activity, however, makes the existence of these beings necessary. Once a thing that is only possible by virtue of itself comes into being, it is necessary by virtue of something else (*wājib al-wujūd bi-ghayrihi*). It is, first of all, the necessary effect of its proximate efficient cause. That cause, however, is itself the necessary effect of other efficient causes, which proceed in a chain of secondary efficient causes from God. Everything that we witness in creation is possible by virtue of itself and necessary by virtue of something else, ultimately necessitated by God.[126]

In the Western philosophical tradition, in which Avicenna became an influential contributor after the translation of his works into Latin during the thirteenth century, the introduction of the synchronic conception of modality is credited to John Duns Scotus (d. 1308). An avid reader of Avicenna, Duns Scotus claimed that the domain of the possible is an *a priori* area of what is intelligible and as such does not have any kind of existence in the outside world. Among his successors in Latin philosophy, this led to a view in which modality lacks an essential connection with time. This disconnect allowed for alternative possibilities at any given time, as well as the development of a notion of possible words, some of them not actualized.

John Duns Scotus, however, was not "the first ever" to employ a synchronic conception of modality, as some Western historians of philosophy assume.[127] Such a view had already been developed in Ash'arite *kalām*. The notion of God as a particularizing agent (*mukhaṣiṣ*), who determines, for instance, when the things come into existence, is an idea that appears in the writings of al-Bāqillānī and of other Ash'arite authors.[128] The idea of particularization (*takhṣīṣ*) implicitly includes an understanding of possible worlds that are different from ours. The process of particularization actualizes a given one of several alternatives. Yet the alternatives to this world—which would be: "X comes into existence at a time different from when X actually comes into existence"—are not explicitly expressed or even imagined. The *kalām* concept of preponderance (*tarjīḥ*), however, explicitly discusses the assumption of possible worlds. The preponderator distinguishes the actual state of being from its possible alternative state of nonbeing. Whereas it is equally possible for a given future contingency to either exist or not exist, each time a future contingency becomes actual, the preponderator decides between an actual world and an alternative world in which that particular contingency is nonexistent. In *kalām*, the idea of preponderance (*tarjīḥ*) already appears in the work of the Mu'tazilite Abū l-Ḥusayn al-Baṣrī in the context of human actions.[129] Abū l-Ḥusayn was a younger contemporary of Avicenna, and he had received a philosophical education. He also developed a particularization argument for the existence of God.[130] Based on these developments within *kalām*, al-Juwaynī was the first Ash'arite who developed a stringent argument for God's existence based on the principle of particularization.[131] In his *Balanced Book* in the *Letter for Jerusalem*, and in his *Scandals of the Esoterics*, al-Ghazālī reproduces versions of this proof. Al-Ghazālī's versions contain strong overtones of Avicenna's ontology: because everything in the world can be perceived as nonexisting, its nonexistence is by itself equally possible as its existence. Existing things necessarily need something that "tips the scales" (*yurajjiḥu*) or preponderates between the two equally possible alternatives of being and nonbeing. God is this "preponderator" (*murajjiḥ*), who in this sense determines the existence of everything that exists in the world.[132]

Avicenna's view of modalities does not break with Aristotle's statistical model, yet he postulates possibility as mental conceivability, thus taking a step toward an understanding of possibility as a synchronic alternative state of affairs. We see one element of such a synchronic alternative in Avicenna's describing God as the "preponderator" (*murajjiḥ*) between the existence of a thing and its alternative of nonexistence. Avicenna's ontology of quiddities, wherein existence depends on a separate act of coming-to-be, fosters the idea of God as a preponderator between being and nonbeing. In Avicenna's major work, *The Healing*, however, the word "preponderance" (*tarjīḥ*) and its derivates do not appear that often. It is much more prominent in one of Avicenna's early treatises on divine attributes. This small work, *Throne Philosophy* (*al-Ḥikma al-'arshiyya*), made a significant impression on al-Ghazālī. When he reports Avicenna's teachings on this subject, for instance, he stresses the idea of preponderance and follows Avicenna's language from his *Throne Philosophy* more than the language of *The Healing.*[133]

Even though the Ashʿarites readily embraced the concept of preponderance, they rejected Avicenna's understanding of the modalities. For al-Ghazālī, Avicenna's lack of distinction between existence in mind (*fī l-dhihn*) and existence in the outside world (*fī l-aʿyān*) removes an important difference: whether possibility and necessity exist in things outside of our mind, or whether they are simply predicates of our judgment. Al-Ghazālī's critique of Avicenna's understanding of the modalities was anticipated by al-Juwaynī's notion of necessity and possibility in his proof of God's existence in the *Creed for Niẓām al-Mulk.* Al-Juwaynī begins his argument there with an explanation of the modalities. Every sound thinking person finds within himself "the knowledge about the possibility of what is possible, the necessity of what is necessary, and the impossibility of what is impossible."[134] We know this distinction without having to study or make further inquiry into the world; it is an impulse (*badīha*) of our rational judgment (*ʿaql*).

> The impulsive possibility that the intellect rushes to apprehend without [any] consideration, thinking, or inquiry is what becomes evident to the intelligent person when he sees a building. This [*scil.* the building] is [simply] a possibility that comes into being (*min jawāz ḥudūthihi*). He knows decisively and offhand that the actual state (*ḥudūth*) of that building is from among its possible states (*jāʾizāt*) and that it is not impossible in the intellect that it had not been built.[135]

The intelligent person (*al-ʿāqil*, here meaning a person with full rational capacity) realizes that all of the features of the building—its height, its length, its form, and so forth—are actualized possibilities that could be different from what they are. The same possibilities apply to the time when the building is built. We immediately realize, al-Juwaynī says, that there is a synchronic alternative state to the actual building. This is what we call, contingency (*imkān*). Realizing that there is such an alternative is an important part of our understanding: "The intelligent person cannot realize in his mind anything about the states of the building other than through a comparison with what is contingent like it (*imkān mithlihi*) or what is different from it (*khilāfihi*)."[136]

Knowledge about the modalities is "on an impulsive rank" (*bi-l-martaba al-badīha*), meaning it is *a priori*: it cannot be derived from any other prior knowledge.[137] This statement is limited to the modalities when they are considered by themselves. Al-Juwaynī realizes that God's creative activity makes all the unrealized possibilities impossible. If considererd on its own, the actual movement of the celestial spheres (*aflāk*) from east to west could be imagined differently. The intellect can imagine that the spheres could move in the opposite direction. Studying the movements in heaven, however, leads to the realization that this possibility is not actualized.

Al-Juwaynī understands possibility as synchronic alternative states to what actually exists. This is different from Avicenna's understanding of possibility and necessity as modes of actualized beings. It also shifts the perspective of the modalities away from what exists in actuality toward what is considered alternative states in the human mind. Al-Ghazālī's critique of Avicenna's modal theory is in no way haphazard but is an outcome of long-standing consideration of modalities developed in Ashʿarite *kalām*.[138]

What Does al-Ghazālī Mean When He Claims That Causal Connections Are Not Necessary?

Once Avicenna's and al-Ghazālī's differing understandings of the modalities are applied to the initial statement of the seventeenth discussion, they change the established meaning of this passage. When al-Ghazālī says that "according to us (*ʿindanā*)," the connection between any given efficient cause and its effect is not necessary,[139] he aims to point out that the connection *could* be different, even if it never will be different from what it is today. For Avicenna, the fact that the connection never was different and never will be different implies that the connection is necessary. Not so for al-Ghazālī. His understanding of modal judgments does not require that any given causal connection was different or will be different in order to be considered possible and not necessary. The possible is that for which the human mind can perceive an alternative state of affairs. For al-Ghazālī, the connection between a cause and its effect is possible—or, to be more precise: contingent (*mumkin*)—because an alternative to it is conceivable in our minds. We can imagine a world in which fire does not cause cotton to combust. Or, to quote the second sentence of the initial statement of the seventeenth discussion:

> It is within divine power to create satiety without eating, to create death without a deep cut (*ḥazz*) in the neck, to continue life after having received a deep cut in the neck, and so on to all connected things. The *falāsifa* deny the possibility of [this] and claim it to be impossible.[140]

Of course, a world in which fire does not cause combustion in cotton would be radically different from the one in which we live. A change in a single causal connection would likely imply that many others would also change. Still, such a world can be conceived in our minds, which means it is a possible world. God, however, did not choose to create such an alternative possible world. He chose to create this world among alternatives.

In the initial sentence of the seventeenth discussion, al-Ghazālī argues against two types of adversaries. First, he argues against those who hold that a causal connection is necessary by itself. This group includes people who claim that any given proximate efficient cause is an independent efficient cause (*fāʿil*) of its effect. This group also includes some natural philosophers who reject secondary causality as well as the Muʿtazilites, who argue that humans create their actions and the immediate effects of them. Al-Ghazālī, however, makes a clear distinction between the teachings of the Muʿtazilite and those of the Avicennan *falāsifa*.[141] The Avicennan *falāsifa* are the second group of adversaries in the seventeenth discussion. Although al-Ghazālī does not argue against the idea of secondary causality in Avicenna, he does reject Avicenna's teaching that the connection cannot be any different from what it is. Being contingent by itself, according to Avicenna, the connection between cause and effect is necessary on account of something else, namely, God's nature. God's nature cannot be conceived any differently from what it is. For Avicenna, there can be no world alternative to the one that exists.

In the initial statement of the seventeenth discussion, al-Ghazālī also claims that "the connection [between cause and effect] is due to the prior decision (*taqdīr*) of God."[142] When he objects to Avicenna and states that these connections are not necessary, al-Ghazālī wishes to express that God could have chosen to create an alternative world in which the causal connections are different from those of this world. Al-Ghazālī upholds the contingency of the world against the necessitarianism of Avicenna. For al-Ghazālī, this world is the contingent effect of God's free will and His deliberate choice between alternative worlds.

While rejecting this necessitarian element in Avicenna's cosmology, al-Ghazālī does not object to the philosopher's concept of secondary causality. Of the two pillars in Avicenna's cosmology—secondary causality and necessitarianism—al-Ghazālī rejects only the latter. In the First Position of the seventeenth discussion, al-Ghazālī uses secondary causality to refute the view that proximate causes are independent efficient causes. In the Second Position, he offers two alternative explanations ("approaches") of prophetical miracles, the first based on occasionalism, the second, on secondary causality and the existence of natures (*ṭabāʾiʿ*). In all this discussion, al-Ghazālī says nothing about whether God actually breaks his habit, meaning the existent laws of nature, when creating the prophetical miracle. For al-Ghazālī, the connection between the cause and its effect is contingent even if God never changes His habits. The sole possibility of His breaking His habit—that we could conceive of God breaking His habit—or just the possibility that He could have arranged the laws of nature differently means that any individual connection between two of His creations is not necessary. Although it is conceivable and therefore possible that God would break his habit or intervene in the assigned function of the secondary causes, an actual break in God's habit is not required for the connections to be contingent.

7

Knowledge of Causal Connection Is Necessary

In the seventeenth discussion of the *Incoherence,* is there a consistent line of argument with regard to causality? After proposing his most radical epistemological criticism in the First Position—that sense perception does not lead to necessary judgments—al-Ghazālī presents in the Second and the Third Positions two alternatives to the Avicennan model of metaphysics and physics. In the First Approach of the Second Position, occasionalism is contrasted with the deterministic cosmology of his opponents. Al-Ghazālī aims to show that a congruent occasionalist model can be a viable alternative to Avicennan metaphysics. He implicitly claims that the *falāsifa* can accept this model and still continue to pursue the natural sciences. The "laws of nature" that, according to the *falāsifa,* govern God's creation may be understood as habitual courses of action subject to suspension, at least in principle. Our human experience, however, has shown us that God does not frivolously break His habit. This insight allows us to equate God's habit with the laws of nature, for all practical purposes. In the natural sciences, we study God's actions and reformulate their habitual course into laws that we justifiably consider, if not necessary, at least stable, unchanging, and permanent.

In the Third Position, al-Ghazālī puts up a far less radical alternative to Avicennan metaphysics and natural sciences. Although not clearly explicated, this theory appears to be a slightly altered version of Aristotelian physics. This physical theory postulates that in addition to the rules of logic, God cannot violate laws of nature that rely on the relationships of implications. Such implications are usually formulated in definitions. Will is defined as existing in a being that has knowledge, for instance, and knowledge is defined as existing in a being that has life. God therefore cannot create will in a being that is lifeless. Equally, God cannot "change the genera" (*qalb al-ajnās*), meaning that He cannot transform a material body into an immaterial being and vice versa. Al-Ghazālī was certainly aware that these three conditions limit God's omnipotence significantly. He here lists what can be viewed as the unchangeable essence of God's creation. And although the laws of nature from among this core group cannot be altered once creation unfolds, God reserves the power to alter others of His habits, such as making water flow uphill or creating life in any given material object, such as a stick.

These two alternative theories to Avicenna's cosmology frame a passage of roughly two pages, which, to the Avicennan, forms the most persuasive part of the seventeenth discussion. In addition to these two alternative cosmological theories (alternative to Avicenna's cosmology), al-Ghazālī defends a slightly modified Avicennan explanation of causal connections in the Second Approach of the Second Position. Here, al-Ghazālī is willing to accept that chains of secondary causes connect every event in creation with the creative activity of the creator. In this part of the seventeenth discussion he clearly accepts the existence of "natures" (*ṭabāʾiʿ*). He requires the Avicennan simply to acknowledge that we lack exhaustive knowledge of the full possibilities of these natures. They might allow causal connections that we have not yet witnessed. The miracles reported in revelation have causes unknown to us. They are not true miracles but mere marvels.

In the *Incoherence*, al-Ghazālī presents what might be called a nominalist criticism of the modalities, in some sense a criticism of human judgments as a whole. Using the parlance of Avicenna, al-Ghazālī implicitly asks whether we can know that any given object that we witness in the outside world is possible by itself (*mumkin bi-dhātihi*) and at the same time is necessitated by something else (*wājib bi-ghayrihi*). Al-Ghazālī rejects Avicenna's assumption that modalities exist in the outside world. This rejection goes to the heart of the Avicennan ontology that regards potentiality as a paradigm that strives to actualize itself. Like Avicenna, al-Ghazālī views human knowledge as a conglomerate of judgments.[1] He agrees with Avicenna that true knowledge is congruent to the outside world and describes it as such. For Avicenna, however, there can be only one true explanation of any given phenomenon in the world. True human knowledge describes the necessary and only way the world is constructed. Demonstration (*burhān*) is the best means to achieve such correct knowledge about the world. Where demonstration is not available, humans choose less perfect means of acquiring knowledge. Al-Ghazālī agrees with Avicenna on the imperfect nature of these means. He realizes, however, that where demonstration cannot be achieved, multiple explanations are compossible, that is one explanation may coexist with another without needing to decide which applies. The inability to demonstrate the unchanging nature of the connection between cause and effect creates a situation in which more than one explanation of causal connections is viable. Only a nominalist position toward human knowledge allows the assumption of two different explanations of a given process as compossible.

Al-Ghazālī's nominalist critique of Avicenna is an important element in the understanding of his cosmology. We must point out that al-Ghazālī was not a nominalist in the sense of his contemporary Roscelin (d. *c.* 1120) or William of Ockham (d. 1347) in the Latin West.[2] These nominalists outspokenly denied any ontological coherence between things and their formal (and universal) representations in our minds. In the Latin dispute about the status of universals—a dispute that lasted from the late thirteenth to the end of the fourteenth centuries—the nominalist criticism was directed against the Aristotelian claim of an eternal and invariant formal level of being that shapes both the individual things in the outside world as well as our knowledge of them. This position, which is known as epistemological realism, essentially maintains that individual things are what they are because of real existing universals. The consistency of our knowledge with the outside world is due to the ontological coherence between the two. Human souls have access to these universals, and their apprehension constitutes our knowledge. In the Latin West, Avicenna was one of the most important proponents of the realist position.

In the Muslim East, the parameters of the dispute on the status of universals were different. Here, the nominalist criticism of Avicenna developed from Ashʿarite occasionalism, as in the case of al-Ghazālī. Yet nominalist positions were not unknown within the discourse of *falsafa* in the East. Justifying his position that the modalities exist only in minds and not in the outside world, al-Ghazālī cites a moderate nominalist view toward human knowledge that were current among the *falāsifa*. He tries to persuade his philosophical readers to accept his position on the modalities by comparing them to universals. According to views held by the *falāsifa* themselves, al-Ghazālī continues, the universals are just concepts in the mind without referring objects (*maʿlūmāt*) in the outside world. The universals do not exist in the outside world:

> What exists in the outside world (*fī l-aʿyān*) are individual particulars that we perceive with our senses and not in our mind. But they are (only) the cause; because the mind abstracts from them intellectual judgments that are empty of matter. Therefore being a color (*lawniyya*) is a single judgment (*qaḍiya*) in the mind (*ʿaql*) similar to blackness or whiteness. One cannot conceive that there exists a color that is neither black nor white nor any other of the colors. In the mind there exists the form of "being a color" without any details; and one says it is a form and it exists in the minds and not in the outside world.[3]

The position referred to here needs not be that of a nominalist. Avicenna himself taught that the perception of individual objects cannot lead to universal judgments.[4] Although admitting that universals have no existence in matter, the Avicennan opponent still holds that they exist in a real and immaterial way in the active intellect, outside of the human mind. Al-Ghazālī uses this argument, however, to advance a distinctly nominalist critique of the position that modalities exist outside of the human mind. We will later see how al-Ghazālī made productive use of some nominalist tendencies within Avicenna's œuvre.[5]

In the methodological introduction to *The Highest Goal in Explaining the Beautiful Names of God*, al-Ghazālī develops a distinctly nominalist theory of semantic relations that combines Ashʿarite notions with philosophical distinctions.[6] It is also apparent, however, that the influence of Avicenna's realist epistemology on him was so strong that he often applies to his own writings a realist concept of the universals.[7] What distinguishes al-Ghazālī from Avicenna, as we will see in the course of this study, is that he remained ontologically uncommitted to the existence of the universals outside of individual human minds. Although the universals may exist as entities in the active intellect, such an existence cannot be demonstrated. The realist understanding of the universals may or may not be true. In the Second Approach of the seventeenth discussion, he counters the realist position with the occasionalist position that human cognitions are the immediate creations of God and are only congruent with the outside world if God wills it.

Some of al-Ghazālī's criticism in his *Incoherence of the Philosophers* centers on questioning the ontological connection between the formal structure of the world and the formal structure of our knowledge. Averroes (d. 595/1198), for instance, who shared Avicenna's realist epistemology, was surprised by al-Ghazālī's effort to defend an occasionalist position with the argument that human knowledge may become disconnected from the world it aims to describe. That cannot be the case, Averroes says, "because the knowledge created in us is always in conformity with the nature of the real thing, since the definition of truth is that a thing is believed to be such as it is in reality."[8] Yet this conformity (*tabaʿ*) is precisely what al-Ghazālī argues against. Since there is no proof of the necessity of the connection between a cause and its effect, there is also no proof of the necessary conformity of our knowledge with the world. The mere possibility of a disconnect between the two proves that there is no formal—and thus necessary—coherence between the world and our knowledge of it.

In a later passage of the *Incoherence*, al-Ghazālī comments on what he does in the seventeenth discussion. This comment appears in the twentieth discussion of the book, on the subject of corporeal resurrection in the afterlife. The *falāsifa* argue that a resurrection of bodies is impossible, as it necessitates the impossible feat of transformation of substances, such as iron transforming into a garment. In his response, al-Ghazālī refers his readers back to the Second Approach of the Second Position in the seventheenth discussion, in which he claims to have already discussed this problem. He argues that the unusually rapid recycling of the matter of the piece of iron into a piece of garment is not impossible. In the Second Approach of the Second Position, al-Ghazālī had argued that the matter that makes up a piece a wood may change in other than its known and usual way from a stick into a serpent. "But this is not the point at issue here," al-Ghazālī continues; the real question is whether such a transformation "occurs purely through [divine] power without an intermediary, or through one of the causes."[9] The question cannot be put more bluntly: does God create such transformations mono-causally—in accord with an occasionalist worldview—or by means of secondary causality?

> Both these two views are possible for us (*kilāhumā mumkinān ʿindanā*) (. . .) [In the seventeenth discussion we stated] that the

> connection of connected things in existence is not by way of necessity but through habitual events, which can be disrupted. Thus, these events come about through the power of God without the existence of their causes. The second [view] is that we say: This is due to causes, but it is not a condition that the cause [here] would be one that is well-known (*maʿhūd*). Rather, in the treasury of things that are enacted by [God's] power there are wondrous and strange things, one hasn't come across. These are denied by someone who thinks that only those things exists that he experiences similar to people who deny magic, sorcery, the talismanic arts, [prophetic] miracles, and the wondrous deeds [done by saints].[10]

The solution al-Ghazālī chose in the seventeenth discussion of his *Incoherence* is thorough and well reasoned, and we will discuss many of its implications in this chapter. One realizes how carefully al-Ghazālī had crafted and considered this position when one sees that al-Ghazālī maintained this position throughout all his later works. All through his life al-Ghazālī remained ultimately undecided as to whether God creates mono-causally and arranges directly in each moment all elements of His creation, or whether God mediates His creative activity by means of secondary causes. Al-Ghazālī accepted both explanations as viable explanations of cosmology.

The Dispute over al-Ghazālī's Cosmology

In a 1988 article, Binjamin Abrahamov attempted to determine al-Ghazālī's position on causality in works written after the *Incoherence of the Philosophers*. Given that the *Incoherence* is a work of refutation in which the author himself admits that his arguments may not represent his real opinion,[11] Abrahamov assessed al-Ghazālī's teachings from works considered closer to his actual teachings. These works include *The Revival of the Religious Sciences, The Book of the Forty*, and al-Ghazālī's commentary on the Ninety-Nine Noble Names. Abrahamov concluded that in these three works, al-Ghazālī uses language that assumes that causes *do* have efficacy on other things. To be sure, it is God who creates the causes and maintains and regulates their influences. Yet in these works, al-Ghazālī suggests that the influence of causes is indeed real and not just an illusion. Once put into place, the causes lead to effects that are themselves desired by God. Abrahamov also noted that in a fourth work of al-Ghazālī, *The Balanced Book on What-To-Believe*, the author uses language that is distinctly occasionalist. Here he maintains that God should be regarded as the immediate creator of each individual event and that if He so wished, He could break His habitual patterns of creation and suspend what we postulate as the laws that govern creation. Given that those works implying a causal theory were written after *The Balanced Book*, Abrahamov suggests that al-Ghazālī changed his mind "but preferred to conceal his true opinion by contradicting himself."[12] In this analysis, Abrahamov follows Leo Strauss in his exegesis of Maimonides

(d. 601/1204). Strauss claimed that when medieval authors such as Maimonides use "conscious and intentional contradictions, hidden from the vulgar," they wished to compel their readers "to take pains to find out the actual meaning," which was often the one that appears least frequently in their writings.[13]

The apparent contradiction observed by Abrahamov had been earlier noted by W. H. T. Gairdner in a 1914 article. Gairdner observed that whereas in some of his works, al-Ghazālī explains God's creative activity by means of secondary causality, creation mediated by other created beings, in other works, he employs explanations that are distinctly occasionalist. Gairdner suggested that al-Ghazālī had published two different sets of teachings, one in works written for the ordinary people (*ʿawāmm*) and a different set of teachings in works that were written for an intellectual elite (*khawāṣṣ*). Whether al-Ghazālī considered these two teachings to be equally true was for Gairdner the "Ghazālī problem."[14] Gairdner supported his view with quotations from Ibn Ṭufayl (d. 581/1185–86) and Averroes, claiming that they had been bothered by the very same problem. Gairdner's article encouraged the widespread assumption in twentieth-century research that in works such as *The Niche of Lights*, al-Ghazālī taught an "esoteric" theology, while in works such as his autobiography or *The Balanced Book*, he accommodated his teachings to the expectation of the target audience and taught occasionalism.[15]

In 1992, Richard M. Frank presented the most thorough study of al-Ghazālī's cosmology to date.[16] Like Abrahamov, Frank bases the bulk of his analysis on the works *The Highest Goal in Explaining the Beautiful Names of God*, *The Book of Forty*, and several books of the *Revival*. Frank also includes *The Niche of Lights*, *Restraining the Ordinary People from the Science of Kalām*, and *The Balanced Book on What-to-Believe*, and was thus able to cover almost the whole Ghazalian corpus. Frank claims that contrary to common opinion, al-Ghazālī teaches (1) that the universe is a closed, deterministic system of secondary causes whose operation is governed by the first created being, an "angel" (or "intellect") associated with the outermost sphere; (2) that God cannot intervene in the operation of secondary causes, celestial or sublunary; and (3) that it is impossible that God has willed to create a universe in any respect different from this one He has created.[17] God governs the universe through intermediaries, and He cannot disrupt the operation of these secondary causes. Frank concluded that whereas al-Ghazālī rejected the emanationism of al-Fārābī and Avicenna, for instance, his own cosmology is almost identical to that of Avicenna. Earlier contributions to the academic debate, Frank points out, had already established that al-Ghazālī accepted some of Avicenna's teachings while rejecting others: "What we have seen on a closer examination of what [al-Ghazālī] has to say concerning God's relation to the cosmos as its creator, however, reveals that from a theological standpoint most of the theses which he rejected are relatively tame and inconsequential compared to some of those in which he follows the philosopher."[18]

Unlike Gairdner or Abrahamov, Frank does not propose that al-Ghazālī presents two different kinds of teachings in different works. He rejects the division of al-Ghazālī's works into esoteric and exoteric.[19] Al-Ghazālī's views on causality in *The Balanced Book on What-to-Believe*, for instance, do not differ from those in his commentary on God's Ninety-Nine Noble Names or in *The Niche of Lights*. Frank implicitly acknowledges that al-Ghazālī used both causalist and occasionalist language in his works. The contradictions that were noted by earlier readers, however, exist only on the level of language and do not reflect substantive differences in thought. When al-Ghazālī uses occasionalist language, Frank claims, he subtly alters the traditionalist language of the Ashʿarite school, making it clear that he does not subscribe to its teachings. Thus, although al-Ghazālī's language in such works as *The Balanced Book* often reflects that of the traditionalist Ashʿarite manuals, his teachings even in that work express creation by means of secondary causality.[20]

Frank's ideas were not unopposed. Michael E. Marmura in particular, who in a number of earlier articles had argued that al-Ghazālī was an occasionalist,[21] rejected the suggestion that al-Ghazālī accepted efficient causality among God's creatures.[22] Other interpreters such as William L. Craig had followed Marmura in their analysis and had maintained that al-Ghazālī "did not believe in the efficacy of secondary causes."[23] Reacting to Frank's suggestion, Marmura conceded that al-Ghazālī makes use of causalist language, "sometimes in the way it is used in ordinary Arabic, sometimes in a more specifically Avicennian / Aristotelian way" and that this usage of language is innovative for the Ashʿarite school discourse.[24] Yet in all major points of Muslim theology, al-Ghazālī held positions that closely followed ones developed earlier by Ashʿarite scholars, such as the possibility of miracles, the creation of human acts, and God's freedom in all matters concerning the creation of the universe.[25] In Marmura's view, al-Ghazālī never deviated from occasionalism, although he sometimes expressed his opinions in ambiguous language that mocked philosophical parlance, likely to lure followers of *falsafa* into the Ashʿarite occasionalist camp.

Marmura does not assume that al-Ghazālī expressed different opinions about his cosmology in different works. In research published since Frank's 1992 study, Marmura focuses on *The Balanced Book* and tries to prove that at least here, al-Ghazālī expresses unambiguously occasionalist positions.[26] Using a passage in the *Incoherence*, Marmura assumes this work to be the "sequel" to that work of refutation, in which al-Ghazālī "affirms the true doctrine."[27] For Marmura, the *Balanced Book* is thus the most authoritative work among al-Ghazālī's writings on theology. Like Frank, he claims that a close reading of all of al-Ghazālī's texts will find no contradictions on the subject of cosmology. Marmura acknowledges that al-Ghazālī uses causalist language that ascribes agency to created objects in the *Revival*, in the *Incoherence*, in the *Standard of Knowledge*, and in other works. Yet such language is used metaphorically, just as we might say "fire kills" without assuming that it has such agency in real terms.[28] Rather, the causal language must be read in occasionalist terms.[29] Al-Ghazālī's use of such words as "cause" (*sabab*) or "generation" (*tawallud*) is only metaphorical, Marmura claims. These terms are commonly used in Arabic, and "it would be cumbersome to have to keep on saying that this is metaphorical usage, or that the reference is to habitual causes and so on."[30] Like Frank,

Marmura is aware of the significant extent to which Avicenna's thought has shaped al-Ghazālī's theology. Marmura sees in al-Ghazālī "a turning point in the history of the Ash'arite school of dogmatic theology (*kalām*)."[31] He adopts many of Avicenna's ideas and reinterprets them in Ash'arite terms. Although al-Ghazālī's exposition of causal connections often draws on Avicenna, the doctrine that he defends is Ash'arite occasionalism.[32]

Both Frank and Marmura deny the possibility that al-Ghazālī showed any uncertainty or may have been in any way agnostic about which of the two competing cosmological theories is true.[33] Frank bemoans al-Ghazālī's failure to compose a complete, systematic summary of his theology.[34] He also believes that there was no notable theoretical development or evolution in al-Ghazālī's theology between his earliest works and his last. This theology is the one Frank had characterized in his *Creation and the Cosmic System*, and it is, in Frank's view, "fundamentally incompatible with the traditional teaching of the Ash'arite school."[35] Rejecting this last conclusion, Marmura does agree that al-Ghazālī held only one doctrine on cosmology and causation. Marmura discusses the passage from the twentieth discussion in the *Incoherence* where al-Ghazālī admits that "both these two views are possible for us."[36] Marmura argued that the evidence from texts such as *The Balanced Book on What-to-Believe* and some textual expressions in the *Incoherence* lead to the assumption that al-Ghazālī was committed only to his first causal theory from the Second Position of the seventeenth discussion, the occasionalist one. The "second causal theory"—that is, the one from the Second Approach of the Second Position, which accepts the existence of natures and assumes that causal relations are not suspended when God creates the miracles—has been introduced merely to win the argument that all miracles reported in revelation are possible; al-Ghazālī was not committed to it.[37]

Recently Jon McGinnis proposed an explanation that reconciles the textual evidence provided by Frank and Marmura to support their mutually exclusive claims. McGinnis believes that al-Ghazālī developed an intermediate position between traditional Ash'arite occasionalism and the *falāsifa*'s theory of efficient causality. For al-Ghazālī, causal processes exist, according to McGinnis, but they are immediately dependent upon a divine, or at least angelic, volitional act. A cause is only sufficient for its effect to occur, according to McGinnis's interpretation of al-Ghazālī, when such a higher volitional act immediately actualizes the cause. Cause and effect react to what might be understood as their natures—thus allowing humans to predict their reactions—but these natures are only passive powers that do not develop any agency or efficient causality by themselves. God or a volitional agent must actualize their passive powers. This volitional agent is the real agent or efficient cause of the causal connection. The actualization is immediate and cannot be mediated by a chain of secondary causes, for instance. According to McGinnis, al-Ghazālī rejected both the occasionalist position of classical Ash'arism as well as the secondary causality of the *falāsifa* and developed a third view that combines elements of these two.[38]

Five Conditions for Cosmological Explanations in the *Incoherence*

When Michael E. Marmura considered the suggestion that al-Ghazālī might actually have held two different explanations of cosmology as compossible, he saw "no compelling reason or textual indication for believing that he is committing the error of thinking that they are."[39] Occasionalism and secondary causality are mutually exclusive, Marmura argues; one denies causal efficacy while the other affirms it. Assuming compossibility in this case, however, does not assume that an event is caused both by an inner-worldly efficient cause *and also* immediately by God. Rather it means—as al-Ghazālī has put it several times in the seventeenth discussion of the *Incoherence*—that God is the creator of the event "either through the mediation of the angels or without mediation."[40] Although God's control over all events in this world is unquestioned, the way He exerts this control is left open.

Still, one might ask, given that occasionalism and secondary causality are so different, how could al-Ghazālī posit that they offer equally convincing theories of God's creative activity? In his *Incoherence,* al-Ghazālī developed certain conditions with which any occasionalist and causalist theory must comply in order to explain adequately both phenomena in the world and God's creative activity as learned from revelation. These conditions are nowhere clearly listed or spelled out, yet they can be inferred mostly from the Second Position of the seventeenth discussion. There, al-Ghazālī tries to convince his readers that a properly conceived occasionalist position as well as a proper view of secondary causality each lead to accepting the prophetical miracles of revelation.

Accepting the miracles reported in revelation is the first of these five conditions. It is not, however, al-Ghazālī's only concern in these passages. He puts drastic words in the mouth of his opponent when he makes him criticize occasionalism's indeterminism. An occasionalist worldview forfeits the possibility of making any assumptions about what is currently happening in places that are not subject to our immediate sense perception, as well as for events in the future. As al-Ghazālī portrays his philosophical adversary saying, occasionalism leads to the assumption of "hideous impossibilities" (*muḥālāt shanīʿa*) that destroy not only the pursuit of the natural sciences but also any coherent understanding of the world.[41] Al-Ghazālī's examples are not chosen—or adopted—without humor, and his readers are clearly left to enjoy the occasionalist position as an object of ridicule.

Creating a coherent understanding of the world that allows assumptions or even precise predictions about what is not immediately witnessed and what will happen in the future was a clear concern of al-Ghazālī and it is the second condition on our list. He would not have accepted an occasionalist explanation of cosmology that violates this criterion. Two other criteria for his cosmology can be taken from other parts of the *Incoherence.* At the end of that work, al-Ghazālī condemns three positions as unbelief (*kufr*). Two of the three positions that he condemns concern cosmological theories, namely, that the world is eternal and that God does not take note of individuals but only knows classes of beings. Since these positions "do not agree with Islam in any respect, and (. . .) none of the Muslim groups believes in it,"[42] any cosmological explanation acceptable to al-Ghazālī must—in a reverse conclusion—acknowledge that the world is created in time and that God knows all His creations both universally and as individuals.

Finally, a fifth condition can be gathered from the pages of the *Incoherence*. In the First Position of the seventeenth discussion, al-Ghazālī denies that fire could be either the efficient cause or the agent (*fāʿil*) of the cotton's combustion. Fire is inanimate and has no action.[43] This argument refers back to the third discussion of the *Incoherence*, in which al-Ghazālī criticizes Avicenna and his followers for their views on God's will. It is true, he says, that the *falāsifa* claim God is the maker (*ṣāniʿ*) of the world as well as its agent or efficient cause (*fāʿil*). In order to be an agent or efficient cause, however, one needs to have both a will and a free choice (*murīd mukhtār*). "We say that agent (*fāʿil*) is an expression [referring] to one from whom the act proceeds together with the will to act by way of free choice (*ikhtiyār*) and the knowledge of what is willed."[44] Here, the *falāsifa* disagree and say that any being can be an agent (*fāʿil*) as long as it is the proximate efficient cause of another being. Fire as the proximate efficient cause of the cotton's combustion may be called its secondary agent.[45]

Al-Ghazālī strongly objects and refuses to accept the terminology of the *falāsifa*. He insists that the word "action" is elliptical for "voluntary action" since an involuntary action is inconceivable.[46] The disagreement is fundamental and its implications are far-reaching. In addition to being the efficient cause of another thing, an agent must thus fulfill three other conditions. He or she must (1) have will or a volition (*irāda*), (2) have a choice (*ikhtiyār*) between alternative actions, and (3) know what is willed.[47] In the *Incoherence*, al-Ghazālī gives the strong impression that humans and other animated beings such as the celestial spheres can be considered agents. Later in his *Balanced Book*, al-Ghazālī clarifies that although humans may fulfill the two first conditions, that is, volition and free choice, the last condition cannot apply to humans since they do not have a full knowledge of what is created when they act.[48] In his autobiography, al-Ghazālī says clearly that the celestial objects, for instance, have no action (*fiʿl*) by themselves, as they are all subject to God's command who employs all of nature according to His will.[49] The same is true for humans, who are subject to God's will and lack this full knowledge. That humans are not agents and that God is the only agent in the universe are prominent motifs in the *Balanced Book* as well as in the *Revival*. Al-Ghazālī's position in the *Incoherence* must be considered dialectical, aiming to convince the *falāsifa* of the rather limited position that inanimate beings can never be considered "agents."[50]

In the *Incoherence*, al-Ghazālī does not present anything that might be considered a philosophical argument as to why he rejects the technical language of the *falāsifa* on this particular point.[51] He simply refers to the common usage of the word "action," seemingly just disagreeing over the choice of language. Al-Ghazālī prefers to use the Arabic word *fāʿil* according to the meaning it has in Muslim theology over its meaning for the Aristotelian philosophers.[52]

Among the *mutakallimūn*, however, language usage was a commonly used tool for establishing *kalām* doctrines. Unlike in *falsafa*, where the terminology was often based on Arabic expressions constructed to parallel Greek words, the Muʿtaziltes established early the habit of invoking common usage of Arabic to support distinct theoretical positions.[53] The Ashʿarites were the heirs to the Muʿtazilites in this approach. Their underlying idea seems to be that language and the particular relationship between words and their referring objects are God's creations. This theory is particularly true for Arabic, the language chosen by God for His revelation. Relying on referential relationships that are not sanctioned by common usage not only is erroneous but also is tampering with the bond that God created between Himself and humans through creating a language that is used by both sides.

Al-Ghazālī accuses the *falāsifa* of obfuscation and of using language that aims to create the impression (*talbīs*) that their God is a true agent. Yet they implicitly reject this position because they deny His will and free choice. In reality, the *falāsifa* teach that God "acts" out of necessity, which means for al-Ghazālī that God does not act at all. The philosophers' God differs from a dead person only inasmuch as He has self-awareness.[54] When the philosophers say that God is the maker (*ṣāniʿ*) of the world, they mean it only in a metaphorical sense.[55] In his *Incoherence of the Philosophers*, al-Ghazālī ridicules Avicenna for attempting to ascribe a will to God while still denying an active desire or deliberation on God's part.[56] This usage, al-Ghazālī says, is a purely metaphorical use of the word "will," and it unduly stretches its established meaning. Al-Ghazālī criticizes Avicenna's teachings as effectively being a denial of the divine attribute of will.[57] In the Third Position of the seventeenth discussion, in which al-Ghazālī discusses rules that not even God can violate in His creation, he clarifies, "we understand by 'will' the seeking after something that is known (*ṭalab maʿlūm*)." Therefore, there can be no will where there is no desire.[58]

For al-Ghazālī, the concept of divine will (*irāda*) on God's part excludes His acting out of necessity.[59] All through the *Incoherence*, al-Ghazālī maintains that God creates as a free agent (*mukhtār*) rather than out of the necessity of His nature. In total, there are thus five conditions for cosmological explanations that can be gleaned from the *Incoherence*. Any viable explanation of cosmology:

1. must include an act of creation from nothing at some point in time;
2. must allow that God's knowledge includes all creatures and all events, universally and as individuals;
3. must account for the prophetical miracles that are related in revelation;
4. must account for our coherent experience of the universe and must allow predictions of future events, meaning that it must account for the successful pursuit of the natural sciences; and
5. must take into account that God freely decides about the creation of existences other than Him.

What would an occasionalist explanation that fulfills these five criteria look like? Any occasionalist cosmology easily fulfills criteria 1, 2, 3, and 5. In the

Incoherence, al-Ghazālī points out that a wrongly conceived occasionalism violates the fourth condition, that of the predictability of future events. As long as one cannot discount that books could be turned into animals, for example, there is no way that an occasionalist explanation can allow or even support the pursuit of the natural sciences. The fourth criterion is fulfilled, however, if the occasionalist assumes that God does not make sudden *ad hoc* decisions about what to create next. In the *Incoherence,* such a conviction is bolstered by the premise that God's actions are strictly habitual. Absurdities such as the one mentioned above will not happen, because they are known to have never happened in the past. We build our knowledge of God's habit from past occurrences that we witnessed ourselves and that others have reported to us. This knowledge enables us to detect and formulate stable patterns in God's habit.

Still, there is no guarantee that an omnipotent God will not frivolously—or rather purposefully—break His habit. The occasionalist believer firmly trusts in God (*tawakkala*) that He will not turn his library into an animal zoo. This is one of the lower degrees of trust in God, writes al-Ghazālī in the thirty-fifth book of his *Revival of the Religious Sciences.* There, he compares the occasionalist believer who has trust in God to someone involved in a legal dispute in court. The claimant puts his confidence in winning the case in the hands of a legal attorney (*wakīl*).[60] The clients of the attorney are well familiar with his habits and how his customary procedures follow regularly after each other (*ʿādātuhu wa-ṭṭirād sunanihi*). The claimant is familiar, for instance, with the attorney's custom to represent his clients without calling them as witnesses. The attorney defends his clients just on the basis of what they have written down in a file (*sijill*). If the client is well familiar with this habit of his attorney and if he truly trusts him, he will assume that the attorney will try to resolve the case based solely on the file and that the attorney will not call upon him in court. The client will thus plan accordingly, preparing a comprehensive file to hand the attorney while also knowing that his attorney will not ask him to testify in court. He can sit calmly and trustingly and await the outcome of the case:

> When he entrusts [his affairs] to him [*scil.* the attorney], his trust is complete (*tamām*) when he is familiar with his [attorney's] customary dealings and his habits and when he acts according to what they require (*wāfin bi-muqtaḍāhā*).[61]

Trust in God, therefore, requires acting in accord with God's habitual order of events. "You understand that trust in God does not require one to give up any kind of planning (*tadbīr*) or action."[62] Rather, it requires arranging one's life patterns to match what we know is God's habit. Someone who is convinced of occasionalism and who has trust in God, for instance, does not need to keep the windows of his library closed simply because he might fear that his books may be turned into birds and fly away. Such a provision is unwarranted, given what we know about God's habits.

Determination by an Unchanging Divine Foreknowledge

Yet there are higher degrees of trust in God (*tawakkul*) that provide the believer with deeper certainty about the strictly habitual character of God's actions. These levels of trust are already hinted at in the seventeenth discussion of the *Incoherence*. There, in the First Approach of the Second Position, in which al-Ghazālī aims to present occasionalism as a viable explanation of physical processes, he suggests that all events in the world have already been determined by God's foreknowledge. In such an occasionalist universe, prophetical miracles can indeed be created: God disrupts His habitual course of action and adapts the knowledge of the witnesses to His disrupted course of action. It seems that in this occasionalist universe, God is not bound by anything. Yet here al-Ghazālī throws in a thought:

> There is, therefore, nothing that prevents a thing from being possible within the capacities of God [but] that it will have already been part of His prior knowledge that He will not do it—despite it being possible at some moments—and that He will create for us the knowledge that He does not do it in that moment.[63]

If God has a pre-knowledge of all events that are to be created in the future, that pre-knowledge not only limits how He will act upon His creation but also determines all His future actions.

The idea of a divine foreknowledge that determines creation was expressed most strongly in the generation after al-Ghazālī in one of the creeds that Ibn Tūmart taught to his Almohad followers. Ibn Tūmart found eloquent ways of expressing God's prior determination of events: "The means of living (*arzāq*) have already been allocated, the works have been written down, the number of breaths have been counted, and the lifespans (*ajāl*) have been determined."[64] Chapter twelve in Ibn Tūmart's *Creed of the Creator's Divine Unity* (*Tawḥīd al-Bāriʾ*) is even more explicit:

> Everything that is preceded by [God's] decision (*qaḍāʾ*) and His determination (*qadar*) is necessary and must become apparent. All created things come out of (*ṣādira*) His decision and His determination, and the Creator makes them appear according to how He determined them in His eternity (*fī azaliyyatihi*). [They follow out of his decree] without addition or diminishing, without alteration of what has been determined, and no change of what has been decided. He generates them without an intermediary and without bestowing them to a cause (*ʿilla*). He has no companion in his originating activity (*inshāʾ*) and no assistant in making [things] exist (*ījād*).[65]

Ibn Tūmart clearly imagines an occasionalist universe in which God "generates without an intermediary and without bestowing [His creations] to a cause" (*awjadahā lā bi-wāsiṭa wa-lā li-ʿilla*). Yet if all future breaths are counted, the future contingencies in such a universe are limited to what is already known to God. God's eternal foreknowledge has already determined the course of the world.

The notion that God knows future events appears already in the Qur'an. Several verses mention that God determines every human's lifespan (*ajal*) and time of death (Q 6:2, 11:3, 14:10, 16:61, etc.). At death, God executes His predetermined decision and "calls home" (*tawaffā*) the person (Q 39:42). Like the time of death, the means of living (or: sustenance, *rizq*) are allocated to the human individuals (Q 11.6, 89:16, 13:26). Finally there is the more general idea, expressed in verses 9:51 and 57:22 of the Qur'an, that nothing will happen to humans that has not been recorded by God. In the prophetical *ḥadīth*, the motif of divine predetermination is even stronger than in the Qur'an. Al-Bukhārī documents a number of versions of a prophetical saying that teaches that while the child is still in the womb, God determines four characteristics for him or her: the sex, the person's redemption or ruin in the afterlife, the sustenance (*rizq*), and the lifespan.[66] Other prophetical *ḥadīths* refer directly to God's pre-knowledge of some future events. One prophetical saying states: "Fifty thousand years before God created the heavens and the earth, He wrote down the measure of the creatures (*maqādīr al-khalāʾiq*)."[67]

In particular, the numerous Qur'anic verses on the set lifespan (*ajal*) of a human have produced much theological speculation. Does a murder override God's determination and cut short the appointed lifespan of the victim, or is the murderer rather the means by which God makes His determination come true?[68] Is only the human time of death predetermined, or does every event have its predetermined time? Indeed, the Qur'an does say that "every nation has its lifespan" (*li-kull umma ajal*, Q 7:34).

Early Sunni Muslim theology centers on opposition to Muʿtazilism, which stressed human freedom rather than the invariable predetermination of their time of death.[69] Sunni theologians, therefore, found it easy to accept predestinarian positions. Al-Ashʿarī, for instance, believed that everything that comes into being is necessarily the will of God; God not only wills the time of a person's death but also the way it comes about. The same is true for a person's sustenance (*rizq*) and—this subject became connected to this discussion in *kalām* literature—the prices (*asʿār*) of things.[70] Al-Ashʿarī's understanding of God's knowledge clearly includes an element of foreknowledge. He taught that "God wills the coming into existence of the thing according to how divine knowledge precedes it (*mā sabaqa bihi al-ʿilm*); and He wills what is known [to Him] to come into existence, and what fails to be known [to Him] not to come into existence."[71] For al-Ashʿarī, however, the subject of divine foreknowledge is somewhat of a side issue in the debate with the Muʿtazila about whether God wills the world's mischief and harm (*sharr*). From his teachings on other subjects, it is clear that al-Ashʿarī did not believe in a universal predetermination of events recorded in God's foreknowledge.[72]

The Nishapurian Ashʿarites make stronger statements about God's foreknowledge, which gradually lead toward the direction of universal predestination. In his *Creed*, al-Isfarāʾīnī requires his followers to believe that God's knowledge "comprises the objects of knowledge in a way that He always knew all of them including their (accidental) attributes and their essences."[73] His colleague ʿAbd al-Qāhir al-Baghdādī clarifies the relationship between God's foreknowledge and His will: whatever God knows will happen is exactly what He wills to happen. God's knowledge represents the decisions of His will: "Whatever God wants to come into existence will come into existence at the time that he wants it to happen (. . .)."[74]

The subject of divine foreknowledge was not one of the major themes in early Ashʿarite literature. Their notion, however, did attract the criticism of Muʿtazilites such as al-Kaʿbī (d. 319/931), who realized that admitting divine foreknowledge destroys human free will and questions God's justice.[75] In the early part of the fifth/eleventh century, his Muʿtazilite colleague Abū l-Ḥusayn al-Baṣrī argued against the determinism of Sunni theologians. These theologians—most probably Ashʿarites—are quoted as saying, "What the divine knowledge knows will occur cannot possibly not occur," and "the divine knowledge that a thing will not exist necessitates that it will not exist."[76] Abū l-Ḥusayn al-Baṣrī's lengthy refutation indicates that this position was the subject of a lively debate between the Ashʿarites and their Muʿtazilite adversaries.

Because knowledge is one of the divine attributes that resides in His essence, all Ashʿarites make the statement that God's knowledge exists from past eternity (*qadīm*) while human knowledge is generated in time.[77] Al-Juwaynī draws the full consequences of this statement. His position on divine knowledge appears to respond to Muʿtazilite and philosophical objections. Avicenna postulated that if God's knowledge is pre-eternal, (*qadīm*), it cannot simply change with each new creation.[78] Al-Juwaynī agrees, teaching that changing knowledge is a characteristic of humans, whose knowledge adapts to a changing reality. To assume, however, that God's knowledge of the world is like human knowledge and contains "cognitions" or "pieces of knowledge" (*ʿulūm*) that generate in time (*ḥāditha*) is implausible. It also violates the consensus of the Muslim scholars, al-Juwaynī says, even amounting to leaving Islam.[79] The pre-eternal character of God's knowledge implies that God's knowledge never changes. It contains all future objects of knowledge, including the "time" when they will be realized.

An adversary may come and say, al-Juwaynī assumes, that in His eternity (*fī azalihi*), God had the knowledge that the world will one day be created. Once the world has been created and continues to exist, there was a new and different object of knowledge. The opponent holds that God's knowledge and awareness of the existence of the world has adapted to this new reality. This opponent maintains that there are new cognitions (*ʿulūm*) in God's knowledge every time there is change. Al-Juwaynī categorically rejects this line of thinking:

> We say: The Creator does not acquire a new awareness (*ḥukm*) that did not exist before. There are no successive "states" (*aḥwāl*) for Him because the succession of states would imply for Him what is implied by the succession of accidents in a body. The Creator is qualified as having only one single knowledge that extends to eternity in

> the past and in the future. This knowledge necessitates for Him an awareness that encompasses all objects of knowledge with all their details. The Creator's knowledge does not increase in number when the objects of knowledge become more. [This is not like in the case of] those cognitions that come about in time, which become more numerous when the objects of knowledge become more numerous. The Creator's knowledge does not become more numerous when there are more objects of knowledge and equally it does not become new when they become new.[80]

When someone learns that Zayd will arrive tomorrow, al-Juwaynī explains, he does not require a new cognition about Zayd's arrival once he has arrived. He knew that all along, strictly speaking. The uncertainty of Zayd's action prior to its actualization, however, requires us humans to form a new cognition once Zayd has arrived. In God's knowledge of His own actions, however, there is no such uncertainty. Knowing that Zayd will arrive at a certain time is identical to knowing the realization of this event; no modification of God's knowledge is needed when the event is actualized.

According to al-Juwaynī, God's knowledge of the world is timeless. It contains a "before" and "after" but does not follow the course of events according to the patterns of past, present, and future. Those events that are currently in the past are to be realized before those that are currently in the future. God knows precisely the succession of events. He knows what has happened in the past, just as He knows—with the same amount of detail—what will happen in the future. His knowledge exists in a timeless realm—"in His eternity," as al-Juwaynī and Ibn Tūmart say—outside our human categories of past and future. Since there are no obstacles to whatever God wills, His knowledge is the result of His will. The two are, however, not identical, nor does God's knowledge determine His will. God's will and His knowledge do not consist of smaller units that could be called volitions or cognitions. God has one eternal will as well as one eternal knowledge.[81]

Divine Foreknowledge in the *Revival of Religious Sciences*

Al-Ghazālī subscribed to al-Juwaynī's understanding of God's knowledge as single and all-encompassing. In a passage that appears in the *Book of the Forty* and in the short creed at the beginning of the second book in the *Revival*, al-Ghazālī uses colorful language to illustrate that God knows every speck on the earth and in the heavens (cf. Q 10:61):

> In the darkest night God knows the crawling of the panther on the solid rock and He senses the movement of the dust-motes in the air. He knows what is hidden and what is apparent. He is aware of the innermost thoughts, the movement of ideas, and the secret fears through a knowledge that is pre-eternal (*qadīm*) and everlasting

> (*azalī*) and He will continue to be characterized by this knowledge in all eternity. His knowledge is not renewed and in its essence does not adapt to the undoing [of earlier arrangements] or to relocation.[82]

If God's knowledge is not renewed by the changing of events, it follows that it has a detailed and determining foreknowledge of the future. In the several creeds that al-Ghazālī wrote during his lifetime, he was somewhat careful not to mention too openly that God predetermines all future events. He is probably most explicit in a brief list of articles of faith at the beginning of the second book in his *Revival*. There, he says:

> God's will is an eternal attribute that He has, which subsists (*qā'ima*) within His essence (*dhāt*) as one of His attributes. By virtue of it He is continuously described as someone who wills in His eternity (*fī azalihi*) the existence of the things in their moments (*fī awqātihā*) that He has determined. They exist in their moments as He wills it in His eternity without one of them coming before or after [He wills it]. Rather, they occur in accordance with His knowledge and His will without change or alteration (*min ghayr tabaddul wa-lā taghayyur*). He has arranged (*dabbara*) the things not by means of a sequence of thoughts [that He has] and nor does He wait for a [specific] time. Therefore, one thing does not distract Him from another.[83]

This passage seems to have been one of the inspirations for Ibn Tūmart's creed.

Yet, although al-Ghazālī requires belief in divine foreknowledge, he does not explicitly say that God's will "in His eternity" predetermines future events in this world, such as the number of breaths that a human will take during his or her lifetime. In his *Letter for Jerusalem*, which follows a few pages after this passage, he is even less explicit on this subject. On divine knowledge, he just says that God's universal knowledge is evident in the detailed arrangement (*tartīb*) of even the smallest things in creation. God paves the way (*raṣṣafa*) for the existence of everything.[84] He then slips into an elaborate argument taken from one of al-Juwaynī's writings. Al-Ghazālī's master is said to have used it, according to al-Murtaḍā al-Zabīdī, against the Muʿtazilite al-Kaʿbī. Al-Kaʿbī claimed that if God had a detailed foreknowledge of future events, it would make His will redundant. Al-Ghazālī then reproduces al-Juwaynī's rebuttal, targeting al-Kaʿbī's accusation that for the Ashʿarites God's knowledge is the same as His will. Al-Ghazālī's counterargument denies al-Kaʿbī's hypothesis that a thing comes into being at the time when God's foreknowledge foresees it, rather than at the time when His will willed it. If that hypothesis were true, al-Ghazālī responds, one could also say that God's foreknowledge would make His power redundant were He to foresee something before enacting it. Rather, al-Ghazālī aims to correct this perception by saying that whereas God's power encompasses all possible creations, His will directs His power to enact one of the possible actions and prevents the alternatives from happening.[85] In the

Revival, however, he fails to clarify the role of divine foreknowledge in this process. He covers this subject in *The Balanced Book on What-to-Believe* in a long chapter about God's will and its relationship to His omnipotence and His foreknowledge.[86] There he adds that divine foreknowledge is not sufficient to replace the will, because "divine knowledge follows that what is known" (*al-ʿilm yatbaʿu al-maʿlūm*), meaning that the decisions of the divine will determine the contents of the divine knowledge. "What is known" (*al-maʿlūm*) to the divine knowledge are the divine acts that God's will has chosen to actualize from among all the acts possible for God's power. The foreknowledge does not affect this decision. The divine attribute of will decides among equally possible alternatives. The attribute of knowledge is true to (*ḥaqqa*) the divine will and takes account of this decision; al-Ghazālī says it "attaches itself" (*yataʿallaqu bi-*) to the decision.[87]

Although al-Ghazālī discusses some of the doctrinal problems of divine foreknowledge in his *kalām* textbook and in the second book of the *Revival* on the creed of Islam, he hardly ever explains its practical consequences for such subjects as cosmology or human actions.[88] This is particularly true of the other books of the *Revival* that are concerned with rectifying human actions (*muʿāmalāt*), in which divine foreknowledge is only mentioned in brief references. Divine predestination and foreknowledge are variously referred to as God's "eternal power" (*al-qudra al-azaliyya*), God's "eternal judgment" (*ḥukm azalī*), or God's "eternal will" (*irāda azaliyya*),[89] yet it is never explained what the "eternal" stands for and what implication it has on God's creation. The reason for al-Ghazālī's reluctance to give his readers a detailed account of God's foreknowledge is didactic. If half-educated people are told that God knows the future, they may draw false conclusions, decline to handle their affairs, and fall into a fatalistic apathy. Al-Ghazālī expresses this danger in several passages of the *Revival*; wishing to guide his readers to good action, he stresses that God will be pleased by some of their actions while detesting others. His readers are exhorted only to perform those actions that will please God and gain them afterlife's reward.

The human's choice stands in an obvious conflict with God's predestination. In at least two passages, al-Ghazālī tries to resolve this conflict, as we will see below. In various other places, however, al-Ghazālī simply rejects any discussion of this conflict. He presents the problem in the familiar terminology of God's decision (*qaḍāʾ*) and His determination (*qadar*). In theological discussions, both terms refer to God's predetermining future events.[90] The subject of divine predestination appears several times in the thirty-second book of his *Revival*, in the discussions of the human's patience and his or her thankfulness to God. Yet al-Ghazālī tries to avoid candid statements about God's all-encompassing predestination, several times shunning his inquisitive readers for questioning God's predetermination of the future:

> Accept God's actions (*ādāb*) and stay calm! And when the predestination (*qadar*) is mentioned, be quite! The walls have ears and people who have a weak understanding surround you. Walk along

> the path of the weakest among you. And do not take away the veil from the sun in front of bats because that would be the cause of their ruin.[91]

"Divulging the secret of predestination" (*ifshāʾ sirr al-qadar*) is simply not allowed.[92] It is best to be silent on this subject and follow the example of the Prophet who, according to al-Ghazālī, said: "Predestination is God's secret, so do not divulge it!"[93] In fact, those who have insight say: "Divulging the secret of God's lordship is unbelief."[94] At times, however, al-Ghazālī himself comes close to disregarding this advice. When he discusses divine predestination, however, he limits himself to saying that God wills all human actions, those that please Him as well as those that He detests, and that He creates both the good and the bad human actions. This distinction is directed against the Muʿtazilite position that God cannot will morally bad actions. Al-Ghazālī leaves no doubt, however, that although God creates all events in the world, the choice between good and bad actions is left to humans, who are all responsible for what they do.

Divine foreknowledge and God's all-encompassing predetermination are important parts of al-Ghazālī's cosmology and his ethics.[95] Understanding that God has such pre-knowledge represents a higher degree of trust in God than relying on conclusions drawn from God's habits. This higher trust in God is closely linked to the proper understanding of divine unity (*tawḥīd*). Indeed, advancing to the higher stages of *tawḥīd* is the root that helps one develop this superior trust in God. Acquiring a correct understanding of God's unity and thus a deep trust in God represents the knowledge—belief in the heart (*taṣdīq bi-l-qalb*) is tantamount to knowledge—that will lead to good and virtuous actions.[96]

Al-Ghazālī's ethics in his *Revival* is premised by the thought that God's will as well as His knowledge are pre-eternal (*azalī*) and have existed long before creation began. They include the first event of creation as well as the last. God already knows whether the crawling panther will catch his prey, and He knows which direction each speck of dust will take in the wind. Most important, if God's knowledge is single and unique, it will also never change. The concept of an unchanging divine foreknowledge has significant repercussions for an occasionalist view of creation. God does not make *ad hoc* decisions about what to create next; His decisions have already been made long before He started acting. In addition, God's decisions are recorded in one of His loftiest creations. All past and future events are contained in the "well-guarded tablet" (*al-lawḥ al-maḥfūẓ*) that sits in a heavenly realm.[97] For al-Ghazālī, the tablet, which is mentioned in verse 85:22 of the Qur'an, represents a blueprint of God's creation and records human actions as well as all other created events.[98] A divine pen has written God's plan for His creation onto this tablet. In his *Decisive Criterion*, al-Ghazālī quotes a canonical *ḥadīth* that identifies this pen, which appears in two enigmatic references in the Qur'an (68:1, 96:4), as God's first creation.[99]

The view that the well-guarded tablet holds the detailed draft for God's creation is widespread in philosophical literature. In Avicenna's *Throne Philosophy* (*al-Ḥikma al-ʿarshiyya*), "the well-guarded tablet" is read as a Qur'anic reference to two different beings: the highest created being as well as the active intellect, both are intellects in the heavenly realm. In the sixteenth discussion of his *Incoherence,* al-Ghazālī reports the philosophical teaching that the well-guarded tablet is a Qur'anic reference to the active intellect. There he criticizes this element of the *falāsifa*'s teaching as unproven and bemoans that the people of religion (*ahl al-sharʿ*) do not understand the well-guarded tablet in this way.[100] Yet the reported positions on the well-guarded tablet are not at all controversial, nor was al-Ghazālī's own view significantly different. He later refers to an important element of the philosophers' teachings that touches on the subject of the well-guarded tablet. In his *Revival,* he explains prophetical divination as a contact between the minds of the prophets and the well-guarded tablet, which here functions equivalently to the *falāsifa*'s active intellect.[101] Sometimes normal people achieve such a contact in their dreams, which may lead to the phenomenon that we today call *déjà vu.* For some time after this dreamtime contact with the active intellect, one remembers the future events one has seen there, and when such an event occurs, one gets the impression that it has happened for the second time. Prophets achieve such a contact and experience of future events while they are awake. In other words, the prophets can "read" future events on the well-guarded tablet, and they report these future events to their followers.[102]

When al-Ghazālī expounds this view in the twenty-first book of his *Revival,* he describes the well-guarded tablet as that thing "which is inscribed with everything that God has decided upon until the Day of Judgment."[103] Here "the well-guarded tablet" does not refer to the active intellect but rather to God's first creation, which is much higher in the celestial hierarchy of intellects. The same categorization applies to a passage in the *Book of the Forty* in which al-Ghazālī quotes approvingly the position of an unnamed scholar as saying that "[God's] decision (*qaḍāʾ*) means that all beings exist on the well-guarded tablet, both in a general way as well as in [their] details."[104] In al-Ghazālī's thought, just as in Avicenna's *Throne Philosophy,* "the well-guarded tablet" refers to both the first creation as well as the active intellect, without clearly distinguishing between these two.

God's unchanging foreknowledge turns an occasionalist explanation of the world into one that fulfills all the five criteria outlined earlier in this chapter. The habitual character of God's creations is no longer understood as a mere routine of God that He may practice on an *ad hoc* basis. Rather, God's habits are inscribed in His foreknowledge. The contingent correlations that we experience in God's universe are the necessary results of a coherent and comprehensive plan of creation that exists from eternity.

Prophetical Miracles and the Unchanging Nature of God's Habit

Al-Ghazālī's occasionalist explanation of the universe includes the conviction that God's decisions follow a habit inscribed in a timeless divine foreknowledge. But how strict is God's commitment to His habit? Does He ever break it? In the *Incoherence,* al-Ghazālī argues that the possibility of a break in God's habit should lead us to acknowledge that the connections between what we call causes and their effects are not necessary. Does God ever actualize this possibility? According to the classical Ashʿarite view, prophetical miracles are breaks in God's habit. Given that the natural scientist studies the lawfulness of God's habits, would prophetical miracles not spoil his or her efforts?

Classical Ashʿarism had already developed an answer to this problem. The effect of a prophetical miracle depends on those witnessing it knowing it to be a miracle. They must be made aware that what they have witnessed is a break in God's habit.[105] Classical Ashʿarite theology recognized several conditions for prophetical miracles that aim at making prior identifications of miracles. According to al-Ashʿarī, a true prophet must announce and describe the miracle that God will perform. He must issue an announcement (*daʿwa*) that God will perform a miracle and a challenge (*taḥaddin*) to those to whom he is sent. Muḥammad, for instance, issued a challenge to his adversaries when he dared them to produce a single *sura* like those contained in the Qur'an (Q 2:23, 10:38). In order for the miracle to be valid and acceptable to his audience, God must perform it exactly the way the prophet earlier describes it.[106]

Al-Juwaynī gives a detailed description of the conditions that are necessary in order to accept a miracle. They include the prophet's announcement and his challenge to those who doubt his prophecy. The goal of these strict conditions was to distinguish a prophetical miracle both from simple marvels and from sorcery. Given that in classical Ashʿarism, the miracle is considered the only way to verify prophecy, much was at stake. The authority of revelation and with it the existence of revealed religion rested on the proper identification of the prophetical miracle and on its distinction from mere coincidence or magic.[107]

Other than in his *Incoherence*, al-Ghazālī writes a few times about prophetical miracles in traditional Ashʿarite terms.[108] Unlike his master al-Juwaynī, however, he does not write about the conditions of the miracle and does not say, for instance, that a miracle must be preceded by a challenge. This is because, unlike his predecessors in the Ashʿarite school, he no longer believes that miracles are the only way, or even a good way, to verify the claims of a prophet. Al-Ghazālī believed that miracles could not be credibly distinguished from marvels and sorcery. In his autobiography, he discusses the case of someone claiming to be a prophet when he performs one of the prophetical miracles that, according to the Muslim tradition, confirmed the prophecy of Jesus. The Qur'an reports that Jesus revived the dead (Q 3.39, 5.110), mirroring chapter eleven in the Gospel of John describing Jesus' reviving Lazarus from his grave. Let's assume, says al-Ghazālī, that someone comes along who pretends to do the same and he announces the performance of this miracle in advance—just as earlier Ashʿarites required him to do. Even if he announces and successfully performs the revivication of an apparently dead person, that would not, according to al-Ghazālī, prove his status as a prophet. Al-Ghazālī justifies his position because the miracle of reviving the dead did not create certain knowledge of Jesus' prophecy. Certain knowledge about Jesus' prophecy is gained through other means. One should not accept people's claims to prophecy just on the bases of so-called miracles. Speaking to those who would follow a pretender purely on the bases of his so-called miracles, al-Ghazālī says:

> Let's assume that your Imam points out to me the miracle of Jesus, peace be upon him, and says: "I will revive your father, and that shall be the proof for me saying the truth." Then he actually revives him and explains to me that he is truly [a prophet]. Yet, how do I know that he speaks the truth? Not all people gained knowledge through the miracle [of reviving a man] that Jesus, peace be upon him, spoke the truth. Rather, the matter was beset with questions and uncertainties that can only be answered by subtle intellectual reasoning. (. . .) That the miracle points towards the veracity [of him who performs it] cannot be accepted unless one also accepts [the existence of] sorcery (*siḥr*) and knows how to distinguish it from a miracle, and unless one acknowledges that God doesn't lead humans astray. It is well known that the question of whether or not God leads us astray is quite difficult to answer.[109]

If prophetical miracles were to create definite knowledge about the claims of prophets, there would be no disagreements among humans as to who is a prophet. Jesus did revive Lazarus, yet the Jews still did not accept his prophecy. The Qur'an (Q 5.110) states that the unbelievers among the Children of Israel considered all miracles performed by Jesus to be mere sorcery (*siḥr*). This is due to it being nearly impossible, al-Ghazālī implies, to distinguish a prophetical miracle from sorcery. While God creates the former to guide people to His revelation, He also chooses to create the latter to confuse and misguide people. Humans are not given the faculty, so goes the implication, to clearly distinguish between the two.

In addition, there is the problem that only a limited number of people would personally witness the miracle, and all other humans would have to believe the viewers' judgment that the miracle was indeed not sorcery. Thus, when deciding whether an event or a text is truly a divine revelation, humans can only practice *taqlīd*; they must accept the positions of other people uncritically. This is quite a horrible thought for al-Ghazālī. In addition, further generations must verify the reports about the miracle and the judgments of its witnesses through impeccable chains of transmission (*tawātur*). This creates a new source of error. Al-Ghazālī was quite skeptical about the value of *tawātur*. Muḥammad's alleged appointment of ʿAlī at Ghadīr Khumm is an example of an event that never happened, according to al-Ghazālī, yet many in the Shiite community still trust its veracity because of its supposedly impeccable chains of transmission. If such a large group of Muslims accepts the historicity of a past event that never actually took place, no community can be immune to error in matters of *tawātur*.[110]

In the *Deliverer from Error*, al-Ghazālī says that only at an advanced stage of his spiritual and intellectual development did he realize that miracles are not the best way of verifying prophecy. After reading Sufi works, he understood there to be a way of distinguishing the true prophet from the false pretender without requiring recourse to a prophetical miracle. Prophets create through their teachings and their revelations effects in the souls of those who witness their prophecy. In the *Book of Forty*, al-Ghazālī describes the outward effect (*athar*) that reciting the Qur'an can have: weeping, breaking into sweat, shivering, getting goose bumps, quivering, and so forth.[111] These physical manifestations will inspire reflection on one's deeds. The direct experience (*dhawq*) of the prophet's positive effects on one's soul is the best indicator for the truth of his mission. This method is quite similar to how we distinguish a true physician from a charlatan or a true legal scholar from someone who only claims to be that. In all these cases we look at the people's work. Does the physician heal the sick? Does the legal scholar solve legal problems? If the answers are positive, we accept their claims. The same should be true for the prophets, who are termed physicians of the soul.[112] If we feel the positive effects of a prophet's work on our souls, we know that we are dealing with a true prophet.[113] This method is superior to those of the earlier Ashʿarites:

> Seek certain knowledge about prophecy from this method and not from the turning of a stick into a serpent or from the splitting of the moon. For if you consider that event by itself, and do not include the many circumstances that accompany this event you may think that it is sorcery (*siḥr*) and imagination (*taḥyīl*). (. . .)[114]

There are certain problems (*asʾila*) with prophetical miracles, al-Ghazālī says later in this passage. The classical Ashʿarite argument that a miracle is a sign for prophecy can easily be countered by arguments "about the problematic and doubtful nature of the miracle."[115] The miracle is only one of many indications of true prophecy, al-Ghazālī says cautiously. This position may have resulted from his reflections on miracles in the seventeenth discussion of the *Incoherence*. It is quite clearly expressed in his *Revival*. Here, al-Ghazālī says that Moses gained many followers by changing a stick into a serpent. Yet these same people later followed the false prophet, "the Samaritan" (*al-Sāmirī*), when he made them build the golden calf while Moses was on Mount Sinai: "Everyone who became a believer by seeing a snake inadvertently became an unbeliever when he saw a calf."[116] For most people, miracles are indistinguishable from sorcery and cannot serve as distinctive markers for prophecy. Avicenna had taught that prophetical miracles and sorcery result from the same faculty (*quwwa*) of the human soul. The prophet applies this capacity with good intentions, while the sorcerer (*al-sāḥir*) applies it with bad ones. Sorcerer and prophet, however, have the same kind of strong soul that can affect their surroundings and make other bodies do their bidding.[117] The essential similarity between prophetical miracles and sorcery is due to their origin in the same faculty (*quwwa*) of the prophet's and the sorcerer's souls. This shared origin makes the two events practically indistinguishable. Because of this essential similarity, al-Ghazālī rejected miracles as a means to verify prophecy, and thus he never discussed the conditions of prophetical miracles in his writing. Yet he nowhere denies that prophets perform miracles and does acknowledge those that are mentioned in revelation.

Al-Ghazālī's view as to what counts as a prophetical miracle also differed markedly from his Ash'arite predecessors' views. In addition to denying that miracles are sufficiently distinguishable from marvels and sorcery, he also rejected the position that they must be a break in God's habit. This direction of thought again has its roots in al-Juwaynī. According to al-Ash'arī, a miracle is defined as "a break in [God's] habit that is associated with a challenge which remains unopposed."[118] Although he quotes the traditional Ash'arite position that prophetic miracles and the wonders (*karamāt*) performed by some extraordinary pious people (*awliyā'*) are "a break in the habit" (*inkhirāq al-'āda*), al-Juwaynī's own position seems to have been more complex. A break in God's habit is indeed a "sign" (*āya*) that can verify a prophet's authenticity. The miracle, however, which al-Juwaynī sees as the only means of verifying prophecy, is no longer described as a break in God's habit but merely as the incapacity of the opponents to respond to the prophet's challenge.[119]

Apart from what he writes in the *Incoherence*, there is no indication that al-Ghazālī ever believed that miracles are a break in God's habit. In his *Balanced Book*, he says that the believer comes to trust the prophet's veracity "through strange things and wondrous actions that break the habits."[120] "Habits" ('*ādāt*)—in plural—seems to refer to the customs of persons or of things in this world, including the habits of the prophets, rather than to God's habit. For example, when the stick is turned into a serpent, the habitual behavior of the stick is broken although God had not changed His habit. This usage of the word "habit" ('*āda*) is already present in the *Incoherence*, in which the *falāsifa*'s position that the prophet has a more powerful practical faculty in his soul is described as "the special character [of the prophet] differs from the habit of the people (*tukhālifu 'ādat al-nās*)."[121]

There are clear indications that al-Ghazālī believed that although "miracles" are extraordinary and often marvelous events, they do not require God to break His customary habit—the laws of nature. In the thirty-first book of his *Revival*, al-Ghazālī says that God creates all things one after the next in an orderly manner. After making clear that this order represents God's habit (*sunna*), he quotes the Qur'an: "You will not find any change in God's habit."[122] This sentence is quoted several times in the *Revival*; in one passage, al-Ghazālī adds that we should not think that God would ever change His habit (*sunna*).[123] The implication is clear: since God never changes His habit, the prophetical miracle cannot be a break in His habit. It is merely an extraordinary occurrence that takes place within the system of the strictly habitual operation of God's actions. Miracles are programmed into God's plan for His creation from the very beginning, so to speak, and they do not represent a direct intervention or a suspension of God's lawful actions.[124] If this was al-Ghazālī's position about prophetical miracles, and I am quite convinced that it was, he nowhere states it explicitly in any of the core works of the Ghazalian corpus. Here, the Second Approach of the Second Position of the seventeenth discussion of the *Incoherence* remains one of the more explicit expressions of this view.[125]

Those who studied with al-Ghazālī or who read his works carefully certainly understood the revolutionary character of his teachings on prophetical miracles. Ibn Ghaylān, the Ghazalian from Balkh, reports with some bewilderment that al-Ghazālī did not oppose the *falāsifa* in their teachings on prophecy and prophetical miracles.[126] Al-Ghazālī's adversaries were more outspoken. In his widely known epistle on why the burning of al-Ghazālī's *Revival* in al-Andalus was justified, al-Ṭurṭūshī complains that regarding prophecy, al-Ghazālī adopted the teachings of the *falāsifa* and particularly those of the Brethren of Purity (*Ikhwān al-ṣafā*ʾ). These philosophers teach, al-Ṭurṭūshī continues, that God does not send prophets; rather, those who develop extraordinarily virtuous character traits acquire (*iktasaba*) prophecy. Al-Ṭurṭūshī is not entirely correct in his characterization of the Brethren of Purity. He is more correct when he says that the *falāsifa* teach that some prophetical miracles are ruses and trickery (*ḥiyal wa-makhārīq*) and that al-Ghazālī agreed with them on this point.[127] Al-Ṭurṭūshī was in close contact with Abū Bakr ibn al-ʿArabi and maybe with other students of al-Ghazālī.

For Avicenna, prophetical insight is caused by the extraordinary character traits of those who become prophets. Prophecy is linked to normal human psychology, and although it is rare, it is indeed a part of the normal course of nature. The origins of Avicenna's teachings on prophecy—and subsequently much of what we find in al-Ghazālī's psychology—lie in the works of Aristotle and his Neoplatonic interpretors, most prominently al-Fārābī.[128] Although the Brethren of Purity shared the Neoplatonic origins of al-Fārābī's and Avicenna's teachings, their presentation of psychology and prophecy is less detailed and well developed.[129] Avicenna's detailed explanation of prophecy certainly influences al-Ghazālī's understanding, and he does reproduce many of its features.[130] Future studies must decide whether the Brethren's psychology also significantly influenced al-Ghazālī, or whether the connection between the two merely resulted from parallel methods of teaching that are only roughly similar.

It is true, however, that the Brethren's work expresses certain mystical notions that also appear in al-Ghazālī but are explicitly expressed neither by al-Fārābī nor by Avicenna. Particularly regarding the inspiration that "friends of God" (*awliyāʾ Allāh*) receive—knowledge similar to revelation but at a lower level—the Brethren's ideas are reminiscent of Sufi concepts.[131] The Brethren, for instance, stress that receiving inspiration (*ilhām*) and revelation (*waḥy*) require the soul's purification from the pollutions of the natural world—a motif prominently expressed by al-Ghazālī in his letter to Abū Bakr ibn al-ʿArabī.[132] In general, the presentation of prophecy in the Brethren's *Epistles* shows closer connections among philosophical teachings, Muslim religious discourse, and Qur'anic passages than we see in al-Fārābī's and Avicenna's more theoretical treatments of prophecy. Unlike the two Aristotelians, who only occasionally back their teachings with an exegesis of verses in revelation, the Brethren frequently engage in figurative interpretations of Qur'anic verses. Al-Ghazālī was inspired by some of their suggestions.[133] Among religious intellectuals, the Brethren's close association with Qur'anic motifs may have created more interest in their work than in al-Fārābī's and Avicenna's work. This, in turn, would make the Brethren of Purity's work more threatening to al-Ghazālī's conservative opponents such as al-Ṭurṭūshī. As he does in his discussion of logics, al-Ghazālī replaced some of the technical language in the psychology of Avicenna with words more familiar to religious scholars that connect more seamlessly to motifs in the Qur'an. Borrowing from Q 38:72, al-Ghazālī frequently uses the word "spirit" (*rūḥ*), where Avicenna would have used the term "intellect" (*ʿaql*).[134] This usage may have made al-Ghazālī's psychological teachings seem closer to those of the Brethren of Purity, who use the term "spirit" frequently, than to those of Avicenna, who uses it only occasionally.

Al-Ghazālī was likely familiar with the *Epistles of the Brethren of Purity*.[135] Some of his cosmological teachings may go back to them, such as equating the heavenly spheres with the "realm of sovereignty" (*ʿālam al-malakūt*) and seeing the human body as a microcosm of the universe.[136] It seems that already during his lifetime, al-Ghazālī was accused of having copied from the *Epistles*. In his autobiography, he implicitly admits that some of his teaching also appear in these treatises, although he denies any influence and argues that the correlation is more or less coincidental. He says that in general, the teachings in the *Book of the Brethren of Purity* (*Kitāb Ikhwān al-ṣafāʾ*)—al-Ghazālī assumes that it was written by a single author—are weak philosophy, based on Pythagoras, and that Aristotle represents a more advanced stage. This work is "the chatter of philosophy" (*ḥashw al-falsafa*), al-Ghazālī adds, and it is false (*bāṭil*). He singles out the *Book of the Brethren of Purity* as an example of a misleading philosophical text, particularly because it aims at appealing to the religious scholars.[137]

Al-Ghazālī's critics, however, continued to associate his position on prophecy with the Brethren. Al-Māzarī al-Imām (d. 536/1141), a Tunisian contemporary of al-Ṭurṭūshī who wrote a polemic against al-Ghazālī, says some students of al-Ghazālī reported that he "constantly cleaved to the *Epistles of the Brethren of Purity*."[138] Al-Māzarī's polemic is unfortunately lost and known only from quotations in later texts, yet his opinions proved to be quite influential among later opponents of al-Ghazālī. In addition to the Brethren of Purity, al-Māzarī attributes the philosophical influence on al-Ghazālī to Avicenna and to Abū Ḥayyān al-Tawḥīdī (d. 414/1023).[139] More than a hundred years after al-Māzarī and al-Ṭurṭūshī, the Sufi philosopher Ibn Sabʿīn (d. *c*. 668/1270) from Ceuta claimed that the teachings presented in four of al-Ghazālī's works on the human intellect, the spirit, and the soul come from the *Epistles of the Brethren of Purity*.[140]

Authors from the Muslim East also understood that on the subject of prophecy, al-Ghazālī got quite close to the *falāsifa*. Ibn Taymiyya, for instance, chastises al-Ghazālī for having followed the "pseudo-philosophers" (*al-mutafalsafa*) in their view that knowledge of prophecy can be verified without someone having witnessed a miracle.[141] Because of al-Ghazālī's teachings on how the souls of the prophets and of "friends of God" (*awliyāʾ*) receive revelation as inspiration and insight from the heavenly spheres, Ibn Taymiyya saw al-Ghazālī as "from the same ilk as the heretical Qarmatians and the Ismāʿīlites." What is more, he complains, al-Ghazālī and others after him, such as Ibn ʿArabī (d. 638/1240), present these views about prophecy as Sufism and claim that it is a deeper truth.[142] Ibn Taymiyya diligently collected the criticism of earlier scholars on this matter, reproducing a long passage from al-Māzarī's lost polemic.[143] Earlier, influential Sunni scholars such as Ibn al-Ṣalāḥ al-Shahrazūrī had already spread al-Māzarī's criticism of al-Ghazālī. In his comments on the latter, Ibn Taymiyya rejects al-Māzarī's suggestion that al-Ghazālī had been influenced by al-Tawḥīdī, but he accepts al-Māzarī's view that al-Ghazālī's position on prophecy is based on Avicenna and the Brethren of Purity.[144] After his teachings on the best of all possible worlds, which will be discussed below, later scholars of Islam found al-Ghazālī's views on prophecy to be most objectionable.

Necessary Knowledge in an Occasionalist Universe

In its practical implications and particularly regarding the pursuit of the natural sciences, the occasionalist universe of al-Ghazālī is indistinguishable from the universe of the *falāsifa*. Both cosmologies assume that events in God's creation are predetermined. Both assume that fire *always* makes cotton combust. Both assume that the laws of nature or God's habit will *always* apply. The distinction between al-Ghazālī's type of occasionalism and the position that God exerts control through secondary causality is limited to the cosmological explanation of causal connections. This question belongs to the realm of metaphysics, teaches al-Ghazālī, and should have no influence on how we respond to God's creative activity. If a person is killed by the blow of a sword to his neck, he writes in his *Standard of Knowledge*, our sense perception recognized that death in this person comes "together with" (*maʿa*) the deep cut (*ḥazz*) in his neck. If this conjunction appears repeatedly, we have no doubt that a cut in the neck and death are connected, and we conclude that one is the cause (*sabab*) of the other.[145] Despite this conjunction, some may indeed doubt the connection; a *mutakallim*, for instance, may claim that the cut is not the cause of death and that God created the cut and death "side by side" (lit. "in the stream," *ʿinda jarayān*). Al-Ghazālī shows little patience with this *mutakallim*. Would he doubt his son's death were he to receive the unfortunate news that his son has a cut in his neck?

> When it comes to the question whether this is an inseparable and necessary connection that cannot be otherwise or whether this is an arrangement according to the normal course of God's habit (*sunnat Allāh*) through the efficacy of God's pre-eternal will which is not affected by change or alteration, [we say:] the question is about the kind of connection not about the connection itself. This should be understood and it should be known that doubting the death of a person who has received a blow to his neck is pure delusion (*waswās*) and that the conviction (*iʿtiqād*) that he is dead is certain (*yaqīn*) and should not be called into question.[146]

If the occasionalist agrees with al-Ghazālī that God's habit is the result of His pre-eternal will (*mashiʾatuhu al-azaliyya*), which "is not affected by change or alteration" (*lā taḥtamilu al-tabdīl wa-l-taghyīr*), the dispute the occasionalist has with a believer in causality is limited to the type of connection between cause and effect. The existence of a direct efficacy of the cause on the effect cannot be demonstrated. Both must agree, however, that the connection itself is inseparable, meaning that the occurrence of the cause (cut in the neck) is *always* concomitant to the appearance of the effect (death).

Richard M. Frank suggested that for al-Ghazālī, connections between what we call causes and their effects are indeed necessary: "Given the actuality of all causal conditions for its occurrence an event comes to be inevitable (*lā maḥāla*) and by necessity (*ḍarūrat^an*)."[147] But how, one must ask, can this conclusion be reconciled with the first sentence of the seventeenth discussion in the *Incoherence* in which al-Ghazālī explicitly says that "according to us" (*ʿindanā*), such connections are not necessary? In his *Balanced Book on What-to-Believe*, al-Ghazālī looks at the same example of a person who received a blow to his neck.[148] That volume's discussion is prompted by the question of whether the murderer cut short his victim's lifespan. Al-Ghazālī's goal is to correctly understand the connection between these two events, the murder and the victim's appointed time of death (*ajal*). He discusses three different ways of how things in this world are connected to one another, the third being the connection between a cause (*ʿilla*) and its effect (*maʿlūl*). By way of a general statement, al-Ghazālī says that in our judgment, the connection of these two is necessary: "If there is only a single cause for the effect and if it has been determined that the cause doesn't exist, it follows from it (*yalzamu min*) that the effect doesn't exist."[149] In this book, al-Ghazālī uses the language of classical Ashʿarism. In the case of the man who has received a cut in his neck, cause and effect are accidents that are connected to one another:

> "Being killed" is an expression for a cut in the neck and that is traced back to certain accidents, namely the movement of the hand of him who holds the sword and other accidents, meaning the cleavages among the atoms in the neck of him who is hit. Another accident is connected with (*aqtarana bi-*) these (accidents), and this is death. If there were no connecting link (*irtibāṭ*) between the cut [in the neck] and death, the denial of the cut would not make the denial of death follow. But these are two things that are created together (*maʿan*) and connected according to an arrangement that follows the habitual course and not according to a connecting link that one of the two has with the other.[150]

The position al-Ghazālī takes in this book is distinctly occasionalist. While by themselves the two events are not connected, they are connected through a habit (*ʿāda*). He does not elaborate as to whose habit this is, and his Ashʿarite readers might assume he means God's habit. Yet in real terms, the habit appears to be that of the creatures, not of God. God may create the two events individually and mono-causally, with each one being considered "a thing autonomously created by God" (*amr^un istabadda al-rabbu*). These two creations, however, always appear together (*maʿan*) and "in a connection according to an arrangement that follows the habitual course" (*ʿalā qtirān bi-ḥukm ijrāʾ al-ʿāda*).

The connection is not of a kind that the first event must be the "generating agent" (*mutawallid*) for the existence of the other. The cut in the neck does not "generate" (*tawallada*) death. Being a cause (*ʿilla*) simply means that, if all other causes of death are excluded, the denial of a cut in the neck makes the denial of death necessary.[151] Cut and death, al-Ghazālī implies, are inseparable, which means the relationship of the corresponding denial of a cut and the denial of death is necessary.[152]

The point al-Ghazālī wishes to make is that in our knowledge, the connection between what we identify as a cause and what we identify as an effect is necessary. Al-Ghazālī uses the Arabic verb *lazima* and its derivates, which indicate both an inseparable connection and a necessary judgment. What we witness is the pure concomitance of two events, grounded in a habit. Al-Ghazālī argues against an understanding of occasionalism that assumes God will break His habit. That, he implies, will not happen. Yet al-Ghazālī needs to be read closely: he nowhere says that the connection between the two events is necessary. He says only that the way our judgment connects these two events is necessary. Here he implicitly reiterates a point already made in the *Incoherence*: necessity is a predicate of human judgments, not a predicate of the outside world.[153] In this passage, the necessary connection is said to exist as a human conviction (*iʿtiqād*):

> He who is convinced (*iʿtaqada*) that the cutting of the neck is a cause (*ʿilla*) of death, and who connects this conviction to his observation that the body of the deceases is sound and that there are no other outside perilous forces involved, is convinced that the denial of the cut and the denial of any other possible cause necessarily means the denial of the effect, because all causes are denied.[154]

In this case, we conclude necessarily that the person whose body we inspect is not dead. To be convinced that there are imminent causes in this world does not mean to say, however, that these causes have a real efficacy toward their supposed effects. Here in his *Balanced Book on What-to-Believe*, al-Ghazālī compares the explanations of causal connection provided (1) by those who posit causality (*ʿinda qāʾilīna bi-l-ʿilal*) and (2) by those of the Sunnis (*ahl al-sunna*) who are convinced that God "is autonomous in the original creation [of events]" (*mustabiddᵘⁿ bi-l-ikhtrāʿ*) and does not allow other creatures to generate (*tawallad*) anything. He says that these two explanations do not differ regarding the conclusions we draw from observing causal connections. Yet on the level of cosmology, there is still a conflict between these positions that is "lengthy," and "most people who plunge into it do not realize its divisive character (*mithāruhā*)." Al-Ghazālī has no interest in engaging with that conflict. Regarding questions as to whether the cutting of the neck causes death or not, he recommends resorting to a simple rule (*qānūn*): one must avoid assuming that something could be generated (*tawallada*) by anything other than God. God creates everything, and in the case of the killed human, it is best to say: what really killed him was the end of his appointed lifespan (*ajal*).[155]

Despite its openly occasionalist language, even in his *Balanced Book*, al-Ghazālī shows no signs that he committed himself exclusively to an occasionalist cosmology. He stresses that the Muʿtazilite explanation of physical events through "generation" (*tawallud*) is wrong. Events in the created world do not simply "generate" from other created beings and certainly not from human decisions. Yet here, as in most of his works, al-Ghazālī wishes to leave open whether these events are created directly by God or are the results of secondary causes. Given that his target readership tends toward the former position, he has no problem stating his position in a language that they will find easy to adopt.

Concomitant Events and Rational Judgments

Al-Ghazālī regarded the reliance on atomism and occasionalism as a viable method to explain God's creative activity, and in some of his works such as the *Balanced Book on What-to-Believe* he succeeds in these explanations. This book was likely written as a textbook of Ashʿarite *kalām* to be used by students at the Niẓāmiyya madrasa in Baghdad. *The Revival of the Religious Sciences*, which al-Ghazālī started composing after he had left the Niẓāmiyya in Baghdad, does not have as distinct a target readership. In this book, al-Ghazālī is not quite as committed to the occasionalist language of the Ashʿarite *mutakallimūn*. Although some books in the *Revival* do use that terminology, most are cast in a more advanced language that tries to give equal justice to both occasionalism and secondary causality. On first reading, these texts appear to employ a distinctly causalist language. At the beginning of the thirty-fifth book, for instance, which discusses belief in God's oneness (*tawḥīd*) and trust in God (*tawakkul*), the author explains the difficulties of developing deep confidence in the reliability of God's habit. Trust in God is difficult to comprehend because many people look exclusively at the causes (*asbāb*) of things, rather than see God's activity. Yet it is wrong to think that causes could stand on their own. This difficulty is expressed in an ambiguous sentence in which al-Ghazālī evidently wishes to remain uncommitted about the true nature of causes. However, he does want to make his readers understand that the common word "cause" (*sabab*) does not mean an independent or absolute efficient cause:

> Basing oneself on the causes (*asbāb*) without viewing them as "causes" (*asbāb*) means to outsmart rationality and plunge into the depths of ignorance.[156]

These "causes" can be either secondary or just an expression of the habitual concomitance of God's immediate creative activity. In neither case do they have independent agency. To assume such independent agency would be the gravest mistake one could make with regard to causes, akin to bringing "polytheism into the idea of God's unity" (*shirk fī l-tawḥīd*). Then again, completely disregarding the causes, defames the Prophet's *sunna* and slanders his revelation (*ṭaʿn fī l-sunna wa-qadḥ fī l-sharʿ*). Qur'an and prophetical *ḥadīth*, al-Ghazālī implies, discuss causes as if they have real efficacy. To understand the true meaning of trust in God, one must balance the conviction that there is only one agent or efficient cause in this world (*tawḥīd*) with rationality (*ʿaql*) and with revelation (*sharʿ*).[157]

Rationality and revelation are the two pillars of verifiable human knowledge. Neither of them provides a decisive answer as to which of the two competing explanations of God's creative activity is correct. Al-Ghazālī implies that neither the Qur'an nor the *ḥadīth* provides a clear statement in favor of either position. This indecisiveness also applies to rationality: in the seventeenth discussion of the *Incoherence*, he aims to show that there is no demonstration that proves the direct and immediate character of the connection between a cause and its effect. These effects may be determined by secondary causes, or the concomitance of them may be determined by God's habitual course of action as he creates each event individually, one by one.

A critical reading of al-Ghazālī must be aware of these ambiguities. If he says that two things are created "side by side" (*ʿalā l-tasāwuq* or *ʿinda jarayān*), this may be due to their being a cause and its effect in a causal chain that has its beginning in God or due to God's immediate arrangement. If things have a "connection" (*iqtirān*) or if there is a "connecting link" (*irtibāṭ*) between two things, their relationship may be either determined by laws of nature or due to God's habitual course of action. Even if something is called a "cause" (*sabab*), the reader of al-Ghazālī cannot be certain that this means "secondary cause." According to al-Ghazālī, this is just the way we talk about our environment, and it would be unwise to jump to conclusions about the cosmological character of the "causes." From this perspective, it is unsurprising that in the great majority of his works, al-Ghazālī promotes a naturalist understanding of "causes." Fire causes ignition, bread causes satiety, water quenches thirst, wine causes inebriety, scammony loosens the bowels, and so forth. The same naturalist understanding applies to the effective existence of natures (*ṭabāʾiʿ*). "A date stone," al-Ghazālī acknowledges in the twenty-second book of the *Revival*, "can never become an apple tree."[158]

In his two works on logics, the *Standard of Knowledge* and the *Touchstone of Reasoning in Logics*, al-Ghazālī discusses how we acquire knowledge of causal connections. Here the nominalist underpinnings of his epistemology become evident. Causal connections are understood through experience or experimentation (*tajriba*). Experimentation represents one of five different means for acquiring certain knowledge, the other four being *a priori* concepts (*awwaliyyāt*), inner sense perceptions (*mushāhadāt bāṭina*), outer sense perception (*maḥsūsāt ẓāhira*), and knowledge that has been reliably reported on other people's authority (*maʿlūmāt bi-l-tawātur* or *mutawātirāt*). In addition to these five sources of certain knowledge (*ʿilm yaqīnī*), there are also types of knowledge that cannot be sufficiently verified and can thus never be used as premises in demonstrations. These are either judgments that immediately appear to be true but that are unverifiable (*wahmiyyāt*) such as "all existence is spatial" or "beyond the boundaries of the world is no vacuum" or notions that are commonly accepted by the majority of the people (*mashhūrāt*), yet verifiable only through other sources, such as judgments about which human actions are morally good or bad.[159]

Al-Ghazālī lists numerous examples of how experience can produce certain knowledge about causal connections. They cover the full range of what is considered causality: fire burns, bread leads to satiety, water quenches thirst, hitting an animal causes it pain, a cut in the neck causes death, and scammony has a laxative effect on one's bowels.[160] These judgments are different from sense perception, al-Ghazālī explains, as they express universal judgments rather than merely individual observations of isolated events. Universality cannot be produced solely by the senses, but it rather must be formed in the human rational capacity (*ʿaql*). Such judgments of experience (*mujarrabāt*) must be based on the repeated sensation of single events in our sense perception.[161] They are a combination of sense perception and rational judgment. Consistent with his criticism in the *Incoherence* that necessity is a predicate of judgments and not of things in the outside world, al-Ghazālī highlights that the universal necessity of these judgments cannot be wholly taken from the outside world. The necessity and universality is due to a "hidden syllogism" (*qiyās khafī*) that combines the multitude of observations into a single judgment. Al-Ghazālī admits, however, that the reason why we acquire certain universal knowledge, rather than just probable or false knowledge, still remains unknown. All we can say is that experience imposes (*awjaba*) upon us either a decisive judgment (*qaḍāʾ jazmī*) or one that we consider valid for the most part (*aktharī*), and that this is by means of a "hidden syllogistic power."[162] This power works on our minds in an inescapable way. In his *Touchstone of Reasoning*, al-Ghazālī gives an example of this hidden syllogistic power:

> If someone who has a painful spot [on his body] pours a liquid over it and the pain goes away, he will not acquire knowledge that the liquid has stopped [the pain] because he will account the disappearance of pain to coincidence.[163] This is similar to when someone reads the Sura "Devotion" (Q 112) once over such a spot and the pain disappears. He would get the idea that the disappearence of [pain] appears by coincidence. If the pain disappears repeatedly [after reading the sura] and on many occasions, however, he acquires knowledge [about such a connection]. Thus, if someone tries it out and reads the sura "Devotion" once the first signs of the illness appear, and every time—or at least in the majority of cases—the pain vanishes, he acquires certain knowledge that [reading the sura "Devotion"] is something that makes the pain vanish, just as he has acquired certain knowledge that bread makes hunger vanish and dust does not make hunger vanish but actually increases it.[164]

Al-Ghazālī invites his readers to consider a situation in which the recitation of the sura "Devotion" (*al-Ikhlāṣ*) and the vanishing of pain at a certain spot repeatedly appear in conjunction. In such a situation we will conclude, he argues, that there is a connection between the two events. What makes us establish such a judgment is not a real causal connection between the two events but simply their concomitant appearance, which is indeed a connection, although not necessarily a causal one.[165] The knowledge that we acquire, however, is that reading the sura causes the pain to go away. Knowledge about what we regard as causal connection is acquired by seeing an inseparable relationship (*talāzum*) between two events and the consecutive and habitual pattern (*iṭṭirād al-ʿādāt*) of their conjunction.[166]

Judgments about causal connections are universal (*qaḍāyā ʿumūmiyya*) and apply to all individuals within a certain species (*jins*). They cannot be attained though sense perception alone, as sense perception (*ḥiss*) can only produce judgments about individual objects (*ʿayn*). All universal judgments that we do not accept from revelation are either *a priori* and primordial or must rely on a syllogism; in the case of experience, the syllogism is hidden and not conscious:

> If you look closely into this you will find that the intellect (*al-ʿaql*) attains these judgments after some sense perception and after their repeated occurrence through the mediation of a hidden syllogism (*qiyās khafī*) that is inscribed in the intellect. The intellect has no cognitive perception (*shuʿūr*) of that syllogism because it does not attend to it and it does not form it in words.[167]

In the First Position of the seventeenth discussion of the *Incoherence*, al-Ghazālī makes his major point on this subject, namely, that without this hidden syllogism, human perception cannot come to universal judgments, including universal judgments about causal connections. In his *Touchstone of Reasoning*, he reminds his readers:

> We have mentioned in the *Book of the Incoherence of the Philosophers* that which alerts [the readers] to the depth of these matters. The gist is that the judgments acquired through experimentation (*al-qaḍāyā l-tajribiyya*) go beyond sense perception.[168]

What exactly makes the judgments of experience go beyond sense perception is not clear: "We cannot say what is the cause (*sabab*) in reaching the perception of this certainty after we know that it is certain."[169] Consequently, the hidden syllogism is nowhere clearly explained. It comes to the fore when a connection between two individual sense perceptions appears so frequently that it cannot be explained as a coincidence. Again in the *Touchstone of Reasoning* he writes:

> The intellect usually says: Were it not for the fact that this cause leads to its [effect], [the effect] would not continuously occur for the most part; and if [the effect] happened by coincidence it would appear [sometimes] and [at other times] not. Consider someone who eats bread and later has a headache while his hunger has gone away. He concludes that the bread satisfies hunger and does not cause the headache because there is a difference between these two effects. The difference is that the headache appears on account of another cause whose connection with the bread is coincidental. Because if it came about through (*bi-*) the bread, [the effect] would appear always together (*maʿa*) with the bread or for the most part, like satiety.[170]

The continuous appearance of one event together (*maʿa*) with the other makes us conclude that the one is the cause of the other. It is worth noting that al-Ghazālī's treatment of experience sees the connections expressed by our judgment as necessary and constituting certain knowledge, even if the underlying sense perceptions concur only "for the most part." There can be no doubt that these kinds of judgments qualify for al-Ghazālī as certain knowledge, despite their nearly-but-not-universal occurrence.[171] In his autobiography, for instance, al-Ghazālī says that the experience (*tajriba*) of the positive effects of a prophet's work on one's soul generates necessary knowledge (*ʿilm ḍarūrī*) of his prophecy.[172] In this case, the judgment of experience is established by the repeated concomitance between performing the Prophet's ritual prescriptions and the positive effects this practice has on one's soul. That resulting judgment, namely, that Muḥammad can effectively heal the soul through his revelation, establishes certainty about prophecy (*yaqīn bi-l-nubuwwa*) and results in belief that equals the power of knowledge (*al-īmān al-qawī l-ʿilmī*).[173]

For al-Ghazālī, the fact that two events always appear together or do so for the most part implies that their concomitance is not coincidental. Once we are convinced that we are not dealing with coincidence, our mind moves toward a necessary judgment about the one being the cause of the other. Talking about the individual sense perceptions that lead to this judgment, al-Ghazālī says that "the cause and the effect always are inseparable (*yatalāzimān*) and if you want you can say 'cause' (*sabab*) and 'effect' (*musabbab*) or if you want you can say 'necessitator' and 'necessitated.' "[174]

Experience (*tajriba*) in Avicenna and in al-Ghazālī

In al-Ghazālī's epistemology, experimentation (*tajriba*) establishes necessary knowledge about causal connections solely from the repeated concurrence of two events. This method stands in striking contrast to the Aristotelian view of how we know about causal connections. In Avicenna's thought, as in most Aristotelian theories of the sciences, the majority of causal connections are the results of active and passive powers in the essences of the cause and the effect. The passive power (*quwwa munfaʿila*) of flammability, for instance, is an essential attribute of cotton that is implied by the fact that it is the product of a plant. All plants and their products are flammable. Equivalently, fire has in its essence the active power (*quwwa fāʿiliyya*) of burning. Once the two come together, inflammation must occur due to the essential nature of these two things. According to Aristotle, we know these essential qualities by witnessing these characteristics in the outside world *and* subsequently inducing their essential nature from the universal forms of cotton and fire. The necessary judgment that "fire burns cotton" is reached not by "experience" (Greek *empeiría*, Arabic *tajriba*) but by "induction" (Greek *epagôgé*, Arabic *istiqrāʾ*). In this case, the human intellect observes a certain process and reaches a necessary conclusion through the assistance or mediation of the separate active intellect when it imprints or illuminates the forms of fire and cotton in the human intellect.[175] That fire has the active power of burning and cotton the passive power of inflammability can only be known through the mediation of the active intellect.[176] We first need to receive the intelligible universal forms of "fire" and "cotton" from the active intellect before we conclude that fire necessarily inflames cotton.

In Avicenna, the individual particulars of a thing are perceived by the senses and stored in the faculty of imagination (*khayāl*). The "light of the active intellect shines upon the particulars" in imagination, and the intelligible universal forms "flow upon" (*yafīḍu ʿalā*) the human soul. The intelligible universal forms are "abstracted" (*mujarrad*) from individually perceived particulars "through the mediation of illumination by the active intellect."[177] In Avicenna, like in Aristotle, the source of our knowledge of the essential active and passive powers of things is not nature and its observation but the separate active intellect. Sensual perception, Avicenna teaches, cannot lead to necessary judgments.[178] It is important to note that induction only works if the active and passive powers that lead to causal connections are part of the essences of the things.[179]

When the active and passive powers that necessitate the causal connection are not part of the essences of the things, Avicenna mandates the use of experimentation (*tajriba*). An example that Avicenna and al-Ghazālī both mention is that in medicine, we witness that scammony causes purgation in the gallbladder. According to Avicenna, the relationship between scammony and the purgation of bile is not due to an active power that is part of the essence of scammony. Rather, the effect is due to an "inseparable accident" (*ʿaraḍ lāzim*) or a proprium (*khāṣṣa*) of scammony, meaning an accident that inheres permanently and is therefore an inseparable part of it.[180] Since the cause of this laxative effect is an accidental characteristic, we cannot know it through induction (*istiqrāʾ*). In this case, experimentation (*tajriba*) leads us to conclude that the accident of causing this laxative effect inheres in scammony. The repeated observation of this connection establishes that there is something either in scammony's nature or just "with it" (*maʿahu*) that causes—at least in our lands, Avicenna adds—purgation of bile.[181]

An important aspect of Avicenna's theory of experience is that it establishes universal judgments not only when the relationship is always (*dāʾim*[an]) observed, but also even in cases in which we only observe that relationship in most cases (*aktharīyy*[an]). The force of necessity in our judgments is considered a syllogism (*qiyās*). "There is a syllogism," Avicenna says, "that is produced in the mind without being perceived."[182] The syllogism, however, is merely the way that the necessity of the judgment is expressed; it cannot be the source of the necessity. In fact, it is not entirely clear what precisely justifies the epistemological leap from an observation of events that likely indicate a relationship to the necessity of a syllogism.[183] Experimentation in Avicenna seems to be based on the underlying assumption that when two things repeatedly happen together, they do so either due to chance or due to necessity. When the two things are just as likely to happen together as not to happen, the repeated observation that they *always* happen together, or in the great majority of cases, justifies the conclusion that they do not happen together by chance (*ittifāqan*).[184] They therefore happen together due to some necessity.

In Avicenna's view, experimentation informs us *that* scammony has a purging effect, yet it does not allow us to conclude *how* this effect occurs. Unlike induction, it does not provide the underlying causal explanation. Experience thus does not provide scientific knowledge (Greek *episteme*, Arabic *ʿilm*) in the strict Aristotelian sense of it being both necessary *and* explanatory.[185] In addition, Avicenna admits that because of its shaky epistemological basis, experimentation does not provide "absolute syllogistic knowledge" but only "universal knowledge that is restricted by a condition."[186] This condition is the methodologically sound application of the judgment. When using experimentation, the scientist must record the variables and background conditions surrounding the observations. Only when experimentation is conducted in this careful way can one be certain that there is a necessary relation between the two events in question. This method often forces the scientist to limit his or her results to the conditions he or she observed, such as when Avicenna says that scammony has the observed effect "in our lands."[187] Limitations, such as the acknowledgment that scammony may not have its purging effect in other climates, are very important in Avicenna's theory of experience. They are a result of the fact that we are only dealing with a cause that is an accident in scammony, and not a part of its essence.[188] Even if all methodological conditions are fulfilled, Avicenna notes, experience is no safeguard against error; and in his work, he further discusses likely mistakes when pursuing experimentation.[189] Nevertheless, experience can provide certain knowledge, albeit of a limited kind.[190]

For Avicenna, experimentation becomes much more important than for earlier Aristotelian theories of knowledge because he believed that induction (*istiqrāʾ*) should always be combined with experience (*tajriba*). At the end of his discussion of experience, Avicenna admits that even induction (*istiqrāʾ*)—usually considered a stronger and more reliable source of knowledge that experimentation—relies on experimentation. Comparing the results of sense perception, of induction, and of experimentation, Avicenna says that unlike sense perception, which just produces individual observations, induction and experimentation both produce universal knowledge. By itself, however, induction produces no more than an "overwhelming assumption" (*ẓann ghālib*), which is not knowledge. The result of induction must be combined with experimentation in order to produce a universal judgment that is not limited by any conditions. Studying nature's connections through experimentation (*tajriba*) is part of the process of obtaining truly universal knowledge from the active intellect. Avicenna says that experimentation is "more reliable" (*ākad*) than induction, and while induction by itself cannot produce certain universal knowledge, experimentation can.[191] By itself, however, experimentation produces universal knowledge, whose universality is limited by the conditions of the underlying observations, meaning, for instance, it is valid where observed, though not necessarily elsewhere.[192]

Jon McGinnis argues that in Avicenna's critique of induction, he moves from a pure Aristotelian position of how we have knowledge of causal connections toward the direction of a more modern epistemology where causal connections are not learned from the universal forms of the active intellect.[193] Avicenna's follower al-Ghazālī went much further on this path. In al-Ghazālī's discussion of the sources of human knowledge, there is a trace of neither induction (*istriqrāʾ*) nor the apprehension from the active intellect of the essential characteristics of things. This epistemology is consistent with al-Ghazālī's nominalist criticism of Avicenna's position on causality. Al-Ghazālī does not distinguish between fire burning cotton or scammony producing a laxative effect: both are examples of a singular type of causal connections. Subsequently, al-Ghazālī does not distinguish between active and passive powers that are either rooted in the essence of things or formed by their concomitant accidents. In fact, al-Ghazālī nowhere mentions the existence of active and passive powers in things.

Causal connections are, for al-Ghazālī, merely the repeated conjunction of two events. Witnessing such events, our rational capacity (*ʿaql*) produces necessary judgments about these connections. Al-Ghazālī's treatment of experience relies heavily on that of Avicenna. The judgments of experimentation (*al-tajribiyyāt*), Avicenna says, "are matters [in the mind] to which credence is given from the side of sense perception through the assistance of a hidden syllogism (*qiyās khafī*)." We have already seen that in al-Ghazālī, the universal judgments provided by experimentation rely on a sequence of sense perceptions in which the connection has been observed either constantly or only for the most part. In both cases, the judgments consist of two elements: the repeated observation that two events occur together and a hidden syllogistic force (*quwwa qiyāsiyya khafiyya*) that merges many observations into one. Like Avicenna, al-Ghazālī also requires experience to be pursued with a certain degree of rigidity. The data from sense perception must be gathered by sound sense organs when the object is close to the senses and when the medium between the senses and its object is dense.[194]

In a long sentence, al-Ghazālī describes the whole process of acquiring knowledge about causal connection through experience, taking account of all aspects of our judgments that two events are causally connected:

> If the [repeated concurrence of two events] were coincidental or accidental and not inseparable (*lāzim*), it would not continue to occur for the most time without variation; so that even if the event that is inseparable (*lāzim*) [from a first event] has not come into existence, the soul (*nafs*) regards the delay of [the second event] from the first as a single occurrence or one that happens rarely (*nādir*[an]), and it would search for a cause (*sabab*) that prevented the [second] event from occurring.
>
> If the individual sense perceptions that occur repeatedly one time after the other are brought together, and the number of occurrences cannot be determined, like the number of authorities (*mukhbir*) in a securely transmitted tradition (*tawātur*) cannot be determined, and if each occurrence is like an expert witness, and if the syllogism (*qiyās*)

> that we mentioned above is combined with it, then the soul grants assent.[195]

In this context, the fact that the soul "grants assent" (*ʿanat al-nafs li-l-taṣdīq*) to the judgment means that the necessity of the connection is established, and it can be used as a premise in demonstrative arguments. If conducted in the right way, experience produces universal and certain knowledge of *all* kinds of causal connections. Unlike Avicenna, al-Ghazālī does not limit the validity of these judgments to certain regions or lands, for instance, or to other circumstances.

It would be false to say, however, that for al-Ghazālī, causal connections are mere mental patterns without correspondence in the real world. The apparent regularity of the connection between what we call a cause and its effect justifies the judgment that scammony causes loosening of the bowels. Although there may be no true causal efficacy on the side of scammony, the regularity of two concomitant events triggers our judgment of causes and effects.[196] Unlike Avicenna, al-Ghazālī never mentions a concomitant laxative accident in scammony, and on some level he pleads ignorant as to whether it really exists. In his cosmology he remains uncommitted to scammony's agency on the loosening of the bowels. The causal inference, however, is not just something the mind puts into the world. The outside world is evidently ordered in a way *as if there were* causal connections. Although the true cause of the regularity of concomitance is uncertain, the fact that they appear together is certain.

Following Avicenna's terminology, however, it would not be correct for al-Ghazālī to say that necessity is solely a feature of our judgments. Necessity, which for Avicenna is identical with temporal permanence, exists when two things *always* appear together; and the latter fact is not denied by al-Ghazālī. Al-Ghazālī's criticism of causality in Avicenna breaks with the statistical interpretation of modal concepts and applies a view of necessity based on the denial of synchronic alternatives. Both agree that the connection between a cause and its effect appears always. For Avicenna, this is synonymous to saying it is necessary. Al-Ghazālī, however, points out that whereas the causal connections we witness in the outside world will always appear in past, present, and future, God could have chosen an alternative arrangement. The possible existence of an alternative means that the connection in the outside world is not necessary.

Making truly necessary connections that allow no alternatives is, according to al-Ghazālī, solely a feature of the human rational capacity (*ʿaql*). Logic is the domain where this rational capacity is applied in its purest form. Al-Ghazālī openly endorsed the logic of the Aristotelians, favoring it over that of the *mutakallimūn*.[197] Averroes and Richard M. Frank questioned how al-Ghazālī could claim to adhere to Aristotelian logic while also subscribing to a cosmology that believes the connection between a cause and its effect is not necessary.[198] In the Aristotelian understanding of logics, the connection between the two premises of a syllogism and its conclusion is that of two causes that are together sufficient and necessary to generate the conclusion. More precisely, it is the combination of the truths of the two premises that causes the conclusion to be true. In the *Touchstone of Reasoning*, a textbook of Aristotelian logics written for students in the religious sciences, al-Ghazālī shares this position. Michael E. Marmura suggested that here, as in other works where he defends Aristotelian logics, al-Ghazālī reinterprets the demonstrative method alongside occasionalist lines without this affecting either the formal conditions that logics must satisfy or its claim for attaining universal certainty.[199] For al-Ghazālī, therefore, the seemingly causal connection between the premises of a syllogism and its effect is just one of those cases where an event, namely, the combination of two true premises, regularly appears concomitantly with another event, namely, the truth of the conclusion. After explaining that any kind of proposition can form the premise of a syllogism, he clarifies in his *Standard of Knowledge* how the conclusion is derived:

> Therefore, those cognitions that are verified and that one has granted assent to are the premises of a syllogism. If they appear (*ḥaḍara*) in the mind in a certain order, the soul (*nafs*) gets prepared for the [new] knowledge to comes about (*yaḥduthu*) in it. For the conclusion comes from God.[200]

We regard the connection between the premises of a syllogism and its conclusion as necessary. Were we not, we could have no trust in rationality and would have to conclude it is mere conjecture. The connection between the premises and the conclusion is of the same kind as the connection that exists between causes and their effects in the outside world. Our assumption about the necessary character of the syllogistic connections in our mind suggests that all causal connections should indeed be considered necessary.[201] This is, in fact, al-Ghazālī's position. In all contexts where the cosmological or epistemological aspects of causal connections are irrelevant, he assumes that *for us* causal connections are necessary. At no point, however, does he call the connection that exists as such between the cause and its effect necessary. Only human judgments about the connections are necessary. Consistent with his criticism in the seventeenth discussion of the *Incoherence*, al-Ghazālī does not assume that causal connections in the outside world are necessary. While they will always happen just as they happen now, they are subject to God's will and thus can be different if He decides to change His arrangement—which we know He never will.

8

Causes and Effects in The Revival of the Religious Sciences

The voluminous *Revival of the Religious Science* (*Iḥyā' 'ulūm al-dīn*) is al-Ghazālī's major work on ethical conduct in the everyday life of Muslims. It is divided into four sections, each containing ten books. With the exception of the first two books, the first section discusses ritual practices (*'ibādāt*), the second, social customs (*'ādāt*), the third, those things that lead to perdition (*muhlikāt*) and should thus be avoided, and the fourth, those that lead to salvation (*munjiyāt*) and should be sought. In the forty books of the *Revival*, al-Ghazālī severely criticizes the coveting of worldly matters, reminding his readers that human life is a path toward Judgment Day and its corresponding reward or punishment. In the first book of his *Revival*, al-Ghazālī says that one cannot expect to achieve redemption in the afterlife without a firm knowledge of this world's causes and effects.[1] Throughout this book, however, he shows no interest in clarifying the ontological character of the connection between what we call a cause and its effects. In the introduction, he says that he wishes to avoid discussions that have no consequences in terms of human actions.[2] This focus on the practical results of human knowledge leads to an attitude in which it suffices to understand that God is the efficient cause of all events, regardless of whether He causes them directly or through the mediation of secondary causes. Nowhere in his *Revival* does al-Ghazālī even so much as hint that there are two competing explanations for God's creative activity. Since in this book, he wishes to give clear and detailed guidance to his readers on how to earn a place in the afterlife, there is no treatment of cosmology. Consequently, causal connections appear in the *Revival* without any scrutiny, just discussed according to how they should be treated in all practical contexts: as necessary connections.

Al-Ghazālī generally sees it as self-evident that the causes that we witness in our daily affairs are themselves only the effects of other causes. This is true for *all* causal connections and thus also true for human actions. Al-Ghazālī's stance on human actions is very simple: like all other events in this world, they are God's creation. This is true not only of the human act itself, but also of all causes that have led to it. A human act is prompted by the human volition (*irāda*), which is itself determined by one or more motives.[3] God creates these motives as well as the volition. The human motive is a judgment that is preceded and determined by two elements: the human's knowledge and his or her desire.[4] Al-Ghazālī discusses the example of a man walking on the street who realizes that a woman is walking behind him; he wishes to see the women and decides that to see her, he must turn around. The motive to turn around is triggered by the knowledge that the woman is there and the desire to see her. This motive may, however, be opposed by a countermotive (*ṣārif*), and thus it may not lead to the volition—and thus also not lead to the action—of turning the head.[5] Humans are not held responsible for their motives, because the motives depend both on the human's knowledge and on his or her desires, two things given to them. Humans are responsible for their volition, however, and thus responsible for those motives that they choose.[6] In his later work, *The Choice Essentials* (*al-Mustaṣfā*), al-Ghazālī clarifies that reason (*ʿaql*) cannot be considered a motive (*dāʿin*). Love of oneself and fear of pain are motives for human actions, and these motives are "dispatched" (*tanbaʿithu*) by the soul (*nafs*). Reason can only be a guide (*hādin*) that shows how best to realize these motives, which themselves can vary in strength.[7] The existence of different motives leads to deliberation (*fikr*) on the side of the human and may also lead to hesitation (*taraddud*). Al-Ghazālī treats the human volition as a causal effect of the motive, with the motive as a causal effect of the human's knowledge combined with his or her desires. The fact that God creates all elements in this causal chain—the human knowledge, the desire, the motive, the volition, and the human action—still does not diminish any of the human's responsibility for his or her actions.

The Creation of Human Acts

Al-Ghazālī explains his view of human actions a few times in his *Revival*, albeit never giving the topic the systematic treatment that would answer all the questions on this subject usually discussed by Ashʿarites. His most illuminating passages can be found in books thirty-one, thirty-two, and thirty-five of the *Revival*. The thirty-fifth book contains a particularly clear passage on how to understand divine unity (*tawḥīd*).[8] Earlier Ashʿarite theologians had differentiated between voluntary and involuntary human actions. When someone has a tremor, for instance, he has no control over certain of his actions and cannot be made responsible for them. The tremor is an involuntary act, a creation of God, similar to other aspects of the outside world that involve no human volition. The human must perform such actions, just as a tree is compelled to move its branches in the wind.

Although voluntary actions are also God's creations, as the Ash'arites stress, they differ in key ways from involuntary ones. With voluntary actions, humans make a decision in their will, and they are individually responsible for their choices. Earlier Ash'arites express the double nature of such actions by saying that humans *acquire* these actions while God *creates* them. The linguistic terms that humans "acquire" or "appropriate" (*kasaba* or *iktisaba*) their actions have their roots in the language of the Qur'an (Q 2:81, 2:134, 5:38) and precede al-Ash'arī. The earliest understanding of these ideas may simply have stressed the idea that humans are responsible for all that they perform, regardless of the cosmological explanation for how these actions are created.[9] With al-Ash'arī and his followers, the understanding of "acquisition" becomes more complex. Most of the Ash'arite theories of human action that precede al-Ghazālī assume that God gives a "temporary power-to-act" (*qudra muḥdatha*) to the human that allows him or her to perform the act that he or she has chosen. This implies that although God creates the action and its results in the outside world, the human is regarded as the agent (*fāʿil*) and the maker of the act.[10]

In his textbook of Ash'arite theology, al-Ghazālī upholds the doctrine that humans have power (they are *qādir*) over their actions, or else the obligations of the religious law would be meaningless.[11] However, the traditional implication that humans are the agents of their actions is incompatible with al-Ghazālī's cosmology in which there is only one agent or efficient cause (*fāʿil*). Understanding God's true nature (*tawḥīd*) includes the realization that there is no agent or efficient cause (*fāʿil*) other than God and that He is the one who creates all existence, sustenance, life, death, wealth, poverty, and all other things that can have a name.[12] The only true agent in this world is God.[13] In the thirty-fifth book of his *Revival*, al-Ghazālī implicitly dismisses the distinction between voluntary and involuntary actions. Opening and closing one's eyelids, for instance, is usually considered a voluntary action. But once a sharp needle approaches the human's eye, the human is compelled to close his eyelids:

> Even if he wanted to leave his eyelids open he couldn't, despite the fact that the compelled closing of the eyelids is a voluntary act. Once, however, the picture of the needle is perceived in his sense perception, the volition to close [the eyelids] appears necessarily and the movement of closing occurs.[14]

The voluntary closing of the eyelids is compelled by a volition (*irāda*), which itself is compelled by perceiving the needle approaching the eye. This is a causal chain in which the human knowledge causes the volition to develop in a certain way, and this volition causes the power-to-act (*qudra*), which causes the action. In classical Ash'arism, the temporarily created power-to-act distinguishes a voluntary human act from an involuntary one. Here in al-Ghazālī's thought, the power-to-act is a mere human faculty,[15] neither singled out from among the basic faculties of human life nor created in any way different from others of God's creation. The power-to-act is simply one link in a chain of secondary causes: "The volition (*irāda*) follows the knowledge, which judges that a thing is pleasing (or: agreeable, *muwāfiq*) to you."[16] The causal chain of knowledge, volition, power-to-act, and action applies to all voluntary human actions. Involuntary actions have a different causal chain, which does not include the human power-to-act, volition, and knowledge. Both types of actions, however, are the result of compulsion (*iḍṭirār*).

In most voluntary actions, the reaction of the human volition is not as immediate as in the case of the needle approaching the eye. A particular subclass of voluntary actions includes those actions that involve a human choice (*ikhtiyār*). Our previous example of the action of closing one's eye when a needle approaches is considered a voluntary action but does not involve a choice. The person whose eye is approached by a needle cannot choose an action that is alternative to closing the eyelid. The human will is compelled to close the eye. Human choice (*ikhtiyār*) means to be able to choose between alternatives. Those actions that involve choice, however, do not differ fundamentally from those performed without it. For al-Ghazālī, choice (*ikhtiyār*) means the human capacity of selecting what appears most agreeable or most beneficial (*khayr*) to us. Often the volition hesitates, and the intellect (*ʿaql*) finds it hard to decide whether something is agreeable or not. In such a case, we deliberate until we decide which actions appears to benefit us most. Once the process of deliberation leads to a clear knowledge about what promises to be best for us, knowledge "arouses" (or: "dispatches," *inbaʿatha*) the volition and thus initiates the part of the causal chain that leads to action. The judgment of the intellect follows what appears best to it, and in this sense, the human action is determined by what the intellect judges as best. This judgment often involves sense perception (*ḥiss*) and our inner sense of imagination (*takhyīl*). All connections in the causal chain between sense perception and human action are considered necessary:

> The motive of the volition (*dāʿiyat al-irāda*) is subservient to the judgment of the intellect and the judgment of sense perception; the power-to-act is subservient to the motive, and the movement [of the limb] is subservient to the power-to-act. All this proceeds from him [*scil.* the human] by a necessity within him (*bi-l-ḍarūra fīhi*) without him knowing it. He is only the place and the channel for these things. As for them coming from him? No and once again no![17]

Given the necessary predetermined character of all human actions, one might think that humans are forced (*majbūr*) to do the actions they perform. Yet that is not the case, al-Ghazālī stresses, as they still have a choice about how to act. Here he implicitly uses al-Fārābī's distinction between two types of necessity. In *The Balanced Book*, al-Ghazālī addresses the question of whether something that is not contained in God's foreknowledge can be created.[18] Viewed by itself (*yunẓaru ilā dhātihi*), every future contingency is a possible event. What the eternal divine will determines, however, is what is necessary, and its alternatives will not happen. A possible future event that is not contained in the divine foreknowledge will never be actualized. Such an event is considered "possible with regard to itself" (*mumkin bi-ʿtibār dhātihi*) yet at the same time "impossible with regard to something else" (*muḥāl bi-ʿtibār ghayrihi*).[19] It is rendered impossible by the divine will and foreknowledge. When the human decides his action—and here we return to the passage in the thirty-fifth book of the *Revival*—he decides between various alternatives that are possible with regard to themselves. He is unaware that all the alternatives that he will eventually reject have already been rendered impossible by the divine will and foreknowledge. Since the divine foreknowledge contains all factors that cause such decisions, it knows what appears most agreeable to the human intellect and thus knows which possible action will be actualized.

The human is a free agent (*mukhtār*) in the sense that he or she is the place (or substrate, *maḥall*) of the free choice (*ikhityār*). Free choice means that humans choose what appears most beneficial (*khayr*) for them; all human actions are motivated by self-interest.[20] Indeed, the human is forced by God to decide his or her own actions that are congruent with his or her self-interest. Responding to one of the oldest disputes of Muslim theology, al-Ghazālī says that one can say that humans lack agency in the sense that they are forced to make a choice (*majbūr ʿalā l-ikhtiyār*). Whereas causal connections in the outside world such as the one between fire and cotton are pure compulsion (*jabr maḥḍ*), and the actions of God are pure free choice (*ikhtiyār maḥḍ*), the actions of the human lie in between these two extremes. This is why earlier scholars decided to name this third category neither free choice nor compulsion. Following the terminology of revelation, al-Ghazālī says, they came to call it "acquisition" (*kasb*). This word is opposed neither to compulsion nor to free choice but "rather, for those who understand, it brings these two together."[21] Al-Ghazālī's novel interpretation of this term "acquisition" thus departs from earlier Ashʿarite teaching.[22]

Al-Ghazālī's teachings on how human acts are generated are quite reminiscent of the *falāsifa*'s teachings in general and of Avicenna's teachings in particular.[23] Avicenna describes human action as triggered by a volition, and this volition is "dispatched" (*mubʿatha*) either by a conviction (*iʿtiqād*) that follows from "an appetitive or irascible imaginative act" or by a rational opinion that follows from an act of cognitive thinking or from the conveying of an intellectual form.[24] These forms come from the active intellect. Whatever happens within the human mind is just a segment in a larger causal chain that begins with God, passes through the heavenly realm, passes through the human mind, and manifests itself in the material world outside our minds. In the thirty-fifth book of the *Revival*, al-Ghazālī includes a rather long parable of an "inquiring wayfarer" (*al-sālik al-sāʾil*) who investigates the cause of a certain written text—a writ of amnesty granted by a king—and follows its causal chain from the paper and the ink, via the human, to the heavenly realm until he reaches God. In this parable, the causes and effects in the material world are called the "world of dominion" (*ʿālam al-mulk*), the part of the chain that happens in the human mind is called the "world of compulsion" (*ʿālam al-jabarūt*), and the part of the causal chain that lies beyond the human in the heavenly realm is called the "world of sovereignty" (*ʿālam al-malakūt*).[25]

Al-Ghazālī's theory of human acts is an original contribution to a centuries-old debate in Muslim theology of how to reconcile God's omnipotence with His justice. If God creates human actions—by means of what appears to us as causal determination—how can He judge human actions and base reward and punishment on that judgment? Again, the answer lies in a simple causal chain. In the thirty-second book of the *Revival,* al-Ghazālī shows divine revelation to be one of the causes that God employs to lead his servants to salvation. The passage starts when an interlocutor asks why humans should ever bother with independent action if all is predetermined, including their fate in the afterlife. If everything is predetermined one might well refrain from doing anything and rest in fatalistic inactivity. Al-Ghazālī's answer focuses on statements of revelation, for the Qur'an and the *ḥadīth* corpus urge humans to act. Both texts contain the imperative "act!"[26] This formulation implies that one will be punished and censured for being disobedient unless one acts. The imperative language triggers a certain conviction in us, with divine words causing (*sabab*) our knowledge that God wants us to act. This knowledge is the cause of a decisive motive (*dāʿiya jāzima*) that propels those who believe in revelation to act and be obedient to God.[27] The motive is the cause for the volition that triggers the movement of the limbs. Thus, divine revelation becomes a cause of good deeds in a human. Al-Ghazālī explains how revelation causes the conviction (*iʿtiqād*) that one is punished for bad deeds and how that conviction causes salvation in the afterlife:

> (. . .) and the conviction [that some humans will be punished] is a cause for the setting in of fear, and the setting of fear is a cause for abandoning the passions and retreating from the abode of delusions. This is a cause for arriving at the vicinity of God, and God is the one who causes the causes (*musabbib al-asbāb*) and who arranges them (*murattibuhā*). These causes have been made easy for him, who has been predestined in eternity to earn redemption, so that through their chaining-together the causes will lead him to paradise.[28]

God's revelation is the cause of the human's fear of punishment in the afterlife. This fear, in turn, causes the human to heed the words of the prophets, which leads to good actions in this world that then causes the believer's redemption in the afterlife.[29] This chain is a further development of al-Juwaynī's notion that God makes a human intelligent and removes obstacles "to make God's path easy for him."[30]

One generation after al-Ghazālī, his follower Ibn Tūmart illustrates how God causes humans to become believers. He traces the human's decision to become a believer in God through a chain of causes and effects to God's prophetical miracle. In his *Creed of the Creator's Divine Unity* (*Tawḥīd al-Bārī*), Ibn Tūmart writes that a Muslim's belief (*īmān*) and piety (*ikhlāṣ*) is accompanied by the knowledge (*ʿilm*) of God's existence and His attributes. The believer's knowledge results from his search (*ṭalab*) for it. This search for knowledge is triggered by a volition (*irāda*), and the volition is the effect of desire and fear. Desire and fear are prompted by what revelation promises regarding reward and punishment in the afterlife (*al-waʿd wa-l-waʿīd bi-l-sharʿ*). Revelation, in turn, takes its authority from the trustworthiness of the

Prophet (*ṣidq al-rasūl*), and the Prophet's trustworthiness is established by the prophetic miracle (*al-muʿjiza*). At the end, this chain of events explaining human belief arrives at God because "the evidence of the miracle is by God's permission (*idhn Allah*)."[31]

Ibn Tūmart's narrative may not concur in all its details with al-Ghazālī's idea of what causes humans to pursue a devout and religious lifestyle.[32] Yet the two agree that the process can be described by a chain of secondary causes, one started and wholly controlled by God. In the thirty-second book of the *Revival,* al-Ghazālī makes his literary interlocutor summarize his own perspective on how human actions are the causes of their own redemption:

> You might say: The gist of this [*scil.* al-Ghazālī's] talk is to say that God has put a purpose (*ḥikma*) into everything. He made some human acts causes (*asbāb*) for the fulfillment of this purpose and for its attaining the objective that is intended in the causes. God (also) made some human actions obstacles to the fulfillment of the purpose.[33]

In all of his works, al-Ghazālī promotes the perspective that God's creation is a perfect conglomeration of causes and effects, with one creation harmoniously dovetailing with the next. In such works as his *Revival of the Religious Science* or in the less well-known *Intellectual Insights* (*al-Maʿārif al-ʿaqliyya*), where the complete harmony of God's creation is elaborated in fine detail, he does not discuss the cosmological nature of causal connection.[34] In these works, it suffices for al-Ghazālī to say that "in actual terms there is only one efficient cause (*fāʿil*) and He is the one who is feared, who is the object of hope, in whom one has trust, and upon whom one relies."[35] In an adaptation of Q 85:16, he says that God is the producer (or the active agent, *faʿʿāl*) of everything that He wills to create.[36] God is "the causer of the causes" or, as Richard M. Frank translates, "the one who makes the causes function as causes" (*musabbib al-asbāb*).[37] Although this term is considered of Avicennan origin, the expression originally used by Avicenna was most probably "cause of causes" (*sabab al-asbāb*).[38] The expression "the one who makes the causes function as causes" (*musabbib al-asbāb*) has a Sufi background and had already been used, for instance, by Abū Ṭālib al-Makkī in his *Nourishment of the Hearts* (*Qūt al-qulūb*).[39] "Cause of causes" expresses the Avicennan position that God is the starting point of all chains of secondary causes and that the relationship between such chains' elements is that of efficient causes to their effect. In contrast to what was likely the Avicennan formula, al-Ghazālī's term avoids committing to an explanation of how the "causes" come about. In al-Ghazālī's *Revival,* God is described as the one who "carries out His custom and binds the effects to causes in order to make His wisdom apparent."[40] All other existences are fully subservient operators (*musakhkharūn*) of Him and lack independence even to move a speck of dust.[41] Using these formulas, al-Ghazālī wishes to leave open whether God's arrangement of "causes" happens by means of secondary causal chains or by creating existences independently, side by side.

The Conditional Dependence of God's Actions

Al-Ghazālī postulates that God created the universe such that what we call an effect *always* exists alongside with what we call its cause. God will always create combustion in a cotton ball when it is touched by fire. In the *Incoherence*, al-Ghazālī argues that the connection between cause and effect is not necessary and could have been constructed differently. In the *Revival*, these connections are described as the result of God's voluntary actions. Al-Ghazālī posits that God's will, which exists from eternity, includes the voluntary decision always to combust a cotton ball if a certain other event—in this case, a close contact with fire—precedes it. In His eternity, God freely decides to limit His creative activity such that humans justifiably conclude that the connection between fire and combustion is an inseparable—and in this meaning: necessary—causal connection.

In the thirty-second book of the *Revival*, al-Ghazālī discusses the concept that humans must be thankful to God. Al-Ghazālī opens the passage with a question of a critical interlocutor who injects that since God is the creator of everything, it is not plausible that humans should be grateful to Him. God does not give anything in particular to His creatures for which they should be thankful. Indeed, God is the creator of all human actions and decisions—including the decision to be grateful to God. After the usual lamentation that this problem belongs to the "mystery of predestination," which he cannot share with his readers, al-Ghazālī explains: the action, which God creates within the human, is the gift for which one should be grateful. If that action is pleasing to God, it will lead to reward in the afterlife: "Your action is a gift from God and inasmuch as you are its place (or: substrate, *maḥall*), He will praise you."[42] The creation of the good action is the first blessing (*niʿma*) of God, and the reward in the afterlife for this very action is a second blessing from Him to the human (*niʿma ukhrā minhu ilayka*). This is again an example for how God has arranged the causes. God's creation of the good action in the human is a cause for His reward in the afterlife. God's first action (creating a good action in a human) is the cause for His second action (rewarding the human in the afterlife). This also applies when God creates thankfulness in a human:

> One of God's two actions is the cause (*sabab*) for the turning of the second action in the direction of what pleases Him. In each case God *has* the gratefulness (*al-shukr*). You are [simply] described as the one who is grateful (*shākir*), and this means that you are the place of the thing that "gratefulness" is an expression of. This doesn't mean that you are the one who brings gratefulness into existence (*mūjid*). Similarly, if you are described as someone who is knowledgeable (*ʿārif wa-ʿālim*), this doesn't mean that you are a creator of the knowledge and the one who brings it in existence. It rather means that you are a place for it and that it has already been brought into existence in you by the Eternal Power (*al-qudra al-azaliyya*).[43]

All causes that lead to salvation in the afterlife are individual acts of God (*fiʿl min afʿāl Allāh*). The causal chain for how God's revelation leads to salvation in the afterlife is characterized as follows: God sends humans a revelation that gives them knowledge about the connection between deeds in this world and redemption in the next. God uses revelation as a secondary cause to create this knowledge in humans. Next, the knowledge of this connection causes a motive (*dāʿiya*) that encourages the obeying of God's imperatives and the performance of good deeds. This motive is also God's creation. The desire to avoid pain in the afterlife and to achieve the pleasures of paradise combined with the knowledge that comes from revelation cause the human motive to act justly and thus please God. Pleasing God will indeed lead to the enjoyment of paradise. God's action of creating pious deeds for the human is the cause of another of God's actions, namely, reward in the next life.

Al-Ghazālī's explanation for how actions in this world lead to reward or punishment in the hereafter is essentially the same as Avicenna's explanation. In his *Pointers and Reminders* (*al-Ishārāt wa-l-tanbīhāt*), Avicenna addresses the question of why God punishes humans if their actions are predetermined. Punishment for one's transgressions, he says, is like a disease that affects the body following gluttony (*nahma*): "Punishment is one of the consequences that past states have led to. The occurrence of these past states and the occurrence of what follows them are both inevitable."[44] Punishment or reward in the hereafter is a causal effect of one's actions in this world. Our good actions in this world are thus the causes of happiness in the next work, al-Ghazālī says, and our bad actions are the causes of distress, just as medicine is the cause of recovery from a sickness and poison the cause of death.[45]

Humans have every reason to be grateful to God, al-Ghazālī argues, since He creates in them the actions that later cause their redemption. Next, al-Ghazālī addresses an objection that he does not explicitly state, although his answer makes the nature of the objection quite evident: if all human actions are in reality God's actions, al-Ghazālī expects his readers to ask, why does He not simply transfer a human into paradise without the whole process of creating knowledge in the human, creating a motive, and creating human actions? If God is truly omnipotent, could He not have made redemption much easier for His creation? Al-Ghazālī answers:

> One of God's acts is the cause (*sabab*) for another; I mean that the first one is the condition (*sharṭ*) for the second. The creation of the body, for instance, is the cause for the creation of the accident (*ʿaraḍ*), since He does not create the attribute before it. The creation of life is a condition for the creation of knowledge and the creation of knowledge is a condition for the creation of volition. All these are from among God's actions and one of them is a cause for the other, meaning that it is a condition. Being a condition means that only a substance (*jawhar*) is prepared to receive the act of life, and only something that lives is prepared to receive knowledge. There is no reception of volition other than by something that has knowledge.

> Therefore, "some of God's actions are a cause for others" means this and it doesn't mean that one of His actions brings the other into existence. Rather [one of God's actions] clears the way for a condition [whose fulfillment is required] for the existence of another of God's actions.[46] If the truth of this is grasped, it elevates to the [higher] stage of belief in God's unity that we have spoken about.[47]

God cannot simply move humans from their cradle into paradise, because the "conditions" of entering paradise are not yet fulfilled when the human is still in the cradle. Entering paradise has a specific cause. Having a cause means one or more conditions must be fulfilled before the creation of the event can take place. Without the fulfillment of these conditions, God cannot create the event. Thus God cannot create someone's entry into paradise unless He has earlier created good deeds in the person. Good deeds, in turn, cannot be created in a human without a prior volition for performing good deeds. The volition requires the prior existence of knowledge. Knowledge, in turn, requires life, and life can only be created in a substance (*jawhar*), be it in a body or in a stable incorporeal entity such as a celestial or human soul.[48] The human's good deeds, his volition, his knowledge, his life, and his substance are all individual elements in a chain of conditions that must be fulfilled before the human can enter paradise. A prophetical *ḥadīth* says that "people will be led into paradise in chains." For al-Ghazālī, this statement expresses the idea that one can only enter paradise "led by chains of causes" (*maqūd bi-salāsil al-asbāb*).[49]

A second passage in al-Ghazālī's *Revival* confirms the view that God's creative activity is limited by rather strict conditions. In this passage from the thirty-fifth book on understanding God's unity (*tawḥīd*), al-Ghazālī rejects the view that knowledge generates (*wallada*) volition, volition generates the human's power-to-act, and this power then generates the movement of the limbs. The reader knows that here al-Ghazālī refers to a Muʿtazilite understanding of the "generation" (*tawallud*) of human acts and their effects. The Muʿtazilite position is wrong, al-Ghazālī stresses: "[t]o say that some of these come into being (*ḥadatha*) from others is pure ignorance, no matter whether one calls it 'generating' (*tawallud*) or anything else." All these events go back to an entity (*maʿnā*) that is known as the "Eternal Power" (*al-qudra al-azaliyya*), and only those who are deeply rooted in knowledge (*al-rāsikhūna fī l-ʿilm*) understand the true nature (*kunh*) of this being.[50] In the next sentence, al-Ghazālī explains some of the workings of the "Eternal Power":

> Some of the objects of this power (*muqdarāt*), however, are arranged so that their coming into being follows others. The arrangement (*tartīb*) is that something conditioned (*al-mashrūṭ*) follows after the condition (*al-sharṭ*). A volition only comes out of (*taṣduru ʿan*) the Eternal Power after knowledge, and knowledge only after life, and life only after there is a substrate for life. And like one cannot say that life is brought into being by the body, which is the condition for life, so [one cannot say this] in the case of all other steps of the arrangement.

> Some conditions are apparent to the ordinary person, but others are only apparent to the elite (*al-khawāṣṣ*), who experience unveiling by the light of the Truth.
>
> In any case, nothing preceding precedes and nothing following follows except by means of right and necessity. This applies to all of God's actions.[51]

According to this passage, the conditioned procession of body, life, volition, and human actions is "by means of right and necessity" (*bi-l-ḥaqq wa-l-luzūm*). Richard Gramlich, in his valuable German translation of books 31–36 of the *Revival*, renders the Arabic word *ḥaqq* (lit. "truth," or also "one's due") in such passages as "laws" or "regulations" (*Gesetzmäßigkeiten*), probably meaning the laws of nature.[52] Although it is not impossible that al-Ghazālī had in mind the lawful character of the arrangement of conditions and the conditioned, it seems a long stretch to extract this meaning from the admittedly highly ambiguous Arabic word *ḥaqq*. More likely, al-Ghazālī means to say that the arrangement follows a rightness that gives each element its allocated due. In Ashʿarite theology, "justice (*ʿadl*) is to put things in their appropriate place."[53] The word "necessity" that follows after this explanation is less problematic in its meaning, though more problematic with regard to what it implies. It suggests that God's actions are the result of an arrangement that works by necessity and leaves no room for alternatives.

In some books of his *Revival*, al-Ghazālī views causes as events that "clear the way" (*mahhada*) for the creation of their effects. The perspective that understands causes as "conditions" for the existence of their effects suggests that God cannot simply create as He wishes, but rather, He must follow a matrix of such conditions. Al-Ghazālī had already put forward a very similar position about conditions for God's creation in the Third Position (*al-maqām al-thālith*) of the seventeenth discussion in the *Incoherence*. Here in the *Revival*, as in his *Incoherence*, al-Ghazālī avoids clarifying the nature of these conditions. This necessity can be either the result of God's choosing or the conditions that are imposed upon God's actions. Al-Ghazālī leaves open the idea whether God Himself chooses such conditions upon His actions or whether they are requirements beyond God's control with which He must comply.

The Conditions of a Creation That Is the Best of All Possible Creations

Assuming that the conditions that apply to God's actions are beyond God's control would mean following Avicenna and accepting that God is not a free agent who cannot choose His actions. Because every causal connection is essentially such a condition and a restriction upon God's actions, adopting the view that God cannot violate causal connections, even if He wanted to, would make the world in which we live necessary while depriving God of all freedom for His actions. For Avicenna, God necessarily acts to establish the best order. Avicenna's position simply does not allow for the world to be any different from this best and necessary order. The divine providence (*al-ʿināya al-ilāhiyya*) that allows for creation results from God being the pure good (*al-khayr al-maḥḍ*) that only emanates the best. The order that follows from God's knowledge is the best order that is possible. For Avicenna, God does not have a particular desire to create the best of all possible worlds; rather He simply cannot help doing so. Everything that He creates is the best of all possible creations.[54]

Al-Ghazālī gives a detailed account of these teachings in the two books in which he reports the position of the *falāsifa*.[55] In the book preserved in MS London, Or. 3126, al-Ghazālī reproduces the relevant passages from Avicenna's *Pointers and Reminders* and from the metaphysics of his *Healing*, while adding his own comments: if one studies the animals and plants and realizes that nature (*al-ṭabīʿa*) cannot generate all these details by itself, one understands that all this must be (*lā maḥāla*) the product of divine providence. The same is true if one evaluates the private interchanges (*muʿāmalāt*) between people. Different people have different habits and different understandings of justice. Divine providence responds to these differences by sending prophets to teach the varied people one true sense of justice. The existence of these and other benefits (*manāfiʿ*) cannot possibly come from any source other than God.[56]

Although these thoughts aim to illustrate Avicenna's teachings, they are not, strictly speaking, part of the latter's doctrine. Observational or empirical evidence of the perfection of God's creation plays next to no role in Avicenna's thought. He merely says that "you cannot deny the wondrous manifestations (*al-āthār al-ʿajība*) in the formation of the world (. . .) all of which do not proceed by coincidence but require some kind of ordering (*tadbīr mā*)."[57] For Avicenna, this arrangement—however perfect it may appear—cannot count as evidence for this world's perfection. The perfection can only be deduced from reflecting on God's knowledge, which is the origin of divine providence. The empirical perception of this world's perfection is a motif of Sufi literature and appears prominently in Abū Ṭālib al-Makkī's *Nourishment of the Hearts* (*Qūt al-qulūb*), among other places. It is also an element of traditional Ashʿarism. For Ashʿarites, the skillfulness (*itqān*) and orderliness (*intiẓām*) of God's creation is a clear sign that God has all-encompassing knowledge.[58] Such arguments based on design and teleological motifs also play an important role in al-Ghazālī's theology.[59] In his *Balanced Book on What-to-Believe*, al-Ghazālī stresses that all of God's creations are skillfully and wisely arranged. Studying God's creation makes one realize how perfectly it is ordered. Here, as in many other places, al-Ghazālī uses the parable of a skillfully handwritten text to point to the many accomplishments of its author and scribe.[60]

In the thirty-fifth book of his *Revival*, al-Ghazālī includes a relatively brief passage in which he also argues that this creation is the best possible creation. The teachings on these two pages became famous for their compressed formula: "There is in possibility nothing more wondrous than what is" (*laysa fī-l-imkān abdaʿ mimmā kān*).[61] This teaching was already seen as controversial in al-Ghazālī's lifetime, and over the following centuries, it stirred a long-lasting debate among Muslim theologians about what exactly al-Ghazālī meant to express here and whether the statement that this world is the best of all possible creations is actually true.[62] Once more, al-Ghazālī failed to be explicit about the theological and philosophical implications of his teachings. This passage in the *Revival* ends with a cryptic statement that the position expressed is a sea of arcane matters in which many have already drowned. Behind it lies the secret of predestination (*sirr al-qadar*) in which the majority of people wonder in perplexity, and those to whom things have been unveiled (*al-mukāshafūn*) are forbidden to divulge the secret.[63] Later, al-Ghazālī commented on this passage in a short explanatory book, *The Dictation on Difficult Passages in the Revival* (*al-Imlāʾ fī ishkālāt al-Iḥyāʾ*), written in response to critics. Here, he confirms the position that this world is the best of all possible creations but hardly adds anything that could clarify the theological background.[64]

Al-Ghazālī took significant parts of this two-page passage on the best of all possible worlds from Abū Ṭālib al-Makkī's Sufi handbook, *The Nourishment of the Hearts*.[65] What interested al-Ghazālī about al-Makkī's earlier text was the apparent orderliness of the world's design that al-Makkī illustrates. Based on these examples, al-Ghazālī posits his theory that this creation is the best of all possible ones, a conclusion not explicitly found in al-Makkī's work. The passage marks the end of al-Ghazālī's explanation of why one must "believe in God's unity" (*tawḥīd*), at which point the text tries to connect God's unity with the idea of "trust in God" (*tawakkul*). The discussion of *tawḥīd* makes clear, al-Ghazālī's literary interlocutor claims, that human actions are not free, but rather they are compelled by the causes (*asbāb*) that determine the human's volition. All events in God's creation, including human actions, are compulsory (*al-kull*^{u} *jabr*^{un}). If this is the case, the interlocutor asks, why does God reward and punish humans for their actions? Since such actions are in reality God's actions, why does God become angry at His own actions? Al-Ghazālī's response refers the reader back to the passage in which he writes that one of God's earlier actions, namely, the action that He creates within a human, is the cause for one of God's later actions, that is, bestowing reward or punishment in the afterlife.[66] Only those who have achieved a high degree of trust in God will understand this aspect of *tawḥīd*.

Complete trust in God, al-Ghazālī continues, results from a firm belief in God's mercy (*raḥma*) and in His wisdom (*ḥikma*). Such belief is itself created by an inquiry into "the one who makes the causes function as causes" (*musabbib al-asbāb*). It would take too long, al-Ghazālī writes, to explain how those to whom truths have been revealed reach their strong level of belief in God's mercy and wisdom. One can only give the gist (*ḥāṣil*) of their method: the one who aims to develop a firm and decisive trust in God believes that, if God had given all humans the understanding of the most understanding among them, the knowledge of the most knowledgeable among them, and the wisdom of the most wise among them, and if He had taught them the secrets of this world and the hereafter, and if He had given them the opportunity to order this world anew, they could not have come up with an arrangement better than or even different from this one, not even by a gnat's wing or a speck of dust.[67]

Al-Ghazālī copied this last long sentence almost verbatim from Abū Ṭālib al-Makkī's book.[68] In al-Makkī's text, however, the sentence has a very different function. He constructs an argument that aims to illustrate the fact that God created this world in accord with human means for understanding it. Al-Makkī wishes to show that God's creation is in a perfect order, as viewed from the perspective of humans. The compatibility between human minds and the order of God's creation gives humans reason to trust the accuracy of their knowledge and their understandings of the world, and it allows them to make predictions regarding future events in this world. According to al-Makkī, trust in God (*tawakkul*) is synonymous with trust in the orderliness of this world, which is a direct result of God's mercy. Al-Makkī writes:

> God carried out this creation according to the arrangement of the minds (*ʿalā tartīb al-ʿuqūl*) and according to the customary notions (*maʿānī l-ʿurf*) and habitual arrangements that come with the well-known causes and familiar mediators according to the yardstick that is imprinted in the minds and that they have been endowed with.[69]

Al-Ghazālī does not reiterate al-Makkī's conclusion that God created this world according to the arrangement of human minds. What fascinated him was the implication that this world is most orderly in its design. As a result, he copied only that part of al-Makkī's text that serves as a fitting illustration for the two facts that this world is created according to a perfect arrangement and that the arrangement is accessible to human understanding. Even the most perfect human minds will perceive nothing but orderliness in the world. For al-Ghazālī, this order is not the result of a simple accord between human minds and God's creation. He comes to a more radical conclusion and says that the order is the best of all possible designs for the world. This is true in absolute terms, not just according to human understanding:

> Everything that God distributes among humans, such as sustenance, life-span (*ajal*), pleasure and pain, incapacity and capacity, belief and unbelief, pious and sinful actions, is all of sheer justice, with no injustice in it, and pure right, with no wrong in it.
>
> Indeed, it is according to the necessary right arrangement (*ʿalā l-tartīb al-wājib al-ḥaqq*) in accord to what should be (*ʿalā mā yanbaghī*) and like it should be (*kamā yanbaghī*) and in the measure in which it should be (*wa-bi-l-qadr alladhī yanbaghī*); and there is in possibility nothing more excellent, more perfect, and more complete than it.[70]

If people live with the impression that their lot in this world is unjust, al-Ghazālī explains, they should wait for the next world to see how they will be compensated for the losses that might be inflicted on them in this world. Those who gain advantages in this world by doing injustice, however, shall have to pay for that in the afterlife.

Imperfections in this world are real, al-Ghazālī says, yet they serve the higher purpose of realizing the most perfect world. In the twenty-second book of the *Revival*, al-Ghazālī says that desire (*shahwa*) and anger (*ghaḍab*) are character traits responsible for much harm in this world. Yet they are necessary because without desire for food and sex, humans could not survive; without anger, they would not be able to defend themselves from those things that threaten their lives.[71] Even the most perfect arrangement for the world includes a certain amount of harm that manifests itself as imperfections that, in turn, point toward the perfect. If there were no sickness, the healthy would not enjoy health. If beasts had not been created, the dignity of man would not have become manifest. Although the punishments in hell may seem like imperfections, they are necessary in order to honor those who will enter paradise and show the righteous the extent of their reward. In a sense, the merits of the righteous are ransomed by the suffering of the unbelievers. This is like saving the health of a person by amputating his gangrenous hand. Perfection and imperfection do not become apparent in absolute terms but only in relation to each other. The perfect, therefore, needs the imperfect in order to demonstrate its perfection: "[God's] generosity and [His] wisdom require the simultaneous creation of the perfect and the imperfect."[72]

The notion that the best of all possible worlds necessarily requires the creation of imperfections comes from philosophical literature. Eric Ormsby, who offers an insightful and detailed analysis of this passage, observed that al-Ghazālī had taken this idea from the works of Avicenna.[73] In his *Pointers and Reminders*, Avicenna writes that it is necessary to create things that are lacking in perfection inasmuch as they are bad or harmful (*sharr*).[74] In order to realize a perfect order, it is also necessary for the good to predominate over the harmful. Yet some harm *must* be there, or else the good would not be able to show its advantages (*faḍīla*). A perfect world, therefore, must contain creations that are absolute evil as well as those in which the evil aspects predominate over the beneficial ones. This is because a small amount of evil preserves (*taḥarraza*) the good creations and safeguards that harmful effects will always be limited. All this is taken into account in God's providence for His creation. God, who according to Avicenna pursues no goals for His creation and has no desires, creates the harmful as if He desires it by accident. One can therefore say that harm enters God's creation by accident, like a disease accidentally affects living beings.[75]

Although harm affects existence accidentally—that is, harm is not necessary for the existence of any kind of world—harm is indeed *necessary* for the realization of a world that is the best of all possible worlds. It is not an undesired side effect of creating the good, but rather it is intrinsic to its establishment. The creation of perfection necessarily requires the simultaneous creation of imperfections for the perfect to exist. Harm is a necessary concomitant of this world's good constitution: "[t]he existence of evil is a necessity that follows from the need for the good."[76] For Avicenna, harm is a privation of perfection, and the most essential privation is the nonexistent (*al-ʿadam*).[77] Something that exists is always better than something that does not exist. Therefore, the fact that God creates this world is a benefit that by itself outweighs many of its privations. If things are affected by harm, they suffer from privation of perfection. Such imperfections manifest themselves as ignorance, for instance, or as physical weakness, deformation, pain, or distress.[78] Harm and evil exist, however, only in the sublunar sphere of generation and corruption, and in that sphere, they affect only individuals and not classes of beings.[79] The heavenly spheres and the universals are perfect and not affected by it. Echoing Aristotle, Avicenna says the harm in the sublunar sphere is insignificant (*ṭafīf*) in comparison to the perfection of the rest of existence.[80]

Al-Ghazālī was evidently impressed by Avicenna's solution to the question of theodicy. In the thirty-second book of the *Revival*, al-Ghazālī mentions the example of a father who forces his infant son to undergo the painful process of cupping in order to heal an illness. This father is more beneficial to the child than his mother who, in her love, wishes to spare him all distress.[81] He elaborates further on this example in his *Highest Goal* (*al-Maqṣad al-asnā*), in which he comments on the divine name "the Merciful" (*al-raḥmān al-raḥīm*). To the objection that God should not be called merciful as long He creates so much poverty, distress, sickness, and harm in His creation, al-Ghazālī responds with a parable:

> A mother cares lovingly for her small child and does not allow that it undergoes cupping, yet the father is insightful (*ʿāqil*) and forcibly treats the child with it. An ignorant person thinks that only the mother is merciful but not the father. An insightful person knows that it is part of the perfection of the father's mercy, his affection, and his complete compassion when he causes pain to the child by making it undergo cupping. [The insightful person also knows] that the mother is an enemy to the child in the guise of a friend. The pain [caused by cupping] is small and yet it is the cause for much pleasure. So it isn't harmful, rather it is good.[82]

This explanation applies to all imperfections and harm in this world. They serve the larger good of preserving the perfections: "There is no harm in existence which does not carry inside some good; were that harm eliminated, the good that it has inside would vanish. The result would be an increase in harm in comparison to what it had before."[83] God referred to this relationship when in a *ḥadīth* He revealed: "[m]y mercy outstrips my wrath." Beneath all this insight, however, lies a secret that revelation cannot fully disclose.[84]

While al-Ghazālī evidently accepts Avicenna's justification of why harm exists in God's creation, he does not accept the metaphysical premise that creating perfection is a necessary result of the divine nature. Al-Ghazālī, for instance, nowhere says that it is in God's nature to create the best creation. Knowledge about the best of all possible worlds is not acquired by reflection on God's attributes. Rather, we know that God creates the best by looking at His creatures. In his *Dictation on Difficult Passages in the Revival,* in which al-Ghazālī apologetically comments on teachings in the *Revival* that prompted opposition among his peers, he devotes little more than one page to the issue of the best of all possible worlds. Here he explains how we know that this creation could not be more perfect:

> If everything that God creates were defective in comparison to another creation that He could have created but didn't create, the deficiency that would infect this existence of His creation would be evident just like it is evident that there are in His [actual] creation particular individuals whom He did create deficient in order to show thereby the perfection of what He creates otherwise.[85]

God creates deficiencies in order to point those insightful humans toward the perfection of His creation. Without the manifest imperfections, the perfection of other creations would simply remain unknown. Imperfect creations draw attention to the perfect ones and make God's perfection obvious:

> Inasmuch as He shows humans His perfection, He points them towards His deficiency; and inasmuch He makes them know His omnipotence, He makes them see His incapacity.[86]

Studying the created beings (*makhlūqāt*) is the only means of knowing that this world is the most perfect. Revelation can only hint at this fact because revealing this world's perfection to the masses of the people would make its perfection void. In his *Dictation*, al-Ghazālī says that the subject of the best of all possible worlds is one of the secrets of worship (*asrār al-ʿibāda*) and cannot be discussed openly. God gives us precisely the right amount of knowledge to enable us to contribute our best actions to this world. The amount of knowledge He gives us is part of the most perfect arrangement of His creation. If people with weak intellects were to become aware that everything is foreseen and in a perfect order, they would draw wrong conclusions and be prompted to perform actions less perfect than those they do without this knowledge. Would God have given those humans destined to enter paradise a way to know their future bliss, for instance, they would never arrive. Such knowledge would lead to bad actions and prevent redemption in the hereafter. The same is true for one who has been told that he will end up in hell. He would make no further effort to restrain his bad passions. It is part of God's perfect arrangement to prevent all but the most learned from gaining knowledge of this world's perfection.[87]

The Necessity of the Conditions in God's Creation

Al-Ghazālī teaches that God chooses to show utmost mercy to His creation and that He creates the best of all possible worlds. He is like the insightful father who chooses to be merciful to his child. Yet, such as the actions of this wise father, God's wise actions can inflict pain upon His creation. It is a sign of wisdom that the world is created with a certain degree of harmfulness intrinsic to it. Even if God freely chooses to follow the wisdom of the plan to create the world, once He decides to create the best possible world, He no longer has a choice about what to create. Among all possible worlds, there is only one that is the best of all possibles. In his *Dictation on Difficult Passages in the Revival*, al-Ghazālī says that God's actions are the result of the free choice (*ikhtiyār*) that this free agent (*fāʿil mukhtār*) has about His actions. Once God chooses to create the most perfect world, however, His actions follow a necessary path that is dictated by wisdom. Al-Ghazālī explains how wisdom (*ḥikma*) determines the divine actions:

> Once God acts, it is only possible for Him to do what is [within] the limit that the wisdom (*al-ḥikma*) requires, of which we know that it is [true] wisdom. God lets us know about this only because we know the channels of His actions and the origins of His affairs and because He verifies that everything which He decided and which He decrees in His creation is by means of His knowledge, and His will, and His power, and that it is of utmost wisdom, of extreme skillfulness, and of the full amount of the creation's generosity. [God lets us know about this] because the perfection of what He creates is a decisive argument and an evident demonstration for His perfection in the attributes of His glory (*jalāl*) that make it necessary to call Him the most glorious (*al-mūjiba li-ijlālihi*).[88]

The divine motive to create the best of all possible worlds explains why God creates this world as it is and why He puts specific conditions on achieving certain benefits. It explains, for instance, why God does not move humans immediately from the cradle to paradise. In the thirty-second book of his *Revival,* al-Ghazālī only partly answers this question. Certain conditions exist, which must be fulfilled for humans to enter paradise. Humans have to perform pious deeds, which in turn require the prior existence of a volition that triggers these deeds. The volition requires the prior existence of knowledge on the part of humans, and so forth. As we have already said, these conditions may also be understood as causal connections. The correct sort of knowledge that an individual has will cause the correct kind of volition, which will cause the correct kind of action to cause entry into paradise. Yet the larger question remains: why do all these conditions—or causal connections—exist? Since God has a universal and detailed pre-knowledge of all events past and future, and since He creates all human actions, why can He not make his chosen people enter paradise even before they experience the hardship of birth and childhood? Why all these complications? Why not simply create human souls and place them into paradise?

For an Ashʿarite, there is no answer to this question and thus no reason to ask. Yet in this particular question, al-Ghazālī clearly goes beyond the Ashʿarite approach and ventures to answer the problem openly, albeit without discussion. The arrangements of the world in which we live are those of the best of all possible worlds. This world cannot be better, because it is already the best possible. It also cannot be worse, because God decided in His mercy not to satisfy Himself with less than what is the best. The arrangement is therefore determined by God's decision to create the best possible world.

In his *Revival,* al-Ghazālī states a few times that God's actions are necessary. One of the most outspoken passages is in the thirty-fifth book, which focuses on *tawḥīd*, shortly before al-Ghazālī writes that this creation is the best possible one. In this passage, creation is described as a necessary process:

> Everything between the heaven and the earth happens according to a necessary arrangement and a binding rightness and one cannot imagine that it would be different from how it happens or different from this arrangement that is found. What comes later comes later only because it waits for its condition. The conditioned (*al-mashrūṭ*) is impossible before the condition (*al-sharṭ*). The impossible cannot be described as being within God's power. Therefore knowledge only comes after the sperm because the condition of life needs to be fulfilled, and volition only comes after knowledge because the condition of knowledge needs to be fulfilled. All this is the way of the necessary (*minhāj al-wājib*) and the arrangement of the rightness (*tartīb al-ḥaqq*). There is no play in it and no coincidence (*ittifāq*); rather all this is through wisdom and ordering.[89]

God's creative activity follows a "necessary arrangement" (*tartīb wājib*) and contains a "binding rightness" (*ḥaqq lāzim*) that cannot be otherwise. The necessity of God's actions exists "through wisdom and ordering" (*bi-ḥikma wa-tadbīr*); wisdom dictates the conditions of the best of all possible worlds, and God chose to abide by its precepts. The necessity of God's creation also appears in the thirty-second book of his *Revival*. In a sentence that we have already quoted above, al-Ghazālī says that "nothing preceding precedes and nothing following follows except by means of the rightness and the necessity (*bi-l-ḥaqq wa-l-luzūm*)."[90] The necessity of God's order is also expressed in the passage where he describes the best of all possible worlds as created "according to the necessary right arrangement" (*ʿalā l-tartīb al-wājib al-ḥaqq*) and "in accord to what should be" (*ʿalā mā yanbaghī*).[91] The necessity in this passage need not be the absolute necessity of Avicenna, but rather a necessity relative to the decision to create the best possible world.

We have thus far given a relatively smooth interpretation of different motifs in the *Revival*. If these interpretations were all that have been proposed, however, al-Ghazāli would not be seen as such a controversial author. The above quoted passage includes at least one formula that cannot be explained by referring to the necessities that spring from the decision to create the best of all possible worlds. Whereas it is plausible that the best order requires that God's actions abide with certain conditions, al-Ghazālī continues and says that any arrangement different from what exists is impossible, and "the impossible cannot be described as being within God's power" (*al-muḥāl lā yūṣafu bi-kawnihi maqdūr*[an]).

There are two ways to understand impossibility in this sentence, a strong way and a weak way. Triggered by this passage, Richard M. Frank proposed these two interpretations.[92] Frank prefers the strong way of understanding impossibility, which suggests that God's actions have to comply with the necessity of God's nature. God *must* follow His generosity (*jūd*); God must create the best of all possible worlds. When al-Ghazālī says, Frank has argued, that God's decisions are made by pure free choice (*ikhtiyār maḥḍ*), he simply means that God is not distracted from choosing what is truly beneficial (*khayr*) for His creation. In reality, however, God cannot help choosing the good, which means that effectively He does not actually choose and cannot make free decisions about His actions. The creation of this world proceeds from His lack of liberty as a necessary act. Reading such strong sense into the words "impossible" and "necessary" assumes that the actions of al-Ghazālī's God—like the God of Avicenna—are determined by His nature. This is the God of the *falāsifa* whose will is identical with His knowledge and His essence.

Al-Ghazālī, however, rejected the idea that creation takes place as a direct and inevitable consequence of God's being. In Avicenna, God's knowledge is the origin of the best of all possible worlds. In al-Ghazālī, however, it is God's will. God chooses to be generous, and this choice is undetermined. God's will is therefore the undetermined determining factor of creation. This idea is expressed forcefully in many of his writings, and Frank acknowledges the importance of this motif in al-Ghazālī's theology.[93] Failing to detach God's will from His knowledge and thus constructing a God who acts out of necessity rather than out of His decisions is al-Ghazālī's main objection against the *falāsifa* in his *Incoherence*.[94] For al-Ghazālī, it is an affront to reason to claim that it is not in God's power to create this world differently from how it is. Because we can easily imagine this world to be larger or smaller, for instance, it is therefore not impossible for it to have been created larger or smaller. The world was possible before it came into existence, and God was never incapable of creating it.[95] In his *Letter for Jerusalem*, al-Ghazālī says that God chooses what He creates among the alternative (*ḍidd*) of not creating it.[96] In the context of Ash'arite theology, al-Ghazālī expresses the divine *liberum arbitrium*—the divine capacity to choose freely—in the tenet that God's will and His knowledge are, like His life, power, hearing, seeing, and His speech, attributes that are not identical to but rather "additional to God's essence" (*zā'id 'alā l-dhāt*).[97] The ubiquity and forceful presentation of this theological motif makes it all but impossible to accept Frank's strong interpretation of why another creation would not be within God's power to create.

The impossibility of any other creation means—according to a second, weaker reading—that the existence of what God does not will to create (that which He knows will never exist) though possible in itself, is actually impossible. This formulation refers to the Farabian distinction between the two types of necessities, restated by al-Ghazālī in his *Balanced Book*. Any future contingency that God knows He will not create is "possible with regard to itself," yet "impossible with regard to something else," meaning impossible with regard to God's foreknowledge. Creating what is not part of God's foreknowledge cannot happen, even if it remains possible in itself. It would turn God's knowledge into ignorance, and that is simply impossible. Therefore, one can say that whatever is not part of God's foreknowledge "is not within God's power to create in the sense that its existence would amount to an impossibility."[98] A creation different from this one is not impossible in absolute terms, as God could have chosen to create it. But it remains impossible relative to the choices God has already made.

9

Cosmology in Works Written after *The Revival*

At various points in his *Revival,* al-Ghazālī describes God's creation as a network of conditions (*shurūṭ*). Only the fulfillment of particular conditions enables God to bring new beings and new events into existence. Humans understand these conditions as causal connections, in the way that God wishes humans to understand conditions; it is God who creates the human's cognitions. Yet causal connections are not the only viable explanation for how God's creation comes about. It may also be the case that God creates the fulfillment of a condition directly and mono-causally and that He produces the event that follows the fulfillment of that condition in the same way. The condition for the combustion of cotton, for instance, is that fire touches it. Al-Ghazālī maintains that God may actually create the touching of the fire to the cotton and the combustion of the cotton as two independent consecutive events. Alternatively, it may be the case that God causally connects the cotton's combustion to the fire's touching it. In both cases, however, God is the ultimate efficient cause of the cotton's combustion. In the first explanation, God would be the immediate cause; in the second, He acts as the only efficient cause at the head of a chain of secondary causes whose effect is the cotton's combustion.

Al-Ghazālī seems to have chosen this uncommitted position quite early in his career. Once he decided that there was no epistemological criterion that could determine which of the two explanations was true, he no longer seemed bothered by the question. He only discusses the equal possibility of these two explanations in his very last work, written shortly before his death, which I will discuss later in this chapter. In most of his writings, however, al-Ghazālī teaches an understanding of the universe in which the cosmological alternative between occasionalism and secondary causality does not appear. For instance, when he teaches that there is a causal connection between the human's knowledge and the way the human acts, such a connection is viable in both kinds of universes. In his *Revival of the Religious Sciences*, al-Ghazālī wishes to convey the understanding that one bears responsibility for one's place in the afterlife and that this care requires focusing on one's actions. However, actions are triggered by a will and motives, which in turn depend on one's knowledge. Consequently, al-Ghazālī wishes his readers to acquire the kind of knowledge that can turn this causal chain toward the right direction. This perspective is different from that of earlier Ash'arites, who taught that fulfilling the prescriptions of Sharī'a can gain one a place in the afterlife. The *Revival*'s underlying assumption is that the right kind of knowledge leads to the development of a good character (*khalq*), which will almost automatically lead to good actions and redemption in the afterlife. The connections between these elements—including the connection between human actions in this world and redemption in the afterlife—may be described as causal.

God's Creation as an Apparatus: The Simile of the Water Clock

The theological notion that God creates and controls everything in His creation through a network of harmoniously interdependent events was more important for al-Ghazālī than committing himself to one specific cosmology. In many books of the *Revival*, the connections between events are referred to as "conditions" (*shurūṭ*); in most books, however, they are referred to as "causes" (*asbāb*) because that is how most readers are familiar with them. God creates a network of causes and effects in order to accomplish a goal, and that network can be likened to an apparatus that produces a certain outcome. In his commentary on the ninety-nine names of God titled *The Highest Goal in Explaining the Beautiful Names of God* (*al-Maqṣad al-asnā fī sharḥ asmā' Allāh al-ḥusnā*) and in his *Book of the Forty* (*al-Arba'īn*), al-Ghazālī introduces a key metaphor and compares God's creation to the apparatus of a clepsydra, or a water clock.

In both books, the water clock is used as an explanatory simile for how God's creation is an expression of His will and how it gives evidence to His wisdom. In *The Highest Goal*, the simile is used to clarify the divine name *al-Ḥakam*, a word that originally referred to God's role as an arbitrator of human actions but that al-Ghazālī uses to refer to God as the holder of absolute wisdom. In the relatively long chapter on the divine attribute of will (*irāda*) in the *Book of the Forty*, al-Ghazālī quotes the water clock passage verbatim from *The Highest Goal*.[1] This latter work must have been composed slightly before the *Book of Forty*. Both works fall in the period after the *Revival* when al-Ghazālī taught at his own small madrasa in Ṭābarān-Ṭūs. They were written some time after 490/1097 and completed before al-Ghazālī began teaching at the Niẓāmiyya madrasa in Nishapur in 499/1106.[2]

When al-Ghazālī introduces the simile of the water clock, he uses motifs familiar from the *Revival*, such as God being the one "who makes all causes function as causes" (*musabbib kull al-asbāb*) and who creates an orderly arrangement of causes (*tadbīr al-asbāb*). Yet, he also presents something new here: the idea that God's arrangement is the "origin of the causes' positioning in order for them to lead to the effects."[3] The causes are positioned so that they "turn toward" (*tawajjaha*) the effects; they are brought in an alignment (*tawjīh*) with the effects.[4]

God has installed (*naṣaba*) the universal causes and their constant and interrelated movements, al-Ghazālī says in this passage, identifying the universal causes (*al-asbāb al-kulliyya*) as the celestial spheres. As elsewhere, al-Ghazālī avoids using technical terms from the lexicon of the astronomers and philosophers and lists the celestial bodies in a language that borrows from the Qur'an and *ḥadīth*. The universal causes are "earth, the seven heavens, the stars, and the spheres"; they are not subject to change and will never cease to be "until what is written is fulfilled" (Q 2:235).[5] Next, al-Ghazālī explains the three divine actions that determine God's creation: God's judgment (*ḥukm*), God's decree (*qaḍāʾ*), and God's predestination (*qadar*). The divine judgment is the initial design of the world; it is "the universal first arrangement and the eternal command (*al-amr al-azalī*), which is like a momentary glance." The divine decree is the concrete creation of the world, the "positioning" (or laying down, *waḍʿ*) of the universal and constant causes, meaning the celestial spheres. Divine predestination (*qadar*) "is the alignment (*tawjīh*) of the universal causes by means of their decreed and calculated movements towards their effects." These effects are "temporal events" (*ḥawādith*) implying—in contrast with the teachings of the *falāsifa*—that all effects, even those in the heavens, are generated in time and will corrupt. The effects are numbered and limited and have a known measure that neither increases nor decreases.[6]

These three steps of divine creation are quite important for al-Ghazālī and appear in other parts of the commentary on the divine names.[7] In order to explain them better, al-Ghazālī presents his reader with a simile (*mithāl*). "Perhaps," he addresses his reader, "you have seen the clock (*ṣandūq al-sāʿāt*) by which the times of prayer are announced." The clock to which he refers is likely not an imaginary one, but rather a real clepsydra with which many of his initial readers were familiar. For those readers who may not have seen it, al-Ghazālī describes the sight:

> There must be in it a device in the form of a cylinder containing a known amount of water and another hollow device placed within the cylinder [floating] above the water with a string attached. One end of the string is tied to this hollow device while its other end is tied to the bottom of a small container (*ẓarf*) placed above the hollow cylinder. In that container is a ball, and below it there is a shallow metal box (*ṭās*) placed in such a way that if the ball falls down from the container it falls into the metal box and its tinkling is heard.
>
> Furthermore, an aperture of a certain size is made in the bottom of the cylindrical device so that the water runs out of it little by little. As the water level is lowered, the hollow device placed on the surface

> of the water will be lowered, thus pulling the string attached to it and moving the container with the ball in it with a movement which nearly tilts it over. Once it is tilted, the ball rolls out of it and falls into the metal box and tinkles. At the end of each hour, a single ball falls.[8]

This water clock is of a quite simple design (figure 9.1). A float swims on the surface of a basin of water, connected by a string to a half-open container above it. Once the water level has fallen to such a degree that the string stretches, the string draws on the container and tilts it to one side. When it is tilted to a horizontal position, a metal ball falls into a metal box and makes a noise. Like most clocks of this period, it did not measure equal hours but rather measured a time span determined by the time of daylight; an "hour" was likely the span

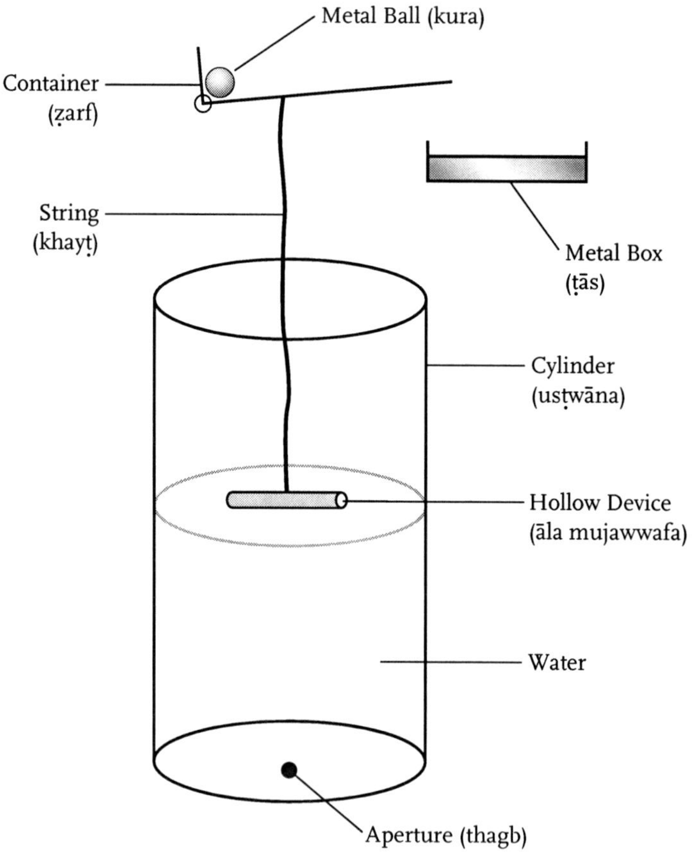

FIGURE 9.1 Al-Ghazālī's water clock (*ṣandūq al-sā ʿa*).

between two prayer times. Since the time intervals vary in length throughout the day and from day to day throughout the year, the clock needed to be set again after each "hour."

Historians have documented the existence of far more advanced water clocks from this time. For instance, an impressive water clock of unknown design is said to have been among the presents Charlemagne received from vassals of Hārūn al-Rashīd in 191/807.[9] When in 478/1085, Castilian troops conquered Toledo in Spain, they were impressed by a large and complex water clock although the new rulers destroyed its mechanism in 538/1133–34 when they began to study it.[10] Because al-Ghazālī expected his readers to know the device to which he refers, we can assume that his water clock likely stood in Ṭūs or in Nishapur. Abū-l Fatḥ al-Khāzinī, a Greek slave by origin, worked in Khorasan as an astronomer at the court of Sanjar and left us a chapter in his book on technical devices on the construction of water clocks. Al-Khāzinī, however, begun his activity in the decade following al-Ghazālī's death, and the water clocks he describes are much more complex than the one sketched out by al-Ghazālī. In his 515/1121–22 book, for instance, we read about a steelyard clepsydra that worked equal hours.[11] The early sixth/twelfth century was a high period for clock-making in Khorasan. Muḥammad al-Sāʿātī (d. 569/1174), for instance, the builder of a famous water clock in Damascus at the Jayrūn Gate, east of the Umayyad Mosque, moved to Damascus from Khorasan in 549/1154.[12]

In al-Ghazālī's clock, the time between the set-up and the falling of a ball is determined by the speed with which the water level in the basin falls. That speed, in turn, is "due to the determination (*taqdīr*) of the size (*saʿa*) of the aperture through which the water flows out; and that is known by way of calculation."[13] The causes thus determine their effects. In the water clock, every effect "is determined when its cause is determined, without increase or decrease."[14] The causal effects of the water clock do not end with the generation of the sound at a calculated and predetermined time. Because this specific clock is used to indicate the times of worship, the sound is a cause for people to perform the prayer. And because praying will ease people's way to redemption in the afterlife, the sound of this clock is one of the causes for bliss in the hereafter:

> Perhaps the falling of the ball into the metal box is a cause for another movement, and this movement is a cause for a third and so on through many steps to the point where remarkable movements are generated by it (*yatawalladu minhu*), determined by some degree of measures. And their first cause is the outflow of water according to a known measure.[15]

Here we should pause and take a closer look at al-Ghazālī's wording. He says that the effects "are generated by" (*tawallada min*) their causes. The language of the "generation" (*tawallud*) of effects appears at least three times in the simile of the water clock. In his earlier works, al-Ghazālī criticized the Muʿtazilites for their usage of the word "generation": those who talk about the "generation" of effects deny both secondary causality and God's direct creation of events, assuming that whatever "generates" an event is its absolute efficient cause. These concerns no longer seem to bother al-Ghazālī in his *Highest Goal.* Here, he uses the term "generation" similar to how he used "cause" in the *Revival.* Michael E. Marmura concluded that al-Ghazālī did not hesitate to use the causal language that is ordinarily used in Arabic, employing common Arabic terms such as "cause" and "generation" for his own meaning, Marmura says, yet with a metaphorical usage.[16] It seems that al-Ghazālī's ideas about what is acceptable language shifted. Although in the *Revival,* he criticizes the usage of the active verb "to generate" (*wallada*) in order to refer to causal connections,[17] he employs the passive "to be generated by something" (*tawallada min*) here in his *Highest Goal.* This is still the language used by Muʿtazilites to describe that humans create their own actions and these actions' immediate results.

In his final step, al-Ghazālī explains what the different elements in the simile of the water clock stand for, and he expounds on its cosmological and theological lessons. The water clock stands for God's universe, which is created in three steps. Here al-Ghazālī comes back to the three divine actions: judgment (*ḥukm*), decree (*qaḍāʾ*), and predestination (*qadar*). The judgment is the arranging (*tadbīr*) of the water clock—we would say its design—by deciding how the device must be constructed in order to achieve a certain effect, the generation of a sound at a certain time. The second step is the "bringing into existence" (*ījād*) of the device and its elements by forming the cylindrical basin containing the water, the hollow vessel on the surface of the water, the string tied to it, the container with the ball in it, and the metal box into which the ball falls. This second step, al-Ghazālī says, is the decree (*al-qaḍāʾ*). Third, the clock maker "must install a cause that necessitates a predetermined movement according to a calculated measure." This he does by making an aperture with a "determined size" (*muqaddarat al-saʿa*) in the bottom of the water basin. Only the constant flow of water from the basin sets the mechanism in motion to lower the water level, straighten the string, tilt the container, set the ball in motion, make it fall, and make a sound at a predetermined time. All these steps come as a result of the first movement, "by a known amount and (pre-)determined measure," which is the speed of the water's flow. This third step, al-Ghazālī says implicitly, is comparable to divine predestination (*al-qadar*).[18]

Whoever understands the working of a water clock has also understood the workings of divine predestination. The heavens and the spheres, the stars and the earth, the sea and the wind are all like such a water clock. They are fuelled by a cause similar to the aperture in the water basin:

> The cause that causes the movement of the spheres, the stars, the sun, and the moon according to a known calculation is like that aperture, which necessitates the flow of water according to a known measure. That the movement of the sun, the moon, and the stars lead to temporal results on earth is similar to the fact that the move-

> ment of the water leads to those movements that result in the ball's falling, indicating that the hour [for prayer] has come.[19]

The determining cause of all movements in the water clock is the size of the aperture in the basin, a cause likened to the determining cause of the celestial movements. These movements have results that are temporal (*ḥawādith*). The effects in this world are causally connected to the cause of the celestial bodies' movements just as the ball's falling is causally connected to the flowing of the water from the basin.

From reading this passage, it is not immediately clear why al-Ghazālī emphasizes the size of the aperture in the water basin. It seems that in this clock, the differing lengths of the "hours" were adjusted by varying the size of the aperture and not the amount of water in the tank, as one would assume. The size of that hole is thus the efficient cause of the movements in the water clock. I will try to give a more modern paraphrase of what al-Ghazālī wishes to convey in this simile. The three steps of judgment, decree, and predestination (*qadar*)—which apply both to the builder of the clock and to God as the builder of the universe—may be best understood in modern terms as designing, building, and supplying with a constant source of energy. The builder of a clock must first make a plan; second, execute this plan and build the clock; and third, set the clock in motion by supplying it with an enduring source of energy. That energy needs to be carefully regulated because only the right amount of energy will produce the desired result. In al-Ghazālī's water clock, the size of the aperture in the water tank regulates the supply of energy. The desired result would not be achieved if the hole were larger or smaller.

Although engineers at the time of al-Ghazālī did not conceptualize energy and its regulation the way we do today, the modern idea of energy input, defined as the requirement of a physical system to perform a certain amount of work, seems to be precisely what al-Ghazālī has in mind by the term *qadar* that he uses for the third stage. When applied to God's actions, the term *qadar* means predestination. The word contains a number of meanings in Arabic, among them, the "ability" or the "power" to perform a certain act. Morphologically, it is also closely connected to words such as "measure" or "quantity" (*qadr*) and to "power" or "capacity" (*qudra*). One may even understand that al-Ghazālī intends to interchangeably call the third step of action either "predestination" (*al-qadar*), "God's decree" (*al-qadr*), or "the measure" (*al-qadr*).[20] These three words cannot be distinguished in the unvocalized Arabic script usually used in manuscripts, and all three can fit in the contexts in which the consonants *q-d-r* appear as one word in this passage. The fourth word, *qudra*, which is related to *qadar*, has a great significance in al-Ghazālī's cosmology. In the thirty-second book of the *Revival*, he says that *al-qudra*, divine power, is used as an approximate expression for the divine attribute from which this world and all new creations originate or emanate. This divine attribute is so noble and lofty that we cannot refer to it with a clear expression in language (*ʿayn wāḍih al-lugha*) and must use a word that can only indicate the full extend of this attribute's majesty.[21]

Cosmology in *The Highest Goal in Explaining the Beautiful Names of God*

With the simile of the water clock, al-Ghazālī portrays the idea that God designs the universe as an apparatus, builds it, and supplies it with what we would call a constant supply of energy. The amount of energy needs to be measured carefully for the apparatus to produce its intended results. In his later discussion of the divine name *al-Ḥakam,* al-Ghazālī writes that the religious benefit (*ḥaẓẓ dīnī*) to be gained from contemplating this name "is to know that from God's side the matter is settled and not to be appealed." The pen that writes all existence is already dry, al-Ghazālī adds. Everything that exists now as well as all that will exist in the future is entirely necessary, being a result of God's initial arrangements in creating this world.[22]

The water clock primarily functions as a simile for the workings of the celestial spheres. Richard M. Frank has observed that in *Highest Goal,* al-Ghazālī borrows significantly more from philosophical teachings than in his earlier works. The universe to which al-Ghazālī compares the water clock is roughly al-Fārābī's and Avicenna's, with its numerous—in al-Fārābī there are nine—celestial spheres that mediate God's creative activity to the lowest sphere below the moon. In his *Revival,* al-Ghazālī proposes the same teachings in less philosophical language. In the thirty-second book on thankfulness, al-Ghazālī explains once again that human actions are truly God's actions, created within humans. The actions that please God advance the objective (*ghāya*) that God pursues with His creation, while those actions that are not pleasing to God are obstacles to realizing that goal.[23] Humans are mistaken when they think that they control their own actions; the actions are rather "of Him who directs your motive (*dāʿiya*)."[24] Nevertheless, most people mistakenly believe that their actions originate in their own selves. Al-Ghazālī here compares these people to boys watching a shadow play, with marionettes made of rags and suspended on fine strings invisible to the boys. The boys do not see the marionette player, and only a few intelligent ones (*ʿuqalāʾ*) among them know that the rags are moved by an outside mover. With the exception of the learned (*al-ʿulamāʾ*), all people are like these boys. They look at people and think they are moving by themselves. The learned know that humans are moved by an outside mover, but they do not know how. Only those who have insight among the learned (*al-ʿārifūn*) and who are firmly rooted in knowledge (*al-rāsikhūn*) understand how the humans are moved. They see the fine strings, thinner than those made by spiders, which reach from the humans into the sky.[25] Al-Ghazālī continues:

> Then they see the beginnings of those strings at the places from where they are suspended and with which they are connected; and they see that these places have handles that are in the hands of the angels, who are the movers of the heavens. They also see how the glances of the heavenly angels are turned towards those who carry the throne while [the angels] expect to receive from them [*scil.* those

> who carry the throne] something from the command (*al-amr*) that is sent down to them from the Lordly Excellence; in order that [the angels] do not disobey God what He orders and do [precisely] what they are ordered.[26]

The angels that move the heavens are the intellects and souls that reside in each of the celestial spheres. There is a twofold division between them; the lower angels act upon the earth, while the higher celestial beings function as the "carriers of the throne" (*ḥamalat al-ʿarsh*).[27] The lower angels receive their commands from these carriers of God's throne. The command (*al-amr*) originates with God and is passed from the higher celestial intellects to the lower ones. In this parable, the lower angels—a group that includes the active intellect—hold the strings that move humans and make them perform their actions.

The parable of humans as marionettes with strings held by heavenly creatures goes back to Plato.[28] In the *Revival,* al-Ghazāli uses this parable to express the same idea as the simile of the water clock: events in the sublunar world are the effects of secondary causes in the heavenly realm. Already in the *Touchstone of Reasoning* (*Miḥakk al-naẓar*), there is a reference to the relationship between events in the sublunar world and their cause in the active intellect. Here, al-Ghazālī considers whether astrology should be classified as a science because it makes valid predictions of the future. Astrology falls in the same class as physiognomy in that it relies on true connections but cannot identify the causes of the future events it predicts, since it is based on the repeated coinciding of two events (*talāzumuhumā*). This coinciding allows us to conclude that both events have the same cause (*ʿilla*), although that cause may be unknown to us. In *Touchstone of Reasoning,* al-Ghazālī encourages his readers to pursue fields of inquiry such as physiognomy and auguring because they lead to exploring the amazing marvels of God's creation. Such marvels can also be found in the connection between the celestial causes and events on earth. When an augur, for instance, counts the red lines in a sheep's shoulder blade and predicts much rain and also much bloodshed that same year, he benefits from the fact that all three events, the red lines, the rain, and the bloodshed among humans, are effects of the same cause in the celestial realm:

> [I]t is not too farfetched that amid the wonders of God's creation there is amongst the celestial causes (*al-asbāb al-samāwiyya*) a single cause (*sabab wāḥid*) that happens to appear in that year. According to a judgment that follows the course of the habit that life has (*bi-ḥukmi ijrāʾi l-ʿādati li-l-ḥayāt*), this [celestial cause] would be a cause (*ʿilla*) for the limbs of the animals and their formation, for the many causes (*asbāb*) of cloud-formation, and for the causes (*asbāb*) of hearts becoming brutish, which are, in turn, the causes (*asbāb*) for fighting, which is the cause for bloodshed.[29]

The heavenly cause itself is unknown to us, yet we witness its effects on many fronts, which allows us to predict future effects that in past occurrences have usually appeared in connection with those effects of this cause that we can already witness. It would be foolish, al-Ghazālī says, to dismiss these causal connections as insubstantial: "Only the ignorant reject this knowledge, people who have no glimmer of the marvels of God's creation and the scope of His power."[30] Yet, as al-Ghazālī writes in the thirty-second book of the *Revival*, those people who know God and know His actions are also aware that the sun, the moon, and the stars are subject to His command.[31] This is a reason why al-Ghazālī bears no objections against astrology as long as it is conducted in this universal way and does not pretend to predict individual events as happening only to certain humans.[32]

In the first book of his *Revival*, however, he describes astrology as a useless science that rarely makes correct predictions about future events. Astrologists have only an incomplete knowledge of the celestial causes; and if they hit it right, it is more due to coincidence than to their insight into the hidden causes. Studying astrology is thus seen as a waste of time.[33] Astrologers are, however, justified in assuming that there are celestial causes for events in the sublunar sphere, even if they can only be incompletely predicted. In a previously quoted passage from the thirty-second book of the *Revival*, al-Ghazāli says it would be unbelief to assume that the stars were by themselves the efficient causes (*fāʿila*) of their effects. Similarly, one must not deny that God governs the stars' movements. In the same passage, al-Ghazālī adds a positive statement:

> The conviction that the stars are causes that have effects that come about on earth, in plants, and in animals by the creation of God—Exalted—is not damaging to religion but it is the truth.[34]

In the *Revival* and in all his subsequent works, al-Ghazālī never doubts the connection between the heavenly bodies and events on earth, describing this connection as causal.[35] Using the figurative language of the *Revival*, he describes the celestial intellects as angels and likens their influence on the sublunar sphere to that of a marionette player on his puppets. Here, he follows his own directive of speaking in signs and symbols. Since the *Revival* is mostly concerned with the actions of humans (*muʿāmalāt*), al-Ghazālī says in its introduction, he will severely limit his exposition of the "knowledge of unveiling" (*ʿilm al-mukāshafa*), of which cosmology is a part. One must not unveil such mysteries in writing, he says, despite the fact that the most sincere people crave this sort of highest knowledge. The learned scholar must follow the example of the prophets and convey this type of knowledge only "through allegory and indication by way of symbolizing and summarizing."[36]

Although he has not completely relinquished these reservations in his commentary on the ninety-nine names of God, they seem to have had less influence on how he expresses himself in that work. In his *Book of the Forty*, al-Ghazālī describes the commentary on the ninety-nine divine names as a work that is more explicit about theoretical knowledge than even the most explicit books in the *Revival*.[37] The *Highest Goal* "knocks at the door of theoretical insight (*maʿrifa*)." Real insights, however, are limited to books that cannot circulate widely and are confined to readers prepared to understand these teachings. Al-Ghazālī particularly recommends the *Highest Goal* to those readers attempting to understand God's actions.[38] Given that most divine names refer to some aspect of the relationship between the Creator and His creation, the subject matter of the *Highest Goal* often veers toward discussing cosmology. Richard M. Frank analyzed al-Ghazālī's cosmology in the *Highest Goal* in a way that allows us to fall back on his results. According to Frank, the cosmology of the *Highest Goal* is largely identical to that of the *Revival*. In the *Highest Goal*, al-Ghazālī is less reluctant to replicate philosophical teachings in plain language, and sometimes he even uses philosophical terminology. In the simile of the water clock, for instance, he describes God as "laying down the universal causes" (*al-asbāb al-kulliya*) so that they will produce certain effects. This is an unmistakable reference to the celestial intellects using standard philosophical terminology.[39] God is "the being necessary by virtue of itself" (*al-mawjūd al-wājib al-wujūd bi-dhātihi*), from which everything whose existence is by itself possible takes its being.[40] Things come into existence by necessity (*bi-l-wujūb*). Everything that is created is both possible by itself and necessary by something else (*al-munkin bi-dhātihi al-wājib bi-ghayrihi*); everything is necessitated by the Eternal Decree.[41] In his *Book of the Forty*, al-Ghazālī adds that God's decree (*qaḍāʾ*) is both the same as His eternal will and the same as His providence for His creation, which is expressed through the order that He creates.[42]

The Highest Goal is only marginally concerned with ethics and thus does not delve as deeply into the nature of human actions as *Revival* does. Yet al-Ghazālī also makes clear here that God creates everything in this world, including human actions. He creates the action as well as the place (or substrate, *maḥall*) that receives the action, which is the human. He also creates the conditions for the action's reception and whatever else contributes to it.[43] God requires humans to "make themselves open" to the outflow of God's mercy upon them, to the creation of beneficial knowledge in them that will lead to praiseworthy actions.[44]

In *The Highest Goal*, al-Ghazālī shows the same ambivalence as in *Revival* with regard to the necessity of the system God creates. God cannot create anything whose conditions for its existence are not fulfilled.[45] If anything were to be changed in God's order, the order itself would become void.[46] If the harmful creations in the world were to be removed, then the good that they produce would be done away with and harm far worse than what currently exists would then come about.[47] Divine liberality (*jūd*) requires the "perfect achievement of the utmost good whose existence is possible."[48] Yet here in the *Highest Goal*, as in many of his other writings, al-Ghazālī distinguishes God's knowledge of how the optimal creation is achieved from God's will to create the optimal. God's actions are not random or coincidental but reflect both his wisdom and his deliberation.[49]

The Niche of Lights: The Philosophers' God as the First Created Being

The *Niche of Lights* (*Mishkāt al-anwār*) is a work from the same period as *The Highest Goal* and *The Book of the Forty*. It was written after the *Revival*, although we cannot say precisely when. Because it is one of al-Ghazālī's most mystical works, earlier chronologies of his output have dated the *Niche of Lights* to the end of his career. It was assumed that during his life, al-Ghazālī developed a progressively stronger inclination toward Sufism, with his most mystical works being his last.[50] Such a supposition, however, is unwarranted, and the *Niche of Lights* could have been composed at any time after 490/1097. We do know that it was composed after *The Highest Goal.*[51]

The *Niche of Lights* is a very rich text in terms of its cosmology, and I will not attempt to give full justice to its complexity. Rather, I will focus on a passage at the very end of the book, known as the Veil Section. Soon after the *Niche of Light* first appeared in print in 1322/1904–5, this section, which is the last of three in that book, inspired sufficient controversy among Western interpreters to the point that they disputed its authenticity.[52] This skeptical position, however, was based on an incomplete view of al-Ghazālī's theology, and today there can be no doubt that all parts of the text of the *Niche of Lights*, as we have it today, are authentically al-Ghazālī's.[53]

The Veil Section at the end of the *Niche of Lights* is not immediately related to the two earlier parts of the book and can be viewed on its own, to a certain degree. Averroes regarded it as the clearest evidence that al-Ghazālī's cosmology continues the tradition of the Aristotelian *falāsifa.*[54] The passage is a commentary on the noncanonical *ḥadīth*: "God has seventy veils of light and darkness; were He to lift them, the august glories of His face would burn up everybody whose eyesight perceives Him."[55] Al-Ghazālī aims to explain the veils of light and darkness that prevent people from grasping who or what God is. He classifies various religious groups according to the kind of veil that prevents them from understanding the true nature of God. In the first division, he discusses those who are veiled by pure darkness (*mujarrad al-ẓulma*), and in the second, those who are veiled "by light along with darkness" (*bi-nūr maqrūn bi-ẓulma*). Both groups are further subdivided. They contain a range of people, from plain unbelievers who hold nature (*ṭab*ʿ) rather than God to be the cause of the world, to heterodox Muslims who believe that God has a bodily form, to Muʿtazilites.

In terms of cosmology, it is most interesting what al-Ghazālī says about the third division, those veiled by pure lights (*maḥḍ al-anwār*).[56] These are people who have gained some insight into God's being. They are again divided into three subgroups that represent different levels of insight into the divine. As noted by Hermann Landolt, this division closely follows the narrative of Abraham's discovery of and ascent to monotheism, as told in Sura 6, verses 75–79 of the Qur'an.[57] Al-Ghazālī introduces this story in an earlier passage from the *Niche of Light.*[58] According to the commentary literature on the Qur'an, the young Abraham grew up in a cave's darkness in order to avoid the persecution of the Mesopotamian king Nimrod.[59] There, he starts searching for his Lord. When he leaves the cave at night, he first sees a star going up in the east and concludes that this is his Lord. Once the star goes down in the west, however, he dismisses that notion. He next sees the moon rising in the east and assumes that this is his Lord. Again, when the moon sets in the west, he rejects this notion. The same happens with the sun: he sees it going up in the morning and thinks it his Lord until it sets in the evening. Finally, Abraham concludes that none of these celestial bodies is his Lord. Rather, the maker of them, the Creator of the heavens and the earth, is his real Lord, and only He should be worshiped.

Abraham's discovery of true monotheism by studying the heavens held great significance for al-Ghazālī, and he refers to it in other works.[60] In the third division of the Veil Section, he compares the three subgroups of scholars who are veiled by pure light to the three false levels of insight that Abraham had gained during his youth. Only a fourth group of people who are not veiled, "those who have arrived" (*al-wāṣilūn*), represents the level of those who truly understand who the Lord is. Only this group has gained a proper understanding of God (*tawḥīd*).

Following the pattern of Abraham's discovery, al-Ghazālī connects the false insight gained by each of the three groups with the celestial being that they assume is "the Lord." These celestial beings come from the ten spheres and their governing intellects as they appear in al-Fārābī's model of cosmology. The fourfold model in this section (three false groups plus one correct) combines philosophical cosmology with doxography or even heresiography.

Al-Ghazālī says the lowest of these three subgroups are people who hold the opinion that the mover of the highest visible heaven, which is the next-to-outermost sphere, the sphere of the fixed stars, is the creator of the world and the "Lord":

> The first among them is a group (*ṭāʾifa*) that knows the meanings of the [divine] attributes properly (*taḥqīqan*) and realizes that the nouns "speech," "will," "power," and "knowledge," and others cannot apply to God's attributes the way that they apply to humans. In their teachings (*taʿrīf*) about God these people avoid using these attributes. When they teach about Him they draw upon the relationship [of God] to the created things just like Moses, peace be upon him, taught about God in his answer to Pharaoh's question: "What is the 'Lord of the Worlds'"? (Q 26:23). These people say the Lord, who is the Holy One and who is exalted over the meanings of these attributes, is the mover of the heavens and the one who governs (*dabbara*) them.[61]

Compared to the groups mentioned earlier in the veil section—those veiled by some kind of darkness—this group has developed a proper understanding of the divine attributes and their transcendence. They understand that the Lord is exalted over all anthropomorphic attributes. When they use words such as "speech," "will," "power," and "knowledge" to describe the Lord; they intend their meaning to transcend the ordinary sense of these words. This passage refers to the polemics between Ashʿarites and Muʿtazilites. The latter are the highest group from the earlier part of those veiled by light and darkness and have just been discussed. Ashʿarites criticized Muʿtazilites for assuming that the human understanding of justice, for instance, is the same as God's understanding. The group described in this passage has gained more insight than the Muʿtazilites and understands that all of God's attributes are transcendent.

When they are pressed by their opponents to explain who is the "Lord of the worlds," they answer what Moses told Pharaoh, namely, that God is "the Lord of the heavens and the earth and all in between" (Qur'an 26:24) and that He is "your Lord and the Lord of your forefathers" (26:26). Pharao asked him about the essence (*māhiyya*) of the divine, al-Ghazālī remarks earlier in the *Niche of Lights*, and Moses responded about the acts of God.[62] Likewise this group draws on God's relation to the created things. Their understanding of divine transcendence leads them to the insight that the Lord is the one who moves and governs the heavens (*muharrik al-samawāt wa-mudabbiruhu*).

The shortcomings of this position are still quite significant, and they are highlighted in the discussion of the next higher group:

> The second group leaves these people behind insofar as it became clear to them that there is multiplicity (*kathra*) in the heavens, and that the mover of each single heaven is a different being that is called an angel, of whom there are many. Their (scil. the angels) relation to the divine lights (*al-anwār al-ilāhiyya*) is the relation of the stars.[63]

The first group incorrectly believes the Lord to be the mover of the next-to-outer sphere, whom they view as the governor (*mudabbir*) of the visible heavens and the cause for the existence of the heavenly bodies. They assume the existence of a single mover of one large heavenly sphere and are unaware of the existence of multiple spheres, each having a mover who may also be called an angel (*malak*).

The last sentence in this passage about the angels' relation (*nisba*) to the divine lights has proved difficult to understand. The sentence is incomplete or at least elliptical, as it analogizes the angels' relationship to the divine lights with the stars' relationship to . . . nothing. A clue to understanding it can be found in al-Ghazālī's *Decisive Criterion* (*Fayṣal al-tafriqa*). There, al-Ghazālī reports that some Sufis interpret the Qur'anic narrative of Abraham seeing the star, the moon, and the sun and identifying them as his Lord in terms of "luminous, angelic substances" (*jawāhir nūrāniyya malakiyya*), to which the words "star, moon, and sun" refer. These substances are purely intellectual and not perceived through the senses, and they have advancing degrees of perfection (*darajāt mutafāwita fī l-kamāl*). The passage ends with the sentence: "The relation (*nisba*) of the amount of differences between one another is like the relation between the star, the moon, and the sun."[64] Here in the Veil Section, the relation of the angels to the divine lights—most probably a reference to God—is the same as the star's relation to the real Lord in Abraham's vision. To complete the elliptical sentence, the words "to the true Lord," or something similar, should be added to its end. Although this sentence remains enigmatic, it is clear that the first level of insight into the divine is likened to the one Abraham reached when in Q 6:76, he erroneously discovers that the star (*al-kawkab*) is his Lord.

In the Veil Section, al-Ghazālī adopts a distinctly philosophical perspective and looks at the world with eyes trained in philosophical cosmology. His first subgroup is defined by its failure to understand the "multiplicity in the heavens." We will see that the next group can be roughly identified with Aristotle and his followers. It seems that when he envisions the first group, al-Ghazālī uses the widespread typology of his time for understanding the history of philosophy, visualizing that group as typifying a pre-Aristotelian stage of philosophy. The supposed failure to recognize multiplicity in the heavens indicates a group that did not believe in the existence of more than one heavenly sphere. The Jewish Aristotelian philosopher Maimonides (d. 601/1204), who wrote three generations after al-Ghazālī, ascribes this view to a group of ancient natural philosophers. In his *Guide of the Perplexed* (*Dalālat al-ḥāʾirīn*), he comments on the cosmology of the earliest generations of philosophers who lived at the times of the Sabians, the pagan polytheists with whom Abraham struggled:

> The utmost attained by the speculation of those who philosophized in those [early] times consisted in imagining that God was the spirit of the sphere (*rūḥ al-falak*) and that the sphere and the stars are the body of which the deity, may He be exalted, is its spirit.[65]

Maimonides refers his readers to Ibn Bājja's (d. 533/1139) commentary on Aristotle's *Physics*. The reference is not entirely clear, since Ibn Bājja neither discusses the cosmology of the early philosophers nor mentions the Sabians. Ibn Bājja, who wrote one generation after al-Ghazālī in the Muslim West, comments on Aristotle's refutation of the teachings of earlier Greek philosophers, most notably Parmenides and Melissos. These early philosophers taught that all that exists is the manifestation of a single unchanging and unlimited principle. There are no real processes in the world, Parmenides taught; rather, what really exists—meaning, what exists on the level of intellectual forms, unaffected by sense perception—is unchanging. Reflecting on Aristotle's writings on these teachings in Book 1 of his *Physics*, Ibn Bājja says that Parmenides and Melissos saw no differences between different existing beings and treated them as if they were all of one kind. This was before the time when Aristotle alerted philosophers to the fundamental difference between beings.[66] But despite Aristotle's attempts to define physics as a science that analyzes processes, the teachings of these earliest philosophers prevailed. The *mutakallimūn*, Ibn Bājja complains, hold basically the same teachings. He implicitly refers to the occasionalism of the Ashʿarites. The *mutakallimūn* reject the existence of natural dispositions (*al-ṭibāʿ*), Ibn Bājja says, and claim that everything consists of atoms. The views of the *mutakallimūn* are not based on any research, writes Ibn Bājja; rather, they have developed these views unsystematically in their internal polemics.[67]

Ibn Bājja's remarks on pre-Aristotelian science are part of a larger tradition of *Physics* commentary in Arabic.[68] In Avicenna's discussion of the *Physics* in his *Healing*, he also connects the teachings of Parmenides and Melissos with the theory of atomism.[69] According to Avicenna's analysis, pre-Aristotelian theories of physics and the opinions of the classical Ashʿarite *mutakallimūn* are erroneous for the same reason: they disregard the substantial differences between beings that underlie Aristotelian physics, such as the difference between a substance and an accident or the difference between composed beings in the sublunar sphere and uncomposed beings in the heavenly spheres. The atomism of the Ash'arite *mutakallimūn*—and by implication, their occasionalism—is just one expression of this disregard for the Aristotelian distinctions between beings. For an Ash'arite occasionalist, all beings consist of indistinguishable smaller parts that are equally close to God's creative activity.

The first group's failure to understand that there is multiplicity (*kathra*) in the heavens may have a more subtle meaning than just the acknowledgment of numerous spheres. Al-Ghazālī may have adopted the Aristotelians' position regarding early theories of cosmology. Pre-Aristotelian cosmology was marred both by the failure to understand that there are numerous spheres as well as by a lack of distinction between different types of beings. The cosmological beliefs of this first group seem to identify them with this early group of philosophers. Additionally, al-Ghazālī may be writing on a naive understanding of occasionalism. Such an occasionalism would simply assume that all things are composed of atoms and accidents and would deny, for instance, the existence of self-subsisting intellects. The failure to understand the "multiplicity in the heavens"—a deliberately unspecific description—may be meant to refer both to an early philosophical approach by pre-Aristotelian thinkers and to a naive occasionalist understanding of the universe.

The fact that the first subgroup in this division has a proper understanding of God's transcendence implies its identification with Ash'arism. According to their own view, the Ash'arites were the only group of Muslim theologians to understand the transcendence of the divine attributes. In his works, al-Ghazālī tirelessly stressed the transcendence of the divine attributes. In the Arabic doxographic tradition, early philosophers also held the view that God is transcendent. The Arabic doxography of pseudo-Ammonius, which was available from the middle of the third/ninth century, reports that pre-Socratic philosophers such as Thales and Pythagoras taught the transcendence of the divine attributes, attributes that neither the human intellect nor the soul is able to comprehend.[70] A generation after al-Ghazālī, al-Shahrastānī repeats these reports in his *Book of Religions and Creeds*.[71] Since this first group is characterized as understanding the transcendence of God's attributes while misunderstanding the composition of the heavens and perhaps also of the world, al-Ghazālī may have been referring either to the Ash'arites or to the pre-Aristotelian philosophers or perhaps even to both.

Compared to this first subgroup, the second subgroup of those veiled by pure light is described as possessing superior insight and believing that the next higher celestial being—the mover of the highest sphere—is their Lord. I have already quoted the passage detailing the group's understanding that there are many heavenly spheres and that each sphere has its own mover. The passage continues:

> Then it became evident to them that these heavens are inside another celestial sphere that moves all the others through its motion once during [every] day and night. They said the Lord is the mover of that celestial body which is furthest away and which envelops all celestial spheres since multiplicity is denied to Him.[72]

In comparison with the first group, this group has an adequate understanding of astronomy and the celestial spheres. Here al-Ghazālī describes the introduction of the *primum mobile*, the outermost starless sphere, as first theorized by Ptolemy.[73] This group's Lord is the mover of the *primum mobile*. Given that there are no physical movements above this sphere, this Lord himself is the highest mover. This group fails to realize that even if there may not be any physical movements beyond the highest sphere, there still are higher beings there. Their failure to acknowledge the existence of beings higher than the mover and governor of the outermost sphere leads them to the false assumption that he, the mover of the outermost sphere, is the Lord of the world.

Again, their error is only pointed out when al-Ghazālī introduces the next higher group:

> The third group leaves these people behind. They say that moving the bodies by way of directly acting upon them (*bi-ṭarīq al-mubāshira*) should be (*yanbaghī an*) [regarded as] a service to the Lord of the Worlds, an act of worship towards Him, and an act of obedience (*ṭāʿa*) towards Him by one of His servants who is called an angel. His [*scil.* the angel's] relation to the pure divine lights is the relation of the moon among the sensory lights.[74]

The members of the second subgroup err when they think that moving the highest sphere is the most supreme task possible to the Lord. In fact, anything that directly moves a physical object cannot be regarded as the truly supreme being. Rather, any being that creates physical movements is just doing a service to the true Lord. The mover of the highest sphere obeys the Lord and worships him by moving the sphere. Using another elliptical reference, al-Ghazālī compares the highest mover to the moon in Abraham's story, connecting this second level of insight into the divine to Abraham's false realization, reported in Q 6:77, that the moon is his Lord.

This second subgroup is characterized by a single conviction, namely, that the unmoved Lord is himself the mover of the highest sphere. This is an unmistakable reference to Aristotle's kinematic proof of God's existence.[75] Al-Ghazālī was well aware of this proof. In the extensive report of his philosophical metaphysics preserved in MS London, British Library Or. 3126, he distinguished between two types of proofs for the existence of God—Aristotle's kinematic proof and Avicenna's proof that God is the being necessary by virtue of itself:

> Know that a group amongst the ancients (*mutaqaddimūn*) argued by way of the contingent for (the existence of) the necessary and by way of the effect for (the existence of) the cause. They started with composed beings. They analyzed them and ascended from there to the elementary things (*basāʾiṭ*) [= celestial beings]. They proved demonstrably that there is nothing that moves without (being moved) by a mover until they ended at a mover who does not move (himself). He is the first mover. The more recent ones (*mutaʾakhkhirūn*) argued by way of the creator for (the existence of) his created beings. They began with the elementary beings then climbed up from them and discovered the necessity of the creator's existence from His existence

> itself. Once they had established this, they established (the existence of) contingent beings through it. They said: "This way to argue is more reliable and nobler, because if we consider the state of being, [we find that] the absolute being (*wujūd muṭlaq*) inasmuch as it is existence, bears witness to Him. So we had no need for the ascent from low to high, because the closest (*awlā*) thing [to mind] is giving evidence to the created things by way of their creator and not giving evidence to Him by way of the created things." This is all good, but the second [method] is better.[76]

This passage is al-Ghazālī's report of Avicenna's position that his proof is superior to Aristotle's proof; it should not be assumed to be al-Ghazālī's own opinion. One source of the report is Avicenna's *Pointers and Reminders*, which is briefly quoted in this passage.[77] The report, however, does demonstrate al-Ghazālī's awareness both of the differences between these proofs and of Avicenna's claim that his proof gives a higher level of insight into God's being. In the London manuscript, al-Ghazālī calls those who use the kinematic proof for God's existence "the ancients." This group seems to be the second subgroup of those who are veiled by pure light. The second subgroup represents the cosmology of Aristotle as al-Ghazālī understands it.

If this identification is correct, the "more recent philosophers," that is, the philosophers who see God as the giver of existence rather than as the first mover, that is, al-Fārābī and Avicenna, are the third and highest group of those who are veiled by pure light. I have already quoted the passage detailing their realization that moving cannot be the most supreme action but is done rather in obedience and as an act of worship to the Lord. The passage continues:

> These people claim that the Lord is the one who is obeyed (*al-muṭāʿ*) by this mover and [they claim that] the Lord, exalted, is the mover of everything by way of the "command" (*al-amr*), not by way of directly acting upon [other things]. Then, there is an obscurity when they try to make the "command" and its essence (*māhiyya*) understood, and this places limits to the deeper understanding. This book cannot not go into that.[78]

The Lord of the second group moves the highest sphere as an act of obedience to the being that this third group considers the true Lord. The Lord of this group is called the "one who is obeyed" (*al-muṭāʿ*). This Lord governs not by causing the movements of lower beings but by giving "the command" (*al-amr*), a vague term that is nowhere explained in this text. Al-Ghazālī blames this group for his own lack of an explanation for what "the command" really is. There are several ways to understand what this "command" might be. The more recent philosophers might, for instance, understand it as the command to exist: "Be!" (Q 6:73). This idea is similar to what al-Fārābī did when he developed Aristotle's causation of motion into a causation of being.[79] Equally, Avicenna characterizes God not as a mover but as the being that bestows existence (*wujūd*) upon His creation. Yet according to al-Ghazālī, even these scholars—al-Fārābī, Avicenna, and their followers—are misguided. The shortcomings of their views are again pointed out when the next group is described. This is the fourth and final group:

> These groups are all veiled by pure light. Only a fourth group are the ones who arrive [at understanding God's oneness]. It has also been disclosed to them (*tajallā lahum ayḍan*) that this one who is obeyed (*al-muṭāʿ*) is characterized by an attribute that is incompatible with pure oneness and utmost perfection; on account of a secret that this book cannot reveal. [It has also been disclosed to them] that the relation of the one who is obeyed [to the real Lord] is the relation of the sun among the sensual lights. Therefore, they have turned their faces from the one who moves the heavens [i.e. the Lord of the second group] and from the one who commands their movements [i.e. the Lord of the third group] to the one who created the heavens and who created the one who gives the command (*al-āmir*) that the [heavens] are moved.[80]

The being that these philosophers consider to be the Lord is himself only the mediator between the real Lord and His creation. Al-Ghazālī compares this version of the Lord, the one who is obeyed (*al-muṭāʿ*), to the sun, comparing this third group to Abraham discovering in Q 6:78 that the sun is his Lord. Al-Ghazālī sees the God of philosophers such as al-Fārābī and Avicenna, who believe that this world emanates from God according to His nature, to simply be a creation of the real God. The real God is the creator of the being that the *falāsifa* consider to be God.

The Cosmology of the "Fourth Group" in the Veil Section of *The Niche of Lights*

The fourth group is the one that possesses true insight into the nature of God. Veiled neither by darkness nor by light, they understand that the philosophers' God is the first creation of the real God. Al-Ghazālī's true cosmology contains two main elements: he first appropriates the cosmology of al-Fārābī, with all its spheres, movers, and its First Being, a cosmology that Avicenna had also adopted. Second and crucially, al-Ghazālī adds to it another layer of creation. For al-Ghazālī, the being that in al-Fārābī and Avicenna's cosmology bestows existence upon others and is obeyed (*muṭāʿ*) by the movers of the spheres is the first creation of the real God. Indeed, the real God does little more than create the one who is obeyed and continuously emanating being into him. The one who is obeyed mediates God's creative activity and converts it into "the command" (*al-amr*), through which creation of the heavens and the earth unfolds.

Only the last of the four groups, "those who have arrived" at a correct understanding of who is the Lord, recognize the created nature of the philosopher's God. Their own insight into the real God is described as follows:

> These people arrived at a being that is exalted above everything that sight has perceived previously. The august glories of His face (*subu-*

> *ḥāt wajhih*)—the First and the Highest—burn up everything that the sight and the insight of the theologians (*al-nāẓirūn*) have perceived since they find in Him someone holy and exalted above everything that we have described before.[81]

This highest level of insight is likened to Abraham's discovery that his Lord is "He who created the heavens and the earth" in Q 6:79. He is the only true existence, and He is the one who truly bestows existence on His creatures. Only "those who have arrived" know of Him and understand that He is the only existence. Among them are a subgroup of those who understand that He is the only one who truly exists. This realization leads to their "annihilation" (*fanāʾ*):

> Then these people divide into smaller groups. Among them is the one for whom everything that he sees is consumed, perishes, and annihilates—but he still remains, observing the beauty and holiness [of God], and observing his own self within His beauty, [a state] that he attained by the arrival at the divine presence (*al-ḥaḍāra al-ilāhiyya*). With regard to these [people], the objects of vision perish, but not he who sees.
>
> Another group who are the elect of the elect pass beyond these. The august glories of His face burn them and the power of glory overcomes them (or: takes control of them). In their selves they are perished and annihilated. No glance at themselves is left to them for they annihilate from themselves. And nothing remains save the One, the Truth. The Qurʾanic verse "everything perishes save His face" (Q 28:88) becomes for them an individual experience (*dhawq*) and a state (*ḥāl*). We referred to this in the first chapter where we mentioned how they apply the word "becoming-one" (*al-ittiḥād*) and how think of it. And this is the [utmost] limit of those who arrive (*al-wāṣilūn*).[82]

Annihilation (*fanāʾ*)—the goal of Sufi practice—is achieved once the believer becomes aware that all being is God, all actions are God's action, and all love is God's love. For al-Ghazālī, annihilation (*fanāʾ*) is not synonymous with a "union" (*ittiḥād*) with God. "Union," al-Ghazālī had said earlier in the book, is only a metaphor for understanding the true meaning of *tawḥīd*, namely the realization that all being is He.[83] In the thirty-second book of the *Revival*, al-Ghazālī had already clarified that when the Sufis say "annihilation of the self" (*fanāʾ al-nafs*), they mean looking at the world through the eye of someone who truly understands divine oneness (*bi-ʿayn al-tawḥīd*). That viewpoint includes the realization that there is nothing in existence other than God (*laysa fī l-wujūd ghayruhu*). It is false to assume that there exists something that is not God. All that exists (*al-wujūd*) is He.[84]

Commenting on a short creed by Ibn Tūmart, in which these Ghazalian teachings are reproduced in an easily comprehensible way, Ignaz Goldziher once remarked that an "air of pantheism" runs through them.[85] For Goldziher, there is here the notion that all things are divine. A more thorough analysis, however, would say that for al-Ghazālī, not all things are divine, but rather the divine is all things. This is not pantheism but rather monism. Alexander Treiger observed that monotheism and monism come very close to each other in al-Ghazālī; monotheism being the view that God is the only existent that is the source of the being for the rest of existents, and monism the idea that God is the only existent at all: "[T]he monistic paradigm views the granting of existence as essentially *virtual* so that in the last analysis God alone exists, whereas the monotheistic paradigm sees the granting of existence as *real*."[86] Treiger concludes that in al-Ghazālī's *Niche of Lights*, both perspectives are present. In some passages, God is the Lord and the Creator and in others, such as the one on the insight of "those who have arrived," God is the only true existent, the other existents possessing only borrowed and metaphorical existence.[87] These two perspectives should not be regarded as being opposed to each other in al-Ghazālī; rather they complement each other. Arriving at true *tawḥīd* means to arrive at a monist perspective of God. This, in turn, includes the monotheist perspective of those levels that represent a less complete understanding of *tawḥīd*.

Monism appears in works other than the *Niche of Lights*. In one of his last works, *The Choice Essentials of the Methods of Jurisprudence* (*al-Mustaṣfā min ʿilm al-uṣūl*), al-Ghazālī discusses how human knowledge is a reflection in the human soul of all the intelligible forms of existence, such as the heavens, the earth, the trees, the rivers, and so forth. Here al-Ghazālī adds:

> Similar, the human soul (*al-nafs al-ādamī*) can be understood as being imprinted with the divine presence (*al-ḥaḍra al-ilāhiyya*] on the whole. The "divine presence" is an expression for the totality of the existences (*jumla al-mawjūdāt*). Altogether they are from (*min*) the divine presence since there is nothing in existence other than God Exalted and His actions.[88]

Those who arrive at a proper understanding of God combine a monist understanding of God's relationship to the world with the monotheism of the *falāsifa*. Most important, they have accepted the philosophical cosmological system. Richard M. Frank gathered evidence that for al-Ghazālī, the celestial intellects are intermediaries (*wasāʾiṭ*) in the transmission of God's blessings to terrestrial beings.[89] Since the Farabian and Avicennan philosophers developed a nearly correct understanding of the one who is obeyed, many elements of their teachings on cosmology are true—but under the condition that it is not God whom they describe in their teachings but the *muṭāʿ*, the highest created being. This near-understanding seems to be the reason why al-Ghazālī writes: "To [the fourth group] it has *also* been disclosed." He implies that the fourth group has accepted many teachings of the third, while integrating their own superior insight that all being *is* God. The third group understands, for instance, that the world is a product of the one who is obeyed (*al-muṭāʿ*) and is created according to his essence. The fourth group refines the understanding of the *falāsifa* and posits that the creative power behind this world is not the essence of the one who is obeyed. The one who is obeyed has no choice of what to create and follows the necessity of His own nature. The true God, however, is not affected by the limitations of the nature of the one who is obeyed, since He is the real originator and exercises his own deliberate choice.

The obeyed one does not act directly upon the rest of the creation but rather acts indirectly via "the command" (*al-amr*). He relies on the mediation of the celestial spheres and their movers to act on the lower spheres, including the sublunar sphere. His acting on all creatures other than himself is by means of "the command" (*al-amr*). The cosmological terminology used in this part of the *Niche of Lights* is both philosophical as well as Qur'anic in its origin. In Sura 81, which starts with a long apocalyptic vision, the Qur'an says that "these are the words of an noble messenger, who holds power with the Lord of the Throne, someone who is of rank (*makīn*), who is obeyed (*muṭāʿ*), and who is also trusted" (Q 81.19–21). The commentary literature identifies this messenger with the archangel Gabriel because he is the one who conveys revelation to the prophets.[90] In his *Decisive Criterion for Distinguishing Islam from Clandestine Apostasy*, al-Ghazālī follows this interpretation and says that the Qur'an refers to Gabriel in many ways, calling him, among other things, "high in rank with the Lord of the Throne" and "the one who is obeyed." This latter phrase is justified because "he is the being that is followed in the rightful actions of some angels."[91] In another passage, al-Ghazālī says that Gabriel, the Holy Spirit (*rūḥ al-quds*), and the Trusted One (*al-amīn*) are all names for the same being, "someone who is obeyed" (*muṭāʿ*).[92] In the Veil Section, al-Ghazālī identifies this being as God's first and most supreme creation. Al-Ghazālī was familiar with the way philosophers used the term *mutāʿ* in their texts. In his report of philosophical teachings preserved in the London manuscript, he includes a chapter from one of Miskawayh's ethical treatises. In that context, which has nothing to do with the above-quoted Qur'anic passage, "someone who is obeyed" (*muṭāʿ*) is a metaphor for the human intellect that governs its domain of the human body as a king reigns over a polity.[93] In philosophical literature, the word *mutāʿ* is an expression for a being that holds absolute authority.[94]

In the case of the word "command" (*amr*), the fusion between Qur'anic terminology and a philosophical reading of revelation is even more apparent. In verse 65:12, the Qur'an says that God created the seven heavens "and through their midst descends the command."[95] In verse 41:12, it is said that after God created the seven heavens, he assigned a command to each of the heavens. Other verses identify God as the one who "governs the command" (*yudabbir al-amr*, Q 13:2). For al-Fārābī, the "world of the command" refers to the world of the highest celestial spheres, just below the "world of lordliness" (*ʿālam al-rubūbiyya*) where the First Principle resides. It is both above the throne and above the "world of creation" (*ʿālam al-khalq*). The "world of the command" is where the pen writes on the preserved tablet (*al-lawḥ al-maḥfūẓ*). The human spirit (*rūḥ*) stems from the "world of the command," and whoever turns from emotions, sense perception, and imagination toward the intelligibles (*al-maʿqūlāt*) will reach the "world of the command," which is the highest part of the "world of sovereignty" (*malakūt*).[96] In a clarifying passage, al-ʿĀmirī explains how the philosophers understood the Qur'anic cosmological metaphors; there he says that they use the term "the command" to refer to the universal forms (*al-ṣuwar al-kulliyya*).[97]

In Avicenna's *Throne Philosophy* (*al-Ḥikma al-ʿarshiyya*), the Qur'anic verse that "You will not find any change in God's habit" (Q 35:43) is explained as referring "to the permanence of the command."[98]

In the standard philosophical lexicon, the "world of command" represents the Platonic concept of an intelligible world of forms beyond the material one. The intelligible world is primarily the world of the celestial intellects, including that of the active intellect that gives humans their universal categories of thought. This is also how al-Ghazālī uses the term in the *Revival*. In the thirty-seventh book of that work, he says: "Every existing thing that is bare of quantity and measure is part of the world of the command."[99] Arent J. Wensinck remarked that "command" is a synonym for the realm of sovereignty (*malakūt*), which in the *Revival* refers to the world of the heavenly intellects, the opposite of the materially created world.[100] In al-Ghazālī's cosmology, the most general meaning of "command" is "the intelligibles." The world of command is the set of universals—or for Avicenna, the quiddities (*māhiyyāt*)—that function as the blueprint for all individual and material creation and that are accessible to the human intellect. "Command" refers to the full set of the classes of beings that make up creation.[101]

In the cosmology of the *falāsifa*, God is the ultimate endpoint of all causal chains. In the *Niche of Lights*, al-Ghazālī does not counter that view, readily accepting that the obeyed one (*al-muṭāʿ*) is the endpoint of all causal chains. If "the command" is a term for the full set of the classes of beings that make up creation, its category also includes the laws of causality. The immaterial universals determine the relationship between individual beings and thus they include the laws of causal connections. These are the "laws of nature"—a phrase nowhere to be found in al-Ghazālī oeuvre—by whose necessity the one who is obeyed governs and creates the world. Yet in this model, the immediate connection between the obeyed one and God seems to be determined by God's free choice rather than by causal necessity. God passes the command (*al-amr*) to the one who is obeyed (*al-muṭāʿ*), meaning that God sets the classes of beings, the quiddities, the universals, and the laws that govern the connections between things in a deliberate act, integrating those settings into the essence of the one who is obeyed, and gives him the power to create the world from his essence.[102] The one who is obeyed turns these settings within his essence into creation by commanding the intellect of the outermost sphere. The one who is obeyed (*al-muṭāʿ*), al-Ghazālī says, is also "the one who gives the command" (*al-āmir*). He commands the intellect in the outermost sphere, who in turn commands the one in the next-to-outermost sphere and so on, until the tenth intellect, the active intellect, the one who controls the sublunar sphere, is reached.

This universe of the Veil Section can be understood as an apparatus similar to that which al-Ghazālī describes with the simile of the water clock. God designs the one who is obeyed, creates him and places him in position, and continues to provide the right amount of "energy" for the apparatus to achieve its intended goals. The apparatus is the whole universe. Creating the one who is obeyed (*al-muṭāʿ*), however, is a sufficient act for God. All other creation follows with necessity from that created being. Establishing the highest creation does indeed imply the creation of all other beings, since they are causal results of this first creation. The first two steps of creation described in the simile of the water clock as judgment (*ḥukm*) and decree (*qaḍāʾ*) are in the Veil Section expressed as the command (*al-amr*) and the creation of the one who is obeyed (*al-muṭāʿ*). The third step of the water clock, during which God provides the energy or power for the apparatus to create the intended results, is not mentioned in this passage. It is made clear, however, that God is the source of all being and provides being for all things that exist. Infusing a carefully calculated amount of being (*wujūd*) into the obeyed one achieves the goals of the creation of this apparatus. This infusion of being seems to be what al-Ghazālī had in mind when he stressed that God's "predestination" (*al-qadar*) will come about "by a known amount and (pre-)determined measure."[103]

Earlier in this book, I proposed a list of five conditions with which any cosmology acceptable to al-Ghazālī must comply.[104] This list was based on al-Ghazālī's criticism of the *falāsifa* in his *Incoherence*. As we have seen, the cosmology of Avicenna and al-Fārābī does not fulfill all five conditions. It fulfills only one of the five, that is, condition number four, which requires that any acceptable cosmology must account for our coherent experience of the universe and allow predictions of future events, meaning that it must account for the successful pursuit of the natural sciences. The cosmology of the *falāsifa* would fail all other four conditions: it would not be able to explain the temporal creation of the world, it would not account for God's knowing all creations individually and as universals, it would be unable to explain all prophetical miracles reported in revelation, and it would not take into account that God freely determines the creation of existences other than Him. The cosmology of the fourth subgroup at the end of the Veil Section, a cosmology that incorporates much of the *falāsifa's* cosmology, would fulfill all five conditions. It would do so, despite the fact that it is based, in fact, on the cosmology of the *falāsifa*.

I will briefly go to the five conditions and explain how this cosmology fulfills them: given that the creation of the one who is obeyed is a deliberate act of the Creator, it is a contingent event that can happen at any time He chooses. The first condition—that of creation in time—is thus fulfilled. In the Veil Section, al-Ghazālī says nothing about the nature and the attributes of the Creator. One may assume, however, that God has knowledge of His creation in a more immediate way than does the God of the *falāsifa*. This detailed knowledge of His creation fulfills the second condition in our list. However, al-Ghazālī nowhere elaborates on this subject, and there are indications that God Himself need not have a detailed knowledge of human actions, for instance. Because al-Ghazālī views salvation in the afterlife as the causal effect of actions in this world, God would only need to include these sorts of causal relations in His creation that he might justly reward or punish humans for their deeds in this world.

The subject of prophetical miracles, the third condition on our list has already been discussed elsewhere in this book. Al-Ghazālī believed that prophetical miracles can be explained as rare effects of causes that are unknown to us. His cosmology acknowledges the existence of all miracles reported in revelation even though it rejects the idea that God is breaking His habit when He creates the miracle. To be sure, he nowhere denies that Moses turned a stick into a serpent, for instance, even if this event was the effect of causes that are as of yet unknown to us. Allowing for the pursuit of the natural sciences is the fourth condition on our list. It is fulfilled once it is clear that all creations below the obeyed one are subject to rules—laws of nature we would say—that result from the essence of the obeyed one (*al-muṭāʿ*) and that will never be suspended.

The fifth and last condition on our list is the most interesting one, and it will make the full merits of the cosmology in the Veil Section evident. Al-Ghazālī demands that an acceptable cosmological theory must acknowledge that God freely decides about the creation of *all* existences other than His. In Avicenna's and al-Fārābī's metaphysics, God creates according to His nature, meaning that He cannot choose the classes of beings, for instance, or the number of individuals in any class that is created. The decision of which beings are possible and which are not is not a matter of God's choice but a result of His nature. In one of his early works Avicenna says that whatever is possible for God to do emanates from Him in a state in which its actual existence has not been determined. That only happens in the second stage, so to speak, when God as the absolute efficient cause becomes the preponderator between the existence and nonexistence of the possibilities (*mumkināt*).[105] Richard M. Frank, who assumes that al-Ghazālī largely adopted Avicenna's cosmology, criticizes him for not discussing the ontological origin of the quiddities and essences. Frank suspects that in al-Ghazālī's cosmology, the origin of what is possible (the *mumkināt*) lies outside of God's power. The possibilities are given to God: "It would seem that for al-Ghazāli, their being possible as possibles is absolute." Either al-Ghazālī was unaware of this metaphysical problem, Frank concludes, "or he was unaware of the seriousness of its theological implications."[106]

Other modern interpreters raised the problem of from where the quiddities come in al-Ghazali's theological system.[107] In Western thought, this has long been seen as a problem of Avicenna's philosophy. If Avicenna's God gives existence to things that are possible by themselves, is He also the one who determines the distinction of what is possible by itself and what is impossible? Gerard Smith and Beatrice H. Zedler have argued that for Avicenna, the realm of the possible is a given that is not determined by any of God's actions. What is possible by itself is just that, meaning its possibility is given by virtue of itself.[108] This prompted David B. Burrell to suggest that for Avicenna, God is a mere demiurge who turns possible beings into actual ones.[109]

The cosmology of the Veil Section suggests that al-Ghazālī understood quite well this problem of Avicenna's cosmology, that the possibles are given at the outset and are not under God's control. Although al-Ghazālī does not discuss it explicitly, it seems clear that the quiddities and possibilities are among those things that God creates with the creation of the one who is obeyed (*al-muṭāʿ*). The quiddities are part of, even identical to, "the command" (*al-amr*) that God creates when He brings His first creation into being. The one who is obeyed (*al-muṭāʿ*) passes the quiddities along the chain of being when he gives "the command" to the lower beings. In this cosmology, God is clearly the creator of everything that exists, including all the possibilities (*mumkināt*). In the simile of the water clock, al-Ghazālī calls the act of designing the apparatus of the world "the judgment" (*al-ḥukm*). Designing the world means determining the quiddities and the possibilities. In this system, determining the precise amount of how much "being" (*wujūd*) is given to the world fine-tunes its effects and determines such things as the number of individuals in each class of being.

An Ismāʿīlite Influence on the Cosmology in the Veil Section?

In a 1991 article, Hermann Landolt suggested that in the Veil Section of *Niche of Lights*, al-Ghazālī adopted Ismāʿīlite cosmological speculation, "to suit his own Sufi world-view."[110] It must be stressed that Landolt's identification of the three subgroups of those veiled by pure light is different from my own. Landolt proposes that the third subgroup represents the Fāṭimid Ismāʿīlites (*al-bāṭiniyya*).[111] This is the subgroup I identify with the followers of Avicenna and al-Fārābī. Landolt's suggestion, though ultimately misleading, I think, points to some interesting parallels between Ismāʿīlite cosmologies of the fifth/eleventh century and al-Ghazālī's own strategy of appropriating Avicenna's cosmology for his own purposes.

God's "command," which is so central in al-Ghazālī's Veil Section, also plays an important role in Fāṭimid Ismāʿīlite accounts of cosmology, particularly in the cosmology of Abū Yaʿqūb al-Sijistānī (d. *c.* 365/975). The Ismāʿīlite cosmology of the fourth/tenth century was heavily influenced by Neoplatonism and interpreted God's divine unity (*tawḥīd*) in a radical way. For these Ismāʿīlite authors, *tawḥīd* meant that God is absolutely transcendent and cannot in any way be part of this world. He is beyond being and beyond knowability. God's absolute transcendence makes it impossible that He causes anything in His creation, since that would require some immanence on His part. His oneness also prevents God from performing more than one single action.

From the early fourth/tenth century on, Ismāʿīlite cosmologies follow a common pattern, one in which God creates a universal intellect by means of His "command" (*amr*). This intellect is the "predecessor" (*al-sābiq*) from which the universal soul, which is also referred to as the "follower" (*al-tālī*), emanates. Matter, form, and the elementary components of the world all emanate from the universal soul.[112] Al-Sijistānī describes creation as a single act of "origination" (*ibdāʿ*), wherein the whole world is put into being. Everything that happens in creation proceeds from this one action: nothing is left out, and nothing can be added or removed at a later time. God issues a single "command," which manifests itself as an intellect. This "command" is the cause of creation. The created intellect exists in a timeless realm. From it emanates soul (*nafs*) and all of those things that are generated and that will eventually corrupt. Al-Sijistānī describes the "command" as something that is uncreated. The "command" is an intermediary (*wāsiṭa*) between God and existence.[113]

In al-Sijistānī's thought, there is no succession of celestial intellects that follows the planetary system. In his cosmology, the "command" transmits or transforms God's creative activity to the first being, and from there, it is further mediated to all other existences. The first intellect mediates creation through the soul to nature and to the material realm.[114] Ḥamīd al-Dīn al-Kirmānī (d. *c.* 411/1021), who was active in the generation after al-Sijistānī and who may have been one of his students, teaches a similar cosmology that adopts the Farabian model of intellects as secondary causes. Unlike the Aristotelians, al-Kirmānī rejects the idea that the highest of these intellects emanates from God, since divine transcendence prevents such a continuing relationship. He also rejected al-Sijistānī's concept of the "command" as a mediator between God and created being.[115] In a single act of origination and creation *ex nihilo* (*ibdāʿ wa-ikhtirāʿ*), God constituted the first intellect, which from then on acted autonomously. Given that God is unknowable, this first intellect is the highest being to which humans can relate, and it is the being that the Qur'an refers to as "God" (*Allāh*). The God of revelation is not a real deity but rather is the true God's first creation. Additionally, this is the being the philosophers and theologians refer to as "God."

The nine other celestial intellects of the Farabian cosmological system and the sublunar world of generation and corruption emanate from this first and universal intellect. Al-Kirmānī retains the philosophical concept that the world is the necessary product of the First Principle (*al-mabdaʾ al-awwal*), which stipulates that the universe emanates according to its essence. However, he adds the idea that this First Principle is, in fact, the first creation (*al-mubdaʿ al-awwal*) of an incomprehensible God. God created this first intellect "in one go" (*dufʿat^{an} wāḥidat^{an}*), under particular circumstances (*kayfiyya*) that cannot be known by humans.[116] For al-Kirmānī, the act of putting into being (*ibdāʿ*) is synonymous with creating the first creature (*al-mubdaʿ al-awwal*).[117] The first creature is also the First Principle of the universe, yet it is not God.[118] All other things follow from the creation of this first being. From the moment of initial creation, the highest being assumes the position of the creator and gives existence to all other beings through the mediation of the other nine celestial intellects and through other secondary causes.

The Ismāʿīlite cosmologies of al-Sijistānī and al-Kirmānī tried to respond to the implication—following from the notion that causes are necessarily related to their effects—that if God were causally related to the world, the latter were a necessary result of Him.[119] Al-Kirmānī, for instance, denied that God is either the agent or the efficient cause (*fāʿil*) of the world. He consciously disagrees with the *falāsifa* when they teach that God is the "first cause" of the world.[120] Al-Kirmānī rejects to declare a causal necessity in the relationship between God and the universe. Ismāʿīlite thinkers allowed causal relations to proceed only from the first intellect downwards. The relationship between the highest intellect and God is not causal. Al-Sijistānī explains it in terms of God issuing a single "command" that leads to the world's creation. In crafting his cosmology, al-Ghazālī found himself in a situation quite similar to al-Sijistānī's and al-Kirmānī's. Avicenna's cosmology accepted the implication that a causal relation between God and His creation precludes deliberative planning on the part of God. In his response to Avicenna, al-Ghazālī avoids casting the relationship between the Creator and His first being as one of cause and effect. Rather, he constructs a relationship that allows *liberum arbitrium* on the side of God. Unlike these Ismāʿīlite thinkers, however, al-Ghazālī never—as far as I can see—elaborates on the relationship between God and "the obeyed one." In al-Ghazālī's thought, "the one who is obeyed"—and not God—issues the "command." This "command" is somewhat different from that of al-Sijistānī, as it is clearly a creation of this world and thus has existence.

Al-Kirmānī's strategy of positioning the God of the Qur'an and of the Aristotelian *falāsifa* as the first creation of the real God may have served as a model for what al-Ghazālī does in the Veil Section of the *Niche of Lights*. When al-Ghazālī writes about the difference between the God of Aristotle and that of Avicenna, he says that Avicenna simply assumed Aristotle's God, the unmoved mover of the highest sphere, to be a created intellect. Avicenna's God transcends this particular intellect and creates it, just as al-Ghazālī does with Avicenna's God. He assumes that Avicenna's understanding of cosmology was limited and that he could only see as far as to "the obeyed one," rather than the creator of this being. This is quite similar to what al-Kirmānī did with the God of the Qur'an. Whether al-Ghazālī knew al-Kirmānī's cosmology is an open question. In his extant refutations of Ismāʿīlite theology—the most important is the *Scandals of the Esoterics* (*Faḍāʾiḥ al-bāṭiniyya*)—he does not refer to a cosmology in which the God of the Qur'an, the Sunni theologians, or the *falāsifa* is regarded as the first creation.[121] His report of the Ismāʿīlīte cosmology is based largely on a stage of their doctrine precededing al-Kirmānī. These teachings were shaped by al-Nasafī (d. 332/943) and al-Sijistānī, with the perfect "intellect" (*ʿaql*) or the "predecessor" (*al-sābiq*), and the imperfect "soul" (*nafs*) or the "follower" (*al-tālī*) standing as the key cosmological agents at a level beneath the totally transcendent God.[122] Al-Ghazālī's report of the Ismāʿīlite cosmology is somewhat confusing since it melds this earlier stage of Ismāʿīlite cosmology with what may indeed be a partial knowledge of al-Kirmānī's cosmology. Al-Ghazālī was, for instance, aware of the Ismāʿilite concept of a totally transcendent God who is neither existent nor nonexistent.[123]

With regard to the earlier stage of Ismāʿilite cosmology, al-Ghazālī seems to have misunderstood that the "intellect" there refers to the totally transcendent deity. Al-Ghazālī mistakenly believed that in Ismāʿīlite cosmology, the "predecessor" (*al-sābiq*) is the very first cause who employs the "follower" (*al-tālī*) as his intermediary (*wāsiṭa*) and that both are considered gods (*ilāhān*). In reality, Ismāʿīlites such as al-Sijistānī saw both the "predecessor" and the "follower" to be intermediaries created and employed by a totally transcendent God.[124] Continuing with this misunderstanding, al-Ghazālī criticizes and condemns the Ismāʿīlites for teaching a dualism of "intellect" and "soul" similar to the light-and-darkness dualism of Zoroastrianism (*al-majūs*).[125] In this part of his critique, he follows earlier Ashʿarites such as ʿAbd al-Qāhir al-Baghdādī.[126] The confusion of the "intellect" with the Ismāʿīlite God, however, does not accord with a brief passage on how the Ismāʿīlite teachings are similar to those of the *falāsifa*. There, al-Ghazālī reports that the Ismāʿīlites—like the *falāsifa*—believe the "intellect" is a creation of the First Principle. A further explication links this passage to al-Kirmānī's Farabian model of cosmology. In his criticism of the Ismāʿīlite cosmology, al-Ghazālī refers the reader to his *Incoherence*, in which he explains its doctrinal problem: in Ismāʿīlite cosmology, the First Being causes the intellect by necessity (*ʿalā sabīl al-*

luzūm) and not through free choice that aims to achieve a certain purpose (*lā ʿalā sabīl al-qaṣd wa-l-ikhtiyār*).[127] In his report on the Ismāʿīlite cosmology, al-Ghazālī tries to fuse two different models, an earlier one by al-Nasafī and al-Sijistānī and a later one by al-Kirmānī, which ultimately meddles elements of both models that do not belong together and thus creates confusion. Despite his claims to have benefited from insider informants, al-Ghazālī did not have enough reliable information on the Ismāʿīlite cosmology to fully penetrate and understand it.

Al-Ghazālī was probably unaware of one of the most significant elements in al-Kirmānī's cosmology, namely, his claim that the God of the Qur'an and the philosophers is not a god at all, but just the first creation of the real and much more transcendent God, who Himself is unable to be in such a close relationship with His creation. Had al-Ghazālī known this, he would have very likely criticized it. We have reason to assume that the higher echelons of the Ismāʿīlite movement tried to keep a tight lid on al-Kirmānī's texts and successfully prevented their dispersion among non-Ismāʿīlite scholars. Few texts were known by the Ismāʿīlites' dogmatic enemies, and al-Ghazālī relied heavily on information passed down from earlier Ashʿarite authors who may have seen some of these texts.[128] We know that al-Ghazālī studied the activities of the Ismāʿīlite missionaries closely, as his works contain reports of the strategies used by these agents. The lively and engaged character of these reports somewhat suggest that these accounts rely on firsthand experience.[129] It is not impossible that al-Ghazālī gained some mediated knowledge either of al-Kirmānī's cosmology or of other Ismāʿīlite cosmologies that applied a similar strategy and that are less well preserved in our sources.

Although both al-Kirmānī and al-Ghazālī describe the philosophical God as a creature of the real transcendent God, there are a number of differences between the cosmologies of these two thinkers. Al-Kirmānī presents varying models of the number of intellects and the spheres that they move. In most places in which he explains the cosmological order, the first intellect is also the unmoved mover of the most outermost sphere, the *primum mobile* (*falak al-aflāk*) that envelops all the other spheres. In one instance, however, the first intellect is not associated with a sphere and is one step removed from the intellect that moves the *primum mobile.*[130] Al-Ghazālī distinguishes between the cosmology of Aristotle and that of Avicenna, putting the God of the latter—whom he terms "the one who is obeyed" (*al-muṭāʿ*)—on a level that transcends physical movement. This first intellect of the Ghazalian cosmology is situated beyond the ten spheres of the Ptolemaic cosmos. In al-Ghazālī, the second intellect is the one that moves the outermost sphere, the *primum mobile.*

More important, al-Kirmānī and al-Ghazālī differ on the attributes of God. Al-Kirmānī applies an almost completely negative theology to God. He is not the creator or the originator; He is not the agent or the cause of the universe.[131] For al-Kirmānī, God is not even a being. Al-Ghazālī rejected negative theologies—even among the Sunni groups—and he vigorously opposed such extreme ones. Al-Ghazālī was convinced that God can be conceived and perceived by humans, albeit only after overcoming much difficulty by education or preparation such as "polishing of the heart." In a parable in the twenty-first book of the *Revival* about a competition between Chinese and Greek painters, a parable later made famous by Niẓāmī (d. *c.* 604/1207) and Jalāl al-Dīn Rūmī (d. 672/1273),[132] al-Ghazālī expresses the opinion that *falsafa* and Sufism are equal ways to comprehend the divine. A king asks a group of Chinese and Byzantine-Greek (*rūmī*) artists each to paint one half of a chamber (*ṣuffa*) in order for him to judge which group does it better. They work independently from one another and cannot see the other group's efforts. When the curtain that separates the chamber is lifted, it turns out that the Greek painters had produced a vivid picture of God's creation using brilliant and shining colors, while the Chinese painters had polished their side so thoroughly that it perfectly mirrored the painting of the Greeks. The king is highly impressed by both groups (figure 9.2).[133]

The Greek painters represent the way of "the philosophers and the scholars" (*al-ḥukamā' wa-l-ʿulamā'*) who comprehend God by acquiring the sciences and obtaining their "picture" (*naqsh*) within their souls, while the "friends of God" (*al-awliyā'*)—meaning the Sufis—perceive God through the manifestation of His splendor upon their polished hearts. In the *Scale of Action* (*Mīzān al-ʿamal*), al-Ghazālī explains this parable and clarifies that the souls of those who have cleansed it from the rusty stains of passions and vices will reflect the true knowledge (*al-ʿulūm al-ḥaqīqiyya*) that is contained in the preserved tablet (*al-lawḥ al-maḥfūẓ*) and in the "souls of the angels" (*nufūs al-malā'ika*), meaning the active intellect and the other separate celestial intellects.[134]

Al-Ghazālī rejected the negative theology of the Ismāʿīlites. In his *Book of the Distinction* (*Fayṣal al-tafriqa*), he reports the Ismāʿīlite position that God is nonexistent (*lā mawjūd*) and is unable to be defined as a single entity (*wāḥid*) or as omniscient. These teachings, al-Ghazālī says, are "clear unbelief."[135] For al-Kirmānī, however, God cannot conceive Himself and thus is also not conceivable by humans.[136] Al-Ghazālī's God is the source of all existence that bestows being on all other beings. He is the creator of the world, who designs all details of this universe according to His free will. He can be conceived in various ways, among them (1) pondering over the sheer fact of existence, like Avicenna did, (2) understanding the marvels of His creation, like the natural scientists do, (3) studying His revelation, like the theologians do, (4) perceiving His splendor in the mirror image of the celestial intellects, like the Sufis do, or, of course, (5) through a combination of all this, like al-Ghazālī did.

Final Doubts about Cosmology: *Restraining the Ordinary People (Iljām al-ʿawāmm)*

Given that the *Niche of Lights* was probably written years after the *Revival* and also after the *Highest Goal,* one might infer that its cosmology reflects a certain development away from al-Ghazālī's uncommitted position regarding the nature of causal connections. Maybe al-Ghazālī had become convinced that truth lies on the side of Avicenna and that the world is governed by secondary causality? Although the subject of causal connections is not discussed in the *Niche of Lights,* it is evident, I believe, that he accepted the cosmology of al-Fārābī and

FIGURE 9.2 A king adoring the two identical paintings of the Chinese and the Byzantine-Greek painters. Miniature illustrating Niẓāmī's *Quintet* (*Khamsah*) by the school of Herat, dated 853/1449–50 (MS New York, The Metropolitan Museum of Art, Gift of Alexander Smith Cochran, 13.228.3, fol. 322a).

Avicenna, including their explanation as to how creation stems from the nature of the being above the unmoved mover of the outermost sphere. There is no clear reference to the occasionalism of the Ashʿarites in the Veil Section. In fact, none of the groups mentioned in the Veil Section can be easily identified with the Ashʿarites. If my identification is correct, all three subgroups of those who are veiled by light are philosophers. Other distinctly Muslim groups, such as the Muʿtazilites, rank below these groups of philosophers. This is an unusual version of a Muslim heresiography, in which Avicenna and al-Fārābī stand only one rank below those who have achieved true insight.

Richard M. Frank had already observed that there is no discernable theoretical development in al-Ghazālī's cosmology between what Frank considers his earliest work on metaphysics and theology and his latest.[137] Although I do not completely agree with Frank what that cosmology is, I concur that there is little or next to no development on this issue between the seventeenth discussion of the *Incoherence* and his later works.[138] The impression given by the *Niche of Lights* that al-Ghazālī eventually accepted the cosmology of the *falāsifa* is shattered by evidence from his very last work, *Restraining the Ordinary People from the Science of Kalām* (*Iljām al-ʿawāmm ʿan ʿilm al-kalām*). Here, al-Ghazālī admits that there are certain things concerning God's creation that we simply cannot know, including whether or not God creates through an intermediary.[139] In this passage, al-Ghazālī aims to convince his readers that even the most experienced Muslim scholar should remain uncommitted on certain issues of metaphysical doctrine, such as whether God creates through the mediation of some creature(s).

There is some evidence that his work *Restraining the Ordinary People* was completed at the beginning of Jumāda II 505 / in December 1111, only a few days before al-Ghazālī died. This is mentioned in a colophon at the end of a manuscript that pretends to be the oldest available manuscript of the text, copied in Shaʿbān 507 / January 1114, roughly two years after al-Ghazālī's death.[140] I was unable to verify the age of this manuscript through an analysis of it, and therefore, the note should be met with at least some amount of suspicion. Both the early date of the manuscript as well as the notice about the dating of the text may have been inserted later in order to increase its marked value. The text of *Restraining the Ordinary People*, however, did also circulate under a second title, *Epistle on the Teachings of the Companions* (*Risāla fī Madhāhib ahl al-salaf*). We may assume that the two titles reflect two different manuscript traditions. A manuscript of this second tradition copied in 836/1433 also mentions that this was al-Ghazālī's last text.[141]

Restraining the Ordinary People is concerned with anthropomorphic descriptions of God that appear in certain verses of the Qur'an and in the prophetical *ḥadīth*. The companions of the Prophet appear in the alternative title of the work because al-Ghazālī wishes to explain how they as the first generation of Muslims understood the anthropomorphic passages in revelation. That does not mean, however, that al-Ghazālī made a turn toward traditionalism during his later life, as his biographer ʿAbd al-Ghāfir al-Fārisī has suggested.[142] *Restraining the Ordinary People* is the work of a rationalist theologian, exploring how the rationalism of the religious elite can be taught to the ordinary people without causing any damage either to their prospects of redemption in the afterlife or to their obeying the religious law that maintains societal order. When someone among the ordinary people is confronted with one of revelation's anthropomorphic verses or reports, he must fulfill seven duties (*waẓāʾif*): (1) he must declare the text holy (*taqdīs*), (2) he must acknowledge its truth (*taṣdīq*), (3) he must acknowledge his incapacity to fully understand it (*iʿtirāf al-ʿajz*), (4) he must keep silent and not ask questions (*sukūt*), (5) he must refrain from rephrasing it in different words (*imsāk*), (6) he must abstain from mentioning his personal opinion about it (*al-kaff*), and (7) he must submit to the authority of the people of knowledge (*taslīm li-ahl al-maʿrifa*).[143]

It must be noted that for al-Ghazālī, the class of "ordinary people" (*ʿumūm al-khalq*) includes many Muslim scholars. He has in mind all those people who have not studied rationalist theology (*kalām*) and who would be unable to present arguments as to why the anthropomorphic descriptions of God in revelation cannot literally be true.[144] The commoners' fifth duty to refrain from rephrasing anthropomorphic passages from revelation implies that they must maintain its original wording and must not paraphrase it. Only learned scholars are allowed to rephrase an anthropomorphic verse or a *ḥadīth* and only under certain conditions. One such condition is when a learned scholar would like to give an explanatory commentary (*tafsīr*) on revelation, including paraphrasing the passage into the Arabic vernacular or into Persian or Turkish. All this is forbidden to the ordinary believer.[145] Additionally, the untrained scholar and the ignorant Muslim must refrain from engaging in metaphorical interpretation (*taʾwīl*), meaning the "explanation of the meaning of the *ḥadīth* after eliminating its literal sense."[146] These things are forbidden whether done by "ordinary people" or in a conversation between a learned scholar (*ʿālim*) and an untrained person. However, if a well-trained scholar (*ʿārif*) engages in such metaphorical interpretation (*taʾwīl*) "in the secret of his heart between him and between his Lord," there is nothing objectionable.[147] This is, in fact, the only occasion when metaphorical interpretation (*taʾwīl*) is allowed. Only someone with a high degree of knowledge might legitimately ponder the meaning of the anthropomorphic descriptions in the Qurʾān and the *sunna*, and he may not convey this to any other than a member of his own class.

This limited permission to interpret gives al-Ghazālī occasion to clarify some parameters for metaphorical or allegorical interpretation (*taʾwīl*) of the revealed text. A well-trained scholar may have three different attitudes (*awjah*) toward what is meant by any given passage of the divine revelation. The first attitude is that he thinks that he has decisive knowledge about (*maqṭūʿ bihi*) what the text intends to convey; the second is that he has doubts about its meaning (*mashkūk fīhi*); and the third is that he has an assumption about the meaning that overwhelms him (*maẓnūn ẓann^{an} ghālib^{an}*). Here, al-Ghazālī distinguishes between the different levels of how strongly one might assent (*taṣdīq*) to a certain proposition. These three different levels of *taṣdīq* are discussed by Avicenna in his *Book on Demonstration*[148] and have influenced other parts of al-Ghazālī's oeuvre, such as his *Book on the Distinction between Islam and Clandestine Apostasy*.[149]

The depth of one's belief in the truth of a certain proposition can lead to varying treatment of the revelatory passage. Again, this is a subject al-Ghazālī has written about in chapters 5–7 of his *Book on the Distinction*.[150] Here in *Restraining the Ordinary People*, he just presents a very short version of these teachings. If one has decisive knowledge about the meaning of a passage or phrase, this meaning becomes part of one's conviction, and one adopts this as part of one's creed (*al-iʿtiqād*). If, however, one has doubts (*shukūk*) about a proposed interpretation, one should push aside the doubtful interpretation and not apply it:

> By no means should one judge about what God and His prophet intend [to convey] in their words by means of a conjecture (*iḥtimāl*), when a similar [conjecture] opposes it and when one cannot tip the scale [between these two conjectures].[151]

In such a case, one must suspend judgment.

The real problem, however, arises with the third attitude, namely, when a scholar is overwhelmed by an assumption (*ẓann*) about the meaning of a passage without having convincing evidence either in favor of or against this proposed allegorical interpretation. In this case, al-Ghazālī says, one must first decide whether the meaning that one is considering is a possible explanation of revelation or whether it is impossible. If the proposed interpretation is impossible, it must be dismissed. The case, however, becomes complicated when its possibility can be proven by a convincing argument, but the well-trained scholar is still reluctant to decide that this is what God intends to convey in revelation. This dilemma, al-Ghazālī says, may well be the case regarding the Qurʾanic verses and the prophetical *ḥadīths* in which it says that God "sat Himself upright on the throne" (Q 7.54, 10.3, etc.) as well as that God is "above" humans (Q 6.18, 16.50, etc.).[152] What al-Ghazālī intends is not to state that these verses may be true in their literal meaning, since valid demonstrations have excluded that from the very beginning.[153] Here, he simply assumes that their literal wording is impossible, and therefore God could not have intended to tell us that He sits on a material throne or that He is spatially above us.[154] The problem rests within the proposed allegorical interpretation itself. It may be unclear to the interpreter what is meant by these verses, particularly if two suggested allegorical interpretations mutually exclude each other. Even a well-trained scholar may hesitate (*taraddada*) to declaim what these verses actually mean.

In the case of God "sitting upright on the throne," the well-trained scholar wonders about God's relationship with the described throne. According to the philosophical interpretation, the throne (*al-ʿarsh*) is a reference to the outermost and highest celestial sphere.[155] Inspired by this reading, al-Ghazālī understands the "throne" to be a possible reference to a being that mediates God's creative activity. Although not explicitly stating the idea, it is quite clear that al-Ghazālī ponders whether the word "throne" in revelation refers to "the one who is obeyed" (*al-muṭāʿ*). He considers interpreting these verses in the Qurʾan to mean:

> [that] with the expression "he sat himself upright on the throne" [God] intends to express [the existence of] a special relationship that

"the throne" has. His relationship would be that God Exalted disposes freely (*yataṣarrafu*) in the whole world and governs the affairs from the heavens down to the earth through the mediation of the throne.[156]

Al-Ghazālī illustrates the relationship between God and His "throne" with a comparison: God may be related to His "throne" in the way that a human's "heart" (*qalb*)—meaning the human soul—is related to the human's brain (*dimāgh*). If a human creates a sculpture or a written text, he or she always needs to have a prior plan (*ṣūra*) in his or her brain. The builder needs to develop a plan in his brain before he can build the house he intends. Thus, one can say that the soul or heart of the human governs its microcosms—that is, its bodily organs—through the mediation of the brain (*bi-wāsiṭat al-dimāgh*).[157] The situation may be similar with God on the level of the macrocosm. Just as humans cannot generate anything without the mediation of the brain, so too God may not create without the mediation of "the throne."

The correspondence between the microcosm of the human body and the macrocosm of the universe is a common motif in al-Ghazālī's works. Although the subject does not appear in Avicenna's works, it is a prominent feature of the *Epistles of the Brethren of Purity*.[158] In al-Ghazālī's works, the correspondence between this universe and the human body is part of the larger theme that, for everything in the "world of perception" (*ʿālam al-shahāda*)—the material world of the sublunar sphere—there is an equivalent in the "world of sovereignty" (*ʿālam al-malakūt*), the realm of the pure ideas that includes the celestial intellects. God created the lower world such that there is a connection (*ittiṣāl*), a relation (*munāsaba*), and most important, an "equivalence" (*muwāzana*) between it and the higher world, and "there is nothing among the things in this world that is not a symbol (*mithāl*) for something in that world."[159] "The lowest is explicatory of the highest," writes al-Ghazālī in his *Jewels of the Qur'an* (*Jawāhir al-Qurʾān*), a work likely written during his years of teaching at his private madrasa in Ṭūs.[160] In his *Highest Goal,* al-Ghazālī compares the whole universe to a single individual. The different parts of the universe are like the limbs of a person. These parts are cooperating and working toward one single goal, which in the case of the universe is the realization of the highest possible goodness.[161] "The whole universe," al-Ghazālī writes in the thirty-second book of his *Revival*, "is like a single person." Just as there is no part of one's body that does not give benefits, so too is there no element in the world that is not beneficial to the overall goal.[162] The idea of the microcosm being equivalent to the macrocosm is already present in al-Ghazālī's early work. When in his *Touchstone of Reasoning* (*Miḥakk al-naẓar*) he introduces the concept that the soul (*al-nafs*) is a self-subsisting entity with no spatial extension, he states that the soul's relationship to the body is equivalent to God's relationship to the universe. And just as the soul is not part of the body, so too is God not part of the universe.[163]

Yet in terms of epistemology, the notion of equating the human body and the whole of God's creation merely indicates that the whole of God's creation is mediated by one single creation. The word "throne" in the Qurʾān only *may* be a reference to something that mediates God's creation. At this point, the epistemological status of the proposed interpretation becomes important. Al-Ghazālī continues:

> Now we may hesitate with regard to asserting this [kind of] relationship that the throne has to God and say: The [relationship between God and the throne] is possible either because it is necessary by itself (*wājib fī nafsihi*) or because God follows regarding this relationship His custom and His habit. The opposite of the relationship is not impossible. This is like in the case where God follows His habit with regard to the human heart.[164]

The relationship between the human heart and the brain in the human microcosm illustrates for al-Ghazālī that the same is *possible* in the macrocosm—yet this possibility says nothing about its actuality. If there is a mediating being in the macrocosm, it is there either because its existence necessarily follows from God's essence or because God had freely chosen to install such a being while always maintaining the power to do otherwise. The first reason for the existence of a mediating being is given by the *falāsifa*, the second by Muslim theologians who assume that God is omnipotent. We should expect al-Ghazālī to choose the second explanation: if there is a mediator, it is there because God habitually lets him mediate, although God could indeed do everything Himself. This is indeed the argument al-Ghazālī makes, although not in a straightforward manner. He begins with the acknowledgment that in the case of the human microcosm, the relationship between the heart and the brain is necessary because God wants it to be necessary:

> Here [*scil.* in the relationship between the human heart and the brain] it is impossible that the heart governs [its body][165] without the mediation of the brain, even if it is within the power of God the Exalted to make it possible without involving the brain. If God's eternal will has foreordained it and if His pre-eternal wisdom (*ḥikmatihi l-qadīma*), which is His knowledge, has made it happen, then its opposite is excluded (*mumtaniʿ*) not because of some shortcoming in [God's] power itself but because that which opposes the pre-eternal will and the eternal foreknowledge (*al-ʿilm al-sābiq al-azalī*) is impossible. Therefore God says: "You shall never find any change in the custom of God." (Q 33:62, 35:43 and 48:23). God's custom does not change because it is necessary. The custom is necessary, because it proceeds from a necessary eternal will (*irāda azaliyya wājiba*), and the product of the necessary is something necessary. What is contrary to the product of the necessary is impossible. It is not impossible in itself but it is impossible because of something else and that is the fact that it would attain to turning the eternal knowledge into ignorance and to prohibiting the effect of the eternal will.[166]

Changing the arrangements of the human microcosm is impossible, but not because the arrangements are the necessary result of human nature. Here again we find al-Fārābī's distinction between two meanings of impossibility. Changing an actual arrangement is not "impossible in itself" (*muḥāl fī dhātihi*) but rather "impossible because of something else" (*muḥāl li-ghayrihi*). A change would contradict God's plan for His creation. Al-Ghazālī calls this arrangement "necessary," not because it could never possibly be changed. "Necessary" here simply means that its final result cannot be changed. The Qur'anic quotation illustrates that the divine plan of creation is considered a "custom" (*sunna*). God has decided, however, never to change His custom, a notion we have already come across. In the quoted Qur'anic verse, God informs humanity that relations of causal concomitance, for instance, will not change and that they are thus necessary regardless of whether or not there is a direct connection between cause and effect. In addition, God's plan is called eternal (*azalī*) and pre-eternal (*qadīm*), two words that in this context stand for the atemporality of God's knowledge. God's knowledge existed before creation started. Finally, God's omnipotence guarantees that whatever He decides will happen, as knowledge of what will happen always coincides with God's plan of creation. God's "customary" decision of what to create and His knowledge are one and the same.

In the context of other works by al-Ghazālī, one would assume that he believes that God makes a free decision about what to create. This theory is suggested at the beginning of the passage where he stresses that God "disposes freely" (*yataṣarrifu*) with regard to His creation and may or may not install a mediating agent.[167] Yet this passage also contains a single sentence that is truly disturbing: God's custom is necessary because it proceeds from a necessary eternal will (*irāda azaliyya wājiba*), as the product (*natīja*) of the necessary is "something necessary" (*wājib*) and its opposite is impossible.[168] Taken at face value, these words say quite explicitly that God's actions and their habitual pattern are by themselves necessary. They proceed not only from a necessitating (*mūjib*) will but also from a will that is itself necessary (*wājib*), a will that is not free but acts in accord with what is by itself necessary.

Richard M. Frank explains the implication of this sentence. Frank draws a parallel with another sentence at the end of the *Standard of Knowledge*. There, al-Ghazālī says that God must be necessary "in all His aspects" (*min jamīʿ jihātihi*). This formula appears again in al-Ghazālī's textbook of Ashʿarite theology, the *Balanced Book*.[169] Avicenna used this phrase to express that God's actions follow with necessity from His essence.[170] If God is necessary "in all His aspects," His essence is by itself necessary, His knowledge is by itself necessary, and His actions are by themselves necessary. Admitting this point implies denying that God is a free agent.[171] These three brief passages—from *Restraining the Ordinary People*, from the *Balanced Book*, and from the *Standard of Knowledge*—pose a challenge for each interpreter of al-Ghazālī. Why would such an accomplished writer as al-Ghazālī, who ceaselessly points out that God's actions are the result of His free will, make such a *lapsus calami*? We must assume that the texts we have are carefully composed and were used as textbooks in teachings sessions. Students and followers may have frequently discussed them before they were made available for copying and would have reacted to inconsistent passages. I will briefly discuss these three passages one by one.

Al-Ghazālī's *Standard of Knowledge* relies significantly on the philosophical teachings preserved in the MS London, Or. 3126. The *Standard of Knowledge* is to some degree a reworking of that report, or at least relies on its same source.[172] According to its own introductory statement, the *Standard of Knowledge* wishes to accomplish two goals: to be a textbook on logic that teaches the syllogistic method, and to acquaint its readers with the technical language of the *falāsifa* so that they will be able to study *The Incoherence of the Philosophers*.[173] The *Standard of Knowledge* straddles the border between being a report of other people's opinions and expressing al-Ghazālī's own views.[174] A closer study of the *Standard of Knowledge* may explain how al-Ghazālī viewed what he posited there concerning God. The passage in question says:

> The being necessary by virtue of itself must be a being that is necessary in all its aspects, to the extent that it is not a substrate of temporary creations, does not change, does not have a delaying will (*irāda muntaẓira*), nor a delaying knowledge (*ʿilm muntaẓir*), and no attribute that delays anything from Its existence. Rather everything that It can possibly have must be present in Its essence.[175]

These teachings are not compatible with those that al-Ghazālī wrote in any work before or after this text. In fact, the passage reads much like an analytical and slightly polemical restatement of Avicenna's position, notwithstanding that the latter believed that God indeed has a will and would not have chosen these specific words on knowledge and will. We might assume this passage is a report rather than al-Ghazālī's own opinion.

The second problematic passage from the *Balanced Book* is less confusing when read in its context. Al-Ghazālī argues that God is not subject to a spatial direction (*jiha*); He is not "above." Were He to be above, the argument goes, one of the six directions would need to be specified and He would be particularized by this one while the five others would not apply to Him. Such particularization requires contingency (*jāʾiz*). Being above negates being below, for instance, and if God were "above," something that particularizes (*mukhaṣiṣ*) would need to have chosen this particular direction. If that were the case, then what particularizes God's direction could not be part of God's essence but must be distinct from it. This is wrong, al-Ghazālī says, since with regard to His place, God is not contingent. Rather He is necessary "from all directions" (*min jamīʿ al-jihāt*).[176] The word *jihāt* here refers to spatial directions and not to "aspects" of God's essence as in the Avicennan formula. Al-Ghazālī wishes to express that all six spatial directions necessarily apply to God. He seems to have chosen these words in a conscious attempt to reject the less literal Avicennan usage of the word "direction" (*jiha*) with regard to God.

Returning to the passage in *Restraining the Ordinary People*, one might speculate that a fatal illness—al-Ghazālī died at age fifty-six—prevented him from putting the necessary care into the composition of this text. When he says that God's will is necessary (*wājiba*), he may have become entangled in the distinction between necessary by itself and necessary by something else and chosen his words carelessly. According to the statements in al-Ghazālī's other works, God's will cannot be necessary by virtue of itself. This would be the position of Avicenna, and al-Ghazālī rejects it in numerous passages of his works. Given, however, that God chooses to create the best of all possible worlds, the will can be considered a more or less necessary effect of combining that choice with God's knowledge about how the best of all possible worlds would look like. The will can thus be considered necessary by virtue of God's knowledge and of God's decision to create the best world.

Apart from this rather confusing sentence, the passage from *Restraining the Ordinary People* stresses God's predetermination of all events in this world and is less concerned with the question of how the divine plan of creation comes about and whether God's will is contingent or necessary. Al-Ghazālī emphasizes that the factual is necessary and cannot be otherwise since God's plan for creation decided matters ages ago in a realm outside of time and in a way that cannot be changed. The argument continues with a return to the macrocosm. Although we have knowledge of the actual situation in the human microcosm, and we know that whatever is actual is also necessary, no such knowledge exists on the level of the macrocosm. Consequently, there is no necessity for the existence of the throne. In general, no necessary conclusions can be drawn with regard to the macrocosm; here, both options are still possible:

> Is the assertion of this [kind of] relationship that God the Exalted has to the throne with regard to the government of the kingdom through the mediation of it—even if it is possible according to the intellect—actual in existence? This is what the theologian (*al-nāẓir*) is hesitant about and maybe he assumes that the relationship between God and the throne does exist.[177]

Regarding God "sitting upright on the throne," the well-trained scholar may ask himself two important but distinct questions. The first question is: is there a relationship between God and the throne in the way that God mediates his creation through the throne? Al-Ghazālī answer is: it is certainly possible that there is such a relationship; but the opposite, namely that there is no such relationship and that the word "throne" refers to something quite different, is also possible. God may mediate His creation through the throne, or he simply may not, and it is impossible for us to decide either way.

This, al-Ghazālī says, is just an example in which the well-trained scholar has developed an assumption about the meaning of a certain term in revelation without any conclusive proof for the truth of the assumption.[178] However, this assumption cannot come from nowhere. In fact, there are always "necessary causes" (*asbāb ḍarūriyya*) for all assumptions (*ẓann*) that cannot simply be washed away.[179] In these cases, the well-trained scholar must adhere to two duties. The first duty is not to console oneself with false tranquility but to be aware of the possibility of error. The scholar should avoid rushing to any conclusions because of such an assumption. His second duty is not to refer to these assumptions as if they were facts, even when he talks with none other than himself. The scholar must realize that he has not been given knowledge about these matters. God reminds us of this when He says in the Qur'ān: "Do not pursue that of which you have no knowledge" (Q 17.36).

Regarding God's governing His creation, there are things of which humans have not been given certain knowledge, neither through clear language in revelation nor by means of demonstrative arguments. If there is no certain knowledge (*qāṭi'*), we only have recourse to speculation, assumption, or conjecture. Thus is the situation with regard to whether God governs his creation immediately or through the mediation of the throne. When al-Ghazālī talks about the proposed figurative interpretation of "the throne," he clearly considered the full apparatus of secondary causality.[180] If the interpretation that there is a throne is correct, al-Ghazālī says at the beginning of this passage, then "God governs the command (*al-amr*) from the heavens down to the earth through the mediation of the throne."[181] "The throne" is not understood just as a single being in the uppermost sphere that mediates God's creation. It is "the one who is obeyed" (*al-muṭā'*) from the Veil Section in the *Niche of Lights*. This being is the first secondary cause according to whose nature all other causes and intermediaries follow. The "throne" thus refers to the whole system of secondary causes and intermediaries as it is known from philosophical literature.

Conclusion

(. . .) *wa-baʿḍuhum qāla bi-l-sabbabiyya fa-stishnaʿūhu*
(. . .) and one of the *mutakallimūn* held the doctrine of causality and in consequence was regarded as abhorrent by them.
—Maimonides, *Guide of the Perplexed*, chapter 1:73

In the introduction to his *Ḥayy ibn Yaqẓān*, Ibn Ṭufayl (d. 581/1185–6) of Guadix in al-Andalus comments on some of al-Ghazālī's books, complaining that none of those that have reached Muslim Spain include the teachings intended for the intellectual elite.[1] Whether or not those books truly exist is an open question for Ibn Ṭufayl. ʿAyn al-Quḍāt al-Hamādhānī (d. 525/1131), who wrote a generation earlier in Iran and who knew the full extent of the Ghazalian corpus, assumed that such books did not even exist. In one of his letters, he posits that because al-Ghazālī feared religious strife (*fitna*), he did not explain the teachings that he intended for the elite in any of his works.[2] Like many readers of the great Muslim theologian, Ibn Ṭufayl and ʿAyn al-Quḍāt felt that al-Ghazālī did not express his teachings in clear terms; in his published books, he left much to be desired.

It is true that no work exists in which al-Ghazālī explains his cosmology in clear and unambiguous terms. Richard M. Frank takes the fact that al-Ghazālī "never composed a complete, systematic, summary of his theology" as an indication, and he doubts whether he had thought his theology through.[3] But when one considers his corpus as a whole, a quite cohesive picture of his theology emerges. Reading al-Ghazālī often requires one to consider interpretations of his work that at first may seem farfetched. One central passage that a critical reader must consider closely is the famous initial statement of the seventeenth discussion from his *Incoherence of the Philosophers*.

That statement has thus far been regarded as one of the most fundamental attacks on the existence of causal connections in the outside world. Al-Ghazālī has been understood as rejecting causal connections and thus denying the laws of nature. Because of his influence on the religious discourse and his legal power as a *muftī*—that is, someone who issues *fatwās*—he has often been made responsible for the assumed decline of the rational sciences after the sixth/twelfth century.[4]

In that famous sentence at the beginning of the seventeenth discussion in the *Incoherence*, al-Ghazālī says that "the connection between what is habitually believed to be a cause and what is habitually believed to be an effect is not necessary according to us."[5] This sentence is not meant to negate the existence of causal connections. A close reading of al-Ghazālī shows that he is merely emphasizing that as a Muslim theologian, he assumes that the connection *could* be different, even if it never was and never will be different. The emphasis here is on the word "necessary." For Avicenna, who applies Aristotle's statistical model of modalities and connects the necessity of a thing to its enduring actuality, a connection that never was different and never will be different is by definition necessary. Al-Ghazālī does acknowledge that causal connections never were and never will be different from how we witness them today. But even if causal connections are inseparable and never change, these connections are still not necessary. The connection between a cause and its effect is contingent (*mumkin*) because we can conceive of an alternative to its actual state. We can imagine an alternative world in which fire does not cause cotton to combust. Of course, such a world would probably be a radically different world from the one in which we live. Still, such a world can be imagined by our minds, which means that it is a possible world. It is thus indeed true that fire does not *necessarily* cause the combustion of cotton.

When he criticizes Avicenna's teaching that any given causal connection is necessary, al-Ghazālī wishes to point out that God could have chosen to create an alternative world in which the causal connections differ from those we know. Al-Ghazālī is indeed willing to accept the Avicennan view that the connection is possible by itself and necessary by something else. This "something else," however, is not the immutable divine nature but God's will, which for al-Ghazālī is distinct from the divine essence (*zā'id 'alā l-dhāt*). In al-Ghazālī's ontology, both possibility by itself and necessity through something else are rooted in God's contingent will.[6] Al-Ghazālī upholds the contingency of the world, in contrast to the necessarianism of Avicenna. For al-Ghazālī, our world is the contingent effect of God's free will and His deliberate choice between alternative worlds. God is not a dreary manufacturer of the world but its accomplished and reflective artisan.

Although he rejects Avicenna's necessarianism, al-Ghazālī has no objections to the philosophers' concept of secondary causality. Our discussion has shown that secondary causality is not a concept alien to Ash'arite occasionalism. The earlier Ash'arites categorically denied necessarian elements in the created world. While they were adamant in their rejection of "natures" (*ṭabā'i'*), they accepted the concept of secondary causality, as in their teachings about

God's creation of human actions through (*bi-*) a created power-to-act (*qudra muḥdatha*).[7] Al-Ghazālī similarly had no problem accepting the secondary causality in Avicenna's cosmology. Throughout his life, al-Ghazālī never attempted to decide *how* God creates the connection between the cause and its effect. What he identifies as causal connections may either be the concomitance of two events that are created individually and whose immediate efficient cause is God, or elements in a chain of secondary causes, in which the ontologically superior element is the immediate efficient cause of the inferior element, the effect. Deciding which of these alternative explanations accurately describes God's control over His universe is impossible, according to al-Ghazālī. When the critical scholar considers the evidence in favor of each view, he may tend toward one of the two options, al-Ghazālī writes in *Restraining the Ordinary People* (*Iljām al-ʿawāmm*). The scholar may thus develop a preference for one explanation. That preference, however, cannot reach the level of certainty (*yaqīn*) and is therefore not knowledge, strictly speaking. God has chosen to withhold that knowledge from humanity.

In both alternative explanations, God is the only efficient cause—or the "agent" (*fāʿil*)—of all events in His creation. Either created beings are not efficient causes at all, or, if they are, their efficacy is only a manifestation of the creator, in whose name they act as intermediates and secondary causes. The connection between cause and effect is in both cases contingent but not necessary. In the case of an occasionalist universe, the contingency between the two events follows from the fact that God *could* change the arrangement of what we call cause and effect at any moment. The concomitance is a mere result of divine habit, and habits can, in principle, be changed. However, God has revealed to humans that He will never change His habit (Q 33:62, 35:43, and 48:23), a revelation confirmed by our experience. Studying the world, we see that the connections between what we call causes and effects are permanent and do not change. Averroes was right when he suspected that every time al-Ghazālī speaks of "God's habit," he means the laws of nature.[8] And although there are no exceptions to the lawful character of God's creation, humans lack complete knowledge of all these laws. Our lack of knowledge becomes evident when we consider prophetical miracles, inexplicable by the standards of the known laws that govern creation but consistent with the yet undiscovered laws of God's creation.

As Michael E. Marmura has observed, al-Ghazālī's thought does contain a first and a second theory of causality.[9] The first theory denies the existence of natures and of active and passive powers, and it denies that what we call a cause is immediately connected to what we call its effect. Instead, the cause and effect are conjoined as two events that regularly appear in sequence. The two events are the direct result of God's will, and their creation is not mediated by any of His creatures. The sequence in which these creations occur manifests God's habit, a habit that He decided never to change. The second causal theory assumes that God mediates His creative activity through His creations, meaning that each of His creations has an unchangeable nature with active and passive powers that determine how this creation will react with others. Every creation in the universe, with its specific nature and its active and passive powers, is the mediated result of God's will, which is the undetermined determining factor of the whole universe.

The fact that al-Ghazālī did not commit himself to either of the two causal theories is an important element of his cosmology. Although both theories offer possible and consistent explanations of God's creative activity, neither of them can be demonstratively proven. Al-Ghazālī accepts the Aristotelians' position that secondary causality is a viable explanation for how God acts upon His creation, but he rejects that the demonstrations they posit indeed prove that theory. This leads to yet another meaning of how the initial sentence of the seventeenth discussion could be understood. Saying that the connection is not necessary means that there is no way for humans to know the connection is necessary. In the human sense perception, "cause" and "effect" are a mere sequence of two events. Only the intellect assigns the role of the "cause" to the first event and that of the "effect" to the second. Although the intellect does that, it still does not know whether cause and effect are directly connected with each other. Whatever we think we know about the true nature of causes and effects does not reach the level of necessary knowledge.

The combination of an occasionalist perspective on God's actions and a causalist perspective regarding events in this world can also be found in Abū Ṭālib al-Makkī's *Nourishment of the Hearts* (*Qūt al-qulūb*). Al-Ghazālī was well aware that this position was different from the one held by earlier Ashʿarites. Most *mutakallimūn*, he says in the first book of the *Revival*, believe that all things come from God, but they fail to pay attention to causes (*asbāb*) and to intermediaries (*wasāʾiṭ*). Although this is a noble position (*maqām sharīf*), it fails to truly understand God's unity (*tawḥīd*) and thus contributes to the *mutakallimūn*'s shortcoming as scholars who focus in their teachings on this world and take little heed of the afterlife.[10] Al-Ghazālī does not explain what he means by saying that the *mutakallimūn*'s opinion "falls short of paying attention" (*taqṭaʿu ltifātihi*) to secondary causes. The *mutakallimūn* may not consider how causes indeed have efficacy on their effects, or they may fail to understand that humans inevitably make causal connections in our understanding of God's creation. For al-Ghazālī, the lack of a demonstration that proves one of the two alternative cosmologies leads to an agnostic position on the type of connection between cause and effect. It also leads to a causalist understanding of these connections in all contexts not related to cosmology and metaphysics. Whatever may be the correct answer to the metaphysical question about the cosmological nature of these connections, it has no bearing on how we deal with these connections in our daily life. Given that God's habit does not change, for all intents and purposes, cause and effect are inseparably conjoined.

For Avicenna, the fact that the conjunction is permanent means that it is necessary. Avicenna follows Aristotle's statistical understanding of necessity, and for him, necessity means that something *always* happens. If two things are always conjoined, their connection is thus necessary. Using an understanding of necessity developed in Ashʿarite theology, al-Ghazālī objects that even permanent connections cannot be considered necessary as long as they could be different. Even if God chooses always to connect the cause with its effect, the possibility of a synchronic alternative to God's action means that this connection is not necessary.

As far as practical human knowledge is concerned, however, al-Ghazālī's position is quite different from his view on the metaphysics of causal connections described above. In human judgments, there is a "hidden syllogistic force" (*quwwa qiyāsiyya khafiyya*) that connects what we identify as the cause with what we identify as its effect. In human judgments, the connection is permanent, and there is no synchronic alternative. Thus in our judgments, the connection between the cause and its effect is necessary. This line of thinking is echoed in the view that the modalities only exist in human judgments, not in the outside world. Although causal connections between events in the outside are not necessary, our knowledge of them is necessary.

It is irrelevant to us whether God's habit manifests itself in the permanent concomitance of certain creations or in chains of secondary causes; either way, we would be unable to tell the difference. We witness a world that is shaped by causes and effects, and we are completely used to referring to these events with the terminology of efficient causality. Indeed, this terminology reflects how God wishes us to refer to these events. All natural processes are governed by necessary causation, as are the movements of the celestial spheres and even human actions. Voluntary human actions are caused by a volition and by its underlying motives. The motives are caused by the human's knowledge and his or her desires; and the human knowledge is the result of various causes, chief among them the influence of the active intellect that governs the sublunar sphere. Redemption or reward in the afterlife is the causal effect of our actions in this world, so that we can say that our fate in the next world is the causal effect of our knowledge in this world. This is why the acquisition of the right kind of knowledge—and acting according to this knowledge—becomes one of the most important tasks for humans in this world.

When it comes to describing the elements of God's creation, their order, and how they interact with one another, al-Ghazālī is willing to accept the teachings of Avicenna and al-Fārābī. The heavens may well consist of nine spheres, each higher sphere being the immediate efficient cause of the lower one. The spheres are of a uniform composition, they move in complete circles, and each sphere receives its movements from a residing mover, an intellect that is caused by the proximate higher intellect. The lowest sphere below the moon is significantly different from the celestial nine spheres. The sublunar sphere is composed of four prime elements (*usṭuqusāt*)—earth, water, fire, and air—and every material being in the sublunar sphere is composed of these four elements. The material beings are individuals from species or classes of beings whose immaterial forms—the quiddities (*māhiyyāt*)—are contained in the active intellect.[11] Creation unfolds from the ontologically superior beings—or in terms of the heavens, the higher ones—to the inferior ones. In a realm defined as ranging from the highest sphere down to the smallest creation on Earth, al-Ghazālī was generally willing to accept the cosmological explanations offered by Avicenna and al-Fārābī.

Unlike these philosophers, however, al-Ghazālī did not assume that the celestial spheres and the four prime elements are pre-eternal. He believed that all came into being at a specific point in time in the past. All things in the universe have been created as the necessary result of the creation of a single being. Al-Ghazālī refers to this being as "the one who is obeyed" (*al-muṭāʿ*). This first being is both the proximate cause of the intellect that moves the outermost sphere and the more remote cause of all other beings within and below that sphere. "The one who is obeyed" (*al-muṭāʿ*), "the throne," (*al-ʿarsh*), and the "well-guarded tablet" (*al-lawḥ al-maḥfūẓ*) are all references to one and the same being, the first creation that then causes the whole universe. In *Scale of Action* (*Mīzān al-ʿamal*), al-Ghazālī writes that the human intellect "flows from" (*yajrī min*) the first intellect, which is God's first creation. The first intelligence is compared to the sun as a source of light.[12] In the *Stairs of Jerusalem* (*Maʿārij al-quds*), the Ghazalian author, who may have been al-Ghazāli himself, refers to this being as the "first creation" (*al-mubdaʿ al-awwal*) and the "holy spirit" (*rūḥ al-quds*).[13] This is also the being that in a prophetical *ḥadīth* is referred to as "the pen" (*al-qalam*) and in an uncanonical *ḥadīth*—which is nevertheless quoted just as often by al-Ghazālī—as "the intellect."[14] Its nature contains all parameters that make this particular world necessary. It passes these parameters to all other creations as forms or as classes of beings, like a treasurer who holds the essences of God's rich resources, meaning his creatures.[15] The classes of beings are intellectual entities, theoretical constructs that determine every material creation. All together, they are referred to as "the command" (*al-amr*). The command passes from the ontologically superior beings to the inferior ones.

God's creation unfolds in three steps: judgment (*ḥukm*), decree (*qaḍāʾ*), and predestination (*qadar*). The first step, judgment, is the planning or drafting of the universe by designing its first creation, the one who is obeyed (*al-muṭāʿ*). The second step, decree, is the creation of this first created being.[16] The third step, predestination, is to provide the first creation with a carefully determined amount of existence (*wujūd*) so that it will cause its intended effects. It is important to note, however, that the relation between God and the obeyed one (*al-muṭāʿ*) is *not* determined by causal necessity. Although all other relations between things in the world may be causally determined, this one relation definitely is not. For al-Ghazālī, God is *not* the cause of the world but its creator. God is a personal agent who freely chooses and who precedes His creation, for instance.[17] The obeyed one receives his particular essence and existence from God and transmits a part of this existence together with the "command" (*amr*) to other beings. The existence of the whole universe follows from this first act of creation according to the plan made in the first step of this process and is realized by creating the obeyed one and providing him with a carefully measured "amount" of existence. The whole universe can be understood as an apparatus designed and maintained in order to achieve certain specific goals.

Al-Ghazālī rejected Avicenna's position that there is no goal (*qaṣd*), pursuit (*ṭalab*), desire (*ārzū*), or intention (*gharaḍ*) present when God creates.[18] God's chosen goal is to achieve the greatest possible benefit for His creation. Given that God is omnipotent and that nothing prevents Him from realizing this goal, the creation of the best possible world is the necessary result of His goal to achieve the best for His creation. In creating the best of all possible worlds, God shows utmost mercy to His creation. It is His mercy that prompts His free decision to create the best possible world. Although al-Ghazālī generally regards this decision as a necessary effect of divine generosity (*jūd*) and compassion (*raḥma*), he also stresses that God exercises free will and chooses between alternatives. David Z. Baneth explained that in al-Ghazālī's cosmology, God's freedom and His necessity become one and the same. The divine will *wills itself to be identical* to divine generosity and thus actualizes the decree to realize the best world order.[19] Studying God's creation and understanding how even the smallest of His creations dovetails with all the others to contribute to the best possible arrangement makes one realize that this is the best of all possible worlds. Harm in this world is a necessary element of creating the best possible world; without harm, the best could never be achieved.

When we examine the Veil Section from *The Niche of Lights*, we see how elegantly al-Ghazali's appropriates Avicenna's cosmology to his own theological system. Here, al-Ghazālī removes God from the sphere of philosophical analysis and assigns to Him a place one step more transcendent than in Avicenna's cosmology. For al-Ghazālī, what Avicenna calls the First Principle is only the first creation of the real God. Avicenna's God is "the one who is obeyed" (*al-muṭāʿ*), meaning the highest intellect that sits one step above the intellect that moves the *primum mobile*, or the highest sphere. Or, if looked at from the perspective of the "lower" world, the sublunar sphere: when Avicenna analyzed the cosmos, he reached only as high as the highest intellect. He did not understand that this intellect is itself only the creation of the real God. As I explained earlier, al-Ghazālī's solution to position the true God one step above Avicenna's First Principle is both elegant and functional.[20] It allows al-Ghazālī to make productive use of Avicenna's cosmology and to expand on its elements, while also allowing al-Ghazālī to reject Avicenna's necessarianism. Whereas Avicenna's God is compelled by principles from a higher ontological plane than His own, al-Ghazālī's God acts freely and chooses the principles of His creation. Additionally because Avicenna's God is a pure intellect, it cannot know the accidents that befall material individuals in the sublunar sphere. In contrast, al-Ghazālī nowhere says that the true God is pure intellect, opening God to the possibility of knowing individuals. In fact, al-Ghazālī remains uncommitted to what God truly is. This is an expression of the Ashʿarite epistemological attitude of "without how-ness" (*bi-lā kayf*) that wished to exempt God's essence and His nature from human rationalist analysis. God's essence and His nature are known to humans only insofar as He reveals knowledge about them in His revelation.

I have already mentioned that when al-Ghazālī gives God's "command" (*amr*) a central position in his cosmology, he is reacting to similar concepts in philosophical literature, mostly of the late fourth/tenth century.[21] The Qur'an uses this word—command (*amr*)—in ways that link it with the different stages of a carefully prepared and well-organized world order.[22] The "command" plays a particularly important role in Ismāʿīlite views of how God created the world. Al-Ghazālī had information on the relatively early stages of Ismāʿīlite cosmology, developed by al-Nasafī and al-Sijistānī, and that may have influenced his own understanding of the "command." Al-Ghazālī lacked, however, enough information on the more complex Ismāʿīlite cosmology of al-Kirmānī to fully penetrate and understand it. For al-Kirmānī, the God of the Qur'an is not a god at all but just the first creation of the real and much more transcendent God, who Himself is unable to be in such a close relationship with His creation. This bears a remarkable resemblance to al-Ghazālī's own technique of adopting Avicenna's God as the first creation of the real God. Yet, the fact that al-Ghazālī is ignorant about this element of Ismāʿīlite cosmology and the many differences between al-Kirmānī's cosmology and al-Ghazālī's appropriation of Avicenna's cosmology make it next to impossible to speak of an Ismāʿīlite influence on al-Ghazālī's cosmology.[23] Rather, al-Ghazālī developed his own adaptation of Avicenna's God as the real God's first creation from an analysis of the relationship between Avicenna's and Aristotle's cosmologies. In the text of MS London, Or. 3126, he gives an account of how Avicenna's proof of God's existence differs from that of Aristotle. That report likely led to the realization that these proofs each reach to different levels on the cosmological ladder of celestial beings, prompting the insight that Avicenna's God is on a higher step on that ladder than the God of Aristotle. Once he understood what Avicenna did to Aristotle's cosmology, it is just a small step toward doing the same to that of Avicenna.

To be sure, this particular move of appropriating Avicenna's God as the real God's first creation may to some degree have been prompted by what al-Ghazālī had discovered on the Ismāʿīlite side.[24] There is, however, no trace of textual evidence for that theory. Except for the *Epistles of the Brethren of Purity* (*Rasāʾil Ikhwān al-ṣafāʾ*), al-Ghazālī probably had no firsthand exposition of Ismāʿīlite cosmology at hand. These *Epistles*, however, do not teach such radical ideas as al-Kirmānī's. They represent moderate Qarmāṭian Ismāʿīlism, and their cosmology is distinct from that of al-Kirmānī, who developed his ideas within the Fāṭimid branch of Ismāʿilism.[25] We earlier discussed the accusations that al-Ghazālī copied his teachings on prophetical miracles from the *Epistles of the Brethren of Purity* (*Rasāʾil Ikhwān al-ṣafāʾ*).[26] There is no question that al-Ghazālī read these epistles and that they influenced his views on distinguishing religious groups in Islam.[27] In his autobiography, al-Ghazālī describes the *Epistles* as a work highly valued by some in the Ismāʿīlite movement.[28] The Fāṭimid and the Nizārī Ismāʿīlite study of the *Epistles* probably only began during al-Ghazālī's lifetime.[29] Later Muslim scholars and critics of al-Ghazālī, however, such as Ibn al-Jawzī, erroneously regarded the *Epistles* as an expression of the official Fāṭimid-Ismāʿīlite propaganda (*daʿwa*).

In chapter seven, I have argued that any resemblance between al-Ghazālī and the *Epistles* is based on a limited number of common motifs and on a common terminology rather than on substantial influence in matters of doctrine. Phrases such as "realm of the unknown and of sovereignty" (*ʿālam al-ghayb wa-l-malakūt*) or "realm of possessing and witnessing" (*ʿālam al-mulk wa-l-shahāda*) come from a distinctly Neoplatonic discourse and do not appear in Avicenna.[30] Earlier generations of Western scholars such as W. H. T. Gairdner, Arent J. Wensinck, or Margaret Smith saw a strong Neoplatonic influence in al-Ghazālī's teachings. If such a strong Neoplatonic influence truly exists, it must stem from the Neoplatonic elements in Avicenna's and al-Fārābī's philosophies as well as in al-Ghazālī's Sufi predecessors. I hesitate to acknowledge the existence of deeper Neoplatonic currents in al-Ghazālī than in these two philosophical thinkers. To be sure, non-Avicennan and non-Farabian philosophy did have its effect on al-Ghazālī. The idea of the human body as a microcosm of the universe, for instance, or the notion that all of nature is a harmonious structure in which every element dovetails with every other are prominent ideas in the *Epistles of the Brethren of Purity* and in al-Ghazālī. Such ideas are not, however, distinctly Neoplatonic.

There is no question that al-Ghazālī was attracted to the writings of pre-Avicennan Arabic philosophers such as Miskawayh and al-Fārābī. His report of the philosophical teachings in metaphysics, preserved in the London manuscript, is an eloquent testimony of this fascination. The same applies to the works of al-Rāghib al-Iṣfahānī and maybe also to those of al-ʿĀmirī. Ibn Taymiyya accepted the opinion of al-Māzarī al-Imām (d. 536/1141), a little known early critic of al-Ghazālī, who claimed that al-Ghazālī based his teachings on Avicenna and on the *Epistles of the Brethren of Purity*.[31] Ibn Taymiyya was probably one of the best-informed critics of rationalism in Islam, and his opinion deserves to be taken seriously. He was certainly right about Avicenna's strong influence on al-Ghazālī. More detailed studies are needed to explore al-Ghazālī's intellectual connection to the Brethren of Purity and to other authors from the second half of the fourth/tenth century.

It seems to me that al-Ghazālī was drawn to the writings of these pre-Avicennan philosophers because they present *falsafa* in a language consciously adapted to the Muslim religious discourse. Whereas Avicenna developed a philosophy that explains Islam and is well suited to it, these earlier *falāsifa* presented their philosophy as an interpretation of Muslim scripture. Unlike Avicenna, they consciously use language that connects to scripture, even modifying their teachings to fit its wording. This attentiveness was certainly attractive to al-Ghazālī. In addition, the *Epistles of the Brethren of Purity* uses allegories, parables, and moralistic stories in order to convey and illustrate its philosophical teachings, a style that al-Ghazālī uses in his *Revival*, in particular. He agreed with the authors of the *Epistles* that literature is a means to promote virtue and to assist people in achieving eternal salvation. Yet, when it comes to the detailed understanding of the universe or of the human soul, for instance, al-Ghazālī seems to have preferred Avicenna's teachings to those of other philosophers. He understands the "realm of the unknown and of sovereignty," for instance, or the "realm of possessing and witnessing" in Avicennan terms, the latter being the sublunar sphere while the former is everything above that, including the active intellect and the concepts contained in it.

There is much room for further studies to explore the ways in which al-Ghazālī's readers in the Islamic tradition made sense of the cosmology in the

Veil Section of the *Niche of Lights*. A casual remark by Averroes suggests that he understood that for al-Ghazālī, God is not the unmoved mover of the *primum mobile* but rather a being ranking one step above him. The mover of the *primum mobile* emanates from God. If that is truly al-Ghazālī's position, Averroes states triumphantly, al-Ghazālī is acknowledging the *falāsifa*'s teachings in metaphysics.[32] Averroes is not entirely correct, however, as al-Ghazālī's God is not one, but two steps above the mover of the first sphere. The radicalism of al-Ghazālī's cosmology seems to have escaped even Averroes. For critics of al-Ghazālī, the Veil Section was one of the most problematic parts of his œuvre. Ibn Ṭufayl quotes the accusation of an unidentified contemporary of his who said that in this passage, al-Ghazālī denied God's oneness (*waḥdāniyya*) and taught that there is multiplicity in God's essence.[33] Even if most readers of al-Ghazālī did not understand the hints and symbols in this enigmatic passage, some sensed that it contained an affinity with Ismāʿīlite teachings. The Ḥanbalite Ibn al-Jawzī (d. 597/1201) was a fierce critic of al-Ghazālī and repeatedly censures him in his book *The Cloaking of Iblīs* (*Talbīs Iblīs*) for his rationalist attitude, his affinity to Sufism, and his carelessness in quoting spurious *ḥadīths*. Commenting on the Veil Section in the *Niche of Lights*, Ibn al-Jawzī reports that the stars, the sun, and the moon, which Abraham saw, refer—according to al-Ghazālī—to lights that are God's veils (*ḥujub Allāh*). This is a misreading of the Qur'anic passage, Ibn al-Jawzī protests, and "this is cut from the same cloth as Ismāʿīlite teachings."[34]

In his 1994 study, Richard M. Frank argued that al-Ghazālī, though belonging formally to the Ashʿarite school (*madhhab*), did not hold the traditional doctrine of the school as his own personal teachings (*madhhab*). Frank concluded that al-Ghazālī's "basic theological system is fundamentally incompatible with the traditional teaching of the Ashʿarite school."[35] In my own conclusion, I argue that al-Ghazālī's undecided position between occasionalism and secondary causality should *not* be seen as a break with Ashʿarism. Indecisiveness is not uncommon in Ashʿarite epistemology. Indeed, it is implied in the "without how-ness" attitude (*bi-lā kayf*) of Sunni theologians toward the nature of God. Arguing that God's transcendence prevents us from fully comprehending His attributes, the Ashʿarites, for instance, objected to Muʿtazilite attempts to explain God's justice by analogizing it to human understandings of justice. One should rather understand that the descriptions of God as "being just" or as "having justice" refer to a different sense of justice than the one we apply to humans. Human reason is only a deficient bridge between the immanent and the transcendent, and it cannot help us understand the divine sense of justice. Additionally, revelation can give only hints that might help humans understand this divine attribute. The indecisiveness of Ashʿarism applies not only to God's attributes but also to questions on the cosmology of the afterlife. Regarding the question of whether atoms cease to exist with the end of this world and are then created anew when resurrection begins, or whether they continue to exist bereft of their previous accidents and are then restored and reassembled into their previous structures (*binya*), al-Juwaynī says that either theory is possible, as revelation gives no information from which to draw a particular conclusion.[36] Ash'arite epistemology developed a nominalist approach to human knowledge; and in that sense, al-Ghazālī is clearly an Ash'arite.

That God is the only agent in this world is a common Ash'arite thesis.[37] Both interpretations of how God acts upon His creation are a conscious attempt to make that particular view compatible with the scientific investigation of the world. Outside of his *Balanced Book on What-To-Believe* (*al-Iqtiṣād fī l-i'tiqād*), al-Ghazālī hardly ever makes a clear statement in favor of occasionalism. He refrains from following his master al-Juwaynī and never says clearly, as al-Juwaynī did, that the power God creates in humans has no effect on its object.[38] Al-Ghazālī also remains uncommitted on the question of whether created powers have efficacy. Instead, he stresses the idea that God controls every aspect of His creation while leaving open how such control is achieved. In a passage from his autobiography typical of this approach, al-Ghazālī writes:

> Nature is forced to operate according to God Exalted; it does not operate of itself but is employed by its creator. The sun, the moon, the stars, and the elemental natures are forced to operate according to His command (*amr*)—none of them has by itself any autonomous activity.[39]

Al-Ghazālī's main goal was to convey both the notion of God's omnipotence *and* the benefit of the natural sciences, of medicine, and of psychology to a readership that may not always understand the subtleties of positions from *kalām* or *falsafa*. Referring to an occasionalist cosmology would not have served the goal of accessibility. References to causes and effects are much more numerous in his works since they conform to commonly held assumptions and do not introduce unnecessary cosmological questions that might be distracting.

Al-Ghazālī's attitude toward other questions that were argued between the Ash'arite *mutakallimūn* and the *falāsifa* is quite similar. Another such question was whether the human intellect is an accident that inheres in the atoms of the human body—a position held by al-Ghazālī's Ash'arite predecessors—or an immaterial self-subsisting substance, as was taught by the *falāsifa*. In this case we have a clear and unambiguous statement by al-Ghazālī, saying that during the ten lunar years between 490 and 500 (1097–1106) he adopted one of these two competing explanations, namely, the one of the human "heart" (*qalb*) as a self-subsisting substance, a teaching he associates with the Sufis and the *falāsifa*.[40] In his earlier books, al-Ghazālī took a more or less agnostic position—similar to his undecided position on how God creates the world. In the first book of the *Revival*, al-Ghazālī refuses to answer which of the two competing views on the soul is correct, stating that this topic does not belong to the "knowledge of human actions" (*'ilm al-mu'āmala*).[41] Throughout the *Revival*, al-Ghazālī uses language that seems to commit sometimes to this and sometimes to the other of the two alternatives.[42] As in the case of the two cosmological alternatives, this leads to passages that can be read quite ambiguously. In the *Revival*, however, al-Ghazālī shows no interest in pursuing any doctrinal conflict between *falsafa* and traditional Ash'arism. His goal is to teach ethics. Both explanations of the character of the human soul offer consistent and noncontradictory explanations of those psychological events that al-Ghazālī refers to in his ethical teachings. What is important is that all Muslims acquire knowledge—and for al-Ghazālī, knowledge includes religious convictions—that lead to good actions. Given that they may already have foregone conclusions or deeply rooted opinions about the nature of the soul, any arguments supporting a contrary position would be counterproductive. In the *Revival*, al-Ghazālī tries to teach good actions without trying to change the convictions of his readership on the nature of the soul.

Unlike Avicenna, al-Ghazālī did not leave a comprehensive account of cosmology that answers all—or at least most—questions about how things come about from God. There is no explanation, for instance, of how the sublunar sphere and its intricate relationship between universal forms and individualizing matter generate from the world of the celestial intellects. It is also unclear whether emanation plays any role in al-Ghazālī's cosmology. In his *Niche of Lights*, he does use emanationalist language,[43] and it is not convincing to argue, as Hava Lazarus-Yafeh did, that for al-Ghazālī, the technical language of emanation in Arabic had lost its emanationalist meaning.[44]

Despite these lacunae in our understanding of al-Ghazālī, there is no indication of a division into esoteric and exoteric teachings where the esoteric would be different or even contradictory to the exoteric. Al-Ghazālī believed that revealing certain teachings to the ordinary people—such as God's complete predetermination of all events, including human actions—can lead to undesirable consequences. This belief results in a reticence to engage his readers on subjects of theology and metaphysics.[45] This reticence is not esotericism but rather the didactic result of al-Ghazālī's view that certain types of knowledge can be harmful to some people.[46] When more than a hundred years ago, W. H. T Gairdner first suggested esotericism in al-Ghazālī, he looked only at a limited amount of text and in doing so missed some of the complexities of al-Ghazālī's cosmology. Al-Ghazālī teaches God's omnipotence and His control over each event in His creation, and he still finds a way to reconcile fully these positions with the cosmological principle of creation through causal chains. Often, assigning esotericism to an author or referring to inconsistencies in a textual corpus is a hermeneutic device to mask the failure of interpreters to understand the texts. The same applies to suggestions that an author may have consciously introduced inconsistencies or contradictions in his works in order to conceal his true position from inattentive readers.[47] Throughout his œuvre, al-Ghazālī constantly reminds his readers how easily humans can fail in their judgments. Failure to understand texts that were written for a very different reader than oneself many centuries before is a natural human shortcoming. This inability, however, is hardly ever acknowledged, but that need not be the case. A good interpretation readily admits the lacunae in its understanding. Only such a frank admission will encourage us to work harder, to read these texts again and again, and to consider new levels of meaning that might reconcile apparent contradictions. Thus, finding such contradictions should lead us to take these texts more—and not less—seriously.

Notes

INTRODUCTION

1. In his *Die klassische Antike in der Tradition des Islam*, 101–55, Felix Klein-Franke reviews a great number of Western contributions that appeared since Edward Gibbon (1737–94) and that make a connection between the fortunes of the Islamic civilization and its ability—or the lack thereof—to integrate fully the ancient sciences.

2. De Boer, *Geschichte der Philosophie im Islam*, 150, 177; Engl. transl. 169, 200.

3. Goldziher, "Stellung der alten islamischen Orthodoxie zu den antiken Wissenschaften," 34, 40.

4. Goldziher, *Die islamische und die jüdische Philosophie des Mittelalters*, 327.

5. Makdisi, "Ash'ari and Ash'arites in Islamic Religious History," 38.

6. Berkey, *The Formation of Islam: Religion and Society in the Near East, 600–1800*, 229–30.

7. Al-Ghazālī, *Tahāfut*, 376.2–10 / 226.1–10. For al-Ghazālī's justification for applying the death penalty in these cases, see my *Apostasie und Toleranz*, 282–91.

8. Munk, in the *Dictionaire des scienes philosophique*, 2:512, and later in his *Mélanges de la philosophie juive et arabe*, 382.

9. Renan, *Averroès et l'averroïsme*, 22–24, 133–36.

10. Goldziher, *Die islamische und die jüdische Philosophie des Mittelalters*, 321.

11. Watt, *Islamic Philosophy and Theology*, 117. In his "Die islamische Theologie 950–1850," 416, a text that he published only in German, Watt further discusses this statement and diminishes much of its thrust.

12. Pines, "Some Problems of Islamic Philosophy," 80, n. 2.

13. Ibid., 80.

14. Bausani, "Some Considerations on Three Problems of the Anti-Aristotelian Controversy Between al-Bīrūnī and Ibn Sīnā," 85. I am grateful to

Mahan Mirza for pointing me to this publication. Bausani's observation fits particularly well with the subject of this book, the development of a nominalist critique of Aristotelianism. It becomes clear that this critique developed in the context of "orthodox" Islamic theology such as Ash'arism and, following al-Ghazālī's influence, became one of the dominating factors in the Islamic discourse on science. In the Latin West, nominalism emerged from a fringe movement that began when translations from the Arabic became available to one that revolutionized the approach to science in the fourteenth century and beyond.

15. Sabra, "The Appropriation and Subsequent Naturalization of Greek Sciences in Medieval Islam: A Preliminary Statement."

16. Endress, "Reading Avicenna in the Madrasa," 392–422, establishes the centrality of these two figures and also offers a survey of more peripheral philosophers after the fifth/eleventh century. On the philosophical commentary-literature of this period, see Wisnovsky, "The Nature and Scope of Arabic Philosophical Commentary."

17. Al-Ghazālī, *al-Munqidh*, 22–23; *Tahāfut*, 15.12–16.4 / 9.6–10.

18. This critique was voiced by Averroes and Richard M. Frank. See p. 212 in this book.

19. For the critique, see al-Ghazālī, *al-Munqidh*, 18–20; for slightly more appreciative comments, see ibid. 25–27.

20. Elsewhere I dealt with the subject of prophecy; see my "Al-Ġazālī's Concept of Prophecy: The Introduction of Avicennan Psychology into Aš'arite Theology." On this subject, see also al-Akiti, "The Three Properties of Prophethood in Certain Works of Avicenna and al-Ġazālī"; and Davidson, *Alfarabi, Avicenna, and Averroes, on Intellect*, 129–44, 149–55. For a study of al-Ghazālī's treatment of the soul in his *Iḥyā'*, see Giannotti, *Al-Ghazālī's Unspeakable Doctrine of the Soul.*

21. See pp. 44, 47 in this book.

22. Frank, *Al-Ghazālī and the Ash'arite School*, 4, 29, 87, 91. Ansari, "The Doctrine of Divine Command," offers the most thorough expression of the opposite view, that is, that of a gradual development in al-Ghazālī's thought from his works of *kalām* (*al-Iqtiṣād fī l-i'tiqād* and *Tahāfut*) to a theology that combines elements of philosophy and mysticsm in his *Mishkāt al-anwār.*

23. See p. 52 in this book.

24. See p. 67.

25. *Al-Iqtiṣād fī l-i'tiqād* is the "right balance" in terms of the teachings presented therein (see Makdisi, "Non-Ash'arite Shafi'ism," 249–50) and also the "balanced middle" with regard to the depth in which it discusses its subject matter (see al-Ghazālī, *Iḥyā'* 1:134.13–16 / 169.16–19; idem, *al-Iqtiṣād*, 215.9–10; and Frank, *Al-Ghazālī and the Ash'arite School*, 71).

26. Watt, "A Forgery in al-Ghazālī's *Mishkāt*?" and "The Authenticity of Works Ascribed to al-Ghazālī," 40–42. For a proper discussion and refutation of Watt's suggestion that the third part of *Mishkāt al-anwār* is a forgery, see Landolt, "Ghazālī and 'Religionswissenschaft,'" 21–29, 62–68.

27. Rosenthal, *The Technique and Approach of Muslim Scholarship*, 22–27.

28. Lazarus-Yafeh, *Studies in al-Ghazzali*, 249–63.

29. Frank, *Creation and the Cosmic System*, 59; idem, *Al-Ghazālī and the Ash'arite School*, 20.

30. In classical Ash'arism, the miracle is the only way the claims of a true prophet can be distinguished from those of an imposter (see Griffel, "Al-Ġazālī's Concept of Prophecy," 101–4). The authority of revelation thus rests on God's performance of prophetical miracles.

31. For a comprehensive report of Frank's and Marmura's interpretations, see pp. 179–82 in this book.

32. Or, as the physicist Steven Weinberg, *The First Three Minutes*, 154, puts it: "It is almost irresistible for humans to believe that we have some special relation to the universe, that human life is not just a more-or-less farcical outcome of a chain of accidents reaching back to the first three minutes, but that we were somehow built in from the beginning."

33. For an explanation of this cosmology, see below, pp. 253–60.

34. Recently, Naṣrullāh Pūrjavādī discovered a text, *al-Kitāb al-Maḍnūn bihi ʿalā ghayri ahlihi*, in which many teachings that al-Ghazālī reports in *Maqāṣid al-falāsifa* are presented as being those of himself; see *Majmūʿah-yi falsafī-yi Marāgha / A Philosophical Anthology from Maragha*, 1–62. The same manuscript (pp. 191–224) also contains one of the numerous versions of *Nafkh al-rūḥ wa-l-taswiya / al-Maḍnūn al-ṣaghīr*, and the *Risāla Fī ʿilm al-ladunī* (pp. 100–120). For the latter, see also the edition of the text from MS Istanbul, Hamidiye 1452, foll. 7b–19b, in ʿĀṣī, *al-Tafsīr al-Qurʾānī wa-l-lugha al-ṣūfiyya*, 182–202.

35. On some occasions I refer in the footnotes to al-Ghazālī (?) *Maʿārij al-quds fī madārij maʿrifat al-nafs*, which is not mentioned in any other work ascribed to al-Ghazālī. The text, however, is very useful, as it explains the background of a number of al-Ghazālī's teachings that appear in his generally accepted works. The work is doubtless of Ghazalian character; see Griffel, "Al-Ġazālī's Concept of Prophecy," 139–42; and al-Akiti, "Three Properties of Prophethood," 196–208. It is also of distinctly Avicennan character; see Janssens, "Le Maʿârij al-quds fî madârij maʿrifat al-nafs: un élément-clé pour le dossier Ghazzâlî-Ibn Sînâ?" Future studies must decide whether it can be truly ascribed to al-Ghazālī. Some of its teachings, such as the notion that God creates without a goal (*gharaḍ*; *Maʿārij al-quds*, 196.12–13) were held by Ibn Sīnā but were rejected by al-Ghazālī in the works that are generally ascribed to him and that are the basis of this study. Yet some classical Muslim scholars such as Ibn Sabʿīn (d. *c.* 668/1269–70) in his *Budd al-ʿārif*, 144.*ult.*–145.4, ascribed the *Maʿārij al-quds* to al-Ghazālī.

36. Ibrahim Agâh Çubukçu and Hüseyin Atay's 1962 edition of *al-Iqtiṣād fī l-iʿtiqād*, based on a comparison of four manuscripts, suffers from a surprisingly large number of misprints, and the list of errors on pp. 269–70, though not complete, should always be consulted. Only after finishing the work on this book, I came across a better edition of *al-Iqtiṣād fī l-iʿtiqād* by Anas Muḥammad ʿAdnān al-Sharafāwī (Jeddah: Dār al-Minhāj, 1429/2008) that compares the edition of Agâh Çubukçu and Atay with two additional manuscripts, one unidentified from the Dār al-Kutub al-Miṣriyya in Cairo and MS Dublin, Chester Beatty Library 3372, copied in 517/1123. An unusually large number of misprints also affects Jamīl Ṣalībā and Kāmil ʿAyyād's edition of *al-Munqidh min al-ḍalāl* (Damascus: Maktab al-Nashr al-ʿArabī, 1939), which compares two manuscripts. Farid Jabre's edition of *al-Munqidh* is based on this text and evens out the misprints. Despite the fact that Jabre does not note the variant readings from Ṣalībā and ʿAyyād's edition, I prefer his edition.

37. These are the words of Muḥyī al-Dīn Ṣabrī al-Kurdī al-Kānīmashkānī (d. after 1357/1938) on the title page of the *editio princeps* of *Kitāb al-Arbaʿīn fī uṣūl al-dīn*, (Cairo: Maṭbaʿat Kurdistān, 1328 [1910]). In the second edition of that work, Ṣabrī al-Kurdī describes in more detail the careful process of establishing the first edition from four different manuscripts in Egypt, Iraq, and Syria and of taking into account the testimony of two further manuscripts for the second edition (Cairo: al-Maṭbaʿa al-ʿArabiyya: 1344 [1925]), 310–11. Ṣabrī al-Kurdī has done pioneering work in bringing books by al-Ghazālī to the printing press and taking care for the reliability of their texts.

38. Al-Ghazālī, *Fayṣal al-tafriqa*, 195.10 / 61.*ult*. Cf. ibid., 190.2–3 / 53.5, 191.8–16 / 55.6–56.2. See also MS Escurial, no. 1130, fol. 84b (copied around 611/1214) and MS Berlin, Wetzstein II 1806, fol. 79b (Ahlwardt no. 2075), which both have *uṣūl al-ʿaqāʾid* in this passage. The latter manuscript was copied around 817/1414 and often contains very original readings, more original than those in MS Istanbul, Şehit Ali Paşa 1712, which, according to its colophon, was copied 508/1115 and which would be the oldest manuscript of the text, copied close to al-Ghazālī's lifetime. The Istanbul manuscript (fol. 66a) has *uṣūl al-qawāʿid* in this passage. This and other apparently less original readings led me to suspect that the colophon in this manuscript is forged.

39. The edition is based on MSS Damascus, Ẓāhiriyya Collection 6469 and 6595. Bījū only occasionally notes variant readings.

40. When one of these early editors prepared more than one edition of a particular text, I work with the latest.

41. Sarkīs, *Muʿjam*, 1409; Badawī, *Muʾallafāt*, 112. It is unclear how the texts in the lithograph editions of the *Iḥyāʾ* that appeared after 1281/1864 in Lucknow (India) and after 1293/1876 in Tehran relate to the one in the early Cairo prints.

42. Bauer, *Dogmatik*, 7. See, however, Richard Hartmann's objection in *Der Islam*, 9 (1919): 263.

43. See the colophon in the four-volume *Iḥyāʾ* print of Būlāq: Dār al-Ṭibāʿa wa-l-Waqāʾiʿ al-Miṣriyya, 1269 [1853], 4:341. On the editor (*raʾīs firaq al-taṣḥīḥ*), see Ziriklī, *al-Aʿlām*, 6:198; and Ṣābāt, *Taʾrīkh al-ṭibāʿa fī-l-Sharq al-Awsaṭ*, 181–82. Later prints almost never mention how the text was established. Muṣṭafā Wahbī, the editor and printer of the second Egyptian edition, claims that he was struck by the odd punctuation (*wuqūf*) in the first printing, compared it with manuscripts, and corrected it. (See the colophon in his four-volume *Iḥyāʾ* print of Cairo: al-Maṭbaʿa al-Wahbiyya, 1282 [1866], 4:469). Muṣṭafā Wahbī was the *maṭbajī* who printed the Ibn Abī Uṣaybiʿa edition of August Müller in 1299/1882. His work led to many complaints and corrections on Müller's side, mainly because Wahbī "changed the reliably established text according to his private ideas of what is correct or beautiful Arabic language." (Müller, *Ibn Abi Useibia*, vii). Uthmān Khalīfa, the Cairene publisher of a four-volume edition printed in 1352/1933, mentions that he took his text from a Būlāq print of 1289/1872–73.

44. See, for example, al-Ghazālī, *Iḥyāʾ*, 4:111.14 / 12:2224.7; 4:302.20 / 13:2490.17. In the first passage, even older prints of Muṣṭafa l-Babī al-Ḥalabī have *mūjid*, which means the mistake was introduced after the 1930s. Cf. al-Zabīdī, *Itḥāf al-sāda*, 9:61.16; 9:385.32.

45. Al-Zabīdī, *Itḥāf al-sāda*. The two editions of this work also add the text from the printed editions of the *Iḥyāʾ* in their margins. Note that the brackets distinguishing the *matn* of al-Ghazālī from al-Murtaḍā al-Zabīdī's commentary are not always set correctly. On al-Murtaḍā al-Zabīdī and his *Iḥyāʾ* commentary, see Reichmuth "Murtaḍā az-Zabīdī (d. 1791) in Biographical and Autobiographical Accounts," 85–87; and Reichmuth's forthcoming book, *The World of Murtada al-Zabidi*, chapter 5.

46. Bauer, *Dogmatik*, 7, compares the two printed versions with MS Berlin, Wetzstein II 19 (Ahlwardt 1680), one of the oldest manuscripts available, which can be dated to 582/1186. He notes that the differences are "less significant than one would expect in a text copied so often." Gramlich's German translation of books 31–36 of the *Iḥyā* notes all variants among the print, the text in al-Zabīdī's commentary, and MS Vienna, Nationalbibliothek 1656, copied in 726/1326.

47. On ʿAbd al-Qādir ibn Shaykh al-ʿAydarūs and his laudatory address on the *Iḥyāʾ*, see Peskes, *Al-ʿAidarūs und seine Erben*, 243–45, index.

48. Farid Jabre uses one of these editions (the one of 1352/1933) for his lexicographical study on al-Ghazālī, *Essai sur le lexique de Ghazali.*

49. These editions were newly typeset from the same stock of fonts. Since the fonts, the size of the paper, and the text remained the same, the differences of pagination between the various editions that al-Ḥalabī produced during the late 1920s and the 1930s are minor. Yet by the end of a volume, they may still add up to three pages between two different editions of this period.

50. Daniel Gimaret, for instance, used this edition in his studies on Ashʿarite theology. It is nicely printed on acid-resistant paper. Given that this is a five-volume edition (the fifth volume contains the texts that were earlier printed in the margins of the four-volume editions), its pagination is not similar to any of the four-volume editions of the 1930s.

51. This edition was used by George F. Hourani in his two articles on the chronology of al-Ghazālī as well as by Hava Lazarus-Yafeh in her *Studies on al-Ghazzali.* I follow their practice and refer to the overall pagination of the edition given at the inside of every page. This edition has been photomechanically reprinted. In the acid-resistant reprint, the folio size is reduced to quarto and the sixteen parts are divided on six volumes.

CHAPTER I

1. Leo Africanus, "Libellus de viris quibusdam illustribus apud Arabes," 262–65.
2. Gavison, *Sefer ʿOmer ha-shikheḥah,* fol. 135*a*; cf. Steinschneider, "Typen," 75.
3. Van Ess, "Neuere Literatur zu Ġazzālī."
4. Humāʾī, *Ghazzālī-nāmah.* The book was written almost twenty years before ʿAbbās Iqbāl Āshtiyānī's edition of al-Ghazālī's letters, *Fażāʾil al-anām.* The second edition of Humāʾī's book, which came out in 1963 and which is richly indexed, has not been further updated and does not refer to Iqbāl Āshtiyānī's edition of the letters or to any other literature that appeared since the publication of the first edition.
5. In 1985, Nakamura, "An Approach to Ghazālī's Conversion," 46–47, rightfully complained that the focus on the *Munqidh* led to a schematic treatment in Western literature, which gave the image of "the eminent orthodox doctor (*ʿālim*) to be reborn as a Sūfī (. . .)."
6. In 2004, Hillenbrand, "A Little-Known Mirror for Princes by al-Ghazālī," 599, for instance, still thought it was impossible to know this date.
7. Al-Baqarī, *Iʿtirāfāt al-Ghazālī aw kayfa arrakha l-Ghazālī nafsahu.* The book appeared in 1943.
8. Al-Ghazālī, *al-Munqidh,* 45.3; al-Subkī, *Ṭabaqāt,* 6:206.7.
9. Cf. for instance, Hillenbrand, "A Little-Known Mirror for Princes by al-Ghazālī," 594; Dabashi, *Truth and Narrative,* xiv, calls the ten years between 488 and 498 "al-Ghazālī's period of doubt uncertainty, and solitude." Michael E. Marmura assumed that al-Ghazālī spent the eleven years after 488/1095 "away from teaching as he became a Ṣūfī" ("Al-Ghazālī," 140), and in the timetable at the beginning of Moosa, *Ghazālī and the Poetics of Imagination,* the author mentions that al-Ghazālī returned to Ṭūs in 493/1100—three years later than he actually did—and lived there "in semiretirement."
10. Krawulsky, *Briefe und Reden,* 42–58. The reader should note that Krawulsky's translation of Hijri dates to Common Era is not always accurate.
11. Ibid., 50. According to Krawulsky, those who contribute original material are ʿAbd al-Ghāfir al-Fārisī, Ibn al-Jawzī, Yāqūt, Ibn Khallikān, al-Isnawī, and Ibn Kathīr. This list seems arbitrary, as al-Dhahabī should certainly be added and Ibn Kathīr be

taken off. It also neglects important historians such as al-Samʿānī and Ibn al-Najjār al-Baghdādī whose direct contributions are lost.

12. The discussion about the character of the *Munqidh* as an autobiography is reviewed in Poggi, *Un classico della spiritualità musulmana*, 16–36. Poggi offers the most comprehensive study on the text of the *Munqidh*, its manuscripts, prints, earlier works that influenced the text, and later works of literature that were influenced by it. Poggi's (ibid, 20–21) suggestion of a connection between Galen's autobiography and the *Munqidh* (earlier, Rosenthal, "Die arabische Autobiographie," 5–8, had already highlighted the influence of Galen's autobiography on Arabic literature) is comprehensively discussed by Menn, "The *Discourse on the Method* and the Tradition of Intellectual Autobiography."

13. On Abū-l Ḥasan ʿAbd al-Ghāfir ibn Ismāʿīl al-Fārisī, see Makdisi, *Rise of Colleges*, 82–83, and Bulliet, *Patricians of Nishapur*, 165–68, and index.

14. Brockelmann, *GAL*, 1:364; *Suppl.* 1:623; see also the decription of his yet unedited commentary on forty selected *ḥadīths* in Ahlwardt, *Handschriften-Verzeichnisse*, 2:210.

15. MS Ankara, Dil ve Tarih Fakültesi Library, İsmail Saib 1544, which contains the second part of the *Siyāq*, is reproduced in a facsimile edition by Frye, *The Histories of Nishapur*, text 2.

16. Al-Ṣarīfīnī (d. 641/1243), *al-Muntakhab min al-Siyāq*, 83–85 = Frye, *The Histories of Nishapur*, text 3, fol. 20a–b. See also the index to the *Siyāq li-taʾrīkh Nīsābūr* and its abridgement by Habib Jaouiche. On the somehow enigmatic relationship of the fragment of the *Siyāq* to the *Muntakhab al-Siyāq*, see Josef van Ess in the preface to Habib Jaouiche's index, pp. vi–vii.

17. Al-Subkī, *Ṭabaqāt*, 1:325.6–9.

18. Compare ibid. 6:204.6–214.3 with Ibn ʿAsākir, *Tabyīn kadhib al-muftarī*, 291.15–296.17 and his *Taʾrīkh madīnat Dimashq*, 55:200.11–204.6. Al-Subkī discusses Ibn ʿAsākir's reason for omitting passages from ʿAbd al-Ghāfir's history in his *Ṭabaqāt*, 6:214.4–11. Al-Subkī's version of ʿAbd al-Ghāfir's report is translated into English by Richard McCarthy, *Al-Ghazali: Deliverance from Error*, 14–19.

19. Al-Subkī, *Ṭabaqāt*, 6:208.4–210.3.

20. Ibid., 6:206.7; al-Ṣarīfīnī, *al-Muntakhab min al-Siyāq*, 84.6.

21. Fragments of al-Samʿānī's report are available in al-Subkī, *Ṭabaqāt*, 6:215.5 ff. and 216–17. There is next to no treatment of al-Ghazālī's life in al-Samʿānī's extant works. Al-Samʿānī does not mention al-Ghazālī in his *Kitāb al-Ansāb*, 10:31, and only marginally in his *al-Taḥbīr fī l-muʿjam al-kabīr*. We know, however, that al-Samʿānī dealt with al-Ghazālī in his lost works, such in as his *Dhayl ʿalā Taʾrīkh Baghdād*. On al-Samʿānī's works, see Rudolf Sellheim's article in *EI2*, 8:1024b, and Brockelmann, *GAL*, 1:329–30; *Suppl.* 1:564–65. On his position among the Shāfiʿite scholars of Khorasan, see Halm, *Ausbreitung*, 84–86.

22. See Griffel, "Al-Ghazālī or al-Ghazzālī? On a Lively Debate Among Ayyūbid and Mamlūk Historians in Damascus," 108. As far as I can see, al-Bayhaqī's Persian work on the history of Bayhaq and its scholars—a district neighboring to Ṭūs—mentions al-Ghazālī only twice in passing (*Taʾrīkh-i Bayhaq*, 79, 235).

23. Ibn ʿAsākir, *Tabyīn kadhib al-muftarī*, 291–306.

24. The *tarjama* on al-Ghazālī in Ibn ʿAsākir's *Taʾrīkh madīnat Dimashq*, 55:200–204, had already been reproduced in Badawī, *Muʾallafāt*, 504–9, and, like the one in *Tabyīn kadhib al-muftarī*, offers no original material. Later historians cite Ibn ʿAsākir with original information about the life of al-Ghazālī that is not included in these two entries. With the recent full edition of Ibn ʿAsākir's *Taʾrīkh madīnat Dimashq* in eighty

volumes, the book becomes available for a much-needed comprehensive search for information on al-Ghazālī and his students in Damascus.

25. Ibn al-Jawzī, *al-Muntaẓam*, 9:55, 87, 168–70.

26. Sibṭ ibn al-Jawzī, *Mir'āt al-zamān*, ed. Hayderabat, 1:39–40; ed. Mecca, 2:548–58. Further notes on al-Ghazālī are in the ed. Mecca, 1:146, 238.

27. Yāqūt, *Mu'jam al-buldān*, 3:560–61. He was the first to include the misleading information that "some say he proceeded to Alexandria and stayed in its lighthouse." Later, al-Subkī's (*Ṭabaqāt*, 6:199.12–3.) mistaken report that al-Ghazālī made his way to Alexandria caused much confusion.

28. Ibn al-Athīr, *al-Kāmil fī l-ta'rīkh*, 10:145, 172, 325, 400.

29. Ibn Khallikān, *Wafayāt al-a'yān*, 4:216–19,; al-Dhahabī, *Siyar a'lām al-nubalā'*, 19:322–46 (largely identical to idem, *Ta'rīkh al-Islām*, vol. 501–20 AH, pp. 115–29); al-Ṣafadī, *al-Wāfī bi-l-wafayāt* 1:274–77; Ibn Kathīr, *al-Bidāya wa-l-nihāya*, 12:137, 149, 173–74; and idem, *Ṭabaqāt al-fuqahā' al-shāfi'iyīn*, 2:533–39. For other, less significant historians of al-Ghazālī, see the anthology by 'Uthmān, *Sīrat al-Ghazālī wa aqwāl al-mutaqaddimīn fīhi*, 84–92, 143–49. See also the reprint of sources in Badawī, *Mu'allafāt*, 471–550. Al-Dhahabī's report is certainly the most interesting as he quotes from scholars who were opposed to al-Ghazālī and who are not mentioned by al-Subkī.

30. For instance, Ibn al-Najjār (d. 643/1245) is quoted as a source of information. The relevant part of his *Dhayl Ta'rīkh Baghdād* is lost, and the excerpts by al-Dimyāṭī, *al-Mustafād min Dhayl Ta'rīkh Baghdād*, 37–38, contain only a brief article on al-Ghazālī.

31. Griffel, "Al-Ghazālī or al-Ghazzālī? On a Lively Debate Among Ayyūbid and Mamlūk Historians in Damascus."

32. Cf. al-Nawawī's (d. 676/1277) extract of Ibn al-Ṣalāḥ's (d. 643/1245) *Ṭabaqāt al-fuqahā' al-shāfi'īyya*, 1:249–64 (= al-Nawawī, *Mukhtaṣar Ṭabaqāt al-fuqahā'*, 267–76) and al-Isnawī (d. 772/1370), *Ṭabaqāt al-shāfi'īyya*, 2:242–44.

33. Al-Subkī, *Ṭabaqāt*, 6:191–389.

34. See Laoust, "La survie de Ġazālī d'après Subkī."

35. Abū 'Abdallāh Muḥammad ibn al-Ḥasan al-Wāsiṭī's work (*al-Ṭabaqāt al-'aliyya fī manāqib al-shāfi'iyya*) is yet unedited. The *tarjama* on al-Ghazālī, however, is edited in al-A'sam, *al-Faylasūf al-Ghazālī*, 153–94; the worklist is on pp. 171–76. Some titles are mentioned twice.

36. Cf. for instance, long articles on al-Ghazālī in the histories by al-Subkī's contemporaries al-Wāsiṭī and al-Yāfi'ī (d. 768/1367), *Mir'āt al-jinān*, 3:145–46, 177–92. A thorough comparison of these two with al-Subkī would yield a systematic picture of the sources that were available to them.

37. Cf., for instance, the texts described by Ahlwardt, *Handschriften-Verzeichnisse*, 9:468–69, and the works used by Ormsby, *Theodicy in Islamic Thought*.

38. Al-Zabīdī, *Itḥāf al-sāda*, 1:6–51, with its center part on al-Ghazālī's life on pp. 7–11. For an earlier commentary, or rather a rewriting of the *Iḥyā'* from a Shiite perspective, see Fayḍ al-Kāshānī's (d. 1090/1679) *al-Maḥajja al-bayḍā fī tahdhīb al-Iḥyā'*. Fayḍ al-Kāshānī was a student and son-in-law of Mullā Ṣadra (d. 1050/1640). His *Maḥajja al-bayḍā* contains no study of al-Ghazālī's life. For a brief survey of its content, see William C. Chittick in *EI2*, 7:476b. Among the more recent Muslim historians that gather earlier material on al-Ghazālī is al-Khwānsārī (d. 1313/1895), *Rawḍat al-jannāt*, 8:3–20, who includes a number of interesting (mostly Shiite and Persian) perspectives.

39. Schmölders, *Essai sur les écoles philosophiques chez les Arabes*. Almost two centuries earlier, a manuscript of the text was already known in Paris (today MS Paris B.N. fonds arabe 1331). In 1697, Barthélemy d'Herbelot paraphrased passages from this manuscript in his *Bibliotheque orientale* 2:66, 693.

40. Macdonald's landmark article, "The Life of al-Ghazzālī," of 1899, for instance, relies mostly on these three sources.

41. Ibn al-Jawzī, *al-Muntaẓam*, 9:168.19.

42. Ibn Khallikān, *Wafayāt al-aʿyān*, 4:218.*peanult*. This sentence also appears in the edition of al-Ṣarīfīnī, *al-Muntakhab min al-Siyāq*, 84.*ult*., which would make ʿAbd al-Ghāfir al-Fārisī its prime source. It is, however, unduly inserted by the editor, and it is not in the facsimile text of the unique manuscript edited by Frye, *The Histories of Nishapur*, text 2, fol. 20b.

43. Al-Ṣafadī, *al-Wāfī bi-l-wafayāt*, 1:277.7–8, mentions it.

44. Al-Ghazālī, *Fażā ʾil al-anām*, 4.16–17; Krawulsky, *Briefe und Reden*, 65. At this time, Sanjar (d. 552/1157) ruled over Khorasan in the name of his brother Sultan Muḥammad Tapar (d. 511/1118), who is also known as Muḥammad ibn Malikshāh and who resided in Isfahan. After Muḥammad Tapar's death, Sanjar would himself become a powerful sultan of the Seljuq Empire.

45. *dowāzdah sāl badīn ʿahd wafāʾ kard*, al-Ghazālī, *Fażāʾil al-anām*, 5.2. The purpose of this letter is to avoid appearing before Sanjar, who had summoned al-Ghazālī; and in order to achieve this, al-Ghazālī doesn't mention the fact that his return to teaching at the Niẓāmiyya madrasa may have already violated his vow in Hebron. This is in line with the view in his *Munqidh*, 48–49, where he implicity rejects the notion that his return to public teaching violated his vow at Hebron. In their respective dating of the letter, the editor Iqbāl Āshtiyānī (in *Fażāʾil al-anām*, 4, n.1) and Krawulsky, *Briefe und Reden*, 14–15, have overlooked this reference.

46. Al-Ghazālī counted periods of his life in lunar years (cf. *al-Munqidh min al-ḍalāl*, 49.17–19). Yet solar calendars were always used for tax purposes and also for the age of people. Cf. Richard Sellheim's discussion of this problem in a book review in *Oriens* 11 (1958): 233–34. Ibn Khaldūn, for instance, informs us that durations in horoscopes were given in solar years (*al-Muqaddima*, 2:164, English trans. 2:224).

47. Al-Ghazālī, *al-Munqidh*, 49.17–19.

48. *wa-qad anāfa l-sinnu ʿalā l-khamsīn*; al-Ghazālī, *al-Munqidh min al-ḍalāl*, 10.11.

49. Ibid., 48–50.

50. Ibid., 49.14–18.

51. I am grateful to Alexander Treiger, who pointed this connection out to me.

52. ʿAlī al-Riḍā was buried in the mausoleum of the ʿAbbāsid caliph Hārūn al-Rashīd, who had died there in 193/809. Shiite contempt for the ʿAbbāsid's grave led to its gradual destruction. The mausoleum became known as that of ʿAlī al-Riḍā.

53. V. Minorsky and C. E. Bosworth, Art. "Ṭūs," in *EI2*, 10:740b–4b; le Strange, *Lands of the Eastern Caliphate*, 388–400; Yāqūt, *Muʿjam al-buldān*, 3:569–60.

54. Al-Ghazālī most likely did not use the name because of its pro-Shiite connotation. Meshed (Mashhad) means "[ʿAlī al-Riżā's] place of martyrdom." In 490/1097, al-Ghazālī referred to Nūqān/Meshed simply as "the site of visitation" (*mazār*, al-Ghazālī, *Fażāʾil al-anām*, 52.12.) On his contemporaries using names such as *Mashhad-i muqaddas-i Riżawī*, see al-Ghazālī, *Fażāʾil al-anām*, 5.16 and 6.6.

55. Al-Subkī, *Ṭabaqāt*, 6:193.10.

56. Abū Ḥāmid Aḥmad ibn Muḥammad al-Ghazālī or Abū Ḥāmid Muḥammad ibn Aḥmad; al-Shīrāzī, *Ṭabaqāt al-fuqahāʾ*, 133. Al-Isnawī, *Ṭabaqāt*, 2:246–47, reports his date of death. See also al-Subkī, *Ṭabaqāt*, 4:87–89, and Griffel, "Al-Ghazālī or al-Ghazzālī? On a Lively Debate Among Ayyūbid and Mamlūk Historians in Damascus," 107–11.

57. Al-Subkī, *Ṭabaqāt*, 6:193.10–194.2; al-Isnawī, *Ṭabaqāt*, 2:242.9–12.

58. *kāna l-Ghazālī yaḥkī hādhā*; al-Subkī, *Ṭabaqāt*, 6:194.3.

59. Al-Dhahabī, *Siyāq*, 19:335.9–17. The student is the unidentified Abū l-ʿAbbās Aḥmad al-Khaṭībī.

60. *taʿallamnā l-ʿilma li-ghayri Llāhi, fa-abā l-ʿilmu an yakūna illā li-Llāh*; al-Ghazālī, *Mīzān al-ʿamal*, 115.13–4 / 343.10–11; *Iḥyāʾ*, 1:71.24–5 / 84.2–3. Cf. al-Subkī, *Ṭabaqāt*, 6:194.3.

61. Al-Subkī, *Ṭabaqāt*, 5:204.9; al-Ṣarīfīnī, *al-Muntakhab min al-Siyāq*, 83 = Frye, *The Histories of Nishapur*, text 3, fol. 20a. Prompted by ʿAbd al-Ghāfir al-Fārisī's information, al-Subkī, *Ṭabaqāt*, 4:91, gives this scholar his own *tarjama* and a full name: Abū Ḥāmid Aḥmad ibn Muḥammad al-Rādhakānī. Cf. also al-Zabīdī, *Itḥāf al-sādā*, 1:19.16, and Halm, *Ausbreitung*, 94.

62. He was the father of Abū l-Azhar al-Ḥasan ibn Aḥmad al-Rādhakānī (d. ca. 530/1135) of Ṭābarān-Ṭūs, who was a scholar. On him see al-Samʿānī, *al-Taḥbīr fī l-muʿjam al-kabīr*, 1:174–75, and idem, *Kitāb al-Ansāb*, 6:29. All we know from al-Samʿānī about the father is his name: Aḥmad ibn Muḥammad al-Rādhakanī. Given the fact that his son grew up in Ṭābarān-Ṭūs, it is likely that he had settled there from the nearby Rādhakān.

63. ʿAbd al-Malik ibn Muḥammad al-Rādhakānī; al-Ṣarīfīnī, *al-Muntakhab min al-Siyāq*, 509 = Frye, *The Histories of Nishapur*, text 3, fol. 96a. ʿAbd al-Ghāfir mentions no connection between this al-Rādhakānī and al-Ghazālī.

64. On Abū l-Qāsim ʿAbdallāh ibn ʿAlī, see al-Subkī, *Ṭabaqāt*, 5:70. Who exactly held the position of the head teacher at the Niẓāmiyya in Nishapur in these years is not known. Cf. Bulliet, *Patricians*, 255. Halm, *Ausbreitung*, 59, thinks it was Shihāb al-Islām ʿAbd al-Razzāq ibn ʿAbdallāh (d. 525/1130), the son of Abū l-Qāsim ʿAbdallāh ibn ʿAlī and the nephew of Niẓām al-Mulk, who is addressed in the anecdote of al-Ghazālī's return from Gurgān. He, however, was born in 459/1066–67 and was probably too young to hold that office during these years. His father, Abū l-Qāsim ʿAbdallāh ibn ʿAlī, however, died in 499/1106, and if he held the position of head teacher at the Niẓāmiyya in Nishapur, it would explain why it became vacant that year, when it was offered to al-Ghazālī.

65. Unlike Makdisi, "Non-Ashʿarite Shafiʿism," 241, 246–47; idem, "Muslim Institutions of Learning," 37; and idem *Rise of Colleges*, 302–3, I see no evidence that the teaching activity at the Niẓāmiyya colleges was limited to *fiqh* and excluded *kalām*. I think there is much evidence to the contrary.

66. *balīda bi-aʿālī Ṭūs*; al-Samʿānī, *al-Ansāb*, 6:28; le Strange, *Lands of the Eastern Caliphate*, 393–94.

67. Niẓām al-Mulk's nephew Shihāb al-Islām ʿAbd al-Razzāq ibn ʿAbdallāh, who has been mentioned in note 64, became the leader of the Shāfiʿites in Nishapur. He also served as vizier of Sultan Sanjar from 511 to 515 (1117–21). Al-Samʿānī, *al-Taḥbīr fi-l-muʿjam al-kabīr*, 1:442–43; al-Subkī, *Ṭabaqāt*, 7:168; Iqbāl Āshtiyānī, *Vizārāt dar ʿahd-i salāṭīn-i buzurg-i saljūqī*, 243–48; Halm, *Ausbreitung*, 59; Klausner, *The Seljuk Vezirate*, 107; Kasāʾī, *Madāris-i Niẓāmiyyah*, 54–55, 99.

68. Al-Subkī, *Ṭabaqāt*, 6:204.9–10. Yāqūt, *Muʿjam al-buldān*, 3:360, gives the distance between Nishapur and Ṭūs as ten *farsakh*.

69. Jabre, "La biographie et l'œuvre de Ghazali," 77, suggests that "Abū Naṣr" was in fact Abū l-Qāsim Ismāʿīl ibn Masʿada al-Ismāʿīlī, an influential teacher of Gurgān who was born in 407/1016–17 and who died in 477/1084–85. He was from a prominent family of Shāfiʿite scholars, and, while in Baghdad, he attracted the attention of Abū Isḥāq al-Shīrāzī, the prominent jurist and theologian who was the first head teacher of the Niẓāmiyya madrasa. On Abū l-Qāsim al-Ismāʿīlī see al-Subkī, *Ṭabaqāt*, 4:294–96, and al-Samʿānī, *al-Ansāb*, 1:243.10–13.

70. A particularly wide-ranging interpretation of this anecdote's significance has been offered by Moosa, *Ghazālī and the Poetics of Imagination*, 90–94. A more sober look is taken by Obermann, *Der philosophische und religiöse Subjektivismus*, 309–11, and Glassen, *Der mittlere Weg*, 79.

71. Ibn al-ʿAdīm, *Bughyat al-ṭalab fī taʾrīkh Ḥalab*, 5:2489–90.

72. It is briefly mentioned in al-Dimyāṭī's *al-Mustafād min Dhayl Taʾrīkh Baghdād*, 38. Cf. also al-Dhahabī, *Siyar aʿlām al-nubalāʾ*, 19:335.

73. Al-Subkī, *Ṭabaqāt*, 6:195.12–13. The question of the highway robber is put slightly more eloquently here: "How can you claim to know what knowledge is contained in these notes when we could have taken them away from you? You have been stripped of the knowledge of those notes and there you are, without any knowledge."

74. *dar daryā-yi ʿulūm-i dīn ghawwāṣī kard*, al-Ghazālī, *Fażāʾil al-anām*, 4.16–17; Krawulsky, *Briefe und Reden*, 65.

75. Al-Ghazālī, *al-Munqidh*, 10.20–11.1.

76. Al-Subkī, *Ṭabaqāt*, 6:208.4–8.

77. On the persecution of the Ashʿarites and the impact of this event on al-Juwaynī and Ashʿarite theology see Griffel, *Apostasie und Toleranz*, 200–215.

78. Glassen, *Der mittlere Weg*, 66ff.

79. Al-Subkī, *Ṭabaqāt*, 6:196.3–6, lists the following subjects in which he surpassed everybody in Nishapur: Shāfiʿite law, differences among the schools of law, disputation (*jadal*), methods of jurisprudence and of theology, and logic: "And he read philosophy (*al-ḥikma wa-l-falsafa*) and became firm in all these subjects."

80. Al-Kiyāʾ al-Harrāsī, *Uṣūl al-dīn*, foll. 27b–62a, and al-Anṣārī, *al-Ghunyā*, foll. 19b–22a, who both also studied with al-Juwaynī in Nishapur, devote much space to refuting the philosophical notion of the eternity of the word. Both understood that this teaching goes back to Aristotle (see in al-Kiyāʾ, fol. 57b; in al-Anṣarī, fol. 20b). Cf. Frank, *Creation and the Cosmic System*, 66.

81. Al-Juwaynī, *al-Shāmil fī uṣūl al-dīn* (ed. Alexandria), 123–342.

82. Ibid., 196–97, 540–41, 618. On these passages, see the remarks on the readings by Frank, *Creation and the Cosmic System*, 17.

83. Al-Juwaynī, *al-Irshād*, 59, 84.

84. Al-Juwaynī, *al-ʿAqīda al-Niẓāmiyya*, 12–13.

85. Ibid., 8–9, 11–12. On this proof and how it differs from Avicenna's proof, see Rudolph, "La preuve de l'existence de dieu," 344–46, and Davidson, *Proofs for Eternity, Creation and the Existence of God*, 187.

86. Jules Janssens and Erwin Gräf suggest that the *Maqāṣid al-falāsifa* was written many years before the *Tahāfut* "by the young al-Ġazzālī in his student days" who "was probably an adept of the (Avicennian inspired) *falsafa*-school of his time" (Janssens, "Al-Ghazzālī and His Use of Avicennian Texts," 48; cf. Gräf in a book review in *ZDMG* 110 [1961], 163.)

87. Al-Ghazālī, *Tahāfut al-falāsifa*, 4.2–9 / 1.11–2.2. On al-Ghazālī's description of this attitude and his analysis of why the followers of the *falāsifa* disregard religion, see Griffel "*Taqlīd* of the Philosophers. Al-Ghazālī's Initial Accusation in His *Tahāfut*."

88. Al-Ghazālī, *Jawāhir al-Qurʾān*, 44.10–46.2; MS Escurial 1130, fol. 14a.

89. Al-Subkī, *Ṭabaqāt*, 16:205.5–7.

90. Al-Ghazālī, *Fażāʾil al-anām*, 4.16–19; Krawulsky, *Briefe und Reden*, 65–66.

91. Abū Bakr ibn al-ʿArabī, *al-ʿAwāṣim min al-qawāṣim*, 57.

92. Al-Ghazālī, *Fażāʾil al-anām*, 12.3–4; Krawulsky, *Briefe und Reden*, 78. The *Mankhūl*, 618, mentions the *Shifāʾ al-ghalīl* by al-Ghazālī, which must have generated in the same period.

93. Al-Ghazālī, *al-Mankhūl*, 618. See Makdisi, *Rise of Colleges*, 244–45, 250. On the many meanings of the word *taʿlīqa*, see Makdisi's *Rise of Colleges*, 114–28.

94. "'You buried me while I am still alive. Can't you wait until I'm dead?' By this he meant to say: Your book outshines mine!" (Ibn al-Jawzī, *al-Muntaẓam*, 9:168–69). Cf. Sibṭ ibn al-Jawzī, *Mirʾāt al-zamān*, ed. Mecca, 548, with the correct amendation. Cf. also al-Dhahabī, *Siyar aʿlām al-nubalāʾ*, 19:335.8, who also understands it this way. Cf. Makdisi, *Rise of Colleges*, 127.

95. Al-Ghazālī, *al-Mankhūl*, 618.9–11.

96. Al-Subkī, *Ṭabaqāt*, 6:205.1–2.

97. Ibid., 6:205.2–4.

98. Ibn al-Jawzī, *al-Muntaẓam*, 9:55.21–23.

99. Ibid. 9:170.12–13. A *qirāt* was the twentieth part of a dinar. Ibn al-Jawzī quotes the *faqīh* Abū Manṣūr Ibn al-Razzāz (d. 539/1144) of Baghdad.

100. Glassen, *Der mittlere Weg*, 131; Halm, *Ausbreitung*, 165. The scholars were Abū ʿAbdallāh al-Ṭabarī (d. 495/1102) and Abū Muḥammad al-Fāmī al-Shīrāzī (d. 500/1107).

101. On the dating of these books see p. 75.

102. Al-Ghazālī, *Fażāʾil al-anām*, 34–35, mentions that Ibrāhīm al-Sabbāk (d. 513 /1119–20) "was for twenty years my companion in Ṭūs, Nishapur, Baghdad, and on the trip to Syria and Hijaz."

103. Ibn al-Jawzī, *al-Muntaẓam*, 9:169.4–5; Glassen, *Der mittlere Weg*, 131.

104. Like that of al-ʿIbbādī (d. 496/1103), an authority on homiletics (*waʿẓ*); Sibṭ ibn al-Jawzī, *Mirʾāt al-zamān*, ed. Hayderabat, 1:5.

105. Together with al-Qaffāl al-Shāshī (d. 507/1114) and Ibn ʿAqīl (d. 513/1119), al-Ghazālī was present during the *bayʿa*-ceremony for the new caliph al-Mustaẓhir in Muḥarram 487 / February 1094; Ibn al-Jawzī, *al-Muntaẓam*, 9:82; Glassen, *Der mittlere Weg*, 132, 158.

106. *akala māla l-sulṭān*. So in his *Fayṣal al-tafriqa*, 197.5/65.3, in which al-Ghazālī complains about people who claim to be Sufis in order to live of the ruler's purse. Part of his vow at Hebron was "no longer to take from the riches of the ruler" (*va-māl-i sulṭān nagīrad*; al-Ghazālī, *Fażāʾil al-anām*, 5.1.).

107. Al-Ghazālī, *Fażāʾil al-anām*, 4.20; Krawulsky, *Briefe und Reden*, 66.

108. Al-Ghazālī, *al-Munqidh min al-ḍalāl*, 18.13–15.

109. Cf. n. 86.

110. Al-Ghazālī, *al-Munqidh min al-ḍalāl*, 33.4–10. Part of the book mentioned, *Mufaṣṣil al-khilāf*, may be preserved in al-Ghazālī, "Jawāb al-masāʾil al-arbaʿ allatī saʾalahā al-bāṭiniyya bi-Hamadān."

111. Daftary, *The Ismāʿīlīs*, 335–38.

112. Bouyges, *Essai de chronologie*, 23–24; al-Ghazālī, *Tahāfut* (ed. Bouyges), ix.

113. Al-Ghazālī, *al-Munqidh min al-ḍalāl*, 28.7–8; *Faḍāʾiḥ al-bāṭiniyya*, 3. The book might be based on the earlier Persian *Ḥujjat al-ḥaqq fī l-raddʿ alā l-bāṭiniyya*, which is lost.

114. Hourani, "Revised Chronology," 293.

115. Glassen, *Der mittlere Weg*, 131–75.

116. Abū Ẓāhir of Arrān (a district in northern Azerbaijan); Juvaynī, *Taʾrīkh-i Jahāngushāy*, 3:204.6.

117. Safi, *The Politics of Knoweldge in Premodern Islam*, 65–74.

118. Ibn al-Jawzī, *al-Muntaẓam*, 9:62.4.

119. Klausner, *The Seljuk Vezirate*, 30.

120. Ibn al-Athīr, *al-Kāmil*, 10:154. There was, however, speculation about the cause of his death. Laoust, *La politique de Ġazālī*, 61, notes that the historians do not agree on the day of al-Muqtadī's death (some say it was four days later), which is an indication that it may not have been made public immedately

121. Juvaynī, *Taʾrīkh-i Jahāngushāy*, 3:206.3–4, 207.9–10. For a detailed narrative of the events shortly before Niẓām al-Mulk's assassination and the long power-struggle afterward, see Laoust, *La politique de Ġazālī*, 56–64, 107–14, and 133–37. On the various conflicts among Seljuq family members during the war of succession after Malikshāh's death, see Claude Cahen's article "Barkyārūk" in *EI2*, 1:1051b–2b. Al-Ghazālī refers to some of these events in a letter he wrote later to Mujīr al-Dīn, the vizier of Sanjar; al-Ghazālī, *Fażāʾil al-anām*, 57–59.

122. Hillenbrand, *1092: A Murderous Year*, 293–94; Glassen, *Der mittlere Weg*, 134–45. Recently, Omid Safi, *The Politics of Knowledge in Premodern Islam*, 74–79, argued that Malikshāh instigated Niẓām al-Mulk's assassination. The evidence he quotes, however, is late (al-Subkī) and does not trump the many voices much closer to the event (Rāwandī, Nīshābūrī, and Rashīd al-Dīn Ṭabīb) who assume that Tāj al-Mulk was behind the murder and that he acted as an agent for Terken Khātūn.

123. Abū Bakr ibn al-ʿArabī, *al-ʿAwāṣim min al-qawāṣim*, 56–57.

124. Ibn al-Athīr, *al-Kāmil*, 10:145–46. The text in Ibn al-Jawzī, *al-Muntaẓam*, 9:62–63, is not altogether clear which of the two parties asked for the provision. Once in Isfahan, Terken Khātūn sent a delegation to the caliph in order to renegotiate the terms of Maḥmūd's appointment. That seems to have led to a mistaken presentation of this episode in Laoust, *La politique de Ġazālī*, 59.

125. Ibn al-Jawzī, *al-Muntaẓam*, 9:62.16–17.

126. Ibn Kathīr, *al-Bidāya*, 12:139.18–19.

127. *lā yajūzu illā mā qālahu l-khalīfa*; Ibn al-Jawzī, *al-Muntaẓam*, 9:63.2–3.

128. On al-Ghazālī's views about the caliphate, see Laoust, *La politique de Ġazālī*, 234–65; F. R. C. Bagley's introduction to *Ghazālī's Book of Council for Kings*, li–lvi; and Binder, "Al-Ghazālī's Theory of Islamic Government."

129. Al-Ghazālī, *Shifāʾ al-ghalīl*, 225.3–4.

130. Hillenbrand, "Islamic Orthodoxy of Realpolitik? Al-Ghazālī's Views on Government," 91.

131. Ibid., 90. See also al-Ghazālī, *Faḍāʾiḥ al-bāṭiniyya*, 173–74. Recently Safi, *The Politics of Knowledge*, 110–24, proposed that there was a shift in al-Ghazālī's political thinking away from the authority of the caliph in his early writings to the authority of the sultan in his later ones. Yet, this is not convincing, since what counted for al-Ghazālī was the *shawka* and not the character of the office that held it. It could, in principle, be held by caliph, sultan, or vizier. See Laoust, *La politique de Ġazālī*, 237–39, 247–52.

132. Majd al-Mulk Abū l-Faḍl al-Qummī al-Balāsānī was a high official (a *mustawfī*) at Berk-Yaruq's court. In 492/1098, he was killed by Berk-Yaruq's generals. Muʾayyad al-Mulk, a son of Niẓām al-Mulk, was first vizier to Berk-Yaruq and later to his rival half-brother Muḥammad Tapar. In 494/1100, Berk-Yaruq executed him with his own hands.

133. Abū l-Fatḥ ʿAlī ibn al-Ḥusayn al-Ṭughrāʾī (d. after 497/1103). He was Sanjar's first vizier and was soon to be replaced by Fakhr al-Mulk. Cf. Krawulsky, *Briefe und Reden*, 32–33; Iqbāl Āshtiyānī, *Vizārāt dar ʿahd-i salāṭīn-i buzurg-i saljūqī*, 195; Klausner, *The Seljuk Vezirat*, 107; Ibn al-Athīr, *al-Kāmil*, 10:180.17.

134. Al-Ghazālī, *Fażāʾil al-anām*, 58.2.

135. Laoust, *La politique de Ġazālī*, 58, already observed that despite having ample opportunity, al-Ghazālī never implied that Niẓām al-Mulk was murdered by an Ismāʿīlite.

136. Al-Ghazālī, *Fażāʾil al-anām.*, 58.17–18.

137. Al-Ghazālī, *Naṣīḥat al-mulūk*, 45.*ult.*–46.6.

138. Ibn Jawzī, *al-Muntaẓam*, 9:170.14–18.

139. Claude Cahen, Art. "Barkyārūk," in *EI2*, 1:1051b–2b, and C. E. Bosworth in *The Cambridge History of Iran. Volume 5*, 109. The *Niẓāmiyya* party changed its allegience and supported Berk-Yaruq's rival, Muḥammad Tapar.

140. Daftary, *The Ismāʿīlīs*, 335–40; Hillenbrand, "The Power Struggle Between the Seljuq and the Ismāʿīlis of Alamut," 206; Bosworth, "The Ismaʿilis of Quhistān."

141. Daftary, *The Ismāʿīlīs*, 343, 354–55.

142. "Esoterics" (*al-bāṭiniyya*) is a pejorative term for Ismāʿīlite-Siites. Ibn al-Jawzī, *al-Muntaẓām*, 9:77.

143. Cf. the trial of a *dāʿī* accused of apostasy conducted by Ibn ʿAqīl in 490/1097 (Griffel, *Apostasie und Toleranz*, 282–83). In 495/1101, even al-Kiyāʾ al-Harrāsī, one of al-Ghazālī's successors in his chair at the Niẓāmiyya, was suspected to be a secret Ismāʿīlite agent. See Makdisi, *Ibn ʿAqīl et la résurgence*, 288–89; Ibn al-Jawzī, *al-Muntaẓam*, 9:129–30; Ibn Khallikān, *Wafayāt*, 3:288–89; al-Subkī, *Ṭabaqāt*, 7:232.

144. Al-Ghazālī, *al-Munqidh*, 36.2–6.

145. Ibid., 36.7–8.

146. Ibid., 36.11–16. Cf. McCarthy, *Al-Ghazali: Deliverance from Error*, 78–79.

147. Al-Ghazālī, *al-Munqidh*, 37.10–12.

148. Ibid., 37.15–16.

149. Ibid., 37.19.

150. Al-Ghazālī, *Fażāʾil al-anām*, 45.5–6. The letter was written after the death of al-Kiyāʾ al-Harrāsī in Muḥarram 504 / July 1110.

151. Al-Ghazālī, *al-Munqidh*, 38.20.

152. Ibid. 37.20–21.

153. Ibid., 37–38.

154. Al-Ghazālī probably could only authorize his brother as a temporary replacement for himself. His appointment of his brother as temporary replacement begs the question to what extent al-Ghazālī really wished to break with the Niẓāmiyya. See Makdisi, "Non-Ashʿarite Shafiʿism," 241. In 489/1096, the chair was given to Abū ʿAbdallāh al-Ṭabarī, who had held it before al-Ghazālī. In 493/1100, al-Kiyāʾ al-Harrāsī took the chair. See Halm, *Ausbreitung*, 165.

155. Later, al-Ghazālī would write a letter of recommendation for him (*Fażāʾil al-anām*, 33–35), in which he mentions that Ibrāhīm accompanied him for twenty years (35.2–3). On Abū Ẓāhir Ibrāhīm ibn Muṭahhar al-Shabbāk, see al-Ṣarīfīnī, *al-Muntakhab min al-Siyāq*, 163 = Frye, *The Histories of Nishapur*, text 3, fol. 36b; al-Subkī, *Ṭabaqāt*, 7:36; Krawulsky, *Briefe und Reden*, 26–27.

156. "Wise master," a Persian title Ibn al-ʿArabī applies to al-Ghazālī in his *Qānūn al-taʾwīl*, 111, 120

157. Ibn al-ʿArabī, *al-ʿAwāṣim min al-qawāṣim*, 23.4–7. Cf. Jabre, "La biographie et l'œuvre," 87–88. The full text of this passage is available on p. 67.

158. Macdonald, "The Life of al-Ghazzālī," 80, 98, suggested that al-Ghazālī had reason to fear the enmity of Berk-Yaruq. Jabre, "La biographie et l'œuvre," 93–94, thought that al-Ghazālī feared assassination by the Ismāʿīlites. The most reliable analysis is offered by Laoust, *La politique de Ġazālī*, 90–105. For other theories, see Abd-el-Jalil, "Autour de la sincerité d'al-Ghazālī"; Sawwaf, *Al-Ghazzali. Etude de la réforme ghazzalienne*, 57–58; Abu-Sway, "Al-Ghazālī's 'Spiritual Crisis' Reconsidered." These theories are discussed in Kojiro Nakamura's article on al-Ghazālī in the *Routledge Encyclopedia of Philosophy*, 4:64. For a thorough discussion of the textual evidence in the *Munqidh* and of earlier interpretations of al-Ghazālī's "conversion," see Poggi, *Un classico della spiritualità musulmana*, 187–210.

159. Al-Ghazālī, *al-Munqidh*, 38.3ff.

160. Not al-Ghazālī's experience of this crisis but rather his very public admission makes it important. Martin Heidegger, for instance, experienced a severe crisis in the spring of 1946 when, because of his earlier Nazi sympathies, he was temporarily stripped of his teaching position at Freiburg University. This crisis, however, never became a prominent part of his biography since he never publicly admitted to it.

161. Van Ess, "Quelques remarques sur le Munqid̲ min aḍ-ḍalāl," 60ff.

162. "The sheikh of the Shāfiʿites in Syria (. . .) The historians say that he was an imam, a resourceful authority (*ʿallāma mufīd*), an expert on the *ḥadīth* and the Qurʾān, an ascetic, noble-minded, pious, and powerful in a way that he had no equal" (al-Yāfiʿī, *Mirʾāt al-jinān*, 3:152.17–20). On him, see Ibn ʿAsākir, *Tabyīn*, 286–87; al-Subkī, *Ṭabaqāt*, 5:351–53; and al-ʿUlaymī, *al-Uns al-jalīl*, 1:297–98.

163. Tibawi, "Al-Ghazālī's Sojourn in Damascus and Jerusalem," 70.

164. Ibn ʿAsākir, *Tabyīn*, 286.10–11; al-Subkī, *Ṭabaqāt*, 5:352.7–8.

165. Menn, "The *Discourse on the Method* and the Tradition of Intellectual Autobiography," 167–68.

166. *māl al-sulṭān wa-ʿummālihi*, al-Ghazālī, *al-Bidāya fī l-hidāya*, 200.13. English transl. in Watt, *Faith and Practice*, 139. The rulers' income and whether one can benefit from it is the subject of a detailed discussion in the fifth chapter of the fourteenth book (*Kitāb al-Ḥalāl wa-l-ḥarām*) of the *Iḥyāʾ* 2:172–80 / 5:890–901.

167. Glassen, *Der mittlere Weg*, 50; Safi, *The Politics of Knowledge*, 101–2; Makdisi, *Rise of Colleges*, 41; Kasāʾī, *Madāris-i Niẓāmiyyah*, 116–17.

168. For the four different categories of *waraʿ*, see al-Ghazālī, *Iḥyāʾ*, 1:31 / 32–33.

169. Ibid., 1.70:7–22 / 94.3–21.

170. Ibn ʿAsākir, *Tabyīn*, 286.11–12; al-Subkī, *Ṭabaqāt*, 5:352–53. The ruler was Tutush ibn Alp-Arslan (d. 488/1095), and the money was from *jizya*.

171. *min ujrat al-naskh*, Ibn al-Jawzī, *al-Muntaẓam*, 9:169.6. The term is unclear as it usually refers to the payment a professional scribe receives for his work. None of the sources mention that al-Ghazālī turned to copying manuscripts, so here the term seems to refer to collecting money for paying scribes to copy and publish books.

172. *pīsh hīch sulṭān narawad va-māl-i sulṭān nagīrad va-munāẓat-i ū taʿaṣṣub nakonad*, al-Ghazālī, *Fażāʾil al-anām*, 5.1, cf. also 45.9–10.

173. Al-Ghazālī, *Iḥyāʾ*, 1:61–70 / 70–81; quote on 1:64.3 / 73.*ult*.

174. Ibn al-Athīr, *al-Kāmil*, 10:172.13–14; al-Dhahabī, *Siyar*, 19:330.7–8. Al-Dhahabī (19:327–28) also reports that al-Ghazālī composed his works *al-Arbaʿīn*, *Qisṭās al-mustaqīm*, and *Miḥakk al-naẓar* in Damascus. He mistakenly assumed that he stayed there for years.

175. Sibṭ ibn al-Jawzī, *Mirʾāt al-zaman*, ed. Hayderabat, 1:171.2–3; al-Yāfiʿī, *Mirʾāt al-jinān*, 3:146.5.

176. Al-Subkī, *Ṭabaqāt*, 6:197.17–18. Tibawi, "Al-Ghazālī's Sojourn in Damascus and Jerusalem," 73–74.

177. Ibn Jubayr, *Tadhkira bi-l-akhbār*, 213–14. Al-Subkī, *Ṭabaqāt*, 6:197.15–16, reports the tale on the authority of al-Dhahabī, who says he has it from Ibn ʿAsākir. Cf. le Strange, *Palestine under the Moslems*, 246, 264. The base of that minaret is part of the remnants from the Roman *temenos* and has largely been unchanged since pre-Islamic times.

178. *fa-qāma Dimashqa sana 489 wa-aqāma bi-hā mudda*; Ibn ʿAsākir, *Taʾrīkh madīnat Dimashq*, 55:200.9

179. Al-Ghazālī, *al-Munqidh*, 38.11.

180. Al-Subkī, *Ṭabaqāt*, 6:199.10–13.

181. Or in the cave under the rock? Al-Ghazālī, *al-Munqidh*, 38.15–16.

182. Al-Ghazālī, *Iḥyā'*, 1:142.21–23 / 1:180.14–16.

183. Al-ʿUlaymī, *al-Uns al-jalīl*, 1:299.14, says that al-Ghazālī "composed several works in Jerusalem." The title of this work contains two indications to Jerusalem: first, the word *quds*, which may stand for *Madīnat al-Quds*, "Jerusalem," and second, the plural of *miʿrāj*, Muḥammad's ascent to heaven on the twenty-seventh of Rajab. Muḥammad is believed to have left from the plateau at the Dome of the Rock, where his footprint is still shown. Incidentally, on 27 Rajab 489 / 22 June 1096, al-Ghazālī was in Jerusalem. Whether or not he wrote this book is unclear.

184. *maḥw al-jāh wa-mujāhadat al-nafs*, Ibn al-Jawzī, *al-Muntaẓam*, 9:169.9.

185. Al-ʿUlaymī, *al-Uns al-jalīl*, 1:299.15–16.

186. Bieberstein/Bloedhorn, *Jerusalem: Grundzüge der Baugeschichte*, 3:200. Properly speaking, the Golden Gate is a double gate with two doors. The north door is known as the *bāb al-raḥma*, the south door as the *bāb al-tawba*. The name "Gate of Mercy," however, also applies to the whole building.

187. In the case of Damascus, the reference to the *zāwiya* to Abū-l Fatḥ Naṣr is unmistakable, as al-Subkī calls it "zāwiyat al-shaykh Naṣr al-Maqdisī."

188. Al-ʿUlaymī, *al-Uns al-jalīl*, 1:298.2–3; 2:34.3–4.

189. Kaplony, *The Ḥaram of Jerusalem 324–1099*, 638–41. Macdonald's claim (in "The Life of al-Ghazzālī," 93) that at the beginning of the twentieth century there was still a *zāwiya* known as *al-Ghazāliyya* in Jerusalem cannot be taken seriously.

190. Burgoyne, *Mamluk Jerusalem*, 49. Al-Malik al-Muʿaẓẓam ʿĪsā was appointed governor of Damascus, including the province of Jerusalem, by his brother al-Malik al-Kāmil in 597/1201. After al-Malik al-Kāmil's death in 615/1218, he became an independent ruler of Syria until his own death in 624/1227.

191. Al-ʿUlaymī, *al-Uns al-jalīl*, 2:34.4–5; Ibn Khallikān, *Wafayāt al-aʿyān*, 3:244.2–3.

192. Ibn Khallikān, *Wafayāt al-aʿyān*, 3:244.2–3. Al-Subkī, *Ṭabaqāt*, 8:327.6, reports that Ibn al-Ṣalāḥ had taught in the *madrasa al-Ṣalāḥiyya* in Jerusalem, that is, the former Crusader Church St. Anne that Ṣalāḥ al-Dīn had converted into a madrasa right after the conquest of 583/1187 (on that institution, cf. Bieberstein/Bloedhorn, *Jerusalem*, 1:217, 3:170–73). These two schools devoted to the memory of Ṣalāḥ al-Dīn should not be confused with the still-existing *khānqāh al-Ṣalāḥiyya*, that is, the former Latin Patriarchat at the ʿAqabat al-Khānqāh close to the Church of the Holy Sepulcher (Bieberstein/Bloedhorn, *Jerusalem*, 2:216–18).

193. I am grateful to Muhammad Ghosheh at the Muʾassasat Iḥyaʾ al-Turāth wa-l-Buḥūth al-Islāmiyya (Center for Heritage and Islamic Research) in Jerusalem for pointing this out to me.

194. Mujīr al-Dīn's report is most probably a reflex on earlier writings about the conversion of the *zāwiya* of Abū l-Fatḥ Naṣr in Damascus into a school that referred in its name to al-Ghazālī. This information was somehow applied to Jerusalem, where Abū l-Fatḥ Naṣr had first taught before he moved to Tyros and Damascus. This then got mixed up with information about a derelict school "al-Nāṣiriyya" above the Golden Gate. The fact that Mujīr al-Dīn refers to this school as the one where al-Ghazālī taught, yet mentions that it was (re)founded by the Ayyūbid al-Mālik al-Muʿaẓẓam (al-ʿUlaymī, *al-Uns al-jalīl*, 2:34.4–5) in 610/1214, is evidence for his confusion.

195. In a letter written in 504/1110, al-Ghazālī mentions that he took his vow at Hebron in the year 489 (al-Ghazālī, *Faḍāʾil al-anām*, 45.8.)

196. See below, pp. 63–64. Cf. al-Ghazālī, *al-Munqidh*, 38.17–18.

197. Al-Subkī, *Ṭabaqāt*, 6:198.1–9. Tibawi, "Al-Ghazālī's Sojourn," 71.

198. Ibn al-Qalānisī (d. 555/1160), *Dhayl Taʾrīkh Dimashq*, 134; Ibn al-Athīr, *al-Kāmil*, 10:185.

199. Al-Ghazālī (?), *Tuḥfat al-mulūk*, 407.11–13. If genuine, this text would be together with ʿAlī ibn Ẓāhir al-Sulamī's (d. 500/1106) *Kitāb al-Jihād*, one of the earliest by a Muslim scholar who calls for *jihād* against the crusaders.

200. Cf. Abū Bakr ibn al-ʿArabī's report below p. 65. The *ribāṭ* or *khānqāh* of Abū Saʿd al-Nīshābūrī (d. 479/1086) was built about twenty years earlier. He left other religious buildings in Baghdad; see le Strange, *Baghdad during the Abbasid Caliphate*, 99–100, and Kasāʾī, *Madāris-i Niẓāmiyyah*, 112–14. Cf. also the valuable map of the quarter surrounding the Niẓāmiyya madrasa in Baghdad shortly before the Mongol invasion of 656/1258, printed at the beginning of Kasāʾī's book.

201. Ibn al-Jawzī, *al-Muntaẓam*, 9:87.7–8.

202. Al-Ghazālī, *Iḥyāʾ*, 1:10.17 / 2.15.

203. Ibid. 1:12.21–23 / 5.4–7. The division mirrors that between the practical sciences (ethics, etc.) and the theoretical sciences (mostly metaphysics) in philosophical literature; see Gilʿadi, "On the Origin of Two Key-Terms in al-Ġazzālī's Iḥyāʾ ʿulūm al-dīn." On the division, see also Lazarus-Yafeh, *Studies*, 357–66.

204. *fiqh ṭarīq al-ākhira*; al-Ghazālī, *Iḥyāʾ*, 1:24.10 / 23.17; about what it entails, see ibid., 1:10.17–18 / 2.15–3.1.

205. Madelung, "Ar-Rāġib al-Iṣfahānī und die Ethik al-Ġazālīs"; Pines, "Quelques notes sur les rapports de l'*Iḥyāʾ ʿulūm al-dīn* d'al-Ghazālī avec la pensée d'Ibn Sīnā." Abrahamov, "Ibn Sīnā's Influence on al-Ghazālī's Non-Philosophical Works," 1–2, gives a report about the secondary literature on philosophical works that have been adapted in the *Iḥyāʾ*. In his article he adds findings from the works of Ibn Sīnā. On the disputed question of when al-Rāghib al-Iṣfahānī lived, see Everett K. Rowson in *EI2*, 8:389b.

206. Al-Ghazālī, *al-Munqidh*, 38.20.

207. Ibn al-ʿArabī. "Shawāhid al-jilla," 311.5; Garden, *Al-Ghazālī's Contested Revival*, 87.

208. Al-Ghazālī congratulates Mujīr al-Dīn (cf. n.133) for his nomination as Sanjar's vizier in 490/1097. This letter mentions al-Ghazālī's "happy affection due to being close to the place of visitation." That refers most probably to the pilgrimage site of Meshed in Ṭūs (al-Ghazālī, *Fażāʾil al-anām*, 52; Krawulsky, *Briefe und Reden*, 147). It certainly does not refer to Baghdad.

209. Zarrīnkūb, *Firār az madrasah*, 109–55.

210. *aṣḥāb al-zawāyā al-mutafarriqūna al-munfaridūna*; al-Ghazālī, *Iḥyāʾ*, 1:99.12–13 / 120.1–2.

211. ʿAbd al-Ghāfir (in al-Subkī, *Ṭabaqāt*, 6:210.4–5) confirms the existence of these institutions at the time when al-Ghazālī gave up teaching in Nishapur. Referring to the earlier period when al-Ghazālī was still teaching at Nishapur, however, the collector of the letters says that he had students in Ṭūs and stayed there in a *khānqāh* (al-Ghazālī, *Fażāʾil al-anām*, 12.15). It is most likely that these institutions were founded when al-Ghazālī returned from Baghdad.

212. Makdisi, *Rise of Colleges*, 161.

213. Ibn al-Jawzī, *al-Muntaẓam*, 9:170.10–11. ʿAbd al-Ghāfir al-Fārisī dates the creation of the *madrasa* and *khānqāh* after the end of al-Ghazālī's teaching at Nishapur (al-Subkī, *Ṭabaqāt*, 6:210.4–5). The letters confirm that they existed earlier.

214. Al-Ghazālī, *Fatrā dar bāra-yi amvāl-i khānqāh*, and idem, *Fatwā ʿAlā man istafāḍa min amwāl ribāṭ ṣūfiyya*. Cf. Safi, *The Politics of Knowledge*, 100.

215. Al-Ghazālī, *al-Munqidh*, 38.21–22.

216. ʿAbd al-Ghāfir al-Fārisī as quoted by al-Subkī, *Ṭabaqāt*, 6:207.2–3. Cf. al-Ṣarīfīnī, *al-Muntakhab min al-Siyāq*, 84 = Frye, *The Histories of Nishapur*, text 3, fol. 20a.

217. Al-Ghazālī, *Fażāʾil al-anām*, 11.15–21; Krawulsky, *Briefe und Reden*, 77.

218. In his *Munqidh*, 49.17–20, al-Ghazālī mentions the two events and says that the period of seclusion (*ʿuzla*) amounted to eleven years. The "twelve years" may be the result of a confusion with a period of that length mentioned in a different letter a few pages earlier in the collection *Fażāʾil al-anām*, 5.2; Krawulsky, *Briefe und Reden*, 66.

219. Al-Ghazālī, *Fażāʾil al-anām*, 10.22. The *khānqāh* is mentioned in another letter on p. 81.21 and in a comment by the collector on p. 12.15.

220. *zāwiya-rā mulāzamat kard*, al-Ghazālī, *Fażāʾil al-anām*, 11.16.

221. Al-Ghazālī, *Fażāʾil al-anām*, 45.10–17; Krawulsky, *Briefe und Reden*, 135–36. Cf. also Brown, "The Last Days of al-Ghazzālī," 95, in which the context of the letter is misrepresented.

222. Al-Subkī, *Ṭabaqāt*, 6:208.4–*ult.*

223. Al-Faḍl ibn Muḥammad al-Fāramadhī; al-Ṣarīfīnī, *al-Muntakhab min al-Siyāq*, 628–9 (= Frye, *The Histories of Nishapur*, text 3, fol. 121a–b); al-Subkī, *Ṭabaqāt*, 5:304–6; Halm, *Ausbreitung*, 94. Fāramadh is one of the villages of Ṭūs.

224. *futiḥa ʿalayhi lawāmiʿun min anwāri l-mushāhada*; al-Subkī, Ṭabaqāt, 5:305. 12–13.

225. *madākhil al-safsaṭa*; al-Ghazālī, *al-Munqidh*, 12–14.

226. Al-Subkī, *Ṭabaqāt*, 6:209.12–15.

227. A more accurate chronology may be given in a brief passage in *al-Munqidh*, 46.14–20, in which the list begins with *falsafa*, followed by Sufism and Ismāʿīlism.

228. Al-Ghazālī, *al-Munqidh min al-ḍalāl*, 48–49.

229. Al-Sukbkī, *Ṭabaqāt*, 6:207.7–11. In his autobiography, al-Ghazālī says that the sultan "issued a binding order to pounce to Nishapur" (*al-Munqidh*, 49.2).

230. Al-Ghazālī, *Fażāʾil al-anām*, 10.10–12.

231. Ibid., 3.9–11.

232. Muḥammad ibn Abī l-Faraj al-Māzarī, who was known as "al-Dhakī" ("the clever one"); on him, see Charles Pellat in *EI2*, 6:943; Garden, *Al-Ghazālī's Contested Revival*, 114–17; Krawulsky, *Briefe und Reden*, 15–16; Ibn al-Jawzī, *al-Muntaẓam*, 9:190; al-Dabbāgh/al-Nājī, *Maʿālim al-īmān*, 3:202–3. He should not be confused with his younger contemporary Abū ʿAbdallāh Muḥammad ibn ʿAlī al-Māzarī (d. 536/1141), who was surnamed "al-Imām." This latter al-Māzarī never left the Maghrib and was a much more respectable scholar than the former. (On him, see *GAL*, *Suppl.* 1:663; Charles Pellat in *EI2*, 6:943, and the sources listed there.) Both al-Māzarīs were highly critical of al-Ghazālī, and al-Māzarī al-Imām wrote a critique of al-Ghazālī's *Iḥyāʾ* with the title *al-Kashf wa-l-inbāʾ ʿalā l-mutarjam bi-l-Iḥyāʾ*. (For the identification of the author, see al-Dhahabī, *Siyar*, 19:330, 340.) Passages from that book are preserved in al-Dhahabī, *Siyar*, 19:330–32, 340–42; al-Subkī, *Ṭabaqāt*, 6:240–58; and Ibn Taymiyya, "Sharḥ al-ʿaqīda al-iṣfahāniyya," 116–19. See also the information on al-Māzarī al-Imām's book collected in al-Zābidī, *Itḥāf al-sāda*, 1:28–29; 179.21–24; 2:411.20–23; 9:442.17–27. The latter passages are translated by Asín Palacios, "Un faqîh siciliano, contradictor de Al Ġazâlî," 224–41.

233. Al-Subkī, *Ṭabaqāt*, 6:207.5–6.

234. Ibid., 6:208.1–2.

235. Ibid., 6:209.14–15; reading *tamarrus* instead of *nāmūs*.

236. Al-Ghazālī, *al-Mankhūl*, 613–18.

237. Al-Shushtarī (d 1019/1610), *Majālis al-muʾminīn*, 2:191; Krawulsky, *Briefe und Reden*, 16.

238. This request comes at the end of the conversation with Sanjar, *Fażāʾil al-anām*, 10.21–22; Krawulsky, *Briefe und Reden*, 75.

239. Al-Ghazālī, *Fażāʾil al-anām*, 10.*peanult.*

240. Ibid., 11.3–4.

241. Ibid., 11.10.

242. In Turūq, south of Ṭūs, on the road to Nishapur; see Krawulsky, *Briefe und Reden*, 219. Sanjar used to pitch his camp there; see Niẓāmī ʿArūḍī, *Chahār Maqāla*, 40.

243. Al-Ghazālī, *Fażāʾil al-anām*, 5.*peanult.*; Krawulsky, *Briefe und Reden*, 67–68.

244. Al-Ghazālī, *Fażāʾil al-anām*, 54–55, Krawulsky, *Briefe und Reden*, 152. This is not the letter to Mujīr al-Dawla that establishes al-Ghazālī's arrival in Ṭūs as 490/1097. On the dating of this letter, see Krawulsky, *Briefe und Reden*, 32–33. I am grateful to Kenneth Garden who pointed me to this letter and its content.

245. Al-Ghazālī, *Fażāʾil al-anām*, 4.10–15; Krawulsky, *Briefe und Reden*, 65.

246. Al-Subkī, *Ṭabaqāt*, 6:210.4–5. Cf. Ibn al-Jawzī, *al-Muntaẓam*, 9:170.9–10; Yāqūt, *Muʿjam al-buldān*, 3:561.7–8.

247. Al-Subkī, *Ṭabaqāt*, 6:210.14–15.

248. See Badawī, *Muʿallafāt*, 112–14; and al-Ḥaddād, *Takhrīj aḥādīth Iḥyāʾ ʿulūm al-dīn*.

249. *anā muzjā l-biḍāʿa fī l-ḥadīth*; al-Wāsiṭī in his *tarjama* edited in al-Aʿsam, *al-Faylasūf al-Ghazālī*, 179.2. Ṭālibī, *Arāʾ Abī Bakr ibn al-ʿArabī l-kalāmiyya*, 1:56, claims he admitted this to his student Abū Bakr ibn al-ʿArabi (who preserved the quote). Ṭālibī's reference, however, cannot be verified.

250. Al-Ghazālī, *Iḥyāʾ*, 1:110.6–111.2 / 134.1–135.5.

251. Al-Anṣārī, *al-Ghunyā fi-l-kalām* and idem, *Sharḥ al-Irshād*.

252. On this institution, see Bulliet, *Patricians of Nishapur*, 124, 230, 251.

253. Kasāʾī, *Madāris-i Niẓāmiyyah*, 99, lists Abū l-Qāsim Salmān ibn Nāṣir al-Anṣārī as a teacher at the Niẓāmiyya in Nishapur right after al-Ghazālī. His biographers are silent about whether he held an office there; see ʿAbd al-Ghāfir al-Fārisī, *al-Siyāq*, in Frye, *The Histories of Nishapur*, text 2, fol. 29b–30a; Ibn ʿAsākir, *Tabyīn kadhib al-muftarī*, 307; al-Subkī, *Ṭabaqāt*, 7:96–99.

254. Al-Ghazālī, *al-Munqidh*, 48–49. ʿAbd al-Ghāfir al-Fārisī devotes a long and eloquent passage to these events that deserves to be closely analyzed. Cf. al-Subkī, *Ṭabaqāt*, 6:207.5–208.3 and 210–11.

255. Al-Ghazālī, *Fażāʾil al-anām*, 37–45. Al-Kiyāʾ al-Harrāsī died on 1 Muḥarram 504 / 20 July 1110. On him, see *EI2*, 5:234 (George Makdisi); Brockelmann, *GAL*, 1:390; *Suppl.* 1:674; Makdisi, *Ibn ʿAqīl et la rèsurgence*, 216–19; ʿAbd al-Ghāfir al-Fārisī, *al-Siyāq*, in Frye, *Histories of Nishapur*, text 2, fol. 72a; Ibn ʿAsākir, *Tabyīn*, 288–89; Ibn Khallikān, *Wafayāt*, 3:286–90; al-Subkī, *Ṭabaqāt*, 7:231–34; Halm, *Die Ausbreitung*, index.

256. Al-Ghazālī, *Fażāʾil al-anām*, 42–45. The original letter was probably written in Arabic. For a fragment of the Arabic version, see MS Berlin, Petermann II 8, p. 126 (Ahlwardt 10070.2). Cf. also Krawulsky, *Briefe und Reden*, 11, 30–31; and Fritz Meier in *ZDMG* 93 (1939): 406–7.

257. Al-Ghazālī, *Fażāʾil al-anām*, 44.16–45.1.

258. Al-Subkī, *Ṭabaqāt*, 6:201.8–12.

259. Al-Abīwardī, *Dīwān*, 2:140.

260. Accoding to ʿAbd al-Ghāfir al-Fārisī, see al-Subkī, *Ṭabaqāt*, 6:211.3. See also Yāqūt, *Muʿjam al-buldān*, 3:561.9–10.

261. In 1915, Diez, *Die Kunst der islamischen Völker*, 82, published a description and the reproduction of a water painting by the Armenian-Iranian artist André Sevruguin (also: Sevrugian, 1894–1996) of the ruins of a large mausoleum in Ṭūs that Diez claimed is the mausoleum of al-Ghazālī. This picture depicts a mausoleum in the midst of Ṭābarān's ruins, which is known as the *Hārūniyya*. For a recent picture of the reconstructed building, see Elton L. Daniel's preface to Field's translation of *The Alchemy of*

Happiness, xl. Local usage mistakenly regards it as the tomb of Hārūn al-Rashīd, who is, however, buried at the site of ʿAlī al-Riḍā in Meshed. There is also a second mausoleum within the former city walls of Ṭābarān, which is the one of Firdawsī (d. 411/1020). Niẓāmī ʿArūḍī, *Chahār Maqāla*, 51, says that Firdawsī was buried in Ṭābarān "outside the gates in a garden." What is today known as Firdawsī's tomb (which is distinct from the *Hārūniyya*) has been lavishly rebuilt in a monumental and modern style during the Pahlevi period. On the various monuments in the vicinity of Meshed, see also Hakami, *Pèlerinage de l'Emâm Rezâ*, 64ff. In 1918, Donaldson, "A Visit to the Grave of al-Ghazzali," reports he found a tombstone in the ruins of Ṭūs that bore al-Ghazālī's name and had been reused in 1007/1598–99 to mark another grave.

262. Al-Subkī, *Ṭabaqāt*, 6:211.5; al-Zabīdī, *Itḥāf al-sādā*, 1:11.17. The *kunya* "Abū Ḥāmid" need not mean (as Smith, *Al-Ghazālī the Mystic*, 57, assumes) that he had a son by the name of Ḥāmid.

263. MS Yale University, Beinecke Memorial Library, Landberg 318, fol. 230a. The *ijāza* is issued by "Muḥammad ibn Muḥammad ibn Muḥammad al-Ghazālī l-Ṭūsī" at some time after the manuscript was copied in 507/1113. Cf. Nemoy, *Arabic Manuscripts in the Yale University Library*, 109, no. 999.

264. In a very brief note in al-Ghazālī, *Fażāʾil al-anām*, 2.9–10; Krawulsky, *Briefe und Reden*, 62. On other scholars with the name *al-Ghazālī* from this period, who were not related to the famous theologian, see Macdonald, "The Name al-Ghazzālī", 21–22; and al-Zabīdī, *Itḥāf al-sāda*, 1:19.

265. Al-Fayyūmī, *Miṣbāh al-munīr fī gharīb al-Sharḥ al-kabīr*, 447 (*sub* gh-z-l). The work is a dictionary of difficult words that appear in ʿAbd al-Karīm al-Rāfiʿī's (d. 623 /1226) commentary to al-Ghazālī's *al-Wajīz*. The history of the Shirwānshāh's is not well known, and their list of kings has lots of lacunae. Cf. Minorsky, *A History of Sharvān and Darband*, 135; and C. E. Bosworth in *EI2*, 11:488–89.

266. Griffel, "On Fakhr al-Dīn al-Rāzī's Life and the Patronage He Received," 339.

267. Ibn al-ʿImād, *Shajarāt al-dhahab*, 7:196. The full name of this scholar and the dearth of information about his background give the impression that this person only pretended to be a decendent of al-Ghazālī. If true, his geneology would imply that al-Ghazālī had both a son and a grandson by the name of Muḥammad.

268. Al-Zabīdī, *al-Muʿjam al-mukhtaṣṣ*, 136; I am grateful to Stefan Reichmuth who pointed me to this reference.

CHAPTER 2

1. Laoust, "La survie de Ġazālī d'après Subkī." See also the list of al-Ghazālī's students in al-Zabīdī, *Itḥāf al-sāda*, 1:44–45.

2. Al-Masʿūdī, *al-Shukūk wa-l-shubah ʿalā l-Ishārāt*. On Sharaf al-Dīn Muḥammad ibn Masʿūd al-Masʿūdī and his works, see *GAL*, 1:474 no. 11 (only in the first edition of 1898); and Shihadeh, "From al-Ghazālī to al-Rāzī: 6th/12th Century Developments in Muslim Philosophical Theology," 153–56.

3. Ibn Ghaylān al-Balkhī, *Ḥudūth al-ʿālam*, 11.18–19.

4. For an overview of Abū l-Futūḥ al-Ghazālī's (d. 517/1123 or 520/1126–27) life and his scholarly œuvre, including the most important secondary literature, see the article by Nasrollah Pourjavady in *EIran*, 10:377–80. On Aḥmad's life, see Aḥmad Mujāhid's introduction to Abū l-Futūḥ al-Ghazālī, *Majmūʿah-yi āṣār-i Fārisī*; Richard Gramlich's introduction to Abū l-Futūḥ al-Ghazālī, *Der reine Gottesglaube*, 1–7; Lumbard, *Aḥmad al-Ghazālī (d. 517/1123 or 520/1127) and the Metaphysics of Love*, 20–128;

Ibn al-Jawzī, *al-Muntaẓam*, 9:260–62; al-Subkī, *Ṭabaqāt*, 6:60–62; and Ibn Khallikān, *Wafayāt al-aʿyān*, 1:97–98.

5. Al-Ghazālī, *Risāla ilā Abū l-Fatḥ al-Damīmī*, 27.10–11; MS Berlin, Petermann II 8, p. 121: "As for preaching, I don't see myself as one of its people because preaching is a (voluntary) alms-tax (*zakāt*) levied on the property (*niṣāb*) of [conducting a pious life] due to other people's preaching (*ittiʿāẓ*), and how can someone who does not have this property pay the tax?"

6. Al-Ghazālī, *Iḥyāʾ*, 1:24.1–4 / 23.6–9.

7. At the beginning of al-Ghazālī, *al-Lubāb min al-Ihyāʾ*, 2.11–14 (1978 edition: 25.5–6), it says: "It had occurred to me during one of my journeys that I extract from my book *The Revival of the Religious Sciences* its kernels." These words are a clear reference to Muḥammad as its author, and he is identified as such in the title. In fact, none of the MSS I saw ascribes the book to Aḥmad. MS Berlin, Wetzstein 99 (Ahlwardt no. 1708), and MSS Princeton, Yahuda 838 and 3717 (Mach no. 2164), attribute the text to Muḥammad. The text of *al-Lubāb min al-Ihyāʾ* appears to be identical to the one in *al-Murshid al-amīn ilā mawʿiẓat al-muʾminīn min Iḥyāʾ ʿulūm al-dīn*, a book ascribed to Muḥammad al-Ghazālī. Brockelmann's identification of Aḥmad as the *Lubāb*'s author (*GAL*, 1:422; *Suppl.* 1:748) follows Ahlwardt, *Handschriften-Verzeichnisse*, 2:313, and seems to be based entirely on Ḥājjī Khalīfa, *Kashf al-ẓunūn*, 1:182–83. On the text, see also Bouyges, *Essay de chronologie*, 135–36; Badawī, *Muʾallafāt*, 114; and Lumbard, *Aḥmad al-Ghazālī*, 122. There also exists a different and shorter excerpt from the *Iḥyāʾ* with the title *Lubb al-Iḥyāʾ* that was authored by neither Aḥmad nor Muḥammad. For this text, see MS Yale University, Beinecke Library, Salisbury 38, foll. 1–45b (Nemoy 797), and MS Berlin, Wetzstein II 1807, foll. 120–46b. (Ahlwardt 1707).

8. Abū l-Futūḥ al-Ghazālī, *al-Tajrīd fī kalimat al-tawḥīd*.

9. One of Aḥmad al-Ghazālī's best-known Persian works, the *Rāz-nāmah* or *Risālah-yi ʿAyniyyah*, is believed to be originally a letter to ʿAyn al-Quḍāt al-Ḥamadhānī. The text is in *Majmūʿah-yi āṣār-i fārisī-yi Aḥmad Ghazzālī*, 175–214.

10. Al-Ghazālī, *Iḥyāʾ*, 1:53.18–19 / 60.19–20.

11. Richard Gramlich in the introduction to Abū l-Futūḥ al-Ghazālī, *Gedanken über die Liebe*, a German translation of the Persian *Kitāb al-Savāniḥ fī l-ʿishq*.

12. The Almoravids conquered the Ṭāʾifa kingdoms in al-Andalus between 445/1053 and 487/1094.

13. ʿAbbās, "al-Jānib al-siyāsī," 218–19; idem, "Riḥlat Ibn al ʿArabī," 61.

14. ʿAbbās, "al-Jānib al-siyāsī," 221.

15. A detailed narrative of Abū Bakr ibn al-ʿArabī's life and his travels is given by Ṭālibī, *Arāʾ Abī Bakr ibn al-ʿArabī*, 1:25–64.

16. Ibn al-ʿArabī, *Qānūn al-taʾwīl*, 92. Fierro, in the preface to her Spanish translation of al-Ṭurṭūshī's *Kitāb al-Ḥawādith wa-l-bidaʿ*, 40, reports that the meeting between the two took place in 486/1093, that is, soon after the arrival of the Ibn al-ʿArabīs in Jerusalem. Abū Bakr studied with al-Ṭurṭūshī his *Mukhtaṣar* of al-Thaʿlabī's (d. 427/1035) Qurʾān commentary during Ramaḍān 487 / September–October 1094 in the al-Aqṣā Mosque of Jerusalem (Ibn al-ʿArabī, *Qānūn al-taʾwīl*, 61; and Ṭālibī, *Arāʾ Abī Bakr ibn al-ʿArabī*, 1:33).

17. *Kitāb Tartīb al-riḥla li-l-targhīb fī l-milla*; Ibn al-ʿArabī, *Shawāhid al-jilla*, 278.3; ʿAbbās, "al-Jānib al-siyāsī," 217.

18. Ibn Ṣāḥib al-Ṣalāt quotes from this book in his *Taʾrīkh al-mann bi-l-imāma*, 258–59; and Ibn al-ʿArabī mentions it in its short title in *al-ʿAwāṣim min al-qawāṣim*, 24.8.

19. Ibn al-ʿArabī, *Shawāhid al-jilla wa-l-aʿyān fī mashāhid al-Islām wa-l-buldān*. The book is apparently an excerpt of the longer *Kitāb Tartīb al-riḥla li-l-targhīb fī l-milla* and

contains the political documents obtained during the trip to the East (Ṭālibī, *Arāʾ Abī Bakr ibn al-ʿArabī*, 27, n, 2; 82). The *Shawāhid al-jilla* is the text in MS Bibliothèque Générale, Rabat,1275 *kāf*, pp.119–40, referred to, for instance, in Griffel, *Apostasie und Toleranz*, 364, n. 21; or in van Ess, "Neuere Literatur," 302. ʿAbbās, "al-Jānib al-siyāsī," 217ff. bases most of his information on Ibn al-ʿArabī's travels on this text and rightfully identifies it as part of the anonymous chronicle *Mafākhir al-barbar* from the eighth/fourteenth century.

20. Ibn al-Arabī, *Qānūn al-taʾwīl*, 107.

21. A. Ben Abdesselem in *EI2*, 10:739a. See also Fierro in the preface to her Spanish translation of al-Ṭurṭūshī's *Kitāb al-Ḥawādith wa-l-bidaʿ*, 40.

22. The text of the letter by al-Ghazālī to Yūsuf ibn Tāsihfīn is preserved in Ibn al-ʿArabī, *Shawāhid al-jilla*, 306–11. See ʿAbbās, "al-Jānib al-siyāsī," 222ff. A short version of this letter is extant in the anonymous *Mafākhir al-barbar* (ed. Lévi-Provençal), 2. The text of al-Ghazālī's *fatwā*, together with Ibn al-ʿArabī's initial request, is in Ibn al-ʿArabī, *Shawāhid al-jilla*, 302–5. Both the question and al-Ghazālī's *fatwā* are also preserved in MS American University of Beirut, Jafet Memorial Library 297.3: G41 iA, pp. 50–56. On al-Ghazālī's *fatwā* in support of the Almoravids, see also the report of Ibn Khaldūn, *al-ʿIbar*, 6:386.

23. Ibn al-ʿArabī, *Shawāhid al-jilla*, 311–12; ʿAbbās, "al-Jānib al-siyāsī," 221.

24. Ibn al-ʿArabī, *al-ʿAwāṣim min al-qawāṣim*, 23.

25. "Wise master," a Persian title Ibn al-ʿArabī applies to al-Ghazālī in his *Qānūn al-taʾwīl*, 111, 120.

26. Ibid., 111.

27. Ibn al-ʿArabī, *Shawāhid al-jilla*, 290–93; ʿAbbās, "al-Jānib al-siyāsī," 227–28.

28. The two passed through Palestine during the early part of the year 492 (November 1098–November 1099), shortly before the hostilities of the First Crusade started there in May 1099. The First Crusade is not mentioned in Abū Bakr's œuvre.

29. Al-Ṭurṭūshī, *Risāla ilā ʿAbdallāh ibn Muẓaffar*, in Ghurāb, "Ḥawla ikhrāq al-Murābiṭīn li-Iḥyāʾ al-Ghazālī," 158–63. See also Fierro in the preface to her Spanish translation of al-Ṭurṭūshī's *Kitāb al-Ḥawādith wa-l-bidaʿ*, 61–64.

30. Al-Dhahabī, *Siyar*, 19:334, 339, 494–96; al-Subkī, *Ṭabaqāt*, 6:240–58; cf. al-Zabīdī, *Itḥāf al-sāda*, 1:28–29; 179.21–24; 2:411.20–23; 9:442.17–27.

31. The list is from Ibn al-ʿArabī's yet unedited *Sirāj al-murīdīn*. ʿAmmār Ṭālibī adds it to his edition of Ibn al-ʿArabī's *al-ʿAwāṣim min al-qawāṣim*, 377–79. He also reproduces parts of the list in his *Arāʾ Abī Bakr ibn al-ʿArabī*, 1:64–65. Works by al-Ghazālī in this book include *al-Mankhūl*, *al-Taʿlīqa*, *Shifāʾ al-ghalīl*, *Miḥakk al-naẓar*, *Miʿyār al-ʿilm*, *Tahāfut al-falāsifa*, and *al-Iqtiṣād fī l-iʿtiqād*. Al-Ghazālī's *Maqāṣid al-falāsifa* is not on this list. It was, however, available to Ibn al-ʿArabī when he later wrote his *al-ʿAwāṣim min al-qawāṣim* (see the list of correspondence between the two books in Ṭālibī, *Arāʾ Abī Bakr ibn al-ʿArabī*, 1:291–92).

32. Ṭālibī, *Arāʾ Abī Bakr ibn al-ʿArabī*, 1:67–68.

33. ʿAbbās, "Riḥlat Ibn al-ʿArabī," 87–88, first made this text available from an Istanbul manuscript.

34. Muḥammad ibn Aḥmad al-Qaffāl al-Shāshī (d. 507/1114) was at this time a teacher at the Tājiyya madrasa in Baghdad. After al-Kiyāʾ al-Harrāsī's death in 504/1110, he taught at the Niẓāmiyya. He was a student of Abū Isḥāq al-Shīrāzī (d. 476/1083) and was known for his traditionalist and less rationalist approach (al-Subkī, *Ṭabaqāt*, 6:70–78, *GAL*, 1:390–91, *Suppl.* 1:674; Makdisi, *Ibn ʿAqīl et la résurgence*, 208–10; Halm, *Die Ausbreitung*, 165, 169).

35. Ibn al-ʿArabī himself lived in Baghdad in the al-Muʿtamidiyya quarter; Ṭālibī, *Arāʾ Abī Bakr ibn al-ʿArabī*, 1:43.

36. I am grateful to Beatrice Gruendler, who assisted in the translation of these verses.

37. Ibn al-ʿArabī, *Qānūn al-taʾwīl*, 111.1–113.6.

38. Ibn Taymiyya, *Darʾ taʿāruḍ*, 1:5.9–10; idem, *Majmūʿ fatāwa*, 4:66.8–10.

39. Al-Ṭurṭūshī, *Risāla ilā ʿAbdallāh ibn Muẓaffar*, 162; see Ghurāb, "Ḥawla ikhrāq al-Murābiṭīn li-Iḥyāʾ al-Ghazālī," 136.

40. Griffel, *Apostasie und Toleranz*, 383.

41. See ʿAmmār Ṭālibī's analysis of the book and the positions defended therein in his *Arāʾ Abī Bakr ibn al-ʿArabī*, 1:89–275.

42. Ibn al-ʿArabī, *al-ʿAwāṣim min al-qawāṣim*,, 23.10–13.

43. Serrano Ruano, "Why Did the Scholars of al-Andalus Distrust al-Ghazālī?"

44. Ibn al-ʿArabī, *al-ʿAwāṣim min al-qawāṣim*, 23–24.

45. *Kitāb Tartīb al-riḥla*. It must be considered lost, cf. p. 63 in this book.

46. In his *Iḥyāʾ*, 1:11.1–2 / 3.2–3, al-Ghazālī refers to the book as "a revival for the religious sciences" (*iḥyāʾ li-ʿulūm al-dīn*); cf. al-Zabīdī, *Itḥāf al-sāda*, 1:59.22.

47. Ibn al-ʿArabī, *al-ʿAwāṣim min al-qawāṣim*, 24.4–11.

48. Rahman, *Prophecy in Islam*, 30–38.

49. Ibn Sīnā, *al-Shifāʾ*, *al-Ṭabīʿiyyāt*, *al-Nafs*, 173.9–174.2.

50. Ibid., 248.9–250.4.

51. Ibid., 200.11–201.9. On these three prophetical capacities in Ibn Sīnā, see Davidson, *Alfarabi, Avicenna, and Averroes on Intellect*, 100–101, 116–23, 139–40; Hasse, *Avicenna's De Anima in the West*, 154–56; Akiti, "Three Properties of Prophethood," 189–95; Rahman, *Prophecy in Islam*, 30–52; and Elamrani-Jamal, "De la multiplicité des modes de la prophetie chez Ibn Sīnā." Al-Ghazālī adapted these in several of his works. See Akiti, "Three Properties of Prophethood," 195–210.

52. Ibn al-ʿArabī, *al-ʿAwāṣim min al-qawāṣim*, 23.13–15.

53. Ibid., 25.6–8. The *Tahāfut*, 274.7–275.1 / 165.3–7 reports a similar example from the teachings of the *falāsifa*. There, walking on an elevated beam is compared to walking on the same beam when it lies on the ground. In the first case, the human falls, in the latter, not. This example is taken from Ibn Sīnā, *al-Shifāʾ*, *al-Ṭabīʿiyyāt*, *al-Nafs*, 200.1–6; and idem, al-*Ishārāt wa-l-tanbīhāt*, 219.13–16.

54. Ibn al-ʿArabī, *al-ʿAwāṣim min al-qawāṣim*, 25.*ult*.–26.3. Read *inbāṭ* for *inbān*.

55. Ibn Sīnā, *al-Ishārāt wa-l-tanbīhāt*, 219–22. On this passage see Hasse, *Avicenna's De anima*, 161–63. Al-Ghazālī copies this passage verbatim into his report of the teachings of the *falāsifa* in MS London, Or. 3126, foll. 283a–284b.

56. Al-Ghazālī, *Fayṣal al-tafriqa*, 191.18–192.5 / 56.5–57.2.

57. *kāna yumayyilu ilā dhālika wa-yastaṭrifuhu*; Ibn al-ʿArabī, *al-ʿAwāṣim min al-qawāṣim*, 93.5

58. Ibid, 232.

59. Ibn al-ʿArabī, *ʿĀriḍat al-aḥwadhī bi-sharḥ Ṣaḥīḥ al-Tirmidhī*.

60. See n, 38 above. Ibn Taymiyya's quotation is already in al-Dhahabī, *Siyar*, 19:327.8–9. Ghurāb, "Ḥawla iḥrāq al-Murābiṭīn li-l-Iḥyāʾ al-Ghazālī," 158, connects it to the unedited *Sirāj al-murīdīn*. Al-Dhahabī, *Siyar*, 19:344, quotes a passage from Ibn al-ʿArabī's yet unedited *Sharḥ al-asmāʾ al-ḥusnā*, in which he also argues against al-Ghazālī's position of the best of all possible words

61. MS Paris, Bibliothèque Nationale, no. 5639 (Fonds Archinard), fol. 138b. How this long text is related to other, shorter versions of *al-Nafkh wa-l-taswiya*—of *al-Maḍnūn al-ṣaghīr* and of *al-Ajwiba al-Ghazāliyya*—requires more study. On the several versions

of *al-Nafkh wa-l-taswiya*, see Lagardère, "A propos d'un chapitre du *Nafḫ wal-taswiya* attribué à Ġazālī," 127–36. Asín Palacios, *Espiritualidad*, 4:164–83, translates the text of the 1309/1891 edition of *al-Maḍnūn al-ṣaghīr* into Spanish.

62. On this dispute in the works of al-Ghazālī, see Frank, *Al-Ghazālī and the Ash'arite School*, 48–67.

63. 'Abbas, "Riḥlat Ibn al-'Arabī," 68–69.

64. Iḥsān 'Abbās worked with an unidentified manuscript at the Moroccan General Library (*al-Khizāna al-'Āmma*) in Rabat. Although there were reports that the film 'Abbās worked with had remained at the library of the American University in Beirut (Film MS:297–3) no such film can be located there today. I am grateful to Bilal Orfali for making inquiries on my behalf. My knowledge on the contents of the text is based on the information given in 'Abbas, "Riḥlat Ibn al-'Arabī," 68–69.

65. This *ḥadīth* goes in full: "The devil runs in the veins of humans and I feared that something of it had spilled into your hearts" (*inna l-shayṭān yajrā min Ibn Ādam majrā l-dam wa-innī khashītu an yaqdhifa fī qulūbikumā shay'an*). See Ibn Māja, *al-Sunan*, ṣiyām 65; similar in al-Bukhārī, *al-Ṣaḥīḥ*, i'tikāf 8, 10. Cf. Wensinck, *Concordance*, 1:342a.

66. MS Cairo, Dār al-Kutub, *majāmi'* 180, foll. 89b–96b. Cf. Bouyges, *Essay*, 59. Badawī, *Ma'allafāt*, 168–71, prints sections of the manuscript. The information in Badawī, *Mu'allafāt*, 16, about manuscripts of this text in Istanbul libraries seems to be erroneous, as I could not find the text in the manuscripts listed. I use the 1984 reprint of Kawtharī's edition. Heer, "Abū Ḥāmid al-Ghazālī's Esoteric Exegesis," 244, and the editor of Ibn Taymiyya's *Dar' ta'āruḍ*, 1:5, n. 3, mention the 1359/1940 print of Kawtharī's edition (Cairo: 'Izzat al-Ḥusaynī), which was not available to me.

67. Al-Ghazālī, *al-Qānūn al-kullī fī l-ta'wīl*, 48–50. The text is translated in Heer, "Abū Ḥāmid al-Ghazālī's Esoteric Exegesis," 244–46.

68. This is also the opinion of Ibn Taymiyya, *Dar' ta'āruḍ*, 1:5.6–9.

69. Gianotti, *Al-Ghazālī's Unspeakable Doctrine of the Soul*, 76–87.

70. See my review of Gianotti's *Al-Ghazālī's Unspeakable Doctrine of the Soul* in *JAOS* 124 (2004): 107–11.

71. Al-Bayhaqī, *Tatimmat Ṣiwān al-ḥikma*, 119. According to al-Bayhaqī, *Tatimmat*, 110–11, Abū l-'Abbās al-Faḍl ibn Muḥammad al-Lawkarī was a student of Bahmanyār ibn Marzubān (d. 458/1066), one of the master students of Avicenna. Al-Lawkarī thus connects the Khorasanian philosophical tradition with Avicenna, and he features in almost all intellectual *isnāds* of Iranian philosophers (see al-Rahim, "The Twelver-Šī'ī Reception of Avicenna"). Unfortunately, nothing is known about his life, and the source for his date of death given by Brockelmann, *GAL*, 1:460, namely, 517/1123–24, is unknown. Al-Lawkarī was still alive in 503/1109 (cf. Badawī, in the preface to his edition of Ibn Sīnā's *al-Ta'līqāt*, 9). On al-Lawkarī, see Dībājī in his introductions to the editions of al-Lawkarī's works *Bayān al-ḥaqq. al-'Ilm al-ilāhī*, 14–15; *Bayān al-ḥaqq, al-Manṭiq*, 67–71; and Reisman, *The Making of the Avicennan Tradition*, index.

72. Modern Meana in Turkmenistan. Mīhana is an alternative pronunciation and is preferred by As'ad's biographers. See, however, C. E. Bosworth's article on "Mayhana" in *EI2*, 6:914b.

73. Abū l-Muẓaffar Manṣūr ibn Muḥammad al-Sam'ānī (d. 489/1096), the grandfather of the historian Abū Sa'd al-Sam'ānī (d. 562/1166), the author of the *Kitāb al-Ansāb*. On Abū l-Muẓaffar, see al-Subkī, *Ṭabaqāt*, 5:335–46; *GAL*, 1:412, *Suppl.* 1:731; and Halm, *Ausbreitung*, 85–86.

74. Guy Monnot in *EI2*, 9:214b.

75. Maḥmūd ibn Muḥammad Tapar ibn Malikshāh was put in charge of Baghdad by his father, Sultan Muḥammad Tapar. When the father died in 511/1118, Maḥmūd first

refused to submit to his uncle Sanjar, who became the supreme sultan and successor to his father. Maḥmūd declined allegiance until he was defeated in 513/1119. That year, Sanjar appeared in Baghdad and reordered its affairs. This event may have prompted Asʿad's departure from the Niẓāmiyya madrasa. Later, Maḥmūd and Sanjar reconciled, and Maḥmūd received the western part of the Seljuq Empire and the title of sultan. He ruled over it until his death in 525/1131.

76. See Ibn ʿAsākir, *Tabyīn kadhib al-muftarī*, 320; Ibn al-Jawzī, *al-Muntaẓam*, 10:13; Yāqūt, *Muʿjam al-buldān*, 3:344; Ibn al-Athīr, *al-Kāmil fī taʾrīkh*, 10:464; Ibn Khallikān, *Wafayāt*, 1:207–8; al-Shahrazūrī, *Nuzhat al-arwāḥ*, 2:57; al-Ṣafadī, *al-Wāfī bi-l-wafayāt*, 9:17–18; al-Subkī, *Ṭabaqāt*, 7:42–43; Krawulsky, *Briefe und Reden*, 18; Halm, *Die Ausbreitung der šāfiʿitischen Rechtsschule*, index; Makdisi, *The Rise of Colleges*, index; and Kasāʾī, *Madāris-i Niẓāmiyyah*, 145–46.

77. The work is referred to as *al-Taʿlīqāt*, *al-Taʿlīqa*, *al-Taʿlīq*, or *ṭarīqa fi-l-khilāf*. On the many meanings of the word *taʿlīqa*, see Makdisi, *The Rise of Colleges*, 114–28. In his *Iḥyāʾ*, 1:60–62 / 68–71, al-Ghazālī expressed severe reservations against the discipline of *khilāf*.

78. Al-Baghdādī, *Kitāb al-Naṣīḥatayn*, fol. 89a.

79. Al-Ṣafadī, *al-Wāfī bi-l-wafayāt*, 21:341.

80. Makdisi, *Rise of Colleges*, 122.

81. *mashhūra (. . .) qalīla al-naẓīr*; Ibn Kathīr, *Ṭabaqāt al-fuqahāʾ al-shāfiʿiyīn*, 2:566.7–8.

82. *wa-lā yaṣilu ilā maʿrifati ʿilmi l-Ghazālī wa-faḍlihi illā man balagha aw kāda yabligha l-kamāla fī ʿaqlih*; al-Subkī, *Ṭabaqāt*, 6:202.7–8.

83. *ke tu-yi madhhab-i kih?* al-Ghazālī, *Fażāʾil al-anām*, 12.15.

84. Juvaynī, *Taʾrīkh Jahāngushāy*, 3:200.8.

85. Al-Ghazālī, *Fażāʾil al-anām*, 3–12; Krawulsky, *Briefe und Reden*, 63–78.

86. Al-Ghazālī, *Fażāʾil al-anām*, 12.15–17.

87. Dawlatshāh, *Taẕkirat al-shuʿarāʾ*, 85.7–11.

88. This Asʿad al-Mayhanī is mentioned by al-Samʿānī, *al-Taḥbīr fī l-Muʿjam al-kabīr*, 1:117–18. His full name is Asʿad ibn Saʿīd ibn Faḍlallāh al-Mayhanī. He was born in 454/1062 and died in 507/1114. His existence resolves the confusion in Krawulsky, *Briefe und Reden*, 18–19; and Humāʾī's *Ghazzālī-nāmah*, 334–35.

89. Dawlatshāh, *Taẕkirat al-shuʿarāʾ*, 85.17.

90. The collection of letters mentions an Asʿad as a Qur'an recitator at the court of Sanjar (*Fażāʾil al-anām*, 6.9). It sometimes also identifies the ruler that al-Ghazālī had an exchange with as "sultan" (ibid., 6.3–8). Sanjar became supreme sultan only after al-Ghazālī's death. During al-Ghazālī's lifetime, he carried the title of a king (*malik*). This mistake may have prompted the misunderstanding that al-Ghazālī had dealings with the supreme sultan of his lifetime, namely Muḥammad Tapar.

91. He wrote *Iljām al-ʿawāmm ʿan ʿilm al-kalām* on this subject.

92. On Abū l-Muẓaffar Aḥmad ibn Muḥammad ibn al-Muẓaffar al-Khawāfī, see al-Ṣarīfīnī, *al-Muntakhab min al-Siyāq*, 146–47 = Frye, *The Histories of Nishapur*, text 3, fol. 35a; al-Samʿānī, *al-Ansāb*, 5:220; Yāqūt, *Muʿjam al-buldān*, 4:486–87, 3:343; Ibn Khallikān, *Wafayāt al-aʿyān*, 1:96–97; al-Subkī, *Ṭabaqāt*, 6:63; and Halm, *Ausbreitung*, 96. He was also a teacher of al-Shahrastānī. In his *Iḥyāʾ*, 1:65.24–26 / 76.3–5, al-Ghazālī warns his readers against taking part in disputations that aim at "silencing one's opponent."

93. His letter of appointment from Sanjar's chancellery, which is unfortunately not dated, is preserved in Muntajab al-Dīn, *ʿAtabat al-kataba*, 6–9; cf. Horst, *Die Staats-*

verwaltung der Großselğuqen, 163. The letter is reprinted in Kasāʾī, *Madāris-i Niẓāmiyyah*, 260–63.

94. Halm, *Ausbreitung*, 250. The two were Abū Manṣūr Muḥammad ibn Muḥammad al-Ṭūsī (d. 567/1171–72) and Abū l-Fatḥ Muhammad ibn Maḥmūd al-Ṭūsī (d. 596/1199–1200). On Muḥammad ibn Yaḥyā's central position in Shāfiʿite intellectual *isnāds*, see Sublet, "Un itinéraire du *fiqh* šāfiʿite," 193.

95. On Abū Naṣr Aḥmad ibn Zirr ibn ʿAqīl al-Kamāl al-Simnānī, see al-Subkī, *Ṭabaqāt*, 6:16–17.

96. Al-Ghazālī, *al-Munqidh*, 49.16. Cf. also the title of his main work: *Ihyāʾ ʿulūm al-dīn*.

97. *Al-Basīṭ fī l-furūʿ fī madhhab al-Shāfiʿī*; it is yet unedited. For a newly discovered text by al-Ghazālī on the *furūʿ* of *fiqh*, see p. 361.

98. Al-Ghazālī, *Iḥyāʾ*, 1:59.9–11 / 68.2–4, mentions *al-Basīṭ* and *al-Wasīṭ*. *Al-Basīṭ* is the earliest of the three works on the Shāfiʿite *furūʿ*. It is referred to in *al-Wasīṭ*, 1:103.3, and in *al-Wajīz*, 1:105.1. On the sources that al-Ghazālī used for the composition of *al-Basīṭ* and *al-Wasīṭ*, see Ibn al-ʿImād, *Shadharāt al-dhahab*, 4:12.18–21.

99. Bouyges, *Essay the chronologie*, 12–13, 49. The chronology is slightly confusing since *al-Wajīz* is also mentioned in *Iḥyāʾ*, 1:196.2 / 260.13, and in *Jawāhir al-Qurʾān*, 27.7, two works that appear to have been published before 495/1101. It is not entirely clear, though, whether in these two passages "*al-wajīz*" truly refers to this book. If so, the passage may anticipate the future completion and publication of *al-Wajīz*. It would not be the only time that al-Ghazālī refers to a future publication.

100. Al-Dhahabī, *Siyar*, 18:340. Al-Ghazālī refers to two of these three works of al-Wāḥidī in *Iḥyāʾ*, 1:57–58 / 67.16–18. On Abū l-Ḥasan ʿAlī ibn Aḥmad al-Wāḥidī and his three Qurʾān-commentaries—*al-Basīṭ*, *al-Wasīṭ*, and *al-Wajīz*—see Saleh, "The Last of the Nishapuri School of Tafsīr"; and Brockelmann, *GAL*, 1:411, *Suppl.* 1:730–31.

101. Al-Ghazālī explains the three set-levels of *iqtiṣār*, *iqtiṣād*, and *istiqṣāʾ* for every science in *Iḥyāʾ* 1:57.21–23 / 66.6–8, and for *kalām* specifically in ibid., 1:134.7–19 / 169.8–*ult.*

102. Muḥammad ibn Yaḥyā also wrote two *taʿlīqāt* on disputation (one titled *al-Intiṣāf fī masāʾil al-khilāf*) that are also lost. On the existing commentaries on al-Ghazālī's *al-Wajīz* and *al-Wasīṭ*, see Brockelmann, *GAL*, 1:424; *Suppl.* 1:752–53.

103. Al-Rāfiʿī, *al-Fatḥ al-ʿazīz fī sharḥ al-Wajīz*.

104. Ibn al-Ṣalāḥ's and al-Nawawī's commentaries are printed in the current edition of al-Ghazālī's *al-Wasīṭ*.

105. On the events of 548/1153 and Muḥammad ibn Yaḥyā's death, see Bulliet, *Patricians*, 76–79, 255; and Ibn al-Athīr, *al-Kāmil*, 11:116–21. On Muḥammad ibn Yaḥyā, see al-Samʿānī, *al-Taḥbīr fī l-muʿjam al-kabīr*, 2:252–53; Ibn Khallikān, *Wafayāt al-aʿyān*, 4:223–24; al-Dhahabī, *Siyar aʿlām al-nubalāʾ*, 20:312–15; idem, *Taʾrīkh al-Islām*, vol. 541–51 AH (vol. 37) 337–39; al-Ṣafadī, *al-Wāfī bi-l-wafayāt*, 5:197; al-Subkī, *Ṭabaqāt*, 6:25–27; and Halm, *Ausbreitung*, 59.

106. Abū Bakr al-Qāsim ibn ʿAbdallāh ibn ʿUmar al-Ṣaffār al-Rīkhī; see Bulliet, *Patricians*, 165, 186, 190; Halm, *Ausbreitung*, 60; al-Dhahabī, *Siyar aʿlām al-nubalāʾ*, 22:109–10; idem, *Taʾrīkh al-Islām*, vol. 611–20 AH (vol. 44), 416–17; and al-Subkī, *Ṭabaqāt*, 8:353. On the sacking of Nishapur and the slaughtering of its inhabitants, see Juvaynī, *Taʾrīkh-i Jahāngushāy*, 1:138–41; and Ibn al-Athīr, *al-Kāmil*, 12:256–57. On the move of Shāfiʿism's center from Iraq and Khorasan to Syria, see Sublet, "Un itinéraire du *fiqh* šāfiʿite."

107. The same who taught Abū Bakr ibn al-ʿArabī; cf. n. 34. On the teachers at the Niẓāmiyya in Baghdad during this time, see Kasāʾī, *Madāris-i Niẓāmiyyah*, 141ff.

108. Al-Mubārak ibn ʿAbd al-Jabbār al-Ṣayrafī was a teacher of *ḥadīth* and was also known as Ibn al-Ḥamāmī and Ibn al-Ṭuyūrī; see al-Samʿānī, *al-Ansāb,* 4:233; idem, *al-Taḥbīr fī l-muʿjam al-kabīr,* 1:570, 2:146; and Ibn al-Jawzī, *al-Muntaẓam,* 9:154.

109. Griffel, "Ibn Tūmart's Rational Proof for God's Existence," 753–56.

110. On al-Ghazālī's qualified endorsement of *al-amr bi-l-maʿrūf wa-l-nahy ʿan al-munkar,* see Cook, *Commanding Right and Forbidding Wrong,* 427–68.

111. A Spanish scholar also translated into Latin the three theological texts by Ibn Tūmart discussed below; see Griffel, "Ibn Tūmart's Rational Proof for God's Existence," 771.

112. *Sifr fīhi jamīʿ taʿāliq al-Imām al-maʿṣūm al-Mahdī al-maʿlūm (. . .) mimma amlaʾahu Sayyiduna al-Imam al-Khalīfa Amir al-Muʾminīn Abū Muḥammad ʿAbd al-Muʾmin ibn ʿAlī.* On the manuscripts and editions, see Griffel, "Ibn Tūmart's Rational Proof for God's Existence," 765–67. Here, we refer to the most reliable edition by ʿAmmār Ṭālibī, published under the title of the first text in this collection, *Aʿazz mā yuṭlab.*

113. On the traditional *kalām* proof for God's existence, see Davidson, *Proofs for Eternity,* 117–53; and Craig, *The Kalām Cosmological Argument,* 3–60 (slightly extended in idem, *The Cosmological Argument,* 48–126).

114. Al-Juwaynī, *al-ʿAqīda al-Niẓāmiyya,* 8–11.

115. Al-Ghazālī, *al-Iqtiṣād,* 24–35; for a sketch of this argument, see Marmura, "Ghazali's *al-Iqtisad fi al-I'tiqad,*" 4–8; and Davidson, *Proofs for Eternity,* 141–46. It is more fully discussed in Craig, *Kalām Cosmological Argument,* 44–49; repeated in idem, *The Cosmological Argument,* 99–104.

116. *min bidāyati l-ʿaqli anna (. . .)*; al-Ghazālī, *Iḥyāʾ,* 1:144.12–13 / 183.5–6; Tibawi, "Al-Ghazālī's Sojourn," 80.26–28, 98.

117. Ibn Tūmart, *Sifr fīhi jamīʿ taʿāliq al-Imām,* 214.7–13.

118. See above p. 30.

119. On al-Ghazālī's proofs from design, see Davdison, *Proofs for Eternity,* 226–27, 234, and below, pp. 221, 226.

120. Ibn Tūmart, *Sifr fīhi jamīʿ taʿāliq al-Imām,* 219.13–14.

121. See below, pp. 236–41.

122. Here I wish to correct the judgment I expressed in my article "Ibn Tūmart's Rational Proof for God's Existence," 777–79.

123. Ibn Tūmart, *Sifr fīhi jamīʿ taʿāliq al-Imām,* 214.5.

124. See pp. 220–21.

125. Garden, *Al-Ghazālī's Contested Revival: Iḥyāʾ ʿulūm al-dīn and its Critics in Khorasan and the Maghrib,* 144–89, 208–23; Griffel, *Apostasie und Toleranz,* 361–66; and idem, "Ibn Tūmart's Rational Proof for God's Existence," 754.

126. Ibn Ṭumlūs, *Madkhal li-sināʿat al-manṭiq,* 12. Cf. Griffel, *Apostasie und Toleranz,* 382–87, 416; idem, "Ibn Tūmart's Rational Proof for God's Existence," 764–65.

127. Griffel, *Apostasie und Toleranz,* 401–62; idem, "The Relationship Between Averroes and al-Ghazālī as It Presents Itself in Averroes' Early Writings."

128. See Opwis, "Islamic Law and Legal Change: The Concept of *Maṣlaḥa* in Classical and Contemporary Islamic Legal Theory."

129. Griffel, *Apostasie und Toleranz,* 354–57; El-Rouayheb, "Was There a Revival of Logical Studies?" 3.

130. El-Rouayheb, "Was There a Revival of Logical Studies?" 4–14; and idem, "Sunni Muslim Scholars on the Status of Logic," 215–16, 226–28.

131. That is at the beginning of the twelfth/eighteenth century; al-Zabīdī, *Itḥāf al-sādā,* 1:179–80. See El-Rouayheb, "Was There a Revival of Logical Studies?" 5.

132. Al-Bayhaqī, *Tatimmat Ṣiwān al-ḥikma*, 109. On the life and works of ʿAyn al-Quḍāt, see his entry in *EIran*, 3:140–43, by Gerhard Böwering. Al-Bayhaqī says ʿAyn al-Quḍāt was a student of ʿUmar al-Khayyām. Both were thoroughly influenced by Ibn Sīnā's ontology. See ʿUmar al-Khayyām's philosophical epistles in al-Khayyām, *Dānishnāmah-yi Khayyāmī*, 324–422; and in the collection of philosophical texts, *Jāmiʿ al-badāʾiʿ*, 165–93. Hamid Dabashi, who in his *Truth and Narrative*, 86 (followed by Safi, *The Politics of Knowledge*, 181), rejects this connection, sees in ʿAyn al-Quḍāt only a Sufi and is largely unaware of the philosophical character of much of his writings. It is quite possible that ʿAyn al-Quḍāt associated himself with ʿUmar al-Khayyām in a similar way to how he associated himself with al-Ghazālī.

133. ʿAyn al-Quḍāt, *Tamhīdāt*, 280–81; Safi, *The Politics of Knowledge*, 172.

134. ʿAyn al-Quḍāt, *Nāmah-hā*, 2:124, 458; Safi, *The Politics of Knowledge*, 173. See above n. 82.

135. *shāgird-i kutub-i ū būdeh-am*; ʿAyn al-Quḍāt, *Nāmah-hā*, 2:316.16. On ʿAyn al-Quḍāt's relationship to Abū Ḥāmid al-Ghazālī, see Pūrjavādī, *ʿAyn al-Qużāt va-ustāẕān-i ū*, 135–79; and Māyil Hirāwī, *Khāṣṣiyyat-i āyinagī*, 8–10, 77–80.

136. ʿAyn al-Quḍāt, *Nāmah-hā*, 1:20–21; Izutsu, "Mysticism and the Linguistic Problem of Equivocation in the Thought of ʿAyn al-Quḍāt Hamadānī," 166–68.

137. ʿAyn al-Quḍāt, *Zubdat al-ḥaqāʾiq*, 6.

138. Ibid., 11–13; Izutsu, "Creation and the Timesless Order of Things," 127–30; Landolt, "Ghazālī and Religionswissenschaft," 55–56. On Ibn Sīnā's proof, see Mayer, "Ibn Sīnā's *Burhān al-Ṣiddiqīn*"; Davidson, *Proofs for Eternity*, 281–310; idem, "Avicenna's Proof of the Existence of God"; and and Marmura, "Avicenna's Proof from Contingency."

139. Al-Ghazālī, *Faḍāʾiḥ al-bāṭiniyya*, 82–83. See Goodman, "Ghazâlî's Argument from Creation," 75–76.

140. Al-Ghazālī, *Tahāfut*, 140–42 / 82–83. Davidson, *Proofs for Eternity*, 366–75; Janssens, "Ibn Sīnā and His Heritage," 4–5; and Goodman, "Ghazâlî's Argument from Creation," 75–85, discuss al-Ghazālī's ambiguity toward the *burhān al-ṣiddīqīn*. Goodman (p. 75) explains why, according to al-Ghazālī, the assumption of an eternal world destroys Ibn Sīnā's argument from contingency.

141. Al-Ghazālī in MS London, Or. 3126, fol. 3a; translated in Griffel, "MS London, British Library Or. 3126," 17.

142. Al-Ghazālī in MS London, Or. 3126, fol. 3a; paraphrasing Ibn Sīnā, *al-Ishārāt*, 146.15–17

143. Frank, *Al-Ghazālī and the Ashʿarite School*, 129, n. 76; Goodman, "Ghazâlî's Argument from Creation," 67, 77–78. Cf. also the comments on the two proofs in al-Ghazālī, MS London, Or. 3126, fol. 3a; translated in Griffel, "MS London, British Library Or. 3126," 17.

144. ʿAyn al-Quḍāt, *Tamhīdāt*, 254–354 (referring to al-Ghazālī, see ibid., 255–56).

145. ʿAyn al-Quḍāt, *Zubdat al-ḥaqāʾiq*, 38.3–4.

146. For these teachings in al-Ghazālī, see *Iḥyāʾ ʿulūm al-dīn*, 4:112.4ff. / 2224.12ff. and 4:120 / 2237. Cf. Gramlich, *Muḥammad al-Ġazzālīs Lehre*, 195–96, 209.

147. Izutsu, "Creation and the Timeless Order of Things," 130–38.

148. ʿAyn al-Quḍāt, *Nāmah-hā*, 2:314–16. This letter by ʿAyn al-Quḍāt to his follower ʿAzīz al-Dīn al-Mustawfī (d. 527/1133) is notable for its comments and for its criticism of al-Ghazālī (ibid., 2:309–31). Cf. Safi, *The Politics of Knowledge*, 174.

149. ʿAyn al-Quḍāt, *Nāmah-hā*, 1:105.

150. ʿAyn al-Quḍāt, *Tamhīdāt*, 349–50; cf. Reisman, *Making of the Avicennan Tradition*, 140.

151. Al-Ghazālī, *Iḥyāʾ ʿulūm al-dīn*, 4:305–22 / 13:2494–518.

152. ʿAyn al-Quḍāt, *Tamhīdāt*, 167. Cf. Safi, *Politics of Knowledge*, 179.

153. ʿAyn al-Quḍāt, *Nāmah-hā*, 1:244; 2:339, 487–88; cf. Safi, *Politics of Knowledge*, 175–76, 182–89.

154. *shayṭān az shayāṭīn-i uns, dushmanī az dushmanān-i Khudā va-rasūl*; Ayn al-Quḍāt, *Nāmah-hā*, 2:375.3.

155. Ibid., 2:58.1

156. Ibn Qasī, *Kitāb Khalʿ al-naʿlayn*. Ibn ʿArabī (d. 638/1240), the famous Sufi, wrote a commentary to this work. On Ibn Qasī and his book, see Goodrich, *A Sufi Revolt in Portugal: Ibn Qasi and his Kitāb khal' al-na'layn* (including an edition of the work on pp. 60–272); and Dreher, *Das Imamat des islamischen Mystikers Abūlqāsim Aḥmad ibn al-Ḥusain ibn Qasī*. Goodrich (pp. 317–18) and Muḥammad Amrānī, the editor of Ibn Qasī's *Khalʿ al-naʿlayn*, suggest that a passage close to the end of that book is copied from al-Ghazālī (?), *Maʿārij al-Quds*, 168–72. The author of this book, however, copied this passage himself from Avicenna (Janssens, "Le Maʿârij al-quds fi madârij maʿrifat al-nafs," 37). Hence, Ibn Qasī might have also adopted this passage from Ibn Sīna, *al-Shifāʾ, al-Ilāhiyyāt*, 348ff.; or idem, *Aḥwāl al-nafs*, 128ff.

157. *bi-ṭṭirāh al-ʿālamayn*; al-Ghazālī, *Mishkāt al-anwār*, 73.9–11 / 161.2–4; cf. *Fayṣal al-tafriqa*, 191.5–6 / 55.3–4. Ibn Taymiyya, *Minhāj al-sunna*, 4:149.5–8, makes a close connection between al-Ghazālī's *Mishkāt* and Ibn Qasī's *Khalʿ al-naʿlayn*.

158. Al-Ghazālī, *Fażāʾil al-anām*, 10–11; Krawulsky, *Briefe und Reden*, 76.

159. Al-Bundarī, *Zubdat al-nuṣra*, 151–52; al-Bayhaqī, *Tatimmat Ṣiwān al-ḥikma*, 109; al-Subkī, *Ṭabaqāt*, 7:128–30; Yāqūt, *Muʿjam al-buldān*, 4:710; al-Ṣafadī, *al-Wāfī bi-l-wafayāt*, 17:540–41; Dabashi, *Truth and Narrative*, 475–536; and Safi, *The Politics of Knowledge*, 189–200.

160. Yāqūt, *Muʿjam al-udabāʾ*, 4:1550–551.

161. Al-Bundarī, *Zubdat al-nusra*, 151.14–15 These words go back to Anūshirwān ibn Khālid (d. around 532/1138) or to ʿImād al-Dīn al-Iṣfahānī (d. 597/1201).

162. Gerhard Böwering in *EIran*, 3:141, lists all the accusations.

163. ʿAyn al-Quḍāt, *Risālat Shakwā l-gharīb*, 9–11.

164. On ʿAyn al-Quḍāt's teachings regarding God's way of knowing the particulars of His creation, see Izutsu, "Mysticism and the Linguistic Problem of Equivocation in the Thought of ʿAyn al-Quḍāt Hamadānī," 163–66; and idem, "Creation and the Timeless Order of Things," 135.

165. ʿAyn al-Quḍāt, *Risālat Shakwā l-gharīb*, 9.14–10.1. See also Landolt, "Ghazālī and *Religionswissenschaft*," 60.

166. Izutsu, "Creation and the Timeless Order of Things"; idem, "Mysticism and the Linguistic Problem of Equivocation in the Thought of ʿAyn al-Quḍāt Hamadānī"; Landolt, "Two Types of Mystical Thought in Muslim Iran: An Essay on Suhrawardī *Shaykh al-Ishrāq* and ʿAynulquẓāt-i Hamadānī," 192–204; and idem, "Ghazālī and *Religionswissenschaft*," 55–60. See also Pūrjavādī, *ʿAyn al-Qużāt va-ustāzān-i* ū; Māyil Hirāwī, *Khāṣṣiyyat-i āyinagī*; and Gerhard Böwering's article in *EIran*.

167. MS Patna (India), Khuda Bakhsh Oriental Public Library (Bankipore), no. 1825. See Riḍwān al-Sayyid's introduction to *al-Asad wa-l-ghawwāṣ*, 8–9. Only after finishing the work on this book I came across another manuscript of the work, MS Yale University, Beinecke Library, Landberg 98 (Nemoy 467), that identifies the author as Abū l-Barakāt Naṣr ibn Salāma al-Dimashqī. I have not been able to identify this person in any of the relevant bio-bibliographical sources and reference works. Brockelmann,

GAL, Suppl. 1:809, was aware of one of the MSS used in the edition but did not identify it as an independent work.

168. Ibn al-Muqaffaʿ, *Kalīla wa-Dimna*, 217–27 / 245–59. There are different recensions of the Arabic text with different chapter arrangements. In Cheikho's edition, the story is the tenth chapter after the introductions. In the oldest MS of the work, which was edited by ʿAzzām and is less representative with regard to the order of the stories, it is the eleventh chapter. The plot of *al-Asad wa-l-ghawwāṣ* takes many elements both from the stories about the lion and the bull and from Dimna's trial in the first and second chapters of *Kalīla wa-Dimna*, 53–124 / 43–124. *Al-Asad wa-l-ghawwāṣ* mentions Ibn al-Muqaffaʿ in its introduction (p. 40.2). Van den Bergh, "Ghazali on 'Gratitude Towards God,'" 92, argues that al-Ghazālī may have been familiar with *Kalīla wa-Dimna*. (On the strength of this argument see below, p. 348, note 81.) As far as I know, al-Ghazālī nowhere mentions Ibn al-Muqaffaʿ in his works. It is true that most MSS and editions of al-Ghazālī's *Fayṣal al-tafriqa* include in its eleventh chapter a derogatory reference to "al-Muqaffaʿ." That, however, is a scribal mistake that happened very early in the manuscript tradition. The original reference is, for instance, in MS Berlin, Wetzstein II 1806, fol. 84b, and refers to "al-Muqannaʿ" ("the veiled one"), a religious propagandist, who appeared in northern Khorasan in 160/777 and who is also referred to in Niẓām al-Mulk's *Siyāsat-namāh*, 252. (See my German translation, *Al-Ġazālī: Über Rechtgläubigkeit und religiöse Toleranz*, 87, 102).

169. *Al-Asad wa-l-ghawwāṣ*, 183.18–19. See also the very helpful German translation of the text by Rotter, *Löwe und Schakal*, 194.

170. *Al-Asad wa-l-ghawwāṣ*, 204–7.

171. Al-Ghazālī, *Risāla ilā Abū l-Fatḥ al-Damīmī*, 27–28; MS Berlin, Petermann II 8, pp. 121–22. The letter was published already during al-Ghazālī's lifetime. The literary technique of having the autobiographic narrator conducting an inner dialogue between himself and his soul appears first in Arabic literature in Burzôye's introduction in Ibn al-Muqaffaʿ, *Kalīla wa-Dimna*, 32–33 / 27–28, and is later used in Sufi literature. The passage in al-Ghazālī is quite reminiscent of the one in Burzôye's introduction.

172. *Al-Asad wa-l-ghawwāṣ*, 70.8–9, cf. 111.2–6.

173. See, for example, al-Ghazālī, *al-Munqidh*, 27.7–8; idem *Iljām al-ʿawāmm*, 10.15–16 / 67.3–4, 18.6 / 77.12; and *Risāla ilā Abū l-Fatḥ al-Damīmī*, 30.10–13, MS Berlin, Petermann II 8, p. 123.

174. Ibn Ṭufayl, *Ḥayy ibn Yaqẓān*, 19.5. The metaphor of scholarship as being a "plunge" (*khawḍ*) into something dangerous is, of course, not limited to al-Ghazālī and appears more often in Arabic religious literature.

175. *dar daryā-yi ʿulūm-i dīn ghawwāṣī kard*; al-Ghazālī, *Fażāʾil al-anām*, 4.16–17; Krawulsky, *Briefe und Reden*, 65. See above pp. 28–9

176. Al-Qushayrī, *al-Risāla*, 1:403. The verse is sometimes attributed to al-Shāfiʿī.

177. *kullu l-ʿadāwati qad turjā izālatuhā* (or: *imātatuhā*), *illā ʿadāwatu man ʿādāka ʿan ḥasad*; al-Ghazālī, Fażāʾil al-anām, 13, and Fayṣal al-tafriqa, 15/128. In *Ayyuhā l-walad*, 49.1–2, the verse is quoted in a general context and not brought in connection with al-Ghazālī's adversaries.

178. *Al-Asad wa-l-ghawwāṣ*, 69.10–70.5.

179. Ibid., 82.8–9, 96–101.

180. The maxim is the general tenor of Book 32 on patience and thankfulness (*al-ṣabr wa-l-shukr*), while the implications on the daily conduct are worked out in the second part of Book 35 on trust in God (*tawakkul*). On the permissible use of astrology, see *Iḥyāʾ*, 4:146 / 2272.

181. *Al-Asad wa-l-ghawwāṣ*, 83.11–17, 92.3–4.

182. Ibid., 199.5–9. Al-Ghazālī, *Iḥyāʾ*, 3:9.4–6 / 1355–56, compares the relationship of the soul with its body to that of a king residing over a city (*madīna*) and a kingdom (*mamlaka*).

183. *Al-Asad wa-l-ghawwāṣ*, 94.1–2. Cf. the saying, "Do not know the truth (*ḥaqq*) by the man [who utters it], rather know the truth [by itself] and you will know its adherents," that al-Ghazālī attributes to ʿAlī ibn Abī Ṭālib and that he quotes several times in his works (see e.g., *al-Munqidh*, 25.16).

184. *Al-Asad wa-l-ghawwāṣ*, 151.9–152.3.

185. Ibid., 91.7–8.

186. Ibid., 187.14–16.

187. Ibid., 167.17–19.

188. Ibid., 193.10–195.8; see also 74.5–10.

189. Al-Ghazālī, *Iḥyāʾ*, 3:69.21–23 / 1442.9–11.

190. Ibid., 3:68.2–5 / 1439–440.

191. Ibid., 1:28–32 / 28–34.

192. Al-Ghazālī, *Kīmyā-yi saʿādat*, 1:5.1–2; *Iḥyāʾ*, 3:78 / 1453–454.

193. Al-Ghazālī, *Himāqat-i ahl-i ibāḥat*, 12.2–9 / 175.9–176.7

194. *dād dar īn dawr bar-andākhteh ast*; Niẓāmī, *Khamsah*, 1:91–3 (*Makhzan al-asrār*, lines 1106–41).

CHAPTER 3

1. Al-Ghazālī, *al-Munqidh*, 18.9–15.

2. Such criticism was voiced, for instance, by al-Māzarī al-Imām (d. 536/1141) in his lost *al-Kashf wa-l-inbaʾ ʿalā mutarjam al-Iḥyāʾ*. The passage is quoted in al-Subkī, *Ṭabaqāt*, 6:240.3–4.

3. Griffel, "MS London, British Library Or. 3126: An Unknown Work by al-Ghazālī on Metaphysics and Philosophical Theology."

4. Janssens, "Le *Dânesh-Nâmeh* d'Ibn Sînâ," 168–77.

5. See, for instance, Bouyges, *Essai de chronologie*, 23–24.

6. Janssens, "Al-Ghazzālī and His Use of Avicennian Texts," 43–45.

7. Al-Ghazālī, *Maqāṣid*, 1:2–3 / 31–32; 3:77 / 385.

8. Janssens, "Al-Ghazzālī and His Use of Avicennian Texts," 45; Griffel, "MS London, British Library Or. 3126," 9–10.

9. Al-Ghazālī, *Tahāfut* 6.6 / 3.5; and idem, *al-Munqidh*, 18.7–10. See Griffel, "*Taqlīd* of the Philosophers," 274–78.

10. Baneth, "Jehuda Hallewi und Gazali," 29, 31; idem, "Rabbi Yehudah ha-Levi we-Algazzali." 313, 315–16.

11. Griffel, "*Taqlīd* of the Philosophers," 286–91.

12. Al-Ghazālī, *Tahāfut*, 4.3–9 / 1.11–2.2; See Griffel, "*Taqlīd* of the Philosophers," 278–86.

13. Al-Ghazālī, *Tahāfut*, 16.8–11 / 9.14–18. See Griffel, "*Taqlīd* of the Philosophers," 287.

14. Al-Ghazālī, *al-Munqidh*, 23.14–15; cf. also the passage 22.21–23.

15. Al-Ghazālī, *al-Qisṭās al-mustaqīm*, 67.11–14; for the opinion that the *falāsifa* took their ethical teachings from the Sufis, see idem, *al-Munqidh*, 24.16–18. (Frank, *Al-Ghazālī and the Ashʿarite School*, 96, believes that al-Ghazālī's identification of the ethical teaching of the *falāsifa* and Sufis is "too transparent a fiction" to have been taken seriously by al-Ghazālī.) According to al-Ghazālī, the ancient physicians also learned their trade from the early prophets (*al-Munqidh*, 45.14).

16. In *Iḥyā'*, 1:46.15–17 / 52.2–5 (= al-Zabīdī, *Itḥāf al-sādā*, 1:226), al-Ghazālī says that only the prophets and the "friends of God" (*awliyā'*) arrive at knowledge of the "metaphysical secrets" (*asrār al-ilāhiyya*), while *falāsifa* and *mutakallimūn* have only an incomplete grasp.

17. Al-Ghazālī, *Faḍā'iḥ al-bāṭiniyya*, 114.11–*ult.*

18. Al-Ghazālī, *Tahāfut*, 252.4–8 / 151.21–152.3. Frank, *Creation*, 83. For al-Ghazālī's view that knowledge "on the configuration (*hay'a*) of the heavens and the stars, their distances, and their sizes, and the way they move" is not demonstrative, see *Mi'yār al-'ilm*, 167.4–7.

19. Al-Ghazālī, *Fayṣal al-tafriqa*, 191.16–192.12 / 56.3–57.8; idem, *al-Munqidh*, 23.17–24.7. See also idem, *Faḍā'iḥ al-bāṭiniyya*, 153.13–154.2; 155.9–11.

20. See, for instance, Munk, *Dictionaire des scienes philosophique*, 2:512, and later in his *Mélanges de la philosophie juive et arabe*, 382, and other scholars quoted in the introduction to this book.

21. Al-Ghazālī, *Tahāfut*, 376.2–10 / 226.1–10.

22. Al-Ghazālī, *Faḍāiḥ al-bāṭiniyya*, 151.17–153.13; idem, *Fayṣal al-tafriqa*, 184.4–5 / 41.5–6; idem, *al-Qānūn al-kullī fī l-ta'wīl*, 44.17–18, 45.1–2; idem, *al-Iqtiṣād*, 249.6–9, 250.5; idem, *Tahāfut*, 376.7–9 / 226.8–9. See Griffel, *Apostasie und Toleranz*, 292–95.

23. Goodman, "Ghazâlî's Argument from Creation," 67–68, 79–82, argues that al-Ghazālī rejected the suggestion of a pre-eternal world so vehemently because, for him, "acceptance of the eternity of the world is inconsistent with belief in the existence of God," and "(. . .) theism itself stands or falls with the doctrine that being once emerged from nothingness."

24. Al-Ghazālī, *al-Iqtiṣād*, 250.3–4. See Griffel, *Apostasie und Toleranz*, 297–93.

25. Al-Ghazālī, *Iḥyā'*, 1:27.6–7 / 27.16–17. Cf. idem, *Fayṣal al-tafriqa*, 195.10–12 / 61–62; idem, *al-Munqidh*, 19.*paenult.*

26. The issue that the *falāsifa* assume the prophets' teachings are false (*takdhīb*) is brought up only once, as far as I can see, in the seventeenth discussion about the *falāsifa*'s denial of a number of miracles that revelation or credible historical reports attribute to the prophets; see al-Ghazālī, *Tahāfut*, 289.11–290.1 / 173.1–3.

27. Griffel, *Apostasie und Toleranz*, 269–70, 295–96.

28. Al-Ghazālī, *Tahāfut*, 377.6–8 / 227.3–5.

29. Al-Ghazālī, *al-Wasīṭ fī l-madhhab*, 6:428–32; idem, *Shifā' al-ghalīl*, 221–24; idem, *Faḍā'iḥ al-bāṭiniyya*, 156–61; idem, *Fayṣal al-tafriqa*, 197.16–7 / 66.2–3; Griffel, *Apostasie und Toleranz*, 285–91; idem, "Toleration and Exclusion," 350–54; Goldziher, *Streitschrift*, 71–73.

30. Griffel, *Apostasie und Toleranz*, 74–82, 92–99. An exception to this, however, existed in the Mālikī school of law.

31. See, ibid. 24–241, 282–91; Griffel, "Toleration and Exclusion"; and idem, "Apostasy" in *EI3*. For a detailed English synopsis of my German book *Apostasie und Toleranz* see Michael Schwarz's review in *Jerusalem Studies of Arabic and Islam* 27 (2002): 591–601.

32. Al-Ghazālī, *Fayṣal al-tafriqa*, 197.15–18 / 66.2–4.

33. Al-Shahrastānī, *al-Milal wa-l-nihal*, 48–49; *Livre des religions et des sects*, 1:242–43. On Abū Mūsā al-Murdār (d. 266/841) and this exchange, see van Ess, *Theologie und Gesellschaft*, 3:136, 5:333–34.

34. On the meaning of *zandaqa* in Muslim legal texts of this period, see Griffel, *Apostasie und Toleranz*, 71–72, 76, 83–89, 134–35, 375–79.

35. On the authority of his cousin, 'Abdallāh ibn 'Abbās (d. 68/687–88), Muḥammad is reported as having said that "whoever changes his religion, kill him!" or "cut off his head!" *man baddala dīnahu fa-qtulhu*, according to Abū Da'ūd, *Sunan*, *ḥudūd* 1; Ibn Māja, *Sunan*,

ḥudūd 2; and al-Bukhārī, *Ṣaḥīḥ, jihād* 149, and *istitāba* 2, or . . . *fa-ḍribū ʿunqahu* according to Mālik ibn Anas, *al-Muwaṭṭaʾ, aqḍiya* 18; cf. Wensink, *Concordance et indices*, 1:153a.

36. Gutas, "Avicenna's maḏhab," 326–34; Janssens, "Ibn Sīnā (Avicenne): un projet 'religieux' de philosophie?"

37. Al-Ghazālī, *Faḍāʾiḥ al-bāṭiniyya*, 146–51; idem, *Fayṣal al-tafriqa*, 187.13–18 / 48.1–8; idem, *al-Munqidh*, 20.15–16; Goldziher, *Streitschrift*, 67–69.

38. Frank, *Al-Ghazali and the Ashʿarite School*, 76–77.

39. Al-Ghazālī, *Fayṣal al-tafriqa*, 127.10–12 / 13.10–14.1.

40. Al-Ghazālī, *Fayṣal al-tafriqa*, 134.4–7 / 25.6–9. In the translation of *sidq, kidhb*, and its derivatives, I follow the analysis of Smith, "Faith as Taṣdīq." For *kidhb* and *takdhīb*, see also *Wörterbuch der klassischen arabischen Sprache*, 1:90–95.

41. See my comments in the introduction to my German translation of the *Fayṣal*, *Über Rechtgläubigkeit und religiöse Toleranz*, 36.

42. Al-Ghazālī, *Iḥyāʾ*, 1:54.9–10 / 62.20.

43. Al-Ghazālī, *Fayṣal al-tafriqa*, 179.20 / 34.8–9. For the *ḥadīth*, see al-Bukhārī, *al-Ṣaḥīḥ, mawāqīt al-ṣalāt*, 11, *iʿtiṣam bi-l-kitāb wa-l-sunna* 3; and Wensinck, *Concordance et indices*, 4:171a.

44. Corbin, *Avicenne et le récit visionaire*, 1:33.

45. Al-Ghazālī, *Fayṣal al-tafriqa*, 182.4 / 36.6–7.

46. Ibid., 175–83 / 27–39.

47. Jackson, *On the Boundaries of Theological Tolerance in Islam*, 49–55; Heath, "Reading al-Ghazālī: The Case of Psychology"; Whittingham, *Al-Ghazālī and the Qurʾān*, 24–27; Kemal, *Philosophical Poetics*, 197–214; Griffel, "Al-Ġazālī's Concept of Prophecy," 121–35; and idem, *Apostasie und Toleranz*, 320–35.

48. Griffel, "Al-Ġazālī's Concept of Prophecy," 129–33.

49. Al-Ghazālī, *Fayṣal al-tafriqa*, 184.2–3 / 41.3–4.

50. Ibid., 184.4–6 / 41.4–7.

51. Ibid., 195.10–16 / 61–62. Dunyā's edition and MS Istanbul, Şehit Ali Paşa 1712, fol. 66a, have *uṣūl al-qawāʿid* instead of *uṣūl al-ʿaqāʾid*.

CHAPTER 4

1. Al-Ghazālī, *Fayṣal al-tafriqa*, 187.2–4 / 47.3–6.

2. Ibid., 184.12–20 / 41.12–43.3.

3. Ibid., 187.5–7 / 47.6–9.

4. Ibid., 187.8 / 47.9–10.

5. Al-Ghazālī, *Faḍāʾiḥ al-bāṭiniyya*, 155.12–14.

6. See, for instance, Marmura, "Al-Ghazali on Bodily Resurrection," 49; idem, "Ghazali's Attitude to the Secular Sciences and Logic," 101; and idem, "Ghazalian Causes and Intermediaries," 91.

7. Al-Ghazālī, *al-Iqtiṣād*, 121–24.

8. *bi-ṭṭirāḥ al-ʿālamayn*; al-Ghazālī, *Mishkāt al-anwār*, 73.9–11 / 161.2–4; and idem, *bi-ttirāḥ al-kawnayn aʿnī l-dunyā wa-l-ākhira*, 70.10–11 / 157.11–12; cf. idem, *Fayṣal al-tafriqa*, 191.5–6 / 55.3–4.

9. The same is true for the other "existences"; once a level of existence is acknowledged, "it includes what comes after it" (al-Ghazālī, *Fayṣal al-tafriqa*, 187.6 / 47.7). Sayeed Rahman made this point in his paper "Are There Two Methods of Interpretation (taʾwīl) in al-Ghazālī's *Fayṣal al-tafriqa* and in his *Mishkāt al-anwār*?" presentated at the annual meeting of the Middle East Studies Association, Boston, November 23–26, 2006. He thus rejects Goldziher's accusation about assumed inconsistencies in al-Ghazālī's

method of Qur'an interpretation (*Richtungen*, 197–207). On the relationship between these two types of Qur'an interpretation in al-Ghazālī, see also Heer, "Abū Ḥāmid al-Ghazālī's Esoteric Exegesis."

10. Al-Ghazālī, *Iḥyā'*, 4:29.16–17 / 2114.6.

11. Ibid., 1:71.5–12 / 73.3–10.

12. Al-Ghazālī reports Miskawayh's theory of the "truthful dream" (*al-manām al-ṣādiq*) as a part of prophecy in MS London, Or. 3126, fol. 253b–55a. Cf. Miskawayh, *al-Fawz*, 133–35. See also Griffel, "MS London, British Library Or. 3126," 19. On Miskawayh's teachings on the soul and his partly reliance on al-Kindī, see Adamson, "Miskawayh's Psychology."

13. On the difference between allegories and symbols with regard to this kind of literature, see Corbin, *Avicenne et le récit visionaire*, 1:34–35.

14. Al-Ghazālī, *Iḥyā'*, 4:29.20–24 / 2114.9–14. See Nakamura, "Ghazālī's Cosmology Reconsidered," 34–35.

15. Al-Ghazālī, *Iḥyā'*, 1:12.24–26 / 5.8–11. This notion is fully developed in al-Ghazālī's *Iljām al-ʿawāmm*, see below, p. 267.

16. Al-Ghazālī, *Iḥyā'*, 1:23.25–24.5 / 23.3–11.

17. Al-Ghazālī, *Faḍā'iḥ al-bāṭiniyya*, 155.14–15.

18. Al-Ghazālī, *Tahāfut*, 356.5–7 / 215.1–2

19. Al-Ghazālī, *Fayṣal al-tafriqa*, 191.19–192.5 / 56.6–57.2.

20. Here I wish to correct my own comments in Griffel, *Apostasie und Toleranz*, 300, n. 24. Jules Janssens rightfully criticizes them in his review in the *Journal of Islamic Studies* 14 (2003): 71.

21. Al-Ghazālī, *Fayṣal al-tafriqa*, 187.8 / 47.9–10, 188.10–11 / 49.8–10.

22. Ibid., 189.7 / 51.2. On the *Miḥakk al-naẓar* and its program, see Frank, *Al-Ghazālī and the Ashʿarite School*, 94–95. On al-Ghazālī's application of Aristotelian logics, see Marmura, "Ghazali and Demonstrative Science," 183–84; idem, "Al-Ghazali's Attitude to the Secular Sciences and Logic," 101–6; idem, "Ghazālī on Ethical Premises"; and Rudolph, "Die Neubewertung der Logik durch al-Ġazālī."

23. Gwynne, *Logic, Rhetoric, and Legal Reasoning in the Qur'an*, 152–89, 203–4; Kleinknecht, "Al-Qisṭās Al-Mustaqīm: Eine Ableitung der Logik aus dem Koran," 167–76.

24. Al-Ghazālī, *Fayṣal al-tafriqa*, 188.13–16 / 49.14–50.3; idem, *al-Munqidh*, 31.3–9.

25. The logical part at the beginning of *al-Mustaṣfā* is essentially an epitome of al-Ghazālī's own earlier textbook of logics, *Miḥakk al-naẓar*.

26. Fakhr al-Dīn al-Rāzī, *al-Munāẓarāt fī bilād mā wara'a l-nahr*, 45–46.

27. On ʿUmar ibn ʿAlī ibn Ghaylān al-Balkhī, see al-Bayhaqī, *Tatimmat Ṣiwān al-ḥikma*, 128; Shihadeh, "From al-Ghazālī to al-Rāzī," 151–53; Michot, "La pandémie Avicennienne," 287–97; and Michot's French introduction as well as Mahdī Muḥaqqiq's Persian introduction to the edition of Ibn Ghaylān's *Ḥudūth al-ʿālam*.

28. See the ninth through eleventh discussions in Fakhr al-Dīn al-Rāzī, *al-Munāẓarāt fī bilād mā wara'a l-nahr*. The date of their meeting can be deduced from the great astronomical conjunction of 29 Jumāda II 582 / 14 September 1186, mentioned on p. 32.5–6. On Muḥammad ibn Masʿūd al-Masʿūdī, see *GAL*, 1:474 (only in the first edition), *Suppl.* 1:817; Rescher, *Development of Arabic Logic*, 176; and Shihadeh, "From al-Ghazālī to al-Rāzī," 153–58.

29. Al-Masʿūdī. *al-Shukūk wa-l-shubah ʿalā l-Ishārāt*.

30. Ibn Ghaylān, *Ḥudūth al-ʿālam*, 11.15–20; cf. ibid., 8.3–15, in which he paraphrases several passages in the first preface of al-Ghazālī's *Tahāfut*.

31. The events are hinted at in Ibn Ghaylān, *Ḥudūth al-ʿālam*, 13.16–17.

32. Ibid., 14–47. It is meant as a refutation of Ibn Sīnā's *Risāla al-Ḥukūma fī-l-ḥujaj al-muthbitīn li-l-māḍī mabdaʾ an zamaniyyan*.

33. Fakhr al-Dīn al-Rāzī, *al-Munāẓarāt fī bilād mā waraʾa l-nahr*, 60.4–5.

34. Ibid., 61.1–2.

35. Ibid., 60.12–13.

36. Griffel, *Apostasie und Toleranz*, 449–60.

37. Ibn Rushd, *Faṣl al-maqāl*, 21.1–4.

38. Ibid., 16.18–19.

39. Fakhr al-Dīn al-Rāzī, *al-Maṭālib al-ʿāliya*, 4:29–33; İskenderoglu, *Fakhr al-Dīn al-Rāzī and Thomas Aquinas*, 69–73.

40. Al-Ghazālī, *Fayṣal al-tafriqa*, 195.10–16 / 61–62. MS Berlin, Wetzstein II 1806, fol. 79b, has probably the correct text of this passage when it describes the three *uṣūl al-īmān* as being, "*al-īmān bi-Llāh bi-l-waḥy li-rasūlin wa-bi-l-yawm al-ākhir.*"

41. Ibid., 191–92 / 56–57.

42. Al-Ghazālī, *Tahāfut*, 376–77 / 226; idem, *al-Iqtiṣād*, 249.*ult*.–250.5; idem, *al-Munqidh*, 24.6–7; see Griffel, *Apostasie und Toleranz*, 277–79.

43. Ibn Rushd, *Faṣl al-maqāl*, 15.13–17.3; 21.11–14. Ibn Rushd mistakenly believed that al-Ghazālī does not allow a judgment of *kufr* in cases in which the consensus of Muslim scholars—but not the outward sense of revelation—is violated. On al-Ghazālī's position regarding this questions (expressed in *Fayṣal al-tafriqa*, 200.6–15 / 71.8–72.3) and Ibn Rushd's mistaken report in his legal works and in the *Faṣl al-maqāl*, see Griffel, *Apostasie und Toleranz*, 430–31, 449–50.

44. Ibn Ghaylān, *Ḥudūth al-ʿālam*. 12.20; see also 14.4.

45. Ibid., 8.3–9.5. For Ibn Ghaylān's taste, however, al-Ghazālī was far too lax and selective toward the many errors of the *falāsifa* when he accepted some of their teachings as true.

46. Al-Ghazālī, *Tahāfut*, 78.3–4 / 46.8–9, explicitly says that he will not bring arguments in favor of the world's creation in time, "as our purpose is to refute their claim that they have knowledge of [its] pre-eternity."

47. Al-Ghazālī, *al-Qisṭās al-mustaqīm*, 41.12–14, 99.8–11; see Kleinknecht, "Al-Qisṭās Al-Mustaqīm: Eine Ableitung der Logik aus dem Koran," 160–61.

48. Al-Ghazālī, *Fayṣal al-tafriqa*, 188.13 / 49.12–13.

49. Al-Ghazali, *Mishkāt al-anwār*, 47.12–15 / 127.10–13. Al-Ghazālī recommends to his readers that they should learn the correct way of pursuing the *ʿaql* from his textbooks *Miʿyār al-ʿilm* and *Miḥakk al-naẓar*.

50. Al-Ghazālī, *Fayṣal al-tafriqa*, 188.10–17 / 49.8–50.3.

51. Al-Ghazālī, *Ḥimāqat-i ahl-i ibāḥat*, 1.5–8 / 153.3–5; 4.2 / 157.4.

52. Al-Ghazālī, *al-Munqidh*, 25–27.

53. Ibid., 26.9–11.

54. Ibid., 27.1–2.

55. Al-Ghazālī, *Fayṣal*, 204.11–12 / 79.7–8.

56. Frank, "Al-Ghazālī on *Taqlīd*," 215–17. Here al-Ghazālī departs from attitudes held by earlier Ashʿarites. Their attitude toward the belief of the masses changes roughly a generation before al-Ghazālī as a result of the Ashʿarites' persecution in Khorasan; see Griffel, *Apostasie und Toleranz*, 200–215. This change prompted al-Ghazālī's revision of the criteria for what counts as *īmān* and what counts as *ʿilm*.

57. Here al-Ghazālī mirrors the attitude of earlier Ashʿarites; see Frank, "Knowledge and *Taqlīd*."

58. Al-Ghazālī, *Iḥyāʾ*, 1:110.6 / 134.1–2.

59. Al-Ghazālī, *Faḍāʾiḥ al-bāṭiniyya*, 17, 73–131; idem, *al-Munqidh*, 29.10–17; Goldziher, *Streitschrift*, 5–6, 38, 52–60. The text of Badawī's edition of the *Faḍāʾiḥ al-bāṭiniyya* should be compared with the quotations from that book in its refutation by the Yemenite Ṭayyibī-Ismāʿīlite *dāʿī muṭlaq* Ibn al-Walīd (d. 612/1215), *Dāmigh al-bāṭil wa-ḥatf al-munādil*. On this book and its author, see Corbin, "The Ismāʿīlī Response to the Polemic of Ghazālī"; Poonawala, *Biobibliography*, 156–61; and Brockelmann, *GAL, Suppl.* 1:715.

60. Al-Ghazālī, *Tahāfut*, 4.3–5.8 / 1.11–2.15; see Frank, "Al-Ghazālī on Taqlīd"; Lazarus-Yafeh, *Studies*, 488–502; and Griffel, "Taqlīd of the Philosophers."

61. Al-Ghazālī, *Tahāfut*, 4.3–4 / 1.11–12. Griffel, "*Taqlīd* of the Philosophers," 282–88.

62. Al-Ghazālī, *Tahāfut*, 13.9–10 / 7.17–18: "Let it be known that (our) objective is to alert those who think well of the philosophers and believe that their ways are free from contradictions (. . .)."

63. Al-Ghazālī, *Ḥimāqat-i ahl-i ibāḥat*, 7.16–17 / 166.8–10 (following Pretzl's and not Pūrjavādī's text); idem, *al-Munqidh*, 29.17–*ult.*; idem, *Fayṣal al-tafriqa*, 133–34 / 22–23. See Goldziher, *Streitschrift*, 19–20. The true prophet is immune from error (*maʿṣūm*).

64. In the century after al-Ghazālī's death, this position is best exemplified by Maimonides and Fakhr al-Dīn al-Rāzī.

CHAPTER 5

1. For attempts in Islamic scholarship to harmonize these Qur'anic narratives with those that appear in the prophetical *ḥadīth*, see Heinen, *Islamic Cosmology*, 61–110.

2. Al-Dhahabī, *Siyar aʿlām al-nubalāʾ*, 15:89; Ibn Khallikān, 4:267–68; and al-Subkī, *Ṭabaqāt*, 3:356–57, relate that after he became detached from Muʿtazilism, al-Ashʿarī confronted his former Muʿtazilite teacher, Abū ʿAlī al-Jubbāʾī (d. 303/915–16), with the "story of the three brothers." It ends in the imagined outcry of one of the three, who led a wicked life, and asks God why He did not let him die early in his life and spare him punishment in the afterlife? Al-Ghazālī tells the same story in his *Iḥyāʾ*, 1:153 / 196–97, and in his *al-Iqtiṣād*, 184–85, without any suggestion that it goes back to al-Ashʿarī. Gwynne, "Al-Jubbāʾī, al-Ashʿarī and the Three Brothers," argues that Fakhr al-Dīn al-Rāzī (d. 606/1210) in his *Tafsīr al-kabīr*, 8:185–86, was probably the first to link this story to al-Ashʿarī. On the story of the three brothers, see also Gardet/Anawati, *Introduction à la théologie musulmane*, 53; and Watt, *Formative Period*, 305.

3. Frank, "The Structure of Created Causality," 20.

4. Ibid., 21, 29.

5. Dhanani, *The Physical Theory of Kalām*; van Ess, *Theologie und Gesellschaft*, 3:224–29, 309–335, 4:450–77; Rudolph/Perler, *Occasionalismus*, 28–51.

6. Gerhard Böwering, Art "Zeit. Islam," in *HWdP*, 12:1223.

7. Rudolph/Perler, *Occasionalismus*, 51–56; Gimaret, *La doctrine d'al-Ashʿarī*, 43–130.

8. Ibn Fūrak, *Mujarrad maqālāt al-Ashʿarī*, 283.17–18.

9. Ibid., 131.7–8.

10. Ibid., 132.23–133.2.

11. See van Ess, *Theologie und Gesellschaft*, 3:116–17, 249; 4:486–88. On Qāḍī ʿAbd al-Jabbār's (d. 415/1025) usage of *khalaqa* in this respect, see Frank, *Al-Ghazālī and the Ashʿarite School*, 44.

12. Ibn Fūrak, *Mujarrad maqālāt al-Ashʿarī*, 131.16–132.6; Gimaret, *La doctrine d'al-Ashʿarī*, 403–9. McGinnis, "Occasionalism, Natural Causation and Science," 445, develops the Ashʿarite argument against the existence of natures in a philosophical language.

13. Ibn Fūrak, *Mujarrad maqālāt al-Ashʿarī*, 134.5–8.

14. Ibid., 176.17; Gimaret, *La doctrine d'al-Ashʿarī*, 459–63.

15. Al-Baghdādī, *Uṣūl al-dīn*, 69; al-Bāqillānī, *al-Tamhīd*, 34–47, 286–87, 300–301; al-Isfarāʾīnī, "al-ʿAqīda," 146.2–4; Gimaret, *La doctrine de al-Ashʿarī*, 408–9; cf. also al-Juwaynī's discussion of *tawallud* in his *al-Shāmil* (ed. Alexandria), 503–6. See Bernand, "La critique de la notion de nature (ṭabʿ) par le kalām"; and Perler/Rudloph, *Occasionalismus*, 60.

16. This is the impression Maimonides (d. 601/1204) gives in his influential report of the occasionalist teachings of the *mutakallimūn* in chapter 73 of the first part of his *Dalālat al-ḥāʾirīn*, 140–41, English translation 1:201–2. Courtenay, "The Critique on Natural Causality," 81–82, reminds us that the occasionalist radicalism of the *mutakallimūn* has often been assumed rather than established. Courtenay points to the significant influence of Maimonides's unsympathetic report of the *mutakallimūn's* occasionalism in the West. On Maimonides' report, see Fakhry, *Islamic Occasionalism*, 25–32, and—pointing out its shortcomings—Rudolph in Perler/Rudolph, *Occasionalismus*, 112–24.

17. We will see that this is also al-Ghazālī's main point against the *falāsifa*.

18. Frank, "The Structure of Created Causality," 30.

19. Ibid., 25–26, 40–41.

20. *waqaʿa bi-qudrati muḥdatha;* Ibn Fūrat, *Mujarrad maqālāt al-Ashʿarī*, 92.6; see also Gimaret, *La doctrine d'al-Ashʿarī*, 390–93.

21. Gimaret, *Théories de l'acte humain*, 92–120.

22. Al-Anṣārī, *al-Ghunya*, fol. 120a.18–19, see below n. 28.

23. Al-Juwaynī, *al-Irshād*, 210.3.

24. Ibid., 210.4–7; see also ibid. 203–25, 230–34. Cf. Gimaret, *Théories de l'acte humain*, 121–22.

25. Al-Juwaynī, *al-ʿAqīda al-Niẓāmiyya*, 30.

26. Humans have no power over the perception of colors. The demand to produce actions while not being capable of it would be like the demand to produce the perception of colors. God makes no such demands. In al-Juwaynī, *al-Irshād*, 203, this is quoted as an argument of his Muʿtazilite adversaries.

27. Al-Juwaynī, *al-ʿAqīda al-Niẓāmiyya*, 32.6–9; also translated in Gimaret, *Théories de l'acte humain*, 122. See also Nagel, *Die Festung des Glaubens*, 227–28, who sees no difference between al-Juwaynī's teachings in *al-ʿAqīda al-Niẓāmiyya* and those in his legal work *Kitāb al-Burhān*.

28. *anna al-qudrata al-ḥādithata lā tuʾaththiru fī maqdūrihā wa-lam yaqaʿ al-maqdūra wa-lā ṣifatan min ṣifātihā;* al-Anṣārī, *al-Ghunya*, fol. 120a.18–19.

29. Al-Juwaynī, *al-ʿAqīda al-Niẓāmiyya*, 32.17, 35.6–7; Gimaret, *Théories de l'acte humain*, 123.

30. Al-Juwaynī, *al-ʿAqīda al-Nizāmiyya*, 34.3–35.5; cf. the translation in Gimaret, *Théories de l'acte humain*, 123.

31. al-Juwaynī, *al-ʿAqīda al-Nizāmiyya*, 32.11.

32. Ibid., 33.13–15.

33. Ibid., 35.9.

34. *al-ḥādithātu kulluhā murādatun li-Llāhi taʿālā;* ibid., 27.9–10.

35. Ibid., 35.10–*paenult.*

36. Ibid., 36.3–4; cf. Nagel, *Die Festung des Glaubens*, 228.

37. Al-Juwaynī, *al-ʿAqīda al-Niẓāmiyya*, 36.5–9.

38. Fakhr al-Dīn al-Rāzī, *al-Tafsīr al-kabīr*, 4:88.5–9 (*ad* Q 2:134); Gimaret, *Théories de l'acte humain*, 124.

39. *inna l-insāna muḍṭarrun fī ṣūrati mukhtār*; Fakhr al-Dīn al-Rāzī, *al-Maṭālib al-ʿāliya*, 9:25.21; 9:57.6–12; idem, *Muḥaṣṣal*, 459.3–4. The sentence goes back to Ibn Sīnā, *al-Taʿlīqāt*, 51.17–18 / 296.7; 53.20 / 108.9. On these earlier appearance of this sentence and a somewhat similar one in the *Rasāʾil Ikhwān al-ṣafāʾ*, 3:294.2–3 / 3:306.22–23, see Michot's introduction to Ibn Sīnā, *Refutation de l'Astrologie*, 69*–71*. On Fakhr al-Dīn al-Rāzī's theory of human actions see Shihadeh, *Teleological Ethics*, 13–44.

40. Gimaret, *Théories de l'acte humain*, 79–128; Frank, "The Structure of Created Causality."

41. The above quotation on p. 129 indicates that he also assumed that "knowledge" (*ʿilm*) causes the human to be knowledgeable (*ʿālim*). See al-Juwaynī, *al-Shāmil* (ed. Alexandria), 302; cf. Nagel, *Die Festung des Glaubens*, 140.

42. Al-Shahrastānī, *al-Milal wa-l-niḥal*, 70.*peanult.*–71.3; idem, *Livre de religions*, 1:327–28.

43. Al-Shahrastānī, *al-Milal wa-l-niḥal*, 71.5–6; idem, *Livre de religions*, 1:328.

44. See above p. 131 (= al-Juwaynī, *al-ʿAqīda al-Nizāmiyya*, 35.10). The theory of "motives" goes back to the Muʿtazilite Abū l-Ḥusayn al-Baṣrī; cf. Madelung, "The Late Muʿtazila and Determinism."

45. Al-Juwaynī, *al-Irshād*, 211.5–11.

46. Al-Juwaynī, *al-ʿAqīda al-Niẓāmiyya*, 38.5–7; cf. Gimaret, *Théories de l'acte humain*, 126.

47. Al-Juwaynī, *al-ʿAqīda al-Niẓāmiyya*, 25–26.

48. Ibn Khaldūn, *al-Muqaddima*, 3:34–35, English translation 3:51–52. Cf. Gardet/Anawati, *Introduction à la théologie musulmane*, 72–76.

49. Wisnovsky, *Avicenna's Metaphysics in Context*, 266. Endress, "Reading Avicenna in the Madrasa," 379, highlights the difference between al-Fārābī and Ibn Sīnā and describes the project of the latter as "set[ting] out to develop philosophy (. . .) as a metaphor of religious knowledge."

50. See, for instance, his report of the philosophers' cosmology in al-Juwaynī, *Irshād*, 234–35, and his comprehensive discussion in *al-Shāmil* (ed. Alexandria), 229–42. The latter passage quotes from an even earlier discussion of philosophical teachings by Abū Isḥaq al-Isfarāʾīnī, which is otherwise lost.

51. Until recently, Abū l-Ḥusayn al-Baṣrī's views on theology were largely unknown. The extant parts of one of his works on theology, *Taṣaffuḥ al-adilla*, have only recently been edited. His follower Ibn al-Malāḥimī (d. 536/1141) reports many of his teachings in his *Kitāb al-Muʿtamad fī uṣūl al-dīn*. On Abū l-Ḥusayn al-Baṣrī's teachings, see Madelung, "Abū l-Ḥusayn al-Baṣrī's Proof for the Existence of God"; idem, "The Late Muʿtazila and Determinism: The Philosopher's Trap"; McDermott, "Abū l-Ḥuṣayn al-Baṣrī on God's Volition"; and Heemskerk, *Suffering in the Muʿtazilite Theology*, 57–59. See also Gimaret's article on him in *EIran*, 1:322–24, and Madelung's article on him in *EI3*.

52. Studies on the innovative aspects of al-Juwaynī's theology are few and far between. See Rudolph, "La preuve de l'existence de dieu chez Avicenne et dans la théologie musulmane"; Gardet/Anawati, *Introduction à la théologie musulmane*, 73–74; Davidson, *Proofs for Eternity*, 187–88, and index; Gimaret, *Théories de l'acte humain*, 120–28; Frank, *Creation and the Cosmic System*, 17–18.

53. Ibn Sīnā, *al-Shifāʾ, al-Samāʿal-ṭabīʿī*, 48.10. Marmura, "The Metaphysics of Efficient Causality," 178–80, deals with the way Ibn Sīnā proves this position.

54. On the importance of causality in Ibn Sīnā, see Bertolacci, "The Doctrine of Material and Formal Causality"; and Wisnovsky, "Final and Efficient Causality in Avicenna's Cosmology and Theology."

55. Wisnovsky, *Avicenna's Metaphysic in Context*, 15; idem, "Final and Efficient Causality," 98.

56. This example is discussed by al-Ghazālī, *Miʿyār al-ʿilm*, 176.5–8.

57. Aristotle, *Physics*, 198a.14–198b.9; idem, *Analytica posteriora*, 94a.20–23. Cf. Johannes Hübner, art. "Ursache/Wirkung," in *HWdP*, 11:377–84. The Arabic terminology reflects the usage of Ibn Sīnā, *al-Shifāʾ*, *al-Ilāhiyyāt*, 194.9.

58. Aristotle, *Metaphysics*, 1050a 21ff.; and idem, *De anima*, 414a 16–17.

59. Wisnovsky, *Avicenna's Metaphysics in Context*, 21–141; Marmura, "The Metaphysics of Efficient Causality."

60. Ibn Sīnā *al-Shifāʾ*, *al-Ilāhiyyāt*, 127.17–128.2; Marmura, "Avicenna on Causal Priority."

61. Ibn Sīnā, *al-Shifāʾ*, *al-Ilāhiyyāt*, 196.14.

62. Ibid., 194.12; Marmura, "The Metaphysics of Efficient Causality," 173–75.

63. Ibn Sīnā, *al-Shifāʾ*, *al-Ilāhiyyāt*, 126.11–15. On Avicenna's notion of essential causality, see Marmura, "The Metaphysics of Efficient Causality," 176–77, 180–81; idem, "Ghazali and Demonstrative Science," 184–86; and idem, "Avicenna on Causal Priority," 67–68.

64. Ibn Sīnā, *al-Shifāʾ*, *al-Ilāhiyyāt*, 31.1–32.3; idem, *al-Najāt*, 225.15–226.5 / 547.12–548.7; McGinnis, "Occasionalism, Natural Causation and Science," 443.

65. Aristotle, *Metaphysics*, 1046a.19–29. On the impact the distinction of active and passive power has on early Muslim theology, see Schöck, "Möglichkeit und Wirklichkeit menschlichen Handels."

66. Ibn Sīnā, *al-Najāt*, 225.5–9 / 547.1–5. The passage is translated in Hourani, "Ibn Sina on Necessary and Possible Existence," 79; and McGinnis, "Occasionalism, Natural Causation and Science," 444. Al-Ghazālī reports this argument in *Tahāfut*, 282.8–283.2 / 169.6–12.

67. Since it moves around the earth once during the (24 hour) day it is also known as the diurnal sphere. The next sphere, that is that of the fixed stars, moves with the speed of one rotation per day minus one rotation in 25,700 years (although the Arab astrologers believed this figure to be in the range of 23,000 years), and the next lower sphere of Saturn moves with the speed of one rotation per day minus one rotation in twenty-nine years, Jupiter with one rotation per year minus one in twelve years, and the sun, for instance, which is situated further below, with the speed of one rotation per day minus one rotation in a year.

68. Al-Fārābī, *Mabādiʾ ārāʾ ahl al-madīna al-fāḍila*, 38.2–3; idem, *al-Siyāsa al-madaniyya*, 31.12.

69. Strictly speaking, the "secondary causes" in al-Fārābī are just the nine celestial intellects above the active intellect; see al-Fārābī, *al-Siyāsa al-madaniyya*, 31–32, 52. In Ibn Sīnā, *al-Shifāʾ*, *al-Ilāhiyyāt*, 360.11–13, the secondary causes are those in the sublunar sphere, whereas the primary ones are the intermediaries (*wasāʾiṭ*) in the heavens.

70. Al-Fārābī, *Mabādiʾ ārāʾ ahl al-madīna al-fāḍila*, 101–5; idem, *al-Siyāsa al-madaniyya*, 31–38. For an analysis of this latter passage, see Druart, "Al-Fārābī's Causation of the Heavenly Bodies"; and Reisman "Al-Fārābī and the Philosophical Curriculum," 56–60.

71. Al-Fārābī, *Mabādiʾ ārāʾ ahl al-madīna al-fāḍila*, 38.8–9.

72. Hasnawi, "Fayḍ," 967–70. On the number of spheres in Ibn Sīnā, see also Janssens, "Creation and Emanation in Ibn Sīnā," 455.

73. Ibn Sīnā, *Risāla Fī sirr al-qadar*, in Hourani, "Ibn Sina's 'Essay on the Secret of Destiny,'" 28.12–14, 31; and in ʿĀṣi, *al-Tafsīr al-Qurʾānī*, 302.13–303.1. The two editions are based on two different manuscripts that the editors compare to the text in an early print. Reisman, *The Making of the Avicennan Tradition*, 140, suggests that *Risāla Fī sirr al-qadar*

was not authored by Ibn Sīnā. He bases his doubts on a "confused argumentation." Reisman alerts us to the fact that some of the smaller works ascribed to Ibn Sīnā may have indeed generated in a Ghazalian intellectual milieu during the sixth/twelfth century. *Risāla Fi sirr al-qadar*, however, seems genuine. The apparent confusion in this epistle results from the difficulty in Ibn Sīnā's works of reconciling human free will with a necessitarian cosmology (see, e.g., Marmura, "Divine Omniscience," 91; or Janssens, "The Problem of Human Freedom in Ibn Sînâ.") Like Al-Ghazālī, Ibn Sīnā avoided being outspoken about the predetermination of all future events and here, like in other of his writings, kept his language elliptic. On that strategy see Gutas, *Avicenna and the Aristotelian Tradition*, 225–34.

74. Marmura, "Some Aspects of Avicenna's Theory of God's Knowledge of Particulars"; idem, "Divine Omniscience," 88–92; Ivry, "Destiny Revisited," 165–68.

75. Belo, *Chance and Determinism in Avicenna and Averroes*, 91–120.

76. Aristotle, *De interpretatione*, 18b.18–25. On this passage and the two major directions of interpretation of why we cannot say which it is, see Hintikka, *Time & Necessity*, 147–78; and Adamson, "The Arabic Sea Battle," 164–67.

77. Al-Fārābī, *Sharḥ al-Fārābī li-Kitāb Arisṭutālīs fī l-ʿIbāra*, 83.13–15; in his English translation, Zimmermann, *Al-Farabi's Commentary*, 76–77, corrects the Arabic text. Cf. Adamson, "The Arabic Sea Battle," 169.

78. Al-Fārābī,*Sharḥ al-Fārābī li-Kitāb Arisṭutālīs fī l-ʿIbāra*, 98.14–19; English translation in Zimmermann, *Al-Farabi's Commentary*, 93.

79. Al-Fārābī,*Sharḥ al-Fārābī li-Kitāb Arisṭutālīs fī l-ʿIbāra*, 98.3–8; English translation in Zimmermann, *Al-Farabi's Commentary*, 92–93.

80. Al-Fārābī,*Sharḥ al-Fārābī li-Kitāb Arisṭutālīs fī l-ʿIbāra*, 99.1–100.13; English translation in Zimmermann, *Al-Farabi's Commentary*, 94–95. Adamson, "The Arabic Sea Battle," 183.

81. There are numerous attempts to interpret what al-Fārābī truly means to say in this passage; see, for example, Marmura, "Divine Omniscience," 84–86; Kogan, "Some Reflections," 96; Leaman, "God's Knowledge of the Future," 25–26; Terkan, "Does Zayd Have the Power Not to Travel Tomorrow"; Wisnovsky, *Avicenna's Metaphysics in Context*, 219–25; and Adamson, "The Arabic Sea Battle," 180–86. For a more complete bibliography on the problem of future contingencies in al-Fārābī, see Adamson's article.

82. See Wisnovsky, *Avicenna Metaphysics in Context*, 219–25; and Kukkonen, "Causality and Cosmology," 39–41.

83. Ivry, "Destiny Revisited"; and Janssens, "The Problem of Human Freedom in Ibn Sînâ," argue that according to Ibn Sīnā, some events in the sublunar world are haphazard and thus not fully determined by God. Goichon, *La distinction*, 162–63; and Michot, *La destinée de l'homme*, 61–64, have argued that there is no contingency in Ibn Sīnā's fully determined system of secondary causes. Marmura, "Divine Omniscience," 91, acknowledges that it remains difficult in Ibn Sīnā's philosophy to reconcile "some of his statements that seem to affirm man's freedom of the will with his necessitarian metaphysics." Belo, *Chance and Determinism*, 55–89, discusses Ibn Sīnā's teachings on this subject and particularly supports Michot's results that, for Ibn Sīnā, all events in the sublunar sphere are fully determined by God.

84. Adamson, "On Knowledge of Particulars," 284–92; Marmura, "Divine Omniscience," 89–91.

85. Ivry, "Destiny Revisited," 166–67; Marmura, "Some Aspects of Avicenna's Theory," 300; and idem, "Divine Omniscience," 81, observe that Ibn Sīnā does not introduce divine foreknowledge "in any precise fashion in his metaphysical writings." For *muṣādamāt* and *taṣādum*, see, for example, Ibn Sīnā, *al-Shifāʾ, al-Ilāhiyyāt*, 359.8–10,

360.11. For how these collisions are still the outcome of a fully determined system, see Belo, *Chance and Determinism*, 110–13; and Ibn Sīnā, *al-Taʿlīqāt*, 131.11–14 / 439.6–10.

86. Ibn Sīnā, *al-Shifāʾ*, *al-Ilāhiyyāt*, 363.4–5.

87. Ibid., 359.18–360.3; idem, *Aḥwāl al-nafs*, 114–21.

88. God's knowledge remains the same before, during, or after the event. Ibn Sīnā, *al-Shifāʾ*, *al-Ilāhiyyāt*, 288–90; idem, *al-Ishārāt wa-l-tanbīhāt*, 182–83; idem, *al-Ḥikma al-ʿarshiyya*, 9.7–15. See Marmura, "Some Aspects of Avicenna's Theory," 301–6; and idem, "Divine Omniscience," 88–89.

89. *ʿalā naḥwin kulliyin*; Ibn Sīnā, *al-Shifāʾ*, *al-Ilāhiyyāt*, 360.13–14.

90. Frank, *Creation and the Cosmic System*, 49, suggests something similar as the meaning of *ikhtiyār* when used by al-Ghazālī, whom he thought was suffering from parallel problems about God's free choice. Certain passages in Ibn Sīnā—for example, *al-Taʿlīqāt*, 51.22–23 / 296.12–15—would support that interpretation. Note also that the term *ikhtiyār* is etymologically related to *khayr* and that God, according to Ibn Sīnā, always creates the best (*al-khayr*) for His creation (the connection between these two words is stressed in *al-Taʿlīqāt*, 50.28–*ult.* / 295.2–4). In his *al-Shifāʾ*, *al-Ilāhiyyāt*, 312.16–18 (=*al-Najāt*, 262.21–23 / 627.4–6), Ibn Sīnā defines *ikhtiyār* as "the intellect's pursuit of what is truly and purely the best." On Ibn Sīnā's use of *ikhtiyār*, see also Goichon, *Lexique de la langue philosophique d'Ibn Sīnā*, 115–16. On Ibn Sīnā applying *ikhtiyār* to God, see his *al-Taʿlīqāt*, 53.22–23 / 108.12–13, in which God is described as the only being who has *ikhtiyār*: "Actions that involve *ikhtiyār* apply in reality to none but the First alone." On these passages from Ibn Sīnā's *al-Taʿlīqāt*, see also the French translations in Michot's introduction to Ibn Sīnā, *Réfutation de l'Astrologie*, 69*–71*.

91. Ibn Sīnā, *al-Ishārāt wa-t-tanbīhāt*, 185.11–13.

92. Ibid., 185.13–16.

93. See, for example, ibid., 185.13–16; or Ibn Sīnā, *Dānishnāmeh-yi ʿAlāʾ-i. Ilāhiyyāt*, 96.1.

94. Ibn Sīnā, *Dānishnāmeh-yi ʿAlāʾ-i. Ilāhiyyāt*, 93.

95. Al-Ghazālī, *Maqāṣid al-falāsifa*, 2:81.9–11 / 235.5–8.

96. Ibn Sīnā, *al-Najāt*, 228.17 / 553.9–10. In his report of the metaphysics of the *falāsifa* in MS London 3126, foll. 197b-198a, al-Ghazālī stresses this element of Ibn Sīnā's teachings more than the philosopher himself had stressed it.

97. Ibn Sīnā, *Fawāʾid wa-nukat*, MS Istanbul, Nuruosmaniye 4894, fol. 242b, lines 30–35; see Yahya Michot's French translation in Ibn Sīnā, *Lettre au vizir Abū Saʿd*, 122*. On this short programmatic text by Ibn Sīnā, which should not be confused with the much more extensive *al-Nukat wa-l-fawāʾid* that is often falsely ascribed to Ibn Sīnā, see Mahdavī, *Fihrist-i nuskhat-hā-yi muṣannafāt-i Ibn Sīnā*, 288 (no. 200).

98. Al-Ghazālī, *al-Munqidh*, 28.*ult.*

99. Ibid. 32.12–18; al-Ghazālī, *Faḍāʾiḥ al-bāṭiniyya*, 79–80; Goldziher, *Streitschrift*, 21–22. Cf. Ibn al-Walīd, *Dāmigh al-bāṭil wa-ḥatf al-munādil*, 1:280–81. In the case of the Ismāʿīlites, this element is their denial of rational arguments (*adilla naẓariyya*), without which they cannot uphold their claim to follow the infallible Imam.

100. Al-Ghazālī, MS London, Or. 3126, foll. 121a–171b and 229b–232b; see Griffel, "MS London, British Library Or. 3126: An Unknown Work," 20.

101. The material in Ibn Sīnā, *al-Shifāʾ*, *Ilāhiyyāt*, 194–205, is paraphrased in foll. 121a–134b of the London MS. Al-Ghazālī quotes and paraphrases the Avicennan texts quite freely and often adds what appear to be his own original comments. At one point he switches to the form of questions and answers (*wa-dhukira hādhā bi-maqāla ukhrā ʿalā wajh al-suʾāl wa-l-jawāb*, fol. 126b) where Ibn Sīnā's text is much less lively. Ibn

Sīnā, *al-Shifāʾ*, *Ilāhiyyāt*, 201–35, appears in a more faithful adaptation of the text on foll. 134a–159a of the London MS.

102. Al-Ghazālī, MS London, Or. 3126, foll. 159b–160b; Ibn Sīnā, *al-Shifāʾ*, *Ilāhiyyāt*, 258.

103. Ibn Sīnā, *al-Shifāʾ*, *Ilāhiyyāt*, 257–59. Cf. Davidson, *Proofs for Eternity*, 339–40.

104. Al-Ghazālī, MS London, Or. 3126, foll. 170b–172b; Ibn Sīnā, *al-Shifāʾ*, *Ilāhiyyāt*, 270–73. Cf. Davidson, *Proofs for Eternity*, 340.

105. The report on the finiteness of the efficient and material causes on foll. 159a–170b is, for instance, not from the corresponding passage in Ibn Sīnā, *al-Shifāʾ*, *al-Ilāhiyyāt*, 262–70, but from another source. Its author doesn't use the same terminology as Ibn Sīnā in his *al-Shifāʾ* and calls, for instance the material cause *ʿilla qābiliyya*, whereas in Ibn Sīnā, it is *ʿilla ʿunsuriyya*. (In the Farabian (?) text *al-Daʿāwa l-qalbiyya*, 9.7, the material cause is called *al-qābil*.)

106. Al-Ghazālī, MS London, Or. 3126, fol. 124a.7–12.

107. Ibid., fol. 241a–247a; quoted passage fol. 241b.4–5. This text is taken from al-Fārābī, *al-Siyāsa al-madaniyya*, 31–38.

108. Al-Ghazālī, MS London, Or. 3126, foll. 230b–231b. The report is based on al-Fārābī, *Mabādiʾ arāʾ ahl al-madīna al-fāḍila*, 100–105. The names of the two uppermost spheres, *falak al-aṭlas* and *falak al-burūj* (instead of *kurat al-kawākib*), are added by al-Ghazālī.

109. Al-Ghazālī, *Iḥyāʾ*, 4:146.7–11 / 2272.10–15. See Marmura, "Al-Ghazālī," 151.

CHAPTER 6

1. Emphasis in the original. *Dictionaire des sciences philosophique*, 2:507–8. This passage was later incorporated in Munk, *Mélanges de philosophie juive et arabe*, 377–78.

2. Marmura, "Ghazali's Attitude to the Secular Sciences," 109. For similar views in recent publications, see, for instance, Moosa, *Ghazālī & the Poetics of Imagination*, 184; or Rayan, "Al-Ghazali's Use of the Terms 'Necessity' and 'Habit.' "

3. This is the prophetic miracle that Moses performed in front of Pharao; cf. *Qurʾan* 7.107, 20.69, 26.32, and 45.

4. Performed by Jesus, see Q 3:49 and 5:111.

5. Al-Ghazālī, *Tahāfut al-falāsifa*, 272.1–5 / 163.18–21; 275.10–11 / 165.17–18.

6. The focus on modalities is prompted by Avicenna's work, yet it also has a predecessor in al-Juwaynī's *al-ʿAqīda al-Niẓāmiyya*, 14–29, with its three chapters, "On What Is Impossible for God" (*Kalām fī-mā yastaḥīlu ʿalā Llāh*), "On What Is Necessary for God" (*Kalām fī-mā yajibu li-Llāh*), and "On What Is Possible for God to Decide" (*Kalām fī-mā yajūzu min aḥkām Allāh*).

7. Al-Ghazālī, *Tahāfut al-falāsifa*, 274.3–275.11 / 164.20–165.18. Kogan, "The Philosophers al-Ghazālī and Averroes on Necessary Connection," 116–20.

8. Al-Ghazālī, *Tahāfut*, 275–76 / 165–66. Kogan, "The Philosophers al-Ghazālī and Averroes," 121–22.

9. The original text expresses these two relations in many more words; cf. Marmura's translation on p. 166, and his comments in "Al-Ghazali on Bodily Resurrection and Causality," 60.

10. Al-Ghazālī, *Tahāfut*, 277.2–278.2 / 166.1–10.

11. Ibid., 270.10–11 / 163.15–16.

12. Lizzini, "Occasionalismo e causalità filosofica," 182.

13. Perler/Rudolph, *Occasionalismus*, 75–77.

14. Ibid., 85–86, 98, referring to al-Ghazālī, *Tahāfut*, 283.9–285.6 / 169.19–170.15 and 292.14–293.4 / 174.120–175.3.

15. Contributions that are based on Ibn Rushd's response to al-Ghazālī in his *Tahāfut al-tahāfut*, 517–542, and Simon van den Bergh's English translation thereof, often take little notice of al-Ghazālī's initial threefold division of his objections.

16. The Third Position (*maqām*) is announced on p. 278.9 / 167.3 but not introduced as such. It starts with the objection on p. 292.2 / 174.9. A helpful analysis of the winding course of the arguments and the "positions" and "approaches" is given by Rudolph in Perler/Rudolph, *Occasionalismus*, 77–105.

17. It is certainly wrong to assume, as Alon, "Al-Ghazālī on Causality," 399, does, that the text is divided into two "philosophical approaches (. . .) called *maqām*, while the religious ones are called *maslak*."

18. Al-Ghazālī, *Tahāfut*, 290.1–7 / 173.6–10; Goodman, "Did al-Ghazâlî Deny Causality," 108.

19. *anna fāʿila l-ikhtirāqi huwa l-nāru faqaṭ*; al-Ghazālī, *Tahāfut*, 278.10 / 167.4.

20. Al-Ghazālī, *al-Munqidh*, 19.4–7; *Tahāfut*, 206.5–207.5 / 123.3–12.

21. Ibid., 377.1–2 / 226.13. On the Muʿtazilte teaching on the generation (*tawallud* or *tawlīd*) of human actions and their effects, see van Ess, *Theologie und Gesellschaft*, 3:115–21, 4:486–88; and Gimaret, *Theories de l'acte humain*, 25–47. Schöck, "Möglichkeit und Wirklichkeit menschlichem Handels," 109–16, discusses in what way the theory of *tawallud* is based on the assumption that natures (*ṭabāʾiʿ*) exist.

22. Al-Ghazālī, *Tahāfut*, 278.13–279.2 / 167.6–8.

23. Ibid., 279.5–11 / 167.12–18.

24. *al-aʿrāḍu wa-l-ḥawādithu allatī taḥiṣalu ʿinda wuqūʿi (. . .) l-ajsām (. . .) tufīḍu min ʿinda wāhibi l-ṣuwar*; ibid., 281.3 / 178.11–13. If this is intended as a paraphrase of Avicenna's position, it is not exactly correct. See Marmura's comment in the notes to his translation on p. 242.

25. Al-Bāqillānī, *al-Tamhīd*, 43.4–9; English translation in Marmura, "The Metaphysics of Efficient Causality," 184–85; see also idem, "Avicenna on Causal Priority," 68; and Saliba, "The Ashʿarites and the Science of the Stars," 82.

26. Al-Ghazālī, *Tahāfut*, 279.3–4 / 167.10–13.

27. Ibid., 280.1–2 / 167.19.

28. Ibid. 279.2 / 167.8–9.

29. Ibn Sīnā, *al-Najāt*, 211.21–22 / 519.7–8: "That from which a thing has its being—without being for that purpose—is the *fāʿil*." Cf. idem, *al-Shifāʾ*, *al-Ilāhiyyāt*, 194.9. See also Goichon, *Lexique de la langue philosophique d'Ibn Sīnā*, 238, 278–79.

30. Al-Ghazālī, *Tahāfut*, 96.11–12 / 56.1–2. Druart, "Al-Ghazālī's Conception of the Agent," 429–32.

31. Among other things, this sentence prompted McGinnis, "Occasionalism, Natural Causation and Science," 449, to argue that al-Ghazālī requires a divine, or at least angelic, volitional act to activate passive dispositions in things. Only this activation allows the connection between cause and effect to materialize. No such act is, however, required.

32. Al-Ghazālī, *Tahāfut*, 281.11 / 167.20.

33. Ibid., 283.4–8 / 169.14–17.

34. Ibid., 283.9–284.6 / 169.19–170.3.

35. Ibid., 283.9 / 169.21.

36. Ibid., 285.12–13 / 170.21–22.

37. Ibid., 286.1–3 / 171.1–2, discusses the example how a prophet knows, through means of divinity, that a person in the future will arrive from a trip. Al-Ghazālī's *al-*

Iqtiṣād, 83–86 (English translation in Marmura, "Ghazali's Chapter on Divine Power," 299–302), discusses the example of Zayd arriving tomorrow and asks whether future contingencies that are not contained in God's pre-knowledge are possible for God to create. For a discussion of this passage and its Farabian background, see pp. 139–40 and 218–19.

38. Courtenay, "The Critique on Natural Causality," 81. On the distinction between God's absolute and ordained power, which developed in thirteenth-century Latin philosophy, see Knuuttila, *Modalities in Medieval Philosophy*, 100.

39. Marmura, "Ghazali's Attitude to the Secular Sciences," 106, 108.

40. Al-Ghazālī, *Tahāfut*, 285.7–12 / 170.17–22.

41. Marmura, "Ghazali and Demonstrative Science," 202–4; Perler/Rudolph, *Occasionalismus*, 86–88; see also Marmura, "Al-Ghazālī's Second Causal Theory," 91, 105–6; and Ibn Rushd, *Tahāfut al-tahāfut*, 531.9–12. Marmura and Rudolph point out that this is nothing new in the Ashʿarite tradition. Already al-Ashʿarī assumed that God creates the human perception (*idrāk*; see Ibn Fūraq, *Mujarrad maqālāt al-Ashʿarī*, 263.7–8) and that our perception corresponds to the world (ibid. 263.5–6).

42. Al-Ghazālī, *Tahāfut*, 286.10–11 / 171.10–11.

43. Ibid., 286.6–7 / 171.7–8.

44. Ibid., 286.12 / 171.12.

45. Marmura, "Al-Ghazālī's Second Causal Theory," 92–95.

46. Al-Ghazālī, *Tahāfut*, 286.12–288.10 / 171.12–172.10; Kukkonen, "Possible Worlds," 497–98.

47. Al-Ghazālī, *Tahāfut*, 291.5–6 / 171–72.

48. Ibid., 270.10–11 / 163.15–16.

49. Ibid., 288.1–3 / 172.2–4.

50. Ibid., 291–92 / 174.7–8.

51. Ibid., 292.2–296.6 / 171.12–177.5. Unlike the earlier two, the beginning of the Third Position is not announced in al-Ghazālī's text.

52. Al-Ghazālī, *Tahāfut*, 292.2–5 / 174.10–13.

53. Ibid. 277.3–4 / 166.2–3; Perler/Rudolph, *Occasionalismus*, 98.

54. Al-Ghazālī, *Tahāfut*, 293.5–7 / 175.5–7; Perler/Rudolph, *Occasionalismus*, 99. Rudolph's interpretation that the third *maqām* concerns what is possible for God to create in the outside world is, for instance, shared by Marmura, "Al-Ghazālī's Second Causal Theory," 103–6; and Goodman, "Did al-Ghazālī Deny Causality?"

55. Al-Ghazālī, *Tahāfut*, 293.8–294.4 / 175.8–19. That will (*irāda*) requires knowledge (*ʿulūm*) is an older Ashʿarite tenet; see al-Juwaynī, *al-Irshād*, 96.12.

56. Goodman, *Avicenna*, 186–87.

57. Goodman, "Did al-Ghazâlî Deny Causality," 118.

58. Al-Ghazālī, *Tahāfut*, 294.4–295.1 / 175.20–176.10.

59. Ibid., 295.1–2 / 176.11–12.

60. Frank, "The Ašʿarite Ontology: I. Primary Entities," 206–8.

61. Goodman, "Did al-Ghazâlî Deny Causality," 105–7, does not make a distinction between the second *maslak* of the second *maqām* and the third *maqām*. He argues that what al-Ghazālī put forward in these two parts is his ultimate position on the issue of causality and that he rejected all others, particularly the occasionalist approach of the first approach in the second *maqām*.

62. Perler/Rudolph, *Occasionalismus*, 101–5. Rudolph (in ibid., 101–2) points to prior discussions within *kalām* literature about the limits of God's omnipotence.

63. Obermann, "Das Problem der Kausalität bei den Arabern," 332–39, and his later, more detailed monograph, *Der philosophische und religiöse Subjektivismus*, 68–85.

64. To my knowledge there is no English-language presentation of Obermann's research despite the fact that he taught in the U.S. (in New York and at Yale) between the time of his migration in 1923 and his death in 1956.

65. Obermann, *Der philosophische und religiöse Subjektivismus*, 73, quoting al-Ghazālī, *Tahāfut*, 37.9–38.2 / 22.1–9. On this example, see also Marmura, "Ghazali and Demonstrative Science," 187.

66. Obermann, *Der philosophische und religiöse Subjektivismus*, 73–74; see al-Ghazālī, *Tahāfut*, 38–39 / 22–23.

67. Obermann, *Der philosophische und religiöse Subjektivismus*, 81, quoting al-Ghazālī, *Tahāfut*, 285.11–12 / 170.20–22.

68. In the early decades of the twentieth century, subjectivism was harshly criticized by philosophers such as Rudolph Carnap and the Vienna Circle. Carnap wanted to establish a purist empiricism, which acknowledges that truth and knowledge are guaranteed through empirical experience of the world and through logical deduction. Other influential thinkers of this time such as Franz Brentano and Edmund Husserl equally bemoaned the "subjectivism" and "anthropologism" of this time.

69. Schaeler, in his review of Obermann's book in *Der Islam* 13 (1923): 121–32, especially 130.

70. Obermann, "Das Problem der Kausalität bei den Arabern," 339; *Subjektivismus*, 85.

71. Obermann, *Der philosophische und religiöse Subjektivismus*, 83–84.

72. Al-Ghazālī, *Tahāfut*, 293.11–13 / 175.11–13.

73. Obermann, *Der philosophische und religiöse Subjektivismus*, 82–83.

74. Ibid., 83, quoting al-Ghazālī, *Tahāfut*, 294.1–4 / 175.16–18.

75. Al-Ghazālī, *Tahāfut*, 293.13–14 / 175.14–15.

76. Ibid., 292.2–5 / 174.10–12.

77. This is what we mean when we say something is contingent: that it is possible but not necessary.

78. Al-Ghazālī, *Tahāfut*, 293.5–6 / 175.5.

79. Ibid., 293.5–7 / 175.5–7. Cf. also al-Ghazālī's earlier definition of impossibility as "conjoining negation and affirmation" (*al-mumtaniʿu huwa l-jamʿu bayna l-nafī wa-l-ithbāt*); ibid. 64.11 / 38.17.

80. Bäck, "Avicenna's Conception of the Modalities," 217–18, 229–31.

81. Aristotle, *De anima*, 431a.1–2.

82. Knuuttila, "Plentitude, Reason and Value," 147. Cf. Hintikka, *Time & Necessity*, 72–80.

83. Al-Ghazālī, *Tahāfut*, 296.4–6 / 177.4–5.

84. Kukkonen, "Plentitude, Possibility, and the Limits of Reason," 555.

85. Al-Ghazālī, *Tahāfut*, 66.8–67.8 / 39.13–40.5; see Kukkonen, "Possible Worlds in the *Tahāfut*," 481.

86. Al-Ghazālī, *Tahāfut*, 80.9 / 47.14–15; 103.6–8 / 60.4–7. For the background to this argument, see Davidson, *Proofs*, 87–88, 352–53.

87. Al-Ghazālī, *Tahāfut*, 67.9–10 / 40.7–8.

88. Aristotle's *Sophistici elenchi*, 166a.22–30.

89. Al-Ghazālī, *Tahāfut*, 66.8–67.8 / 39.15–40.5.

90. Al-Ghazālī, *Tahāfut*, 70.10–71.1 / 42.2–5; Kukkonen, "Possible Worlds in the *Tahāfut*," 482.

91. Street, "Faḫraddīn ar-Rāzī's Critique," 102–3.

92. Bäck, "Avicenna's Conception of the Modalities," 229–31; see also Wisnovsky, *Avicenna's Metaphysics*, 248.

93. Al-Ghazālī, *al-Maqṣad*. 31.15–32.3; Frank, *Creation*, 13.

94. Kukkonen, "Possible Worlds in the *Tahāfut*"; Dutton, "Al-Ghazālī on Possibility." Dutton's article appears to have been written contemporaneous to Kukkonen's article. Although he lists Kukkonen's article in his footnotes, Dutton does not refer to its parallel content. The fact that al-Ghazālī criticizes Ibn Sīnā's concept of the modalities had been pointed out in earlier literature such as Zedler, "Another Look at Avicenna," 517.

95. Al-Ghazālī, *Tahāfut*, 69.5–7 / 41.6–7.

96. Ibn Sīnā, *al-Najāt*, 220.2–5 / 536.4–6; idem, *al-Shifāʾ, al-Ilāhiyyāt*, 137.8–9; cf. Aristotle, *Metaphysics*, 1032a.20.

97. Al-Ghazālī, *Tahāfut*, 70.10–71.1 / 42.2–5. See Kukkonen, "Possible Worlds in the *Tahāfut*," 488; Dutton, "Al-Ghazālī on Possibility," 27.

98. Al-Ghazālī, *Tahāfut*, 343.4–13 / 207.5–14. See Davidson, *Alfarabi, Avicenna, and Averroes, on Intellect*, 152–53.

99. Kukkonen, "Possible Worlds in the *Tahāfut*," 488–89; idem, "Plentitude, Possibility, and the Limits of Reason," 543.

100. Al-Ghazālī, *Tahāfut*, 74.11–12 / 44.13–14.

101. Ibid., 74.6–75.10 / 44.8–45.3.

102. Dutton, "Al-Ghazālī on Possibility," 27–29, 40–5.

103. Gimaret, *La doctrine d'al-Ashʿarī*, 30.

104. McGinnis, "Occasionalism, Natural Causation and Science," 445.

105. Frank, "The Non-Existent and the Possible in Classical Ashʿarite Teaching," 1–4.

106. Knuuttila, "Plentitude, Reason, and Value," 145.

107. Hintikka, *Time & Necessity*, 63–72, 84–86, 103–5, 149–53; Knuuttila, *Modalities in Medieval Philosophy*, 1–38.

108. Knuuttila, "Plentitude, Reason, and Value," 145.

109. Street, "Faḫraddīn al-Rāzī's Critique," 104–5.

110. While possibility is defined as the opposite of impossibility and might therefore include the necessary, contingency excludes both impossibility *and* necessity.

111. Al-Fārābī, *Kitāb Bārī armīniyās ay al-ʿIbāra*, 71.1–5; English translation in Zimmermann, *Al-Farabi's Commentary*, 247. Knuuttila, *Modalities in Medieval Philosophy*, 114.

112. Bäck, "Avicenna's Conception of the Modalities," 231; Rescher, *Temporal Modalities*, 8, 37–38.

113. Ibn Sīnā, *al-Shifāʾ, al-Ilāhiyyāt*, 148–49.

114. Bäck, "Avicenna's Conception of the Modalities," 232.

115. Ibn Sīnā, *al-Shifāʾ, al-Manṭiq, al-Qiyās*, 21.6–12.

116. Ibid., 30.10–12.

117. Bäck, "Avicenna's Conception of the Modalities," 232–36.

118. Ibn Sīnā, *al-Shifāʾ, al-Manṭiq, al-Qiyās*, 21.10–12.

119. Craemer-Ruegenberg, "Ens est quod primum cadit in intellectu," 136; Rescher, "Concept of Existence in Arabic Logic," 72–73. See also Black, "Avicenna on the Ontological and Epistemological Status of Fictional Beings."

120. Bäck, "Avicenna on Existence," 354, 359–61. On the principle that the nonexistent (*al-maʿdūm*) cannot be an object of predication, see Ibn Sīnā, *al-Shifāʾ, al-Ilāhiyyāt*, 25.14–16.

121. Ibn Sīnā, *al-Shifāʾ, al-Manṭiq, al-Qiyās*, 21.9.

122. Ibn Sīnā, *al-Shifāʿ, al-Manṭiq, al-Madkhal*, 15.1–15.

123. Ibn Sīnā, *al-Shifāʾ, al-Ilāhiyyāt*, 22.11–13; 27.18–29.10. The same in al-Fārābī, *Sharḥ al-Fārābī li-Kitāb Arisṭutālīs fī l-ʿIbāra*, 84.3–5; English translation in Zimmermann, *Al-Farabi's Commentary*, 77–78.

124. Bäck, "Avicenna's Conception of the Modalities," 241.

125. See above pp. 141–43.

126. Ibn Sīnā, *al-Shifāʾ, al-Ilāhiyyāt*, 29–34; idem, *al-Najāt*, 224–28 / 546–53. Davidson, *Proofs*, 290–93; idem, "Avicenna's Proof of the Existence of God as a Necessarily Existent Being"; Wisnovsky, "Avicenna and the Avicennian Tradition," 105–27; Hourani, "Ibn Sina on Necessary and Possible Existence."

127. Normore, "Duns Scotus's Modal Theory," 129. On Duns Scotus's modal theory, see also Knuuttila, *Modalities in Medieval Philosophy*, 138–49, 155–57.

128. Al-Bāqillānī, *al-Tamhīd*, 23.13–16; al-Baghdādī, *Uṣul al-dīn*, 69.2–7; al-Juwaynī, *al-Irshād*, 28.3–8; idem, *Lumaʿ fī qawāʿid*, 129.3–6; idem, *al-Shāmil* (ed. Alexandria), 262–65; Davidson, *Proofs*, 159–61, 176–80.

129. Abū l-Ḥusayn al-Baṣrī taught that each time a human considers an act, he or she is equally capable of performing and not performing it. The human's motive is the preponderator (*murajjiḥ*) between these two equally possible alternatives. See Madelung, "Late Muʿtazila and Determinism," 249–50.

130. See the excursus in Ibn al-Malāḥimī's *Kitāb al-Muʿtamad*, 169.9–172.18, in which he reports Abū l-Ḥusayn's argument in favor of God's existence. See also Madelung, "Abū l-Ḥusayn al-Baṣrī's Proof for the Existence of God," 279–80. On the particularization argument and on God as the preponderator (*murajjiḥ*), see Craig, *Kalām Cosmological Argument*, 10–15; repeated in idem, *The Cosmological Argument*, 54–59; and Davidson, "Arguments from the Concept of Particularization."

131. Al-Juwaynī, *al-ʿAqīda al-Niẓāmiyya*, 11.9–13.2. See also idem, *al-Shāmil* (ed. Alexandria), 263–65; and idem, *Lumaʿ fī qawāʿid*, 129–31. Ibn Rushd, *al-Kashf ʿan manāhij*, 144–47, analyzes al-Juwaynī's *murajjiḥ* argument for God's existence and says it is based on similar premises as Ibn Sīnā's proof. On al-Juwaynī's proof and how it differs from Ibn Sīnā's, cf. Rudolph, "La preuve de l'existence de dieu," 344–46. See also Davidson, *Proofs*, 161–62, 187; Saflo, *Al-Juwaynī's Thought*, 202.

132. Al-Ghazālī, *Iqtiṣād*, 25–26, *Iḥyāʾ*, 1:144–45 / 183–84 (= Tibawi, "Al-Ghazālī's Sojourn," 80–81, 98–99); idem, *Faḍāʾiḥ al-Bāṭiniyya*, 81–82; cf. Ibn al-Walīd, *Dāmigh al-bāṭil wa-ḥatf al-munādil*, 1:284–86. On the arguments, see also the literature mentioned on p. 313, n. 140.

133. On the various titles under which Ibn Sīnā's *al-Ḥikma al-ʿarshiyya* was known, see Mahdavī, *Fihrist-i nuskhat-hā-yi muṣannafāt-i Ibn Sīnā*, 75–76 (no. 61). I largely follow Gutas, *Avicenna and the Aristotelian Tradition*, with regard to the titles of works by Ibn Sīnā and the titles' English translations. Preponderance appears in Ibn Sīnā, *al-Shifāʾ, al-Ilāhiyyāt*, 233.4, 303.2, 303.9–11, 335–36. Al-Ghazālī, *Tahāfut*, 23.3–4 / 13.9–10, reports that the *falāsifa* say without a preponderator (*murajjiḥ*), there would be no existence. In the versions of the proof of God's existence in his *al-Shifāʾ, al-Ilāhiyyāt*, 31–32; and al-*Najāt*, 236–37 / 570–71; Ibn Sīnā uses the word *takhṣīṣ* but not *tarjīḥ* or *murajjiḥ*. The same argument in *al-Ḥikma al-ʿarshiyya*, 2–3, however, mentions *tarjīḥ*. Ibn Rushd, *al-Kashf ʿan manāhij*, 144–45, also reports this proof as involving a *murajjiḥ*, not a *mukhaṣiṣ*.

134. Al-Juwaynī, *al-ʿAqīda al-Niẓāmiyya*, 8.*peanult*.–9.1.

135. Ibid., 9.4–7.

136. Ibid., 9.9–10.

137. Ibid., 10.1–2.

138. At this point, the role of the Muʿtazilite Abū l-Ḥusayn al-Baṣrī and his views on *tarjīḥ* are unclear. He may have had a significant influence on al-Juwaynī's and on al-Ghazālī's understanding of the modalities. Soon after al-Ghazālī, Maḥmūd ibn

Muḥammad al-Malāḥimī (d. 536/1141)—one of Abū l-Ḥusayn al-Baṣrī's followers who lived in Khwarezm—wrote a refutation of *falsafa*. This book, *Tuḥfat al-mutakallimīn fī-l-radd ʿalā l-falāsifa*, is currently being edited by Wilferd Madelung.

139. See the translation on p. 149.

140. Al-Ghazālī, *Tahāfut*, 278.2–5 / 167.10–12.

141. Based on a brief note in al-Ghazālī's *fatwā* at the end of the *Tahāfut*, 377.2–3 / 226.12–3; Marmura, "Al-Ghazali on Bodily Resurrection," 48; and "Ghazali's Chapter on Divine Power in the *Iqtiṣād*," 280 assumes that for al-Ghazālī, the causal theories of the Muʿtazila and the *falāsifa* are identical. In the seventeenth discussion, these two causal theories are clearly distinguished and treated differently.

142. Al-Ghazālī, *Tahāfut*, 278.1 / 167.8–9.

CHAPTER 7

1. Al-Ghazālī, *Iḥyāʾ*, 4:305.4–5 / 2494.5–6.

2. On the subject of efficient causality, Ockham taught that the necessity of the connections between the cause and its effect cannot be demonstrated. Nevertheless, he considered the necessity of this connection to be present in human knowledge. See Adams, *William Ockham*, 2:741–98. On his modal theory, see Knuuttila, *Modalities in Medieval Philosophy*, 145–57.

3. Al-Ghazālī, *Tahāfut*, 74.11–75.4 / 44.12–18.

4. Ibn Sīnā, *al-Shifāʾ*, *al-Manṭiq*, *al-Burhān*, 44.11–12; *al-Najāt*, (ed. Dānishpazhūh) 169–70. The passage is missing from Ṣabrī al-Kurdī's edition of Ibn Sīnā's *al-Najāt*.

5. See below pp. 205–12. On nominalist tendencies in Ibn Sīnā, see McGinnis, "Scientific Methodologies in Medieval Islam," 325–27.

6. Al-Ghazālī, *al-Maqṣad*, 15–59; see Gätje, "Logisch-semasiologische Theorien," 162–68.

7. See for instance, the parable of the "inquiring wayfarer" in the thirty-fifth book of the *Ihyāʾ*, in which the "pen," that is, the active intellect, "writes" knowledge on the "spread-out tabled" in the human soul (*Iḥyāʾ*, 4:310.22–312.1 / 2502.12–2504.3). On this parable, see below, p. 219. There are numerous distinctly "realist" comments in the works of al-Ghazālī, such as in the first book of the *Iḥyāʾ*, 1:120.7–16 / 148.5–16, in which he says that knowing is effectively "remembering" (*tadhakkur*) the forms or ideas that humans are taught in their primordial disposition (*fiṭra*). See also a passage in his *al-Mustaṣfā*, 1:80.7–8 / 1:26.12: "(. . .) therefore the [human] intellect can be compared to a mirror in which the forms of the intelligibles are imprinted according to how they really are (*ʿalā mā hiya ʿalayhā*), and I mean by 'forms of the intelligibles' their essences (*ḥaqāʾiq*) and their quiddities (*māhiyyāt*)." Or the *Mishkāt al-anwār*, 67.15–6 / 153.3–4: "If there are in the world of sovereignty luminous, noble, and high substances, which are referred to as 'the angels,' from which the lights emanate upon the human spirits (. . .)."

8. Ibn Rushd, *Tahāfut al-tahāfut*, 531.11–13; English translation by van den Bergh, *Averroes' Tahafut*, 1:325: "Knowledge" always implies truth—falsehood is not considered knowledge."

9. *bi-mujarradi l-qudra min ghayri wāsiṭa aw bi-sababin min al-asbāb*; al-Ghazālī, *Tahāfut*, 369.5 / 222.6–7.

10. Ibid., 369.6–370.1 / 222.7–14.

11. Ibid., 13.10–12 / 7.17–19.

12. Abrahamov, "Al-Ghazālī's Theory of Causality," 91.

13. Strauss, *Persecution and the Art of Writing*, 68–74.

14. Gairdner, "Al-Ghazālī's Mishkāt al-Anwār and the Ghazālī Problem," 153.

15. On esoteric and exoteric writing in al-Ghazālī (though with little reference to the question of his cosmology), see Lazarus-Yafeh, *Studies*, 349–411.

16. Frank, *Creation and the Cosmic System: Al-Ghazâlî & Avicenna.* Frank presents these results first in his article "Al-Ghazālī's Use of Avicenna's Philosophy." Later, in his "Currents and Countercurrents," 126–34, he revisits the subject again and adds new insights.

17. Cf. Frank's own synopsis of his conclusions in his *Al-Ghazālī and the Ashʿarite School*, 4.

18. Frank, *Creation and the Cosmic System*, 86.

19. Frank, *Al-Ghazālī and the Ashʿarite School*, 87.

20. Ibid., 31–37. Frank is highly critical of al-Ghazālī's ability—or willingness—to express himself clearly. On certain subjects, al-Ghazālī "fudges the issue (. . .) in a fog of traditional language," "tends to weasel," "buries the real issue under a cloud of dialectical obfuscation," and offers "somewhat inconclusive rigmarole" (Frank, *Al-Ghazālī and the Ashʿarite School*, 49, 89–90). Frank's analysis of al-Ghazālī's language has been criticized by Ahmad Dallal in his "Ghazālī and the Perils of Interpretation," 777–87. Dallal sees a certain philological sloppiness in Frank's treatment of al-Ghazālī's texts that jumps to preconceived and often untenable conclusions.

21. Marmura, "Ghazali and Demonstrative Science"; idem, "Al-Ghazālī's Second Causal Theory in the 17th Discussion of the Tahāfut"; and idem, "Al-Ghazālī on Bodily Resurrection and Causality in Tahafut and the Iqtisad."

22. Marmura, "Ghazālian Causes and Intermediaries," 92–93.

23. Craig, *Kalām Cosmological Argument*, 45–46; repeated in idem, *The Cosmological Argument*, 101. The position of Craig and Marmura was generally accepted up to 1992.

24. Marmura, "Ghazālian Causes and Intermediaries," 89.

25. Ibid., 91, 93–97, 99–100.

26. Marmura, "Ghazali's Chapter on Divine Power in the *Iqtiṣād*."

27. In *Tahāfut*, 78.4–7 / 46.9–12, al-Ghazālī mentions *Qawāʿid al-ʿaqāʾid* as the title of the book that affirms the true teachings (*ithbāt madhhab al-ḥaqq*). *Qawāʿid al-ʿaqāʾid* is the title of the second book of the *Iḥyāʾ*. The first part of that book also circulates in manuscripts as an independent work under this title. Marmura, "Ghazali's *al-Iqtisad fi al-i'tiqad:* Its Relation to *Tahafut al-Falasifa* and to *Qawa'id al-Aqa'id*," makes the convincing case that the title *Qawāʿid al-ʿaqāʾid* in the *Tahāfut* refers, in fact, to *al-Iqtiṣād fī l-iʿtiqād*.

28. Marmura, "Ghazālian Causes and Intermediaries," 96.

29. Marmura, "Ghazali and Demonstrative Science," 193.

30. Marmura, "Ghazālian Causes and Intermediaries," 97.

31. Marmura, "Ghazali's Attitude to the Secular Sciences," 100.

32. Marmura, "Ghazali and Ashʿarism Revisited," 93, 108.

33. Marmura expressed that explicitly ("Ghazali and Demonstrative Science," 183); Frank never considered that option as far as I can see.

34. Frank, *Al-Ghazālī and the Ashʿarite School*, 3, 100–101. Marmura believes this is available in *al-Iqtiṣād fī l-iʿtiqād*.

35. Frank, *Al-Ghazālī and the Ashʿarite School*, 4, 29, 87, 91.

36. *kilāhumā mumkināni ʿindanā*; see above pp. 178–9.

37. Marmura, "Al-Ghazālī's Second Causal Theory," 86, 96–98, 101–7; idem, "Ghazali on Bodily Resurrection and Causality," 50, 59–65.

38. McGinnis, "Occasionalism, Natural Causation and Science in al-Ghazālī."

39. Marmura, "Ghazali's Second Causal Theory," 97.

40. Al-Ghazālī, *Tahāfut*, 279.2 / 171.8; 279.11 / 171.16; 289.4–5 / 176.15.
41. Ibid., 283.9–285.6 / 173.16–174.14.
42. Ibid., 376.7–10 / 230.6–9.
43. Ibid., 279.2 / 167.8–9.
44. Ibid., 96.11–97.1 / 56.1–3.
45. Ibid., 98.1–2 / 56.16–7.
46. Gyekye, "Al-Ghazālī on Action," 90.
47. *maʿa l-ʿilmi bi-l-murādi* and *huwa ʿālimun bi-mā arādahu*; al-Ghazālī, *Tahāfut*, 96.11–12 / 56.2–3; 100.2–3 / 58.1–2.
48. Al-Ghazālī, *al-Iqtiṣād*, 87.*ult.*–88.2; Marmura, "Ghazali's Chapter on Divine Power," 304; Druart, "Al-Ghazālī's Conception of the Agent," 437.
49. Al-Ghazālī, *al-Munqidh*, 23.11–13.
50. The suggestion that al-Ghazālī developed his views on this subject is not truly convincing. Al-Juwaynī, *Irshād*, 110.3, had already clarified that there is only one agent in this world, which is God.
51. Gyekye, "Al-Ghazālī on Action," 84–88, reviews the arguments and discusses their philosophical underpinnings.
52. Al-Ghazālī argues that in comparison to animate beings, inanimate ones are called agents only by way of metaphor (*Tahāfut*, 98.13–99.6 / 57.8–14); this argument stands mute in light of his whole œuvre, since in the *Iḥyāʾ*, he makes clear that even animate beings cannot be considered agents in the true sense of the word. Here he says that calling a human an agent is only by means of a metaphor (*Iḥyāʾ*, 4:320.12–16 / 2516.4–9).
53. Sabra, "*Kalām* Atomism as an Alternative Philosophy," 207–9.
54. Al-Ghazālī, *Tahāfut*, 120.13 / 71.1.
55. Ibid., 134.5 / 79.12. Druart, "Al-Ghazālī's Conception of the Agent," 428–32.
56. On this element of Ibn Sīnā's teachings, see above pp. 142–3.
57. Al-Ghazālī, *Tahāfut*, 157.1–5 / 92.3–6; 161.6–7 / 95.18–19; 163.2–5 / 96.1–4.
58. Ibid., 293.*ult.* / 175.14.
59. Gyekye, "Al-Ghazālī on Action," 88.
60. Al-Ghazālī, *Iḥyāʾ*, 4:322 / 2518–19.
61. Ibid., 4:325.*ult.* / 2523.12–13; cf. al-Zabīdī, *Itḥāf al-sāda*, 9:465.18–19; and Gramlich, *Muḥammad al-Ġazzālīs Lehre*, 558.
62. Al-Ghazālī, *Ihyā*, 4:326.6–7 / 2523.20–21.
63. *wa-yakūnu qad jarā fī sābiqi ʿilmihi an lā yafʿalahu. . .*; al-Ghazālī, *Tahāfut*, 286.8–10 / 171.9–10. The passage is composed of a full sentence at the beginning plus two *ḥāl* sentences that qualify the first. I follow Marmura's suggestion and see the first *ḥāl* as an objection.
64. Ibn Tūmart, *Sifr fīhi jamīʿ taʿāliq al-Imām al-maʿṣūm al-Mahdī*, 220.6–7; cf. Nagel, *Im Offenkundigen das Verborgene*, 109; and Griffel "Ibn Tūmart's Rational Proof," 779–80.
65. Ibn Tūmart, *Sifr fīhi jamīʿ taʿāliq al-Imām al-maʿṣūm al-Mahdī*, 219.16–20.
66. Al-Bukhārī, *al-Ṣaḥīḥ*, *qadar* 1; *anbiyāʾ* 1; cf. also *tawḥīd* 28 and *badʾ al-khalq* 6. Cf. Wensinck, *Concordance et indices*, 1:22a–b. See Watt, *Free Will and Predestination*, 18; and van Ess, *Zwischen Ḥadīṯ und Theologie*, 1–32.
67. Muslim, *al-Ṣaḥīḥ*, *qadar* 2. Cf. Wensinck, *Concordance et indices*, 5:319a. See Watt, *Free Will and Predestination*, 17.
68. Al-Ghazālī discusses this question in *al-Iqtiṣād*, 222–25. The passage is discussed below, pp. 202–4. He comes down on the latter side, saying that someone always dies "at the time of" (*bi-*, also meaning: "through") his appointed time of death (*ajal*). This

is the usual language applied by Ashʿarites to that question; cf. Gimaret, *La doctrine d'al-Ashʿarī*, 423–28. For al-Ghazālī, this means that the *ajal* and the death are always created concomitantly, just like a cause and its effect.

69. Watt, *Free Will and Predestination*, 135. On the various positions on divine predetermination taken by Muʿtazilites, see ibid., 61–92; van Ess, *Theologie und Gesellschaft*, 4:492–500; and Gimaret, *La doctrine d'al-Ashʿarī*, 424–28.

70. Ibn Fūrak, *Mujarrad maqālāt al-Ashʿarī*, 135–39; Gimaret, *La doctrine d'al-Ashʿarī*, 423–32.

71. Ibn Fūrak, *Mujarrad maqālāt al-Ashʿarī*, 74.12–13; see also idem, 45.15–17; 98.8–11; and Frank, *Creation*, 70.

72. Gimaret, *La doctrine d'al-Ashʿarī*, 393–95, 411–22.

73. Al-Isfarāʾīnī, "al-ʿAqīda," 134.4–5; see also ibid., 162, fragm. 72.

74. Al-Baghdādī, *Uṣūl al-dīn*, 145.10–12.

75. See below p. 191. Al-Fārābī's proposed solution to the dilemma between human free will and divine predestination (pp. 139–40) can also be understood as a reaction to the debate between Muʿtazilites and their Sunni opponents.

76. Abū l-Ḥusayn al-Baṣrī, *Taṣaffuḥ al-adilla*, 116.9–10; 118.14. See also the editors' introduction on pp. xviii–xix.

77. Ibn Fūrak, *Mujarrad maqālāt al-Ashʿarī*, 11.21; al-Bāqillānī, *al-Tamhīd*, 29–30; al-Baghdādī, *Uṣūl al-dīn*, 8.5–6; al-Juwaynī, *al-Irshād*, 13.14–16.

78. On Avicenna's notion of a single eternal divine knowledge and how it contains individual events such as the eclipse of celestial body, see pp. 138–41.

79. *khurūj min al-dīn*; al-Juwaynī, *al-Irshād*, 96.3–7.

80. Ibid., 98.1–8. See Paul Walker's English translation of this passage in al-Juwaynī, *A Guide to Conclusive Proofs*, 56.

81. Al-Juwaynī, *al-Irshād*, 13.14; 94.14. Avicenna also taught that God's knowledge is timeless. On God's knowledge in classical Ashʿarite texts, see Frank, "The Non-Existent and the Possible in Classical Ashʿarite Teaching," 7–16.

82. Al-Ghazālī, *Iḥyāʾ*, 1:124.18–21 / 155–56; idem, *al-Arbaʿīn*, 5.13–6.2 / 5.7–11.

83. Al-Ghazālī, *Iḥyāʾ*, 1:125.1–4 / 156.12–15; corrected according to al-Zabīdī, *Itḥāf al-sāda*, 2:28–29, who adds *ṣifa azaliyya lahu* in the first sentence. See the translation and discussion of this passage in Frank, *Creation*, 53. It is also in al-Ghazālī, *al-Arbaʿīn*, 6.13–7.2 / 6.7–11.

84. Al-Ghazālī, *Iḥyāʾ*, 1:148.9–11 / 188.13–16. Tibawi, "Al-Ghazālī's Sojoun," 84.26–29, 105.

85. Al-Ghazālī, *Iḥyāʾ*, 1:148.16–20 / 189.1–5; Tibawi, "Al-Ghazālī's Sojourn," 85.6–11, 105. See the commentary by al-Zabīdī, *Itḥāf al-sāda*, 2:141.3ff.

86. Al-Ghazālī, *al-Iqtiṣād*, 101–13.

87. Ibid., 101.9–102.7.

88. One might compare this with Fakhr al-Dīn al-Rāzī, *al-Maṭālib al-ʿāliya*, 9:57.7–12 (also in idem, *Muḥaṣṣal*, 459.6–7), who concludes from the existence of an all-encompassing divine foreknowledge that human actions are predetermined and "compelled" (*majbūr*).

89. Al-Ghazālī, *Iḥyāʾ*, 4:111.5 / 2224.8 and 4:317.17 / 2511.*ult.*: *qudra azaliyya*; ibid. 4:12.17 / 2091.4: *hukm azalī*; ibid. 4:30.23 / 2115.21: *irāda azaliyya*. Compare with this, e.g., *al-Maqṣad*, 145.6, in which "God's foreknowledge" (*sābiq ʿilmihi*) is clearly spelled out.

90. Al-Ghazālī, *Iḥyāʾ*, 4:120.22–4 / 2237.6–7.

91. Ibid., 4:121.3–5 / 2237.16–18. See al-Zabīdī, *Itḥāf al-sāda*, 9:74; and Gramlich, *Muḥammad al-Ġazzālīs Lehre*, 209.

92. Al-Ghazālī, *Iḥyā*ʾ, 4:111.8–9 / 2224.*ult.*

93. *al-qadaru sirru Llāhi fa-lā tafshūhu*; ibid., 4:440.4–8 / 14:2680.14. On the non-canonical sources of this *ḥadīth*, see al-ʿIrāqī's notes on the prophetical sayings quoted in the *Iḥyā*ʾ; Gramlich, *Muḥammad al-Ġazzālīs Lehre*, 209; idem, *Nahrung der Herzen*, 2:172; and Michot, in his introduction to Ibn Sīnā, *Lettre au vizir*, 121*.

94. *qāla l-ʿārifūna: ifshāʾu sirru l-rubūbiyya kufr*; al-Ghazālī, *Iḥyā*ʾ, 4:306.23 / 2499.15–16. Cf. al-Makkī, *Qūt al-qulūb*, 2:90.19–20: "Some of those who have knowledge of the implications of *tawḥīd* say: (. . .) The Lordship is a secret, revealing it would forfeit prophecy; and prophecy is a secret, revealing it would forfeit knowledge." Cf. Gramlich, *Muḥammad al-Ġazzālīs Lehre*, 522. Al-Ghazālī discusses the saying in his *al-Imlā*ʾ, 44–46 / 3075–78. In *Iḥyā*ʾ, 1:128–34 / 161–74, he discusses *in extenso* which elements of the creed should not be discussed in public.

95. Cf. Marmura, "Ghazali and Ashʿarism Revisited," 105. Frank, *Creation and the Cosmic System*, 45; idem, *Al-Ghazālī and the Ashʿarite School*, 19.

96. Al-Ghazālī, *Iḥyā*ʾ, 4:305.2–6 / 2494.3–7. Al-Ghazālī promises to discuss just as much of *tawḥīd* as is necessary to develop the right kind of *tawakkul*. It is the third among four degrees of *tawḥīd* that forms the basis of a sound *tawakkul*.

97. Ibid., 3:24.20 / 8:1376.16.

98. Cf. also Q 56:78 and 80:13–16. For the range of views of how the *lawḥ maḥfūẓ* has been understood by Muslims, see Daniel Madigan, "Preserved Tablet," in *EQ*, 4:261–63; van Ess, *Theologie und Gesellschaft*, 4:617–30.

99. Al-Ghazālī, *Fayṣal al-tafriqa*, 182–83 / 37–38. Note that this is not "the pen" of the thirty-fifth book of the *Iḥyā*ʾ that writes on the human tablet. The *ḥadīth* that the first creation is the pen is for instance reported by al-Tirmidhī, *Jāmiʿ al-ṣaḥīh, tafsīr sūrat 68*.

100. Al-Ghazālī, *Tahāfut*, 258–61 / 155–57.

101. On Avicenna's teachings on prophecy, see p. 68.

102. Al-Ghazālī, *Iḥyā*ʾ, 3:24.17–22 / 1376.12–18; 4:241.12–13 / 2406.15–16. Cf. ibid., 4:217.20–22 / 2374.5–7. On the Avicennan influence on al-Ghazālī's understanding of the *lawḥ al-maḥfūẓ*, see Pines, "Quelques notes sur les rapports de l'*Iḥyā*ʾ *ʿulūm al-dīn* d'al-Ghazālī avec la pensée d'Ibn Sīnā," 14–16. According to al-Ghazālī's report of the *falāsifa*'s teachings, the imaginative faculty (*quwwa mutakhayyila*) of the prophets can see in the *lawḥ al-maḥfūẓ* "the forms of future particular events imprinted in it" (al-Ghazālī, *Tahāfut*, 273.8–10 / 164.1–2).

103. *alladhī huwa manqūshun bi-jamīʿi mā qaḍā Llāhu bihi ilā yawmi l-qiyāma*; al-Ghazālī, *Iḥyā*ʾ, 3:24.20–24 / 8:1376.16–17.

104. Al-Ghazālī, *al-Arbaʿīn*, 11.10–12 / 11.4–6. The quotation is from a "Sharḥ li-l-Maṣābīḥ" by an unidentified scholar whom he calls "*al-imām mawlānā ʿAlāʾ al-Dīn*." See Frank, *Creation and the Cosmic System*, 21, 45.

105. Al-Ghazālī hints to this position in the seventeenth discussion of the *Tahāfut* when he writes: "(. . .) the cognitions [that the miracle is among the way God acts habitually] slip away from the [people's] hearts and God does not create them" (*Tahāfut*, 286.7–8 / 171.7–8).

106. Ibn Fūrak, *Mujarrad maqālāt al-Ashʿarī*, 176.16–20; 177.10–15. Gimaret, *La doctrine d'al-Ashʿarī*, 459–60; Gardet, *Dieu et la destinée*, 193–94; Antes, *Prophetenwunder in der Ašʿarīya*, 37–39.

107. Al-Juwaynī, *al-Irshād*, 307–15. For the classical Ashʿarite views on prophecy and its verification, see Gimaret, *La doctrine d'al-Ashʿarī*, 453–63; Gardet, *Dieu et la destinée de l'homme*, 193–204; Antes, *Prophetenwunder in der Ašʿarīya*, 29–46; Griffel, "Al-Ġazālī's Concept of Prophecy," 101–3.

108. See Antes, *Prophetenwunder in der Aš'arīya*, 95. In his *kalām* compendium *al-Iqtiṣād*, 198–99, al-Ghazālī writes about prophetical miracles in a very traditional way, teaching that miracles establish the veracity (*ṣidq*) of the prophets without clearly stating that miracles are a break in God's habit. See also his *al-Risāla al-Qudsiyya* (= *Iḥyāʾ*, 1:154.*ult.*/ 198.14–15) where he says, "Whereas the physician's truthfulness is known through experience and the prophet's truthfulness is known through miracles (. . .)" (Tibawi, "Al-Ghazālī's Sojourn," 91.29–30, 117).

109. Al-Ghazālī, *al-Munqidh*, 32.5–11.

110. Al-Ghazālī, *al-Mustaṣfā*, 2:154–55 / 1:138.16–17; idem, *Faḍāʾiḥ al-bāṭiniyya*, 133–36; Weiss, "Knowledge of the Past: The Theory of *tawātur* According to Ghazālī," 93, 95.

111. Al-Ghazālī, *Arbaʿīn*, 64.1–4. The passage is translated in Gianotti, *Al-Ghazālī's Unspeakable Doctrine*, 156.

112. Al-Ghazālī, *Ḥimāqat-i ahl-i ibāḥat*, 9.17–18 / 171.3.

113. See Griffel, "Al-Ġazālī's Concept of Prophecy: The Introduction of Avicennan Psychology into Aš'arite Theology," 138–44; and Frank, *Al-Ghazālī and the Ashʿarite School*, 67–68.

114. Al-Ghazālī, *al-Munqidh*, 44.1–3; cf. the English translation by McCarthy, *Deliverance from Error*, 86.

115. Al-Ghazālī, *al-Munqidh*, 44.5–7.

116. Al-Ghazālī, *Iḥyāʾ*, 4:315.9–10 / 2508.18–19. The story of the pseudo-prophet al-Sāmirī and how he misled the Israelites to build the golden calf is told in Q 20:83–98.

117. Ibn Sīnā, *al-Ishārāt wa-l-tanbīhāt*, 220.15–221.2. Al-Ghazālī copied this passage in his report of philosophical teachings, MS London, British Library, Or. 3126, fol. 284a. Cf. also al-Ghazālī (?), *Maʿārij al-Quds*, 165.12–13. See above pp. 68–69; and al-Akiti, "Three Properties of Prophethood," 191.

118. Ibn Fūrak, *Mujarrad maqālāt al-Ashʿarī*, 157.4.

119. Al-Juwaynī, *al-Irshād*, 307–8, 314.9–12. According to ibid., 312.3–5, and to al-Juwaynī, al-*Shāmil* (ed. Tehran), 96–97, the "ahl al-ḥaqq" hold that miracles and *karamāt* are breaks in God's habit.

120. Al-Ghazālī, *Iqtiṣād*, 6.14.

121. Al-Ghazālī, *Tahāfut*, 289.11–12 / 173.1–2. That position is repeated, for instance, in al-Ghazālī's letters to Abū Bakr in al-ʿArabī, see p. 69.

122. *lā tajidu fī sunnati Llāhi tabdīlan*, Q 33:62, 48:23 (. . .*taḥwīlan*, in Q 35:43); cf. also Q 30.30: *lā tabdīla li-khalqi Llāh*. Al-Ghazālī, *al-Iḥyāʾ*, 4:8.4–5 / 2084.*ult.*

123. Al-Ghazālī, *Iḥyāʾ*, 4:12.17–18 / 2091.4–5. For other occurrences, see ibid., 4:30.23 / 2115.20–21; 4:58.9–11 / 2151–52; 4:370.4 / 2586.6–7. For Ibn Sīnā's understanding of this Qur'anic verse, see *al-Ḥikma al-ʿarshiyya*, 15.16–17.

124. Frank, *Creation and the Cosmic System*, 59; idem, *Al-Ghazālī and the Ashʿarite School*, 20. For evidence that this position is the one that underlines the whole discussion in the seventeenth discussion of the *Tahāfut*, see Bahlul, "Miracles and Ghazali's First Theory of Causation," 139–41. Marmura, "Ghazali on Demonstrative Science," 196, 200–201; and idem, "Ghazali and Ashʿarism Revisited," 105, maintains that for al-Ghazālī, miracles are a break in God's habit.

125. Al-Ghazālī was most explicit in his *Munqidh*, which was noted by many of his later critics. See, for instance, the remark in Fakhr al-Dīn al-Rāzī's *Muḥaṣṣal*, 491.11–12. On al-Ghazālī's subtle technique of including Ibn Sīnā's three properties of prophecy in his *Munqidh*, see al-Akiti, "Three Properties of Prophethood," 197–99. Al-Ghazālī's position about the verification of prophecy in the *Munqidh* has been a controversial subject among Western interpreters. For reports about the literature, see Poggi, *Un classico*

della spiritualità musulmana, 239–45; and Griffel, "Al-Ghazālī's Concept of Prophecy," 105, n. 12.

126. Ibn Ghaylān, *Ḥudūth al-ʿālam*, 8.19–22.

127. Al-Ṭurṭūshī, *Risāla ilā ʿAbdallāh ibn Muẓaffar*, 160.8–161.5; see Ghurāb, "Ḥawla ikhrāq al-Murābiṭīn li-Iḥyāʾ al-Ghazālī," 136.

128. Al-Akiti, "Three Properties of Prophethood," 194–95; Davidson, *Alfarabi, Avicenna, and Averroes, on Intellect*, 58–63, 116–17.

129. The Brethren present their teachings on prophecy in the thirty-fifth, forty-sixth, and forty-seventh epistles in *Rasāʾil Ikhwān al-ṣafāʾ*, 3:231–48 / 3:227–42, 4:123–96 / 4:61–144. On prophetology in the Brethren, see Marquet, *La philosophie des Iḫwān al-Safāʾ*, 477–508; idem, "Révélation et vision véridique"; Goldziher, *Richtungen*, 186–96; Giese, "Zur Erlösungsfunktion des Traumes"; and al-ʿAbd, *al-Insān fī fikr Ikhwān al-ṣafāʾ*, 254–73.

130. Al-Akiti, "Three Properties of Prophethood," 195–210; Davidson, *Alfarabi, Avicenna, and Averroes, on Intellect*, 141–42; Griffel, "Al-Ġazālī's Concept of Prophecy."

131. Baffioni, "From Sense Perception to the Vision of God," 230–31. Cf also Baffioni's study on the Brethren's terminology with regard to the *awliyāʾ Allāh*: "An Essay on Terminological Research in Philosophy."

132. *Rasāʾil Ikhwān al-ṣafāʾ*, 3:246.15–17 / 3:240.21–23; German translation in Diwald, *Arabische Philosophie und Wissenschaft*, 202. On al-Ghazālī's letter, see above pp. 68–69. Ibn Sīnā's teachings on this subject are hardly different. He also says that "purification" (*tazkiya*) leads the prophets and *awliyāʾ* to their perfection (*al-Ishārāt wa-l-tanbīhāt*, 220.15–18).

133. See, for example, Whittingham, *Al-Ghazālī and the Qurʾān*, 68–69.

134. Ibn Sīnā, however, did not shy away from using the word *rūḥ* himself. See the *ʿaql qudsī* and *al-rūḥ al-qudsiyya* in *al-Shifāʾ*, *al-Ṭabīʿiyyāt*, *al-Nafs*, 248–49, and compare these teachings to al-Ghazālī's use of *al-rūḥ al-qūdsī* in *Mishkāt al-anwār*, 51–52 / 133.10–12, 77.13–15 / 166.9–12, 81.4–10 / 170–71.

135. Gimaret, *Le livre de Bilawhar et Būḏāsf*, 37–38, for instance, makes the case that the allegory of the king who confuses his bride with a corpse in al-Ghazālī's *Kīmiyā-yi saʿādat*, 1:105–6, goes back to the *Rasāʾil Ikhwān al-ṣafāʾ*, 4:212–14 / 4:162–64. Diwald, *Arabische Philosophie und Wissenschaft*, 7, mentions numerous connections between the *Rasāʾil* and works of al-Ghazālī. She promises to present their analysis in a volume that unfortunately never came out,

136. See below pp. 269–71, 219.

137. Al-Ghazālī, *al-Munqidh*, 26.5–17; 27.2–6, 33.19–22. The *Ikhwān al-ṣafāʾ* are not mentioned in the *Tahāfut*.

138. *lahu ʿukūfun ʿalā Rasāʾili Ikhwāni l-ṣafāʾ*; al-Subkī, *Ṭabaqāt*, 6:241.7; and al-Zabīdī, *Itḥāf al-sāda*, 1:28.22. See also Ibn Taymiyya, "Sharḥ al-ʿaqīda al-iṣfahāniyya," 116.19. These passages seem to be quotations from the lost *Kitāb al-Kashf wa-l-inbāʾ ʿalā mutarjam bi-l-Iḥyāʾ* by Abū ʿAbdallāh Muḥammad ibn ʿAlī al-Māzarī al-Imām. For a Spanish translation of the passage, see Asín Palacios, "Un faqîh siciliano, contradictor de Al Ġazâlî," 227.

139. Al-Subkī, *Ṭabaqāt*, 6:241.7–15; and al-Zabīdī, *Itḥāf al-sāda*, 1:28.22–28. Ibn Taymiyya, "Sharḥ al-ʿaqīda al-iṣfahāniyya," 116.19–117.9.

140. Ibn Sabʿīn, *Budd al-ʿārif*, 144.*ult.*–145.4. Ibn Sabʿīn mentions the twenty-first book of the *Iḥyāʾ* (*Sharḥ ʿajāʾib al-qalb*), *Maʿārij al-quds fī madārij maʿrifat al-nafs*, *Mishkāt al-anwār*, and "*Kīmiyāʾ al-saʿāda*." It is unlikely that Ibn Sabʿīn read the Persian *Kīmiyā-yi saʿādat*, so the latter book is most probably the Arabic *Kīmiyāʾ al-saʿāda*, a reworked summary of the *Iḥyāʾ*. Bouyges, *Essay*, 136–37; and Badawī, *Muʾallafāt*, 275–76, have

questioned al-Ghazālī's authorship of this book. On the passage by Ibn Sabʿīn, see Akasoy, *Philosophie und Mystik*, 230–31, 323. On the negative reaction to al-Ghazālī in the Muslim West, see also Serrano Ruano, "Why Did the Scholars of al-Andalus Distrust al-Ghazâlî?"

141. Ibn Taymiyya, "Sharḥ al-ʿaqīda al-iṣfahāniyya," 111.12.

142. Ibn Taymiyya, *Minhāj al-sunna*, 4:148:33–149–21.

143. Ibn Taymiyya, "Sharḥ al-ʿaqīda al-iṣfahāniyya," 115–18.

144. Al-Māzarī al-Imām may have brought up al-Tawḥīdī's name because in his *al-Imtāʿ wa-l-muʾānasa*, 2:11–18, he reports a dispute in the workshop of the copyists at Basra. There, Abū Sulaymān Muḥammad ibn Maʿshar al-Bīstī al-Maqdisī, one of the initial authors of the *Rasāʾil Ikhwān al-ṣafāʾ*, claimed that prophets heal sick people and that the healthy souls of those who practice philosophy (*aṣḥāb al-falsafa*) are in no need of prophecy. Stern, "Authorship of the Epistles," 369, observes that this goes beyond what is taught in the *Rasāʾil* and that "al-Maqdisī, in the heat of dispute, let slip from his mouth opinions which were usually restricted to the inner circle of adepts." Al-Maqdisī's position has more than once been misattributed to al-Tawḥīdī; cf., for instance, Moosa, *Ghazālī and the Poetics*, 155.

145. Al-Ghazālī, *Miʿyār al-ʿilm*, 122.11–20; MS Vatican, Ebr. 426, fol. 128b. This example appears more often in al-Ghazālī's work—see above p. 172—and in many editions, the word *ḥazz* ("incision, notch") is mistakenly rendered as *jazz* ("cutting off"). This led to the false impression, reproduced by most interpreters, that al-Ghazālī here talks about decapitation. The Judeo-Arabic manuscript, in which the letters *ḥāʾ* and *jīm* are very distinct, has *ḥazz*. Already in Bouyges's critical edition of the *Tahāfut*, 277.7, 278.3–4 (= 166.6, 166.11 in Marmura's edition) it is clear that *ḥazz* is the *lectio difficilior* and should have been adopted. This is also true for the discussion in *al-Iqtiṣād*, 223.12–14, which is dealt with below on p. 202, and which clarifies that the *ḥazz* leads to "cleavages (*iftirāqāt*) among the atoms in the neck of him who is hit."

146. Al-Ghazālī, *Miʿyār al-ʿilm*, 123.8–11; MS Vatican, Ebr. 426, fol. 129a. I am reading *ḥuzzat raqabatuhu* according to the MS. This passage is discussed in Marmura, "Ghazali and Demonstrative Science," 195–96; Frank, *Al-Ghazālī and the Ashʿarite School*, 18; and Dallal, "Al-Ghazālī and the Perils of Interpretation," 783.

147. Frank, *Creation and the Cosmic System*, 38.

148. The *Miʿyār al-ʿilm* was most probably written in the same period right after the *Tahāfut* and before the *Iḥyāʾ ʿulūm al-dīn*. The following passage is also discussed in Marmura, "Al-Ghazali on Bodily Resurrection and Causality," 68–70; and Fakhry, *Islamic Occasionalism*, 62–63.

149. Al-Ghazālī, *al-Iqtiṣād*, 223.8–9. Marmura, "Al-Ghazali on Bodily Resurrection and Causality," 69, suggests that the "single cause" here is understood to be God, which would change the understanding of this passage. That interpretation, however, is not viable. It would allow for what can only be an absurd assumption for al-Ghazālī that if God is regarded as the only cause of death, He could not exist. In the whole passage God is nowhere mentioned as a cause (*ʿilla*). Here al-Ghazālī talks about what we usually regard as proximate causes of events such as death. The passage focuses on human knowledge of causal connections and not on the creation of them.

150. Al-Ghazālī, *al-Iqtiṣād*, 223.12–224.1.

151. *lazima min intifāʾihi intifāʾu l-mawt*; ibid., 224.3.

152. *al-mawtu amrun istabadda l-rabbu taʿālā bi-ikhtirāʿihi maʿa l-ḥazz*; ibid., 224.7–8.

153. See above p. 152.

154. Al-Ghazālī, *al-Iqtiṣād*, 224.8–10.

155. Ibid., 224.11–225.1.

156. Al-Ghazālī, *Iḥyāʾ*, 4:302.19 / 2490.15–16. Reading *taghbīr fī wajh al-ʿāql* instead of *taghyīr* according to al-Zabīdī, *Itḥāf al-sāda*, 9:385.30. Gramlich, *Muḥammad al-Ġazzālīs Lehre*, 515–16, in his otherwise meticulous German translation renders *asbāb* as "secondary causes," which leads to undue conclusions.

157. Al-Ghazālī, *Iḥyāʾ*, 4:302.19–20 / 2490.16–17; read *ʿaql* instead of *naql* following al-Zabīdī, *Itḥāf al-sāda*, 9:385.32.

158. Al-Ghazālī, *Iḥyāʾ*, 3:72.11 / 1445.15–16. In the first book of the *Iḥyāʾ*, 1:118.1–119.3 / 145.7–146.16, he clarifies that certain parts of the *ʿaql* are part of the human nature (*ṭabʿ*), among them the instinctive capacity to distinguish "the possibility of the possibilities from the impossibility of what is impossible (*jawāz al-jāʾizāt wa-stiḥālat al-mustaḥīlāt*)."

159. This list of seven sources follows the division in al-Ghazālī, *Miḥakk al-naẓar*, 47–52 (and subsequently *al-Mustaṣfā*, 1:138–46 / 1:44–46). See Weiss, "Knowledge of the Past," 100–101. In the *Miʿyār al-ʿilm*, 121–25, the division is slightly different and excludes reliably reported knowledge (*mutawātirāt*). In *Miʿyār al-ʿilm*, 125–35, there are three kinds (*aṣnāf*) of noncertain knowledge, which are further divided in many subdivisions, most of them discussed in quite an amount of detail. In the *Iḥyāʾ*, 1:103.5–7 / 124.18–20, al-Ghazālī includes *tawātur*. There, the four categories of certain knowledge are: (1) *a priori* knowledge and knowledge established through (2) *tawātur*, (3) experimentation (*tajriba*), and (4) *burhān*.

160. Al-Ghazālī, *Miʿyār al-ʿilm*, 122.12–15; idem, *Miḥakk al-naẓar*, 50.1–6;

161. *ḥukmu l-ʿaqli bi-wāsiṭati l-ḥissi wa-bi-takarruri l-aḥsāsi marratan baʿḍa ukhrā*; al-Ghazālī, *Miḥakk al-naẓar*, 50.1–12; and idem, *al-Mustaṣfā*, 1:141.2–12 / 1:45.10–16. For very similar lists of causes and their effects, see *Miʿyār al-ʿilm*, 122.13–15; and *Maqāṣid al-falāsifa*. 1:47.19–48.1 / 103.4–8. Cf. Frank, *Al-Ghazālī and the Ashʿarite School*, 18.

162. *quwwa qiyāsiyya khafiyya*; al-Ghazālī, *Miʿyār*, 122.16–18.

163. *idh yaḥtamilu anna zawālahu bi-l-ittifāq* only in the parallel passage from *al-Mustaṣfā*.

164. Al-Ghazālī, *Miḥakk al-naẓar*, 50.13–51.1; and *al-Mustaṣfā*, 1:142.2–8 / 1:45.16–46.2.

165. Bahlul, "Miracles and Ghazali's First Theory of Causation," 146–47, observes correctly that in al-Ghazālī, there is no difference between causal connections and "accidental connections," that is, those not representing causal influences.

166. Al-Ghazālī, *Miḥakk al-naẓar*, 51.9–10; idem, *al-Mustaṣfā*, 1:142.14–15 / 1:46.4. Note that al-Ghazālī's language assumes that the things itself have such habits; he does not speak of God's habit.

167. Al-Ghazālī, *Miḥakk al-naẓar*, 51.1–3; *al-Mustaṣfā*, 1:142.9–11 / 1:46.2–3.

168. Al-Ghazālī, *Miḥakk al-naẓar*, 51.11–12; *al-Mustaṣfā*, 1:142.*ult.*–143.1 / 1:46.4–5. Cf. *Miʿyār al-ʿilm*, 123.4–5. See Marmura, "Ghazali and Demonstrative Science," 195; and idem, "Ghazali's Attitude to the Secular Sciences," 107–8.

169. Al-Ghazālī, *Miʿyār al-ʿilm*, 122.16; reading "*ḥuṣūlu idrāki dhālika l-yaqīn*" according to MS Vatican, Ebr. 426, fol. 128b.

170. Al-Ghazālī, *Miḥakk al-naẓar*, 51.4–9 (reading *iqtirānuhu* in line 8); cf. *al-Mustaṣfā*, 1:142.11–13 / 1:46.3–4.

171. Marmura, "Ghazali and Demonstrative Science," 195, remarks that al-Ghazālī's use of certainty in connection with the result of experimentation is somehow ambiguous. I see no such ambiguity.

172. Al-Ghazālī, *al-Munqidh*, 54.1–5.

173. Ibid., 43.12–*ult.* / 44.5–11. See Griffel, "Al-Ġazālī's Concept of Prophecy," 104, 141.

174. *mūjib wa-mūjab*; al-Ghazālī, *al-Munqidh*, 70.8–9.

175. Davidson, *Alfarabi, Avicenna, and Averroes, on Intellect*, 83–94; McGinnis, "Scientific Methodologies in Medieval Islam," 312–13.

176. Aristotle, *Categories*, 2a.35–2b.6.

177. *bi-tawassuṭi ishrāqi l-ʿaqli l-faʿʿāl*; Ibn Sīnā, *al-Shifāʾ, al-Ṭabīʿiyyāt, al-Nafs*, 235; see Hasse, "Avicenna on Abstraction," 53–58; and McGinnis, "Making Abstraction Less Abstract," 173–76, 180.

178. Ibn Sīnā, *al-Shifāʾ, al-Manṭiq, al-Burhān*, 44.11–12; McGinnis, "Scientific Methodologies," 313. Experience (*tajriba*) in Ibn Sīnā is also dealt with in a brief passage in his *al-Najāt*, 61 / 113–14 (see also pp. 169–70, but only in Dānishpazhūh's edition), and a passage in his *Risālat al-Ḥukūma fī-l-ḥujaj al-muthbitīn li-l-māḍī mabdaʾ*^{an} *zamaniyy*^{an}, 134.18–135.6, which are both translated in Pines, "La conception de la conscience de soi," 255–57.

179. McGinnis, "Scientific Methodologies," 314–15.

180. Ibn Sīnā, *al-Shifāʾ, al-Manṭiq, al-Burhān*, 45.15–18, 46.4; McGinnis, "Scientific Methodologies," 317, 320.

181. Ibn Sīnā, *al-Shifāʾ, al-Manṭiq, al-Burhān*, 46.5–7.

182. *hunāka qiyāsun yanʿaqidu fī l-dhihni bi-ḥaythu lā yushʿaru bih*; Ibn Sīnā, *Risālat al-Ḥukūma fī-l-ḥujaj*, 134.23; see also *al-Shifāʾ, al-Manṭiq, al-Burhān*, 161.19; 46.11; 46.20. Janssens, "'Experience' (*tajriba*) in Classical Arabic Philosophy," 56.

183. McGinnis, "Scientific Methodologies," 318–19, argues that although induction attempts to engender a necessary judgment through the enumeration of positive instances, experimentation is based at least in part on the absence of falsifying instances. This requires, as McGinnis admits, that observation "for the most part" not include a falsification and that an exception be extremely rare, perhaps observed only once or twice. Janssens, "'Experience' (*tajriba*) in Classical Arabic Philosophy," 54, objects that this interpretation has no basis in the text and is simply too modern.

184. Ibn Sīnā, *al-Shifāʾ, al-Manṭiq, al-Burhān*, 46.2.

185. Aristotle, *Analytica posteriora*, 71b.9–12; McGinnis, "Scientific Methodologies," 321; Janssens, "'Experience' (*tajriba*) in Classical Arabic Philosophy," 55.

186. *ʿilm kullī bi-sharṭ*; Ibn Sīnā, *al-Shifāʾ, al-Manṭiq, al-Burhān*, 46.20–23; McGinnis, "Scientific Methodologies," 323; Janssens, "'Experience' (*tajriba*) in Classical Arabic Philosophy," 57–58.

187. Janssens, "'Experience' (*tajriba*) in Classical Arabic Philosophy," 58.

188. Ibid., 57–59.

189. Ibn Sīnā, *al-Shifāʾ, al-Manṭiq, al-Burhān*, 47.11; McGinnis, "Scientific Methodologies," 324–27.

190. Janssens, "'Experience' (*tajriba*) in Classical Arabic Philosophy," 59.

191. Ibn Sīnā, *al-Shifāʾ, al-Manṭiq, al-Burhān*, 161.19–*ult.*; Janssens, "'Experience' (*tajriba*) in Classical Arabic Philosophy," 60.

192. Ibn Sīnā, *al-Shifāʾ, al-Manṭiq, al-Burhān*, 48.14–*ult.*; Janssens, "'Experience' (*tajriba*) in Classical Arabic Philosophy," 59–62.

193. McGinnis, "Scientific Methodologies in Medieval Islam," 326–27.

194. Al-Ghazālī, *Miʿyār al-ʿilm*, 122.9–10.

195. Ibid., 122.18–123.1; MS Vatican, Ebr. 425, fol. 128b. Cf. also a parallel passage in *Maqāṣid al-falāsifa*, 1:48.2–3.

196. Marmura, "Ghazali and Demonstrative Science," 196.

197. See above p. 116.

198. Ibn Rushd, *Tahāfut al-tahāfut*, 522.8: "*fa-man rafaʿa l-asbāba fa-qad rafaʿa l-ʿaql*"; Frank, *Al-Ghazālī and the Ashʿarite School*, 17. See also Marmura, "Ghazali and Demonstrative Science," 183–85; and idem, "Ghazali's Attitude to the Secular Sciences," 105.

199. Marmura, "Ghazali and Demonstrative Science," 193.

200. *fa-l-natījatu min ʿinda Llāhi taʿālā*; al-Ghazālī, *Miʿyār al-ʿilm*, 119.8–10; MS Vatican, Ebr. 426, fol. 127a. Marmura, "Al-Ghazali and Demonstrative Science," 194, points to parallels in Ibn Sīnā, in which the conclusion of an argument is an emanation from the active intellect.

201. Kukkonen, "Causality and Cosmology," 33–34.

CHAPTER 8

1. Al-Ghazālī, *Iḥyāʾ*, 1:27.3–5 / 27.11–13. For a synopsis and an index of subjects in the *Iḥyāʾ*, see Bousquet, *Ih'ya ʿouloûm ad-dîn ou vivication des sciences de la foi*.

2. Al-Ghazālī, *Iḥyāʾ*, 1:12.21–23 / 5.4–7.

3. The position that the human act is the causal effect of a motive (*dāʿin* or *dāʿiya*) goes back to the Basran Muʿtazilite Abū l-Ḥusayn al-Baṣrī and to Ibn Sīnā. In Ashʿarite *kalām*, it appears already in al-Juwaynī and had a considerable influence on later Ashʿarite thought, particularly on Fakhr al-Dīn al-Rāzī. On Abū l-Ḥusayn's theory of action, see Gimaret, *Théories de l'acte humain*, 59–60, 124–26, 130–31, 143; Shihadeh, *The Teleological Ethics of Fakhr al-Dīn al-Rāzī*, 25–29; Madelung, "Late Muʿtazila," 250–56; and McDermott, "Abū l-Ḥusayn al-Baṣrī on God's Volition." On translating *irāda* when it applies to humans as "volition" rather than as "will," see Frank, *Creation*, 32–34.

4. Heer, "Moral Deliberation," 166–68.

5. Al-Ghazālī, *Iḥyāʾ*, 3:53.19–20 / 1417.12–13; Heer, "Moral Deliberation," 166, 168.

6. Heer, "Moral Deliberation," 170.

7. Al-Ghazālī, *al-Mustaṣfā*, 1:196.6–9 / 1:61.12–14.

8. Al-Ghazālī, *Iḥyāʾ*, 4:315.11–318.4 / 2508.21–2512.11; the perspective in this passage is distinctly causalist. Another passage in the thirty-first book on *tawba* (ibid. 4:7.19–9.7 / 2084.11–2086.10) uses more occasionalist language. See also a passage in the thirty-second book on *shukr* (ibid. 4:111.7–112.18 / 2223.21–2225.18). The parable of the wayfarer to God in the thirty-fifth book (ibid. 4:307.*ult.*–314.6 / 2498.11–507.5) also includes an explanation of the compelled character of human actions. On al-Ghazālī's theory of human action in the *Iḥyāʾ*, see Frank, *Creation*, 23–27, 31–37; idem, *Al-Ghazālī and the Ashʿarite School*, 42–47; Gimaret, *Théories de l'acte humain*, 130–32; Marmura, "Ghazali and Ashʿarism Revisited," 102–10; Heer, "Moral Deliberation"; Gardet, *Dieu et la destinée de l'homme*, 74–77; and Abrahamov, "Al-Ghazālī's Theory of Causality," 88–90. On the more Ashʿarite formulation of the same theory in the *Iqtiṣād* and other works, see Marmura, "Ghazali's Chapter on Divine Power in the *Iqtiṣād*"; Gimaret, *Théories de l'acte humain*, 129–30; Gyekye, "Ghazâlî on Action"; and Druart, "Al-Ghazālī's Conception of the Agent."

9. Schwarz, "'Acquisition' (kasb) in Early Kalām."

10. See the explanation of al-Ashʿarī's theory of human action above on p. 128. On *kasb* in early Ashʿarite theology, see also Gimaret, *Théories de l'acte humain*, 79–128, esp. 84–85; Watt, *Formative Period*, 189–94; and Abrahamov "A Re-examination of al-Ashʿarī's Theory of kasb."

11. Al-Ghazālī, *al-Iqtiṣād*, 86.*ult.*–87.3; Marmura, "Ghazali's Chapter on Divine Power," 303; Druart, "Al-Ghazālī's Conception of the Agent," 436.

12. Al-Ghazālī, *Iḥyāʾ*, 4:307.14–18 / 2497.19–22; 4:314.24–25 / 2508.3–4.

13. Analyzing the less explicit work *al-Iqtiṣād*, Druart, "Al-Ghazālī's Conception of the Agent," 439, concludes that humans are "agents only in a metaphysical way."

14. Al-Ghazālī, *Iḥyāʾ*, 4:316.5–7 / 2509.*paenult.*–2510.2; al-Zabīdī, *Itḥāf al-sāda*, 9:421.

15. Frank, *Creation*, 33–34.

16. (. . .) *anna l-irāda tabiʿa l-ʿilma alladhī yaḥkumu bi-anna l-shayʾa muwāfiqun laka*; al-Ghazālī, *Iḥyāʾ*, 4:316.10–11/ 2510.7.

17. Ibid. 4:317.4–6 / 2511.8–10; corrected to *al-kullu yaṣduru minhu* according to al-Zabīdī, *Itḥāf al-sāda*, 9:422.10.

18. *inna khilāfa l-maʿlūmi maqdūr?* al-Ghazālī, *al-Iqtiṣād*, 83–86. Marmura, "Ghazali's Chapter on Divine Power," 299–302. See above p. 192. See also al-Juwaynī, *al-Shāmil* (ed. Alexandria), 375–76.

19. Al-Ghazālī, *al-Iqtiṣād*, 85.1–3; 85.5–7; Marmura, "Ghazali's Chapter on Divine Power," 301.

20. *kār-i khalq-i hama ba-rāy khwesh ast*; al-Ghazālī, *Ḥimāqat-i ahl-i ibāḥat*, 9.3 / 169.13.

21. Al-Ghazālī, *Iḥyāʾ*, 4:317.6–11/ 2511.10–16.

22. Fakhr al-Dīn al-Rāzī, *al-Tafsīr al-kabīr*, 4:88.5–9 (*ad* Q 2:134), already ascribes this position to al-Juwaynī and his *al-ʿAqīda al-Niẓāmiyya*. He adds that this position is close to that of Abū l-Ḥuṣayn al-Baṣrī.

23. See Ibn Sīnā, *al-Shifāʾ*, *al-Ilāhiyyāt*, 360.6–9, 362.16–19. On Ibn Sīnā's teachings about the generation of human acts, see Michot's introduction to Ibn Sīnā, *Réfutation de l'astrologie*, 59*–75*; Belo, *Chance and Determinism*, 115–17; and Janssens, "The Problem of Human Freedom in Ibn Sînâ." Al-Ghazālī gives a colorful report of these views in his *Maqāṣid al-falāsifa*, 2:82.4–*paenult.* / 236.3–23, where he discusses, as he does in many works that present his own teachings, the example of how writing is caused. On the Avicennan influence on al-Ghazālī's theory of human action, see Frank, *Creation*, 24–25; and Marmura, "Ghazali and Ashʿarism Revisited," 107. Van den Bergh, "Ghazali on 'Gratitude Towards God,'" points towards the Stoic origins of these teachings.

24. Ibn Sīnā, *al-Shifāʾ*, *al-Ilāhiyyāt*, 133.13–15. See also the detailed discussion of human action in chapter 6.5 of the *Ilāhiyyāt*, 220–35. The word "motive" (*dāʿin* or *dāʿiya*) appears in Ibn Sīnā's *Ilāhiyyāt* a few times, saying that God has no motive (233.4–6, 303.11) or that the actions of humans are guided by motives (223.9, 230.12, 372.18). It is very prominent in certain passages in Ibn Sīnā, *al-Taʿlīqāt*, 50–51, 53 / 108, 295–97. On these passages about the generation of human actions in Ibn Sīnā, see Michot's introduction to Ibn Sīnā, *Réfutation de l'astrologie*, 68*–75*.

25. Al-Ghazālī, *Iḥyāʾ*, 4:307.*ult.*–314.6 / 2498.11–507.5. A brief version of the parable is in *Iḥyāʾ*, 4:103.5–*ult.* / 2213.4–*ult.*; and in *al-Arbaʿīn*, 241.4–242.9 / 220.5–221.5. In this parable, al-Ghazālī offers a view of human knowledge in which the "pen" in the *ʿālam al-malakūt* writes on a blank tablet in the human soul. This is an application of philosophical ideas based on Aristotle, *De anima*, III.5. Al-Ghazālī's "pen" is the active intellect that writes knowledge on what is in Aristotle the "erased tablet" (the *tabula rasa*) within the individual human soul (Aristotle, *De anima*, 430a.1–2). In *al-Ḥikma al-ʿarshiyya*, 12.4–5, Ibn Sīnā identifies the active intellect with "the pen" and the soul of the prophet with "a tablet." On al-Ghazālī's parable, see Gimaret, *Théories de l'acte humain*, 131; Nakamura, "Ghazālī's Cosmology Reconsidered," 40–43; Gianotti, *Al-Ghazālī's Unspeakable Doctrine*, 152–55. On the active intellect in al-Ghazālī's *Mishkāt al-anwār*, see Abrahamov, "Ibn Sīnā's Influence on al-Ghazālī's Non-Philosophical Works," 8–12. On the terms *malakūt*, *jabarūt*, and *mulk* in al-Ghazālī and in previous authors, see Nakamura, "Ghazālī's Cosmology Reconsidered"; Davidson, *Alfarabi, Avicenna, and Averroes on Intellect*, 119, 133–35; Lazarus-Yafeh, *Studies*, 503–22; Frank, *Creation*, 19; and Wensinck, "On the Relationship Between al-Ghazālī's Cosmology and His Mysticism."

26. The imperative "act!" (*iʿmalū*) appears numerous times in the Qur'an (e.g. Q 6.135). From the canonical *ḥadīth* corpus, al-Ghazālī quotes: "Act! because everything has been made easy if it has been created for you" (*Iḥyāʾ*, 4:111.18 / 2224.12). For

this *ḥadīth*, see al-Bukhārī, *al-Ṣaḥīh*, *qadar* 4; or Ibn Māja, *Sunan*, *muqaddima* 10; cf. Wensinck, *Concordance*, 7:364b. The theological implications of this *ḥadīth* are discussed in van Ess, *Zwischen Ḥadīṯ und Theologie*, 39–47; and Gramlich, *Muḥammad al-Ġazzālīs Lehre*, 194–95.

27. Al-Ghazālī, *Iḥyā'*, 4:111.21–23 / 2224:17–19. That human acts are prompted by a motive (*dā'in* or *dā'iya*) goes back to the Basran Mu'tazilite Abū l-Ḥusayn al-Baṣrī but has also been taught by Ibn Sīnā. See above notes 3 and 24. The usage of *jāzima* ("decisive") is entirely Avicennan. For Ibn Sīnā's distinction between a decisive volition (*irāda jāzima*) and an inclining volition (*irāda mumīla*), see Marmura, "The Metaphysics of Efficient Causality," 183; and idem, "Avicenna on Causal Priority," 70.

28. Al-Ghazālī, *Iḥyā'*, 4:112.5–8 / 2225.2–6. See al-Zabīdī, *Itḥāf al-sāda*, 9:62.17–21, who has *yussira* instead of *tayassara* and *taqūdahu* instead of *yaqūdahu*. Cf. Gramlich, *Muḥammad al-Ġazzālīs Lehre*, 195.

29. See also al-Ghazālī, *Iqtiṣād*, 6.10–7.6; on this passage, see van Ess, *Erkenntnislehre*, 338.

30. See above p. 133. Cf. Q 80.20 "(. . .) then [God] makes the path easy for him (. . .)"; Al-Subkī, *Ṭabaqāt*, 3:386.5–6, already remarked that al-Ghazālī's theory of human action is identical both to that of al-Juwaynī and to that of the Mu'tazila. See Gimaret, *Théories de l'acte humain*, 129.

31. Ibn Tūmart, *Sifr fīhi jāmi'*, *ta'ālīq al-Imām*, 214.1–5. This chain also appears in the text "al-Kalām 'alā l-'ibāda"; ibid., 205–6.

32. See, however, a somewhat similar passage in *Iqtiṣād*, 6.10–7.6.

33. Al-Ghazālī, *Iḥyā'*, 4:119.14–15 / 2235.9–11. Cf. al-Zabīdī, *Itḥāf al-sāda*, 9:72.8–11; and Gramlich, *Muḥammad al-Ġazzālīs Lehre*, 207.

34. In the *Iḥyā'*, see most of all the *Bayān fī kayfiyyat al-tafakkur fī khalq Allāh* at the end of the thirty-ninth book on meditation (*tafakkur*); al-Ghazālī, *Iḥyā'*, 4:540–57 / 2822–44; Ormsby, *Theodicy*, 45–51. On al-Ghazālī's *al-Ma'ārif al-'aqliyya*, see Cabanelas, "Un opusculo inédito de Algazel: El libro le las intuicones intelectuales."

35. Al-Ghazālī, *Iḥyā'*, 4:318.17–18 / 2513.9–10. Cf. ibid., 4:323.14 / 2520.4; 4:307.17–22 / 2497.21–2498.3.

36. Ibid., 1:124.*paenult.* / 156.7; 1:148.16 / 189.1; Tibawi, "Al-Ghazālī's Sojourn," 85.6, 105.

37. Frank, *Creation*, 18. The term appears numerous times in al-Ghazālī's *Iḥyā'*; see, e.g., 1:104.20 / 126.6; 4:58.9 / 2151.*peanult.*; 4:120.21 / 2237.4; 4:136.11 / 2258.11; 4:149.23 / 2277.16; 4:321.4 / 2516.20; 4:355.11–13 / 2565.6–8. For the synonymous *mudabbir al-asbāb* ("the one who orders [or: governs over] the causes"), see ibid., 4:340.22 / 2545.23. See also al-Ghazālī's *Tahāfut*, 65.4 / 38.22, 182.9 / 107.19; and *al-Maqṣad*, 116.13. In the *Munqidh*, 49.20, al-Ghazālī refers to God as the *muqallib al-qulūb*, "the one who changes the hearts," meaning the one who determines people's opinions and moods.

38. For the Avicennan background of *musabbib al-asbāb*, see Janssens, "Filosofische Elementen in de mystieke Leer," 341–42; and Frank, *Creation*, 18. The term appears at least twice in works of Ibn Sīnā, in *al-Ḥikma al-'arshiyya*, 9.7, and *al-Shifā'*, *al-Ilāhiyyāt*, 2.16. Based on a close study of the manuscript evidence, Bertolacci, *Reception of Aristote's Metaphysics*, 489, rejects the reading of *musabbib al-asbāb* in Ibn Sīnā's *al-Shifā'*, *Ilāhiyyāt*, 2.16, and corrects it to *sabab al-asbāb*. The Latin translation confirms this and translates *causa causarum* (Ibn Sīnā, *Avicenna latinus. Liber de philosophia prima*, 1:3.2). A similar correction might be necessary for the poorly edited *al-Ḥikma al-'arshiyya*. Al-Juwaynī, *al-Irshād*, 235.3–4, reports that the philosophers say God is the *sabab al-asbāb*. In his *Tahāfut*, 102.4 / 59.10, al-Ghazālī reports roughly the same (cf. also ibid., 97.1 / 56.2). It is possible that al-Ghazālī's prominent use of *musabbib* had an influence on

the Avicennan manuscript tradition and prompted some copyists to change the original Avicennan *sabab* to *musabbib*.

39. Al-Makkī, *Qūt al-qulūb*, 1:209.8; 2:11.9; German translation in Gramich, *Die Nahrung der Herzen*, 2: 97, 317; see Frank, *Creation*, 18. Al-Ghazālī's strategy of combining a causalist view of events in this world with a possible occasionalist perspective on God's actions seems to go back to al-Makkī; see for example his chapter on *asbāb* and *wasā'iṭ* in *Qūt al-qulūb*, 2:10–15; German translation in Gramlich, *Die Nahrung der Herzen*, 2:315–29.

40. *musabbibu l-asbābi ajrā sunnatahu bi-rabṭi l-musabbabāti bi-l-asbābi iẓhāran li-l-ḥikma*; al-Ghazālī, *Iḥyā'*, 4:355.3 / 2564.16–17.

41. Ibid. 4:307.20 / 2498.2. The usage of the verb *sakhkhara* is Qur'anic; see for example Q 13:2, 16:14.

42. Al-Ghazālī, *Iḥyā'*, 4:111.11 / 2224.3–4.

43. Ibid. 4:111.12–15 / 2224.5–8. Ḥalabī's edition has *mūjid* instead of *mūjib*.

44. Ibn Sīnā, *al-Ishārāt wa-l-tanbīhāt*, 188.17–19. Cf. Marmura, "Divine Omniscience," 91–92. Abrahamov, "Ibn Sīnā's Influence on al-Ghazālī's Non-Philosophical Works," 14–16, deals with Ibn Sīnā's influence on some of al-Ghazālī's views about reward in the hereafter.

45. Al-Ghazālī, *al-Maqṣad al-asnā*, 98.4–6.

46. *bal [huwa] mumahhidu sharṭi l-ḥuṣūl li-ghayrihi.*

47. Al-Ghazālī, *Iḥyā'*, 4:111.23–112.3 / 2224.19–*ult*. Cf. Frank, *Creation*, 26.

48. On al-Ghazālī's understanding of *jawhar*—which does not concur with the earlier Ash'arites' understanding of *jawhar* as an atom—see Frank, *Al-Ghazālī and the Ash'arite School*, 48–67.

49. Al-Ghazālī, *Iḥyā'*, 4:112.11–12 / 2225.9–10. On the *ḥadīth*, see al-Bukhārī, *al-Ṣaḥīḥ*, *jihād*, 144; and Wensinck, *Concordance*, 2:501a.

50. Al-Ghazālī, *Iḥyā'*, 4:317.17–20 / 2511.*paenult*–2512.1; *al-rāsikhūna fī-l-'ilm* is taken from Q 3:7 in which—according to al-Ghazālī's interpretation of this verse—those "deeply rooted in knowledge" are identified as the scholars who know the meaning of difficult passages from revelation. On al-Ghazālī's understanding of Q 3:7, see Griffel, *Apostasie und Toleranz*, 448.

51. Al-Ghazālī, *Iḥyā'*, 4:317.20–25 / 2512.3–6. See al-Zabīdī, *Itḥāf al-sāda*, 9:423.20–26, who lacks the word *azaliyya* in the third sentence.

52. Gramlich, *Muḥammad al-Ġazzālīs Lehre*, 542–43.

53. Al-Isfarā'īnī, "al-'Aqīda," 168, fragm. 94. Cf. al-Ghazālī, *al-Maqṣad*, 105.17: "God arranges them (*scil.* all things) in their appropriate places and thereby He is just (*'adl*)." See Frank, *Creation*, 64–65, and particularly 56–57, where he discusses the meaning of *ḥaqq* in a very similar passage to the one we are looking at. On *ḥaqq*, cf. a passage in al-Ghazālī, *al-Iqtiṣād*, 102.2–3, on the relationship between God's will and his foreknowledge: "The divine foreknowledge is true to (*ḥaqqa al-'ilm*) [the decision of the divine will] and contains them as they are."

54. See above pp. 141–43.

55. Al-Ghazālī, *Maqāṣid al-falāsifa*, 2:82.*paenult*.–85.2 / 236.24–239.3; MS London, Or. 3126, foll. 237b–240a. The text in *Maqāṣid* is a free adaptation of Ibn Sīnā, *Dānishnāmah-yi 'Alā'-i, Ilāhiyyāt*, 95–97, that illustrates Ibn Sīnā's teachings with original examples. For the Avicennan texts used in the passage of the London MS, see Griffel, "MS London, British Library, Or. 3126: An Unknown Work," 15.

56. Al-Ghazālī, MS London, Or. 3126, foll. 238a.8–238b.9.

57. Ibn Sīnā, *al-Shifā'*, *al-Ilāhiyyāt*, 339.4–8; *al-Najāt*, 284.12–13 / 668.14–16, quoted in al-Ghazālī, MS London Or. 3126, fol. 238a.5–8.

58. Al-Juwaynī, *al-Shāmil* (ed. Alexandria), 621.21–22 and 622.3–8 (with the example—also used by al-Ghazālī—that a well-written manuscript gives necessary evidence to the skills of the scribe); al-Makkī, *Qūt al-qulūb*, 2:35–36. The motif that God's creation is skillfully arranged goes back to the Qur'an, in which in various contexts it says that God's creation contains "signs (*āyāt*) for those who can understand." The idea that certain divine attributes show in His creation is also Qur'anic. The usual proof for God's oneness and unity in *kalām* literature is by mutual hindrance (*tamānuʿ*). It appears already in Q 23:91, 17:42–3: Because this creation shows no signs of the activity of more than one divine creative force, which would necessarily compete with and hinder one another, there is only one God. See also the story of Abraham's conversion to monotheism in Q 6:75–79, discussed below.

59. Kukkonen, "Plentitude, Possibility, and the Limits of Reason," 545–46; Davidson, *Proofs*, 226–27, 234; Goodman, "Ghazâlî's Argument from Creation," 69. On arguments from design in the thirty-second book of the *Iḥyāʾ*, see van den Bergh, "Ghazali on 'Gratitude Towards God,'" 86–88, 97–98.

60. Al-Ghazālī, *al-Iqtiṣād*, 99–100.

61. Ormsby, *Theodicy in Islamic Thought*, 37, 75. For a discussion of al-Ghazālī's position, see ibid., 39–74; Frank, *Creation*, 60–66; and idem, *Al-Ghazālī and the Ashʿarite School*, 20–21.

62. The debate is documented and analyzed in Ormsby, *Theodicy in Islamic Thought*, 92–265, and al-Zabīdī, *Itḥāf al-sāda*, 9:434–60.

63. Al-Ghazālī, *Iḥyāʾ*, 4:322.1–3 / 2518.5–8; English translation in Ormsby, *Theodicy in Islamic Thought*, 40–41. Cf. al-Zabīdī, *Itḥāf al-sāda*, 9:433. These words seem to be inspired by the beginning paragraph of Ibn Sīnā's *Risāla Fī sirr al-qadar*, in which all elements in al-Ghazālī's passage (the obscurity of the matter, the deep sea, the prohibition to teach it to the *ʿāmma*) also appear. See the text of the *Risāla Fī sirr al-qadar* in Hourani, "Ibn Sīnā's 'Essay on the Secret of Destiny,'" 27–31; and in ʿĀṣi, *al-Tafsīr al-Qurʾānī*, 302–5.

64. Al-Ghazālī, *al-Imlāʾ fī ishkālāt al-Iḥyāʾ*, 50–51 / 3083–85; MS Yale, Landberg 428, pp. 55–56. For an English synopsis of this passage, see Ormsby, *Theodicy in Islamic Thought*, 75–81. The available editions of *al-Imlāʾ fī ishkālāt al-Iḥyāʾ*, printed at the end or on the margins of many editions of the *Iḥyāʾ ʿulūm al-dīn* and on the margins of al-Zabīdī, *Itḥāf al-sāda*, 1:192–204, are of very poor quality, with some textual passages (likely some lines in the underlying manuscript) missing. They also represent a recension of the text that is not original. In this recension, the order of the *fuṣūl* does not match the description of the contents given by al-Ghazālī at the beginning of the work on pp. 19–20 / 3038–39. A more original recension and a much better text is available in manuscripts such as MS Yale, Landberg 428 (once owned by al-Murtaḍā al-Zabīdī), and, with slight variations in the order of the text, MS Berlin, Petermann II 545 (Ahlwardt 1714).

65. On al-Makkī's *Qūt al-qulūb*, see the very helpful German translation by Richard Gramlich, *Die Nahrung der Herzen*, which includes notes and an analytical index.

66. See above p. 222.

67. Al-Ghazālī, *Iḥyāʾ*, 4:321.1–16 / 2516.17–2517.13. English translation in Ormsby, *Theodicy in Islamic Thought*, 38–39.

68. Al-Makkī, *Qūt al-qulūb*, 2:35–36. Cf. Ormsby, *Theodicy in Islamic Thought*, 41, 45, 81.

69. Al-Makkī, *Qūt al-qulūb*, 2:35.*paenult.–ult*. English translation in Ormsby, *Theodicy*, 58; German translation in Gramlich, *Die Nahrung der Herzen*, 2:396; and idem, *Muḥammad al-Ġazzālī's Lehre*, 549.

70. *wa-laysa fī-l-imkāni aṣlan aḥsanu minhu wa-lā atamma wa-lā akmala*; al-Ghazālī, *Iḥyā'*, 4:321.16–18 / 2517.13–16. Cf. al-Zabīdī, *Itḥāf al-sāda*, 9:430.18–26. See the English translation in Ormsby, *Theodicy in Islamic Thought*, 39.

71. Al-Ghazālī, *Iḥyā'*, 3:73.10–13 / 1446–47.

72. Al-Ghazālī, *Iḥyā'*, 4:321.20–26 / 2517.18–2518.2. See the English translation in Ormsby, *Theodicy in Islamic Thought*, 40, and his commentary on pp. 64–69.

73. Ormsby, *Theodicy in Islamic Thought*, 257.

74. Ibn Sīnā, *al-Ishārāt wa-l-tanbīhāt*, 186.5–6. Ormsby, *Theodicy in Islamic Thought*, 257, says that according to Ibn Sīnā, harm appears *accidental* when good is created. This is, however, a misunderstanding that seems to be based on Ibn Sīnā's wording in *al-Ishārāt*, 186.1 and 187.1–3. Frank, *Creation*, 61, shares this misunderstanding. Creating good, however, *necessarily* requires the creation of harm.

75. Ibn Sīnā, *al-Ishārāt wa-l-tanbīhāt*, 185–87. On Ibn Sīnā's teachings on harm or evil (*sharr*), see Steel, "Avicenna and Thomas Aquinas on Evil," 173–86.

76. Ibn Sīnā, *al-Shifā',al-Ilāhiyyāt*, 342.4–5. Steel, "Avicenna and Thomas Aquinas on Evil," 179–81.

77. Ibn Sīnā, *al-Shifā', al-Ilāhiyyāt*, 340.11; idem, *al-Najāt*, 285.11 / 670.17. Steel, "Avicenna and Thomas Aquinas on Evil," 174–77. This seems to be a premise not shared by al-Ghazālī. In his *Imlā'*, 50.10–11 / 3083.20–21; MS Landberg 428, p. 55.20, al-Ghazālī counters the objection that the idea of the best of all possible worlds is incompatible with the position of the world's creation in time. His brief response makes sense only if existence is not regarded as better than nonexistence. On the apparent incompatibility of the best of all possible worlds and creation in time, see Ormsby, *Theodicy*, 76–77; and Frank, *Creation*, 66.

78. Ibn Sīnā, *al-Shifā', al-Ilāhiyyāt*, 339.13–15; *al-Najāt*, 284.18–19 / 669.9–10.

79. Ibn Sīnā, *al-Shifā', al-Ilāhiyyāt*, 341.8–10; *al-Najāt*, 286.4–7 / 672.9–13. The position that species are unaffected by harm does not seem to have been shared by al-Ghazālī, who considers the species of beasts (*bahā'im*) harmful (*Iḥyā'*, 4:321.24–25 / 2518.1).

80. Ibn Sīnā, *al-Shifā', al-Ilāhiyyāt*, 341.8–9; *al-Najāt*, 286.5 / 667.9–10. Cf. Aristotle, *Metaphysics*, 1010a.25–30.

81. Al-Ghazālī, *Iḥyā'*, 4:124.21–125.2 / 2242.17–2243.2. It is here that al-Ghazālī says: "An ignorant friend is worse than an insightful foe." Van den Bergh, "Ghazali on 'Gratitude Towards God,'" 92, remarks that al-Ghazālī "may have read it" in *Kalīla wa-Dimna*. Van den Bergh's article has no references, and as far as I am aware, there is no such sentence in that work.

82. Al-Ghazālī, *al-Maqṣad al-asnā*, 68.1–5.

83. Ibid. 68.6–8.

84. Ibid. 68.15–16; 69.15–16. For the *ḥadīth*, see al-Bukhārī, *al-Ṣaḥiḥ*, *tawḥīd* 15, 22, 28, 55; or Muslim ibn al-Ḥajjāj, *al-Ṣaḥīḥ*, *tawba* 14–16. Cf. Wensinck, *Concordance*, 4:526a.

85. Al-Ghazālī, MS Yale, Landberg 428, p. 56.4–7. In the printed text in *al-Imlā'*, 50.16–18 / 3084.6–8, and in the margins of al-Zabīdī, *Itḥāf al-sāda*, 1:201, this sentence is corrupted. See also the translation in Ormsby, *Theodicy*, 78, based on MS Berlin, Petermann II 545, fol. 16b.

86. Al-Ghazālī, *al-Imlā'*, 50.20–21 / 3083.10–11; idem, MS Yale, Landberg 428, p. 56.9–10. Cf. the text in the margins of al-Zabīdī, *Itḥāf al-sāda*, 1:201.

87. Al-Ghazālī, *al-Imlā'*, 50.23–51.7 / 3084.14–3085.5. This passage is not in MS Yale, Landberg 428. It is this reasoning that likely lies behind al-Ghazālī's decision only to write about the world's perfection in two comparatively brief passages in his *Iḥyā' 'ulūm al-dīn* and in his *al-Imlā' fī ishkālāt al-Iḥyā'*. The subject is not explicitly discussed

in other books of the *Iḥyāʾ* circle, such as *al-Arbaʿīn* or *Kīmiyāʾ-yi saʿādat*. Al-Ghazālī, however, alludes to it in *Tahāfut al-falāsifa*, 289.4–6 / 172.17; al-*Iqtiṣād*, 165–66; *al-Maqṣad al-asnā*ʾ, 47.12–13, 68.6–8, 105–6, 81.12–13, 109.8–15, 152.11–13; and probably many other passages of his works.

88. Al-Ghazālī, *al-Imlā*ʾ, 50.13–16 / 3084.2–6. Corrected according to MS Yale, Landberg 428, pp. 55.*ult*.–56.4, which varies in the following readings: *wa-l-yataḥaqqiqa*; *wa-anna dhālika ʿalā ghāyati l-ḥikma*; and *burhān*[an] *wāḍiḥ*[an].

89. Al-Ghazālī, *Iḥyāʾ*, 4:317.*ult*.–318.4 / 2512.12–16. Cf. al-Zabīdī, *Itḥāf al-sāda*, 9:423–24. This passage is translated and its language discussed in Frank, *Creation*, 56–61. See also Gramlich's German translation in *Muḥammad al-Ġazzālīs Lehre*. 543.

90. See p. 225.

91. See p. 228.

92. Frank, *Creation*, 55–63.

93. Ibid., 69.

94. This is most forcefully expressed in al-Ghazālī, *Tahāfut*, 96–103 / 56–60; 155–60 / 91–94. See also above, pp. 184–85.

95. Al-Ghazālī, *Tahāfut*, 64.5–66.6 / 38.12–39.13. Logically impossible means "conjoining negation and affirmation" (*al-jamʿ bayna l-nafī wa-l-ithbāt*).

96. Al-Ghazālī, *Iḥyāʾ*, 1:148.16–18 / 189.1–3; Tibawi, "Al-Ghazālī's Sojoun," 85.6–7, 105. Kukkonen, "Possible Worlds," 480, concludes that al-Ghazālī's innovations to the philosophy of Ibn Sīnā "have their root in the idea of God freely choosing (arbitrating) between alternatives equal to him."

97. Al-Ghazālī, *al-Iqtiṣād*, 129–39. This category came to be used by the Nishapurian Ashʿarites, most probably in conscious response to the *falāsifa*'s teachings. ʿAbd al-Qāhir al-Baghdādī's *Uṣūl al-dīn*, 117–18, 121–22, refers to a group of divine names that are derived from God's essence and to a second group that "are derived from an attribute that He has residing within Him" (*mushtaqq min ṣifa lahu qāʾimat*[an] *bihi*). A third group is derived from God's actions. On al-Baghdādī's division, see Gimaret, *Les noms divins en Islam*, 107–8. Ibn Sīnā already refers to this concept and says in *al-Ḥikma al-ʿarshiyya*, 9.9, that God's knowledge is not *zaʾid ʿalā l-dhāt*. On al-Ghazālī's conception of the divine attributes, see Frank, *Creation*, 47–52.

98. *ghayru maqdūrin ʿalā maʿnā anna wujūdahu yuʾaddī ilā stiḥāla*; al-Ghazālī, *al-Iqtiṣād*, 85.8–86.4. Marmura, "Ghazali's Chapter on Divine Power," 301–2. On this sense of necessity in al-Ghazālī, see Kukkonen, "Causality and Cosmology," 41–42.

CHAPTER 9

1. Al-Ghazālī, *al-Maqṣad*, 98.6–102.14; idem, al-*Arbaʿīn*, 13.6–18.*ult*. / 12.5–14.*ult*. The passage in *al-Arbaʿīn* is often left out from manuscripts of the book. In fact, among the five manuscripts I looked at—two in Berlin (Sprenger 763 and 941; see Ahlwardt, *Handschriften-Verzeichnisse*, nos. 1715–716), two in Princeton (Yahuda Collection, nos. 3893 and 4374; see Mach, *Catalogue*, no. 2161), and one at Yale's Beinecke Library (Arabic MSS suppl. 425)—none contained this passage, and two of them (Sprenger 763 and Yahuda 3893) have heavily abbreviated sections on God's will (*irāda*), opening the possibility that al-Ghazālī published more than one version of *al-Arbaʿīn*. Cf. the English translation of the passage in *al-Maqṣad* in Burrell and Daher, *The Ninety-Nine Beautiful Names of God*, 86–88. On this passage, see Frank, *Creation and the Cosmic System*, 42–44; and Abrahamov, "Al-Ghazālī's Theory of Causality."

2. On the dating of these two works, see Bouyges, *Essai de chronologie*, 46–47, 50–51; and Hourani, "Revised Chronology," 298–99. Hourani observes that in his

al-Mustaṣfā, 1:5.3–4 / 1:4.2, al-Ghazālī makes it clear that *Jawāhir al-Qurʾān* was composed before "the return to teaching." In *al-Arbaʿīn*, 305.9–10 / 264.13–14, he presents that book as the concluding part of *Jawāhir al-Qurʾān*.

3. *aṣlu waḍʿi l-asbābi li-tatawajjaha ilā l-musabbabāt*; al-Ghazālī, *al-Maqṣad*, 98.9–10; *al-Arbaʿīn*, 13.7 / 12.5–6.

4. Al-Ghazālī, *al-Maqṣad*, 98.7, 109.14–15.

5. Ibid., 98.12; idem, *al-Arbaʿīn*, 13.9–10 / 12.9–10.

6. Al-Ghazālī, *al-Maqṣad*, 98.16–99.1; idem, *al-Arbaʿīn*, 13.13–14.2 / 12.11–14.

7. Al-Ghazālī, *al-Maqṣad*, 79–81; see the analysis in Gimaret, *Les noms divins*, 282, 287.

8. Al-Ghazālī, *al-Maqṣad*, 99.4–13; idem, *al-Arbaʿīn*, 14.5–14 / 12.16–13.4.

9. Einhard, "Annales. 741–829," 194.22–28.

10. Hill, "The Toledo Water-Clock of c.1075."

11. Al-Khāzinī, *Mīzām al-ḥikma*, 153–66. On al-Khāzinī and his descriptions of water clocks, see Hill, *Arabic Water-Clocks*, 47–68.

12. Muḥammad ibn ʿAlī ibn Rustum al-Saʿātī; Ibn Abī Uṣaybiʿa, *ʿUyūn al-anbāʾ*, 2:183–84; Ibn Jubayr, *Tadhkira bi-l-akhbār*, 216–17; Hill, *Arabic Water-Clocks*, 69–74. Soon after this period, at the beginning of the seventh/thirteenth century, Ibn al-Razzāz al-Jazarī wrote his famous *Kitāb fī Maʿrifat al-ḥiyal al-handasiyya* for a local ruler in northern Syria. On the even more sophisticated water clocks in that book, see Donald Hill's annotated translation of al-Jazarī, *The Book of Knowledge of Ingenious Mechanical Devices*, 17–93.

13. Al-Ghazālī, *al-Maqṣad*, 99.15–16; idem, *al-Arbaʿīn*, 14.15–16 / 13.5–6.

14. *kullu dhālika yataqaddaru bi-taqadduri sabbabih*; al-Ghazālī, *al-Maqṣad*, 100.2–3; idem, *al-Arbaʿīn*, 15.2 / 13.9.

15. Al-Ghazālī, *al-Maqṣad*, 100.3–6; idem, *al-Arbaʿīn*, 16.11–16 / 13.9–12.

16. Marmura, "Ghazālian Causes and Intermediaries," 97.

17. Al-Ghazālī, *Iḥyāʾ*, 4:317.15 / 2511.*paenult*.

18. Al-Ghazālī, *al-Maqṣad*, 100.7–101.3; idem, *al-Arbaʿīn*, 15.7–16.6 / 13.13–*ult*.

19. Al-Ghazālī, *al-Maqṣad*, 101.9–14; idem, *al-Arbaʿīn*, 16.11–16 / 14.5–8.

20. Janssens, "The Problem of Human Freedom in Ibn Sînâ," 118, notes the same ambiguity of "measure" and "predestination" in Ibn Sīnā's usage of *qadar/qadr*. For Ibn Sīnā's teachings on *al-qaḍāʾ wa-l-qadar*, see Belo, *Chance and Determinism*, 113–19.

21. *ṣifat*[un] *ʿanha yaṣduru l-khalqu wa-l-ikhtirāʿ*; al-Ghazālī, *Iḥyāʾ*, 4:119.21–*ult.*/ 2235.19–2236.3.

22. Al-Ghazālī, *al-Maqṣad*, 103.1–5; see Kukkonen, "Possible Worlds," 485.

23. Al-Ghazālī, *Iḥyāʾ*, 4:119.14–19 / 2235.9–15.

24. Ibid., 4:122.13–14 / 2239.5.

25. Ibid., 4:122.18–22 / 2239.11–22. Like Gramlich, *Muḥammad al-Ġazzālī's Lehre*, 211–12, in his German translation, I follow al-Zabīdī, *Itḥāf al-sāda*, 9:76.16, who has *illā l-ʿulamāʾ* instead of *bi-l-nisbat al-ʿulamāʾ*.

26. Al-Ghazālī, *Iḥyāʾ*, 4:122.22–25 / 2239.22–2240.2; again following al-Zabīdī, *Itḥāf al-sāda*, 9:76.22, who adds the word "glances" (*abṣāra*) to the second sentence.

27. For this identification and division, see also Frank, *Creation*, 43–44.

28. Plato, *Laws I*, 644d–646a. In Plato, the heavenly creatures are gods who control the affections of humans. On the connection between these two texts, see van den Bergh, "Ghazali on 'Gratitude Towards God,'" 85.

29. Al-Ghazālī, *Miḥakk al-naẓar*, 72.11–16; cf. the English translation in Frank, *Al-Ghazâlî and the Ashʿarite School*, 127.

30. Al-Ghazālī, *Miḥakk al-naẓar*, 72.16–17.

31. This sentence is missing from the 1967/68 edition of the *Iḥyāʾ*, 4:103.13, but it is in the 1937/39 edition (2213.14–15) as well as in al-Zabīdī, *Itḥāf al-sāda*, 9:50.13–14. See almost the same sentence in *Iḥyāʾ*, 1:104.10–17 / 126.6–19; and in *al-Munqidh*, 23.11–13, where he adds that the same is true for the whole nature (*al-ṭabīʿa*).

32. Al-Ghazālī's attitude here is certainly influenced by that of Ibn Sīnā, who accepted the general idea of astrology, namely that events are predetermined by celestial causes, but rejected the way it was conducted by the astrologers because too many causes are involved for humans to keep track of them. See Michot in his introduction to Ibn Sīnā, *Réfutation de l'astrologie*, 76*–79*.

33. Al-Ghazālī, *Iḥyāʾ*, 1:48–49, 57.9–13 / 54–55, 65.12–15.

34. Ibid., 4:146.13–15 / 2272.18–19; al-Zabīdī, *Itḥāf al-sāda*, 9:118.28–29. See Marmura, "Al-Ghazālī," 151, who inexplicably omits the last part of the sentence. On al-Ghazālī's position toward astronomy and astrology, see the brief sketch by Saliba, "The Ashʿarites and the Science of he Stars," 85–87.

35. See also Davidson, *Alfarabi, Avicenna, and Averroes, on Intellect*, 133–38, 151.

36. *bi-l-ramzi wa-l-īmāʾi ʿalā sabīli l-tamthīli wa-l-ijmāl*; al-Ghazālī, *Iḥyāʾ*, 1:12.23–*ult.* / 5.7–10; al-Zabīdī, *Itḥāf al-sāda*, 1:63.

37. These are books thirty-two (*al-Ṣabr wa-l-Shukr*), thirty-six (*al-Maḥabba wa-l-Shawq*), and the first part on *tawḥīd* in thirty-five.

38. Al-Ghazālī, *al-Arbaʿīn*, 25.4–10 / 23.1–7; see Marmura, "Ghazali and Ashʿarism Revisited," 100.

39. Al-Ghazālī, *Maqṣad*, 98.17, 102.6; Frank, *Creation*, 22, 42.

40. Al-Ghazālī, *Maqṣad*, 47.12–13, 50.5–6; Frank, *Creation*, 69.

41. Al-Ghazālī, *Maqṣad*, 137.9–11; Frank, *Creation*, 55.

42. Al-Ghazālī, *al-Arbaʿīn*, 11.14–15 / 11.7–9.

43. Al-Ghazālī, *al-Maqṣad*, 125.9–10; Frank, *Creation*, 22.

44. Al-Ghazālī, *al-Maqṣad*, 83.15–16; Frank, *Creation*, 38.

45. Al-Ghazālī, *al-Maqṣad*, 125.2–12.

46. Ibid., 81.17–18, 152.11–15; Frank, *Creation*, 60–61.

47. Al-Ghazālī, *al-Maqṣad*, 68.7–9; Frank, *Creation*, 61.

48. Al-Ghazālī, *al-Maqṣad*, 152.13; Frank, *Creation*, 60.

49. Al-Ghazālī, *al-Maqṣad*, 81.20; Frank, *Creation*, 48.

50. The argument is reproduced by Bouyges, *Essai de chronologie*, 65, who finds it sound but not fully convincing. See, for instance, Gairdner, "Al-Ghazālī's Mishkāt al-anwār," 121. In the introduction to his English translation of *The Niche of Lights*, xviii–xxi, Buchman recently gave a vocal expression of this mistaken view on the chronology of al-Ghazālī's work. See also Watt, "Authenticity of the Works," 44, who dates the *Mishkāt* to the very end of al-Ghazālī's life in what he calls the *dhawq* period.

51. In the *Mishkāt*, 46.11 / 126.3, 56.15 / 139.1, al-Ghazālī mentions the twenty-first book of the *Iḥyāʾ* and *al-Maqṣad al-asnā*. See Bouyges, *Essai de chronologie*, 65–66; and Hourani, "Revised Chronology," 299. The latter puts the case clearly when he says that there "is no way to attain further accuracy on the date of the *Mishkāt* between *Jawāhir al-Qurʾān* and al-Ghazālī's death (. . .)."

52. Watt, "A Forgery in al-Ghazālī's *Mishkāt*?" I discuss the secondary literature on the veil section in my forthcoming article, "Al-Ghazālī's Cosmology in the Veil Section of His *Mishkāt al-Anwār*."

53. Not all available editions reproduce a reliable text. Particularly the *editio princeps* by Aḥmad ʿIzzat and Farajallāh Zakī al-Kurdī (Cairo: Maṭbaʿat al-Ṣidq, 1322 [1904–5]), which provided the text for several reprints, cannot be relied on.

54. Ibn Rushd, *al-Kashf ʿan manāhij*, 183.*ult.*–184.3; idem, *Tahāfut al-tahāfut*, 117.6–8.

55. See, for instance, Ibn Fūrak, *Kitāb Mushkil al-ḥadīth*, 183.

56. For earlier attempts to identify the groups mentioned in the Veil Section, see the important contributions by Landolt, "Ghazālī and 'Religionswissenschaft,'" 31–62; Ansari, "The Doctrine of Divine Command," 36–37; and Gairdner's introduction to his translation, *Al-Ghazzālī's Mishkāt al-Anwār*, 5–8.

57. Landolt, "Ghazālī and 'Religionswissenschaft,'" 39.

58. Al-Ghazālī, *Mishkāt al-anwār*, 67–68 / 154.

59. Al-Ṭabarī, *Jāmiʿ al-bayān* (ed. Cairo), 11:480–83; Fakhr al-Dīn al-Rāzī, *al-Tafsīr al-kabīr*, 13:47; al-Ghazālī, *Fayṣal al-tafriqa*, 190.18–*paenult.* / 54.9–13. The information about Abraham's youth comes from Rabbinic literature (*Bereshith Rabbā*, 38; *Talmud Nedārīm*, 32; etc.).

60. Cf., for instance, al-Ghazālī, *Fayṣal al-tafriqa*, 190.4–191.3 / 53.6–55.3.

61. Al-Ghazālī, *Mishkāt*, 90.13–*ult.* / 183.11–*ult.*

62. Ibid., 68.10–16 / 155.3–8.

63. Ibid., 91.1–3 / 184.1–3.

64. Al-Ghazālī, *Fayṣal al-tafriqa*, 190.7 / 53.10–11.

65. Maimonides, *Dalālat al-ḥāʾirīn*, 375.23–25; English translation, 2:515.

66. Ibn Bājja, *Sharḥ al-samāʿ al-ṭabīʿī*, 16.8–17.1. Cf. Aristotle, *Physics*, 184b–188a.

67. Ibn Bājja, *Sharḥ al-samāʿ al-ṭabīʿī*, 17.5–10.

68. On Aristotle's teachings in *Physics* I, 2, and 3, 184b–188a, and the commentaries of John Philoponos (d. *c.* 570), Abū ʿAlī Ibn al-Samḥ (d. 418/1027), and Ibn Bājja, and Ibn Rushd's middle and long commentary on this passage, see Lettinck, *Aristotle's Physics and Its Reception in the Arabic World*, 38–53, 71, 78–82.

69. Ibn Sīnā, *al-Shifāʾ, al-Ṭabīʿiyyāt, al-Samāʿ al-ṭabīʿī*, 26.8–9.

70. Rudolph, *Doxographie des Pseudo-Ammonios*, 50.6–9, 51.2–6.

71. Al-Shahrastānī, *al-Milal wa-l-niḥal*, 2:254.3–6, 265.17–*ult.*; French translation, 2:181, 200. Al-Shahrastānī provides a three-page commentary on Abraham's discovery of his Lord (Q 6.75–79) within his doxographic treatment of the Sabians, the pagan polytheists of antiquity. He interprets the full Qur'anic passage as a historic account of how Abraham defeated the "followers of the structures" (*aṣḥab al-hayākil*), a subgroup of the Sabians, by realizing and pointing out that the "structures" (*hayākil*) they believe in, that is, the seven celestial bodies, are moved and governed by a superior power. See al-Shahrastānī, *al-Milal wa-l-niḥal*, 2:247–48; French translation, 2:164–66.

72. Al-Ghazālī, *Mishkāt al-anwār*, 91.4–6 / 184.3–6.

73. See above pp. 136–37.

74. Al-Ghazālī, *Mishkāt al-anwār*, 91.7–10 / 184.7–10.

75. Aristotle's kinematic proof for God's existence is developed in *Physics*, 256a–259a, and *Metaphysics*, 1071b–73a. He uses the idea that every movement is necessarily the effect of a mover to argue that, since there are evidently movements and thus movers in the world, there must be somebody or something that has caused the very first movement. "First" here is not understood in a temporal meaning but rather in an ontological one. Cause and effect must exist simultaneously. Thus God who is the cause of the first movement and of the prime mover exists coeternally with this world.

76. Al-Ghazāli, MS London, British Library, Or. 3126, foll. 3a–b.

77. Ibn Sīnā, *al-Ishārāt wa-l-tanbīhāt*, 146.15–17; cf. Griffel, "MS London, British Library Or. 3126," 17.

78. Al-Ghazālī, *Mishkāt al-anwār*, 91.10–12 / 184.10–12.

79. Druart, "Al-Fārābī: Metaphysics."

80. Al-Ghazālī, *Mishkāt al-anwār*, 91.13–92.1 / 184.13–18.

81. Ibid., 92.2–4 / 184.18–*ult.*

82. Ibid., 92.5–13 / 185.1–7.

83. Ibid., 57.*paenult.*–58.5 / 141.3–9, 92.12–13 / 185.7–8; Davidson, *Alfarabi, Avicenna, and Averroes, on Intellect*, 140.

84. Al-Ghazālī, *Iḥyāʾ*, 4:107.12–108.4 / 2218.18–2219.15.

85. Goldziher, "Materialien zur Kenntnis der Almohadenbewegung," 72, 83: "eine pantheistische Nuance." The article is reprinted in Goldziher's *Gesammelte Schriften*, 2:191–301. Gairdner, "Al-Ghazālī's Mishkāt al-anwār," 152, saw al-Ghazālī "perpetually trembling on the edge of the pantheistic abyss" (discussed in more detail in Gairdner, *Al-Ghazzālī's Mishkāt al-Anwār*, 34–41).

86. Treiger, "Monism and Monotheism in al-Ghazālī's *Mishkāt al anwār*," 1.

87. Ibid., 14–16.

88. Al-Ghazālī, *al-Mustaṣfā*, 1:81.8–11 / 1:27.1–2. On the Ashʿarite background of this passage, see Sabra, "*Kalām* Atomism as an Alternative," 220–21.

89. Frank, *Al-Ghazālī and the Ashʿarite School*, 39–42.

90. Al-Ṭabarī, *Jāmiʿ al-bayān* (ed. Būlāq), 30:51.9; Fakhr al-Dīn al-Rāzī, *al-Tafsīr al-kabīr*, 31:72–73.

91. *bi-ʿtibār kawnihi matbūʿan fī ḥaqqi baʿḍi l-malāʾika*; al-Ghazālī, *Fayṣal al-tafriqa*, 182.*ult.* / 38.7–9. The words *fī ḥaqq* in this sentence are quite unclear. They might express the regular and lawful nature of the angel's actions. This is how Gramlich translated these words in other contexts. See p. 225.

92. Al-Ghazālī, *Iḥyāʾ*, 4:121–22 / 2238.12–14.

93. Al-Ghazālī, MS London, Or. 3126, foll. 252b–253b: "Chapter five on the fact that the intellect is by nature a king who is obeyed (*malik muṭāʿ bi-l-ṭabʿ*)." This chapter is copied from Miskawayh, *Kitāb al-Fawz al-aṣghar*, 130–33.

94. Cf. also al-Ghazālī, *Tahāfut*, 250.1 / 150.13: "the celestial angels are obedient to God."

95. *yatanazzalu l-amru baynahunna.*

96. Al-Fārābī, *Fuṣūṣ al-ḥikam*, 61–62, 71–72, 81–82. Cf. Alon, *Al-Fārābī's Philosophical Lexicon*, 13–14. In Ibn Sīnā, *malakūt* stands for the realm of the immaterial souls of the celestial spheres without their material bodies; see Davidson, *Alfarabi, Avicenna, and Averroes, on Intellect*, 119, 133.

97. Al-ʿĀmirī, *Kitāb al-Fuṣūl fi-l-maʿālim al-ilāhiyya*, ed. Khalīfāt 84.21–22 / ed. Wakelnig 364.11–12. See also *Rasāʾil Ikhwān al-ṣafāʾ*, 3:234.1/ 3:238.20–21; cf. Diwald, *Arabische Philosophie und Wissenschaft*, 179–80. On the philosophical identification of *al-amr* with the universal forms, see Wakelnig, *Feder, Tafel, Mensch*, 161–62, 392.

98. *ilā dawām al-amr*; Ibn Sīnā, *al-Ḥikma al-ʿarshiyya*, 15.16–17. On this verse, see also pp. 198–99.

99. Al-Ghazālī, *Iḥyāʾ*, 3:473.2 / 2017.23–24.

100. Wensinck, "On The Relationship Between Ghazālī's Cosmology and His Mysticism," 199–201.

101. Al-Ghazālī (?), *Maʿārij al-quds*, 203–5, offers more detailed explanations about the nature of the *amr* and the *muṭāʿ*. Ansari, "The Doctrine of Divine Command," 38–41, translates and analyzes this passage.

102. Gairdner's assumption that the *muṭāʿ* is a demiurge or a "vicegerent" ("Al-Ghazālī's Mishkāt al-Anwār," 141–43; and idem, *Al-Ghazzālī's Mishkāt al-Anwār*, 10–25) is not justified. The first intellect has no autonomy and is—in the parlance of the *falāsifa*—simply the first secondary cause even if the relationship between it and God is not strictly causal.

103. Al-Ghazālī, *al-Maqṣad*, 100.7–101.3; idem, *al-Arbaʿīn*, 15.7–16.6 / 13.13–*ult*. See n. 18 above.

104. See above p. 185.

105. Ibn Sīnā, *al-Ḥikma al-ʿarshiyya*, 14.6–10. God knows which of the possibilities will become actual; see ibid., 9.11–14. Before they become truly existent, the quiddities exist, according to Ibn Sīnā, in a state of "divine existence" (*wujūd ilāhī*); see *al-Shifāʾ, al-Ilāhiyyāt*, 156.6–8. On the ontological status of the quiddities in Avicenna, see Bäck, "Avicenna's Conception of the Modalities," 246–48.

106. Frank, *Creation*, 62–63, 65, 84.

107. Kukkonen, "Possible Worlds," 495–96;

108. Smith, "Avicenna and the Possibles," 346–47, 357; Zedler, "Why Are the Possibles Possible?" 115–18; eadem, "Another Look at Avicenna," 513–19. The problem is best put by Aimé Forest in *La structure métaphysique*, 153, 161: "Chez Avicenne les possibles s'offrent éternellement à l'action divine, parce qu'ils ne sont pas constitutés comme tels par sa volonté. Dieu pense nécessairement sa propre nature, sa liberalité n'est que l'acquiescement à cet ordre universel des choses qu'il ne constitue pas. (. . .) [L]es essences sont possible avant d'être, elles sont constituées indépendamment de la volonté divine et de la Sagesse."

109. Burrell, *Freedom and Creation*, 34–37, 43–46.

110. Landolt, "Ghazālī and 'Religionswissenschaft,'" 46. Landolt is rather unspecific about which elements of Ismāʿilism al-Ghazālī adopted and how they served his own teachings.

111. Ibid. 43–46.

112. Halm, *Kosmologie und Heilslehre*, 14; cf. Wilferd Madelung, in his article on "Ismāʿiliyya" in *EI2*, 4:203–4.

113. Walker, *Early Philosophical Shiism*, 82–86; idem, "The Ismā'ilīs," 81–84; and idem, "The Ismaili Vocabulary of Creation."

114. Walker, *Early Philosophical Shiism*, 95–106; Halm, *Kosmologie und Heilslehre*, 83–85.

115. Walker, *Ḥamīd al-Dīn al-Kirmānī*, 85.

116. De Smet, *La quietude de l'intellect*, 110–53, 159–76, 187–99; Walker, "The Ismā'īlīs," 84–89.

117. *al-ibdāʿu lladhī huwa l-mubdaʿu al-awwal*; al-Kirmānī, *Rāḥat al-ʿaql*, 174.14–15; 176.*ult*.–177.1, 254.2.

118. Ibid., 158.10–3.

119. De Smet, *La quietude de l'intellect*, 138–40; Walker, *Ḥamīd al-Dīn al-Kirmānī*, 84–85.

120. Baffioni, "Contrarity and Similarity in God," 19.

121. Wilferd Madelung surveys al-Ghazālī's works on the Ismāʿīlites in *EIran*, 10:376–77.

122. On these earlier cosmologies, particularly the teachings of Muḥammad ibn Aḥmad al-Nasafī see Walker, *Early Philosophical Shiism*, 55–60; idem, "The Ismāʿīlīs," 78–79, Halm, *The Fatimids and Their Tradition of Learning*, 51–53; idem *Kosmologie und Heilslehre*, 53–91; and Daftari, *The Ismāʿīlīs*, 240–45.

123. Al-Ghazālī, *Faḍāʾiḥ al-bāṭiniyya*, 39.5–10; cf. Ibn al-Walīd, *Dāmigh al-bāṭil*, 1:134.15–20.

124. Al-Ghazālī, *Faḍāʾiḥ al-bāṭiniyya*, 38.9–13; cf. Ibn al-Walīd, *Dāmigh al-bāṭil*, 1:131.15–18. See also Landolt, "Ghazālī and '*Religionswissenschaft*,'" 44. For the source of this common misunderstanding among non-Ismāʿīlite authors, see Halm, *Kosmologie und Heilslehre*, 79–80.

125. Al-Ghazālī, *Faḍāʾiḥ al-bāṭiniyya*, 40.8–10: cf. Ibn al-Walīd, *Dāmigh al-bāṭil*, 1:143.*paenult.*–144.2. On a certain similarity of the early Ismāʿīlite teachings with Manichaeism, see Daftari, *The Ismāʿīlīs*, 244.

126. Al-Ghazālī reports the teachings of the "bāṭiniyya" on metaphysics in his *Faḍāʾiḥ al-bāṭiniyya*, 38–40; cf. Ibn al-Walīd, *Dāmigh al-bāṭil*, 1:131–46; and Goldziher, *Streitschrift*, 44–45. He accuses the Ismāʿīlites of teaching a dualism of *ʿaql* and *nafs* as first and second divine beings. For similar criticism, see al-Baghdādī's report of the Ismāʿīlite teachings in his *Farq bayna l-firaq*, 316.11–15. Al-Baghdādī quotes these teachings from al-Nasafī's lost *Kitāb al-Maḥṣūl*. Walker, "The Ismaili Vocabulary of Creation," 79, connects al-Baghdādī's report to al-Sijistānī's teachings.

127. Al-Ghazālī, *Faḍāʾiḥ al-bāṭiniyya*, 40.10–17; cf. Ibn al-Walīd, *Dāmigh al-bāṭil*, 1:144.4–10.

128. Al-Baghdādī, *al-Farq bayna l-firaq*, 316.11–12, quotes from al-Nasafī's *Kitāb al-Maḥṣūl*, which became available to non-Ismāʿīlites (see Walker, *Early Philosophical Shiism*, 55–56). Apparently, al-Ghazālī relied heavily on al-Bāqillānī's lost book, *Kashf al-asrār wa-hatk al-astār*. In his *Iḥyāʾ*, 3:179.14–15 / 900.4–5, he takes the rare step of acknowledging his reliance on that work. Al-Zabīdī, *Itḥāf al-sāda*, 6:122.10, however, omits this passage, which prompted Goldziher, *Streitschrift*, 16, to assume it is an interpolation. Goldziher, however, bases his argument on grounds that are not all convincing. On al-Bāqillānī's book, see also al-Subkī, *Ṭabaqāt*, 7:18.1–5.

129. See, for instance, al-Ghazālī, *al-Munqidh*, 33.11–16; or idem, *Qawāṣim al-bāṭiniyya*, 33–34. In his *Munqidh*, 29.2–6, al-Ghazālī mentions that "one of his colleagues" (? *wāḥid min aṣḥābī*) attached himself to al-Ghazālī after he had professed the Ismāʿīlite teachings. The former Ismāʿīlite informed him of their arguments.

130. Al-Kirmānī, *Rāḥat al-ʿaql*, 279; de Smet, *La quietude de l'intellect*, 279–81.

131. De Smet, *La quietude de l'intellect*, 120–38.

132. Niẓāmī, *Khamsah*, 5:401–4 (*Sharafnāmah*, lines 5111–53). In what is probably its most well-known version in the West in Rūmī's *Mas̱navī*, 1:213–15, English translation 2:189–90 (book 1, lines 3467–99), the roles of the Greek and Chinese painters are reversed from that in al-Ghazālī and Niẓāmī, and the competition is no longer unresolved. The Greeks are declared winners, thus expressing the superiority of mystical experience over acquired knowledge.

133. Al-Ghazālī, *Iḥyāʾ*, 3:28.17–29.3 / 1382.9–22; the story also appears in *Mīẓan al-ʿamal*, 37–38 / 225–26. Moosa (*Ghazālī and the Poetics*, 254–55) and Soucek ("Niẓāmī on Painters and Painting," 14) understand the text as if the Chinese painters are judged superior over the Greek and that al-Ghazālī thus favored the Sufi method. That is, however, not expressed anywhere in al-Ghazālī's texts. The equality of both ways is, in fact, stressed in Niẓāmī's version, which seems to be directly inspired by al-Ghazālī's version. In Niẓāmī, the Greek is the best in painting (*naqsh*) and the Chinese the best in polishing (*ṣaql*) (line 5153), yet both achieve results that are absolutely indistinguishable from each other and equivalent.

134. Al-Ghazālī, *Mīzān al-ʿamal*, 37.18–38.1 / 226.6–8. Al-Ghazālī uses the word "heart" as a synonym to the philosophical usage of the word "soul" (*nafs*); see Griffel, "Al-Ġazālī's Concept of Prophecy," 142, n. 132.

135. *kufr ṣurāḥ*; al-Ghazālī, *Fayṣal al-tafriqa*, 198.5–8 / 66–67. See also *Faḍāʾiḥ al-bāṭiniyya*, 39.5–11; cf. Ibn al-Walīd, *Dāmigh al-bāṭil*, 1:134.15–20.

136. Baffioni, "Contrariety and Similarity in God," 4.

137. Frank, *Al-Ghazālī and the Ashʿarite School*, 91.

138. There seems to have been a development about what al-Ghazālī considered a *fāʿil*. See pp. 184–85.

139. The passage is also discussed by Frank in *Creation and the Cosmic System*, 44–45, 75–77, and in *Al-Ghazālī and the Ash'arite School*, 19. For Frank, this passage is a major textual evidence supporting his conclusion that al-Ghazālī's God has no free choice in his creation and creates out of necessity, without free will (*Creation and the Cosmic System*, 75). Frank, however, reports and translates only a small part of a longer argument.

140. MS Istanbul, Şehit Ali Paşa 1712, fol. 32b. See Bouyges, *Essai de chronology*, 81, in which the date of the colophon on fol. 32b is incorrect.

141. *wa-yuqālu innahā ākhiru muʾallafātihi*; MS Berlin, Petermann II 690, fol. 1a. See Ahlwardt, *Handschriften-Verzeichnisse*, 2:527–28 (no. 2301).

142. See above pp. 56–57.

143. Al-Ghazālī, *Iljām al-ʿawāmm*, 3 / 53–54. The text of this work, which is poorly edited, has been compared to the manuscripts MSS Istanbul, Şehit Ali Paşa 1712, foll. 1a–32b; and Berlin, Petermann II 690. The Istanbul manuscript has been edited by Sāmiḥ Dughaym (Beirut: Dār al-Fikr al-Lubnānī, 1993), and the edition reproduces the original folio division in its margins.

144. Al-Ghazālī, *Iljām al-ʿawāmm*, 13 / 70.

145. Ibid., 8–9 / 64–65.

146. *bayānu maʿnāhu baʿda izālati ẓāhirih*; ibid., 10.12–13 / 66.*ult*. The two MSS have *ẓāhirihi*.

147. Al-Ghazālī, *Iljām al-ʿawāmm*, 11.9–10 / 68.5–6.

148. Ibn Sīnā, *al-Shifāʾ*, *al-Manṭiq*, *al-Burhān*, 51.2–52.8.

149. Griffel, "Al-Ghazālī's Concept of Prophecy," 124.

150. Al-Ghazālī, *Fayṣal al-tafriqa*, 41–60 / 184–94.

151. (. . .)*min ghayri tarjīḥ*; al-Ghazālī, *Iljām al-ʿawāmm*, 11.11–13 / 68.8–9.

152. Ibid., 11.9–11 / 68.5–8 (*farq* in line 6 is to be amended by *fawq*). Cf. *Faḍāʾiḥ al-bāṭiniyya*, 155.11–12.

153. Al-Ghazālī, *Iljām al-ʿawāmm*, 155.12–14.

154. Ibid., 11.18–20 / 68.16–18.

155. The *Rasāʾil Ikhwān al-ṣafāʾ*, 2:22.10–3 / 2:26.13–5, teaches that the next to outer sphere is the "bearing (or the throne)" or the "pedestal" (*kursī*, cf. Q 2:255) and the outermost sphere is the "throne." According to al-ʿĀmirī, *Kitāb al-Fuṣūl fi-l-maʿālim al-ilāhiyya*, ed. Khalīfāt 84.*ult*–87.1 / ed. Wakelnig 364.12–13, the *falāsifa* use the word "throne" to refer to the "straight sphere (*al-falak al-mustaqīm*) and the *primum mobile* (*falak al-aflāk*)," which is the outermost sphere. On the identification of *al-ʿarsh* as the highest being and the starless sphere, see Heinen, *Islamic Cosmology*, 77–81, 83; and Wakelnig, *Feder, Tafel, Mensch*, 167, 392. In his *Tahāfut*, 261.3–8 / 157.1–7, al-Ghazālī reports and criticizes the teachings of the *falāsifa* with regard to the "preserved tablet" (*al-lawḥ al-maḥfūẓ*) and the "pen" (*al-qalam*), but he never mentions those with regard to "the throne." See Frank, *Creation*, 45.

156. Al-Ghazālī, *Iljām al-ʿawāmm*, 11.20–23 / 68.19–21.

157. Ibid., 11.*ult*. / 68.23.

158. See particularly the twenty-sixth and thirty-fourth epistles of *Rasāʾil Ikhwān al-ṣafāʾ*, 3:3–36 / 2:456–79, 3:211–26 / 3:212–30. Cf. Nasr, *Introduction to Islamic Cosmological Doctrines*, 66–74, 96–104.

159. Al-Ghazālī, *Mishkāt al-anwār*, 67.5–7 / 153.9–11; 66.8 / 152.16. This idea is also present in *Rasāʾil Ikhwān al-ṣafāʾ*, 3:3–4 / 2:456–57.

160. *al-adnā bayyina ʿalā l-aʿlā*; al-Ghazālī, *Jawāhir al-Qurʾān*, 51.5; Ormsby, *Theodicy*, 45.

161. Al-Ghazālī, *al-Maqṣad*, 152.11–13; Frank, Creation, 60.

162. Al-Ghazālī, *Iḥyāʾ*, 4:146.6–7 / 2272.9–10.

163. Al-Ghazālī, *Miḥakk al-naẓar*, 124.14–16.

164. Al-Ghazālī, *Iljām al-ʿawāmm*, 11.*ult.*–12.3 / 68.*ult.*–69.2.

165. *lā yumkinuhu min al-tadbīr*, following MS Berlin, Petermann II, 690, fol. 6b.

166. Al-Ghazālī, *Iljām al-ʿawāmm.*, 12.3–9 / 69.2–9.

167. The usage of *taṣarrafa* in this context also appears in al-Ghazālī (?), *Maʿārij al-quds*, 198.14–15: *taṣarrufu l-ādamiyyu fī ʿālimihi aʿnī badanahu yashbihu taṣarrufa l-khāliqi fī l-ʿālami l-akbar.*

168. *natījatu l-wājibi wājibun (not: wājibatun)*; ibid., 12.7 / 69.8; following the MSS.

169. Al-Ghazālī, *Miʿyār al-ʿilm*, 221.21–22; idem,*al-Iqtiṣād*, 43.3; Frank, *Creation*, 72, 81–82. Frank points to another passage in *al-Iqtiṣād*, 78.4–9, but the necessity there is of the sort that follows as a consequence of earlier creations; it is necessity from something else, a sort of necessity that does not pose the kind of problems currently discussed.

170. Cf. Ibn Sīnā, *al-Shifāʾ*, *al-Ilāhiyyāt*, 328.2; *al-Najāt*, 228.17 / 553.9–10.

171. The implications of the statement that God is *wājib al-wujūd min jamīʿ jihātihi* are clearly spelled out in al-Ghazālī, MS London, Or. 3126, fol. 198b.

172. Griffel, "MS London, British Library Or. 3126," 21–29. See also Marmura, "Ghazali and Demonstrative Science," 184, 189.

173. Al-Ghazālī, *Miʿyār al-ʿilm*, 25.1–26.10.

174. Frank, *Al-Ghazālī and the Ashʿarite School*, 93–94; Marmura, "Ghazali and Demonstrative Science," 192.

175. Al-Ghazālī, *Miʿyār al-ʿilm*, 221.21–222.1. The corresponding passage in MS London, Or. 3126, contains a long discussion of God's essence in book 6, chapter 1 of that text (foll. 197b–207a).

176. Al-Ghazālī, *al-Iqtiṣād*, 42.11–43.3.

177. Al-Ghazālī, *Iljām al-ʿawāmm*, 12.9–12 / 69.9–12.

178. Ibid., 12.10–11 / 69.11.

179. Ibid., 12.13–14 / 69.14–15.

180. In al-Ghazālī (?), *Maʿārij al-quds*, 198–99, the author explains that while in the microcosm of the human body, production is mediated first through the animal soul (*al-rūḥ al-ḥayawānī*) and then through the brain (*al-dimāgh*), so too does God create in the macrocosm "through the mediation of moving the heavens and the planets." He specifies that the relationship between the human heart and the brain is equivalent to the relationship between the throne (*al-ʿarsh*) and the stool or pedestal (*al-kursī*). The senses (*al-ḥawāss*) are to the human what the angels, that is, the celestial intellects, are to God.

181. Al-Ghazālī, *Iljām al-ʿawāmm*, 11.20–22 / 68.19–21.

CONCLUSION

1. Ibn Ṭufayl, *Ḥayy ibn Yaqẓān*, 16.*ult.*–17.10.

2. ʿAyn al-Quḍāt, *Nāmah-hā*, 1:79.7–10. More precisely, ʿAyn al-Quḍāt says that al-Ghazālī did not explain the level of meaning in the Qur'an that is geared to the intellectual elite.

3. Frank, *Al-Ghazālī and the Ashʿarite School*, 100–101.

4. For a recent example in more popular literature about the history of science, see Steven Weinberg's comments on al-Ghazālī in his review article, "A Deadly Certitude." In what is partly a response to Weinberg, Robert Irwin, "Islamic Science and the Long Siesta," largely accepts Weinberg's view of al-Ghazālī's devastating effect,

a view that is indeed widespread among current historians of the sciences and of philosophy.

5. Al-Ghazālī, *Tahāfut*, 277.2–3 / 166.1–2.

6. Kukkonen, "Possible Worlds," 496, 499.

7. See above pp. 128–32.

8. Ibn Rushd, *Tahāfut al-tahāfut*, 523.2–16, 531–32, complains about al-Ghazali's use of the word "habit" (*ʿāda*). If al-Ghazālī means that existing things have a "habit," he should use "nature" instead since "habit" is only applicable to animate things. If the "habit" exists only in our judgment, he should instead use the world "intellect" (*ʿaql*) because that is the agent (*fāʿil*) of such a habit.

9. Marmura, "Al-Ghazālī's Second Causal Theory." 86. Marmura, however, believed that al-Ghazālī is committed only to the first theory that represents an occasionalist view of causal connections.

10. Al-Ghazālī, *Iḥyāʾ*, 1:50.7–10 / 56.11–14; al-Zabīdī, *Itḥāf al-sādā*, 1:236–37, with *lam yaqūmū bihi* as a variant to *lam yataṣiffū bihi.*

11. In al-Ghazālī (?), *Maʿārij al-quds*, 14.16–*ult*, the active intellect is described as "the substrate of the cognitions, of revelation, and of inspiration (*ilhām*)." The active intellect is also referred to as "the spirit" (*al-rūḥ*) as well as "the pen" and "the creation that flows out of God's command (*al-mubdaʿ al-ṣādir min amr Allāh*).

12. Al-Ghazālī, *Mīzān al-ʿamal*, 107.8–15 / 331.1–13.

13. Al-Ghazālī (?), *Maʿārij al-quds*, 16.1–2. The Arabic *mubdaʿ al-awwal* or maybe even *mabdaʿ al-awwal* ("first invention") seems almost a pun on the philosophical name for the same being, *al-mabdaʾ al-awwal* ("the first principle").

14. Ibid., 15.5–7; al-Ghazālī, *Fayṣal al-tafriqa*, 182.6–12 / 36.8–37.7; idem, *Iḥyāʾ*, 1:115.17–18 / 142.1–2. The *ḥadīth* that the first creation is the pen is reported by al-Tirmidhī, *Jāmiʿ al-ṣaḥīh, tafsīr sūrat 68*. See Wensinck, *Concordance*, 1:135a. The *ḥadīth* that this first creation is the intellect is not considered sound. An even longer version of the spurious *ʿaql-ḥadīth*, containing a short cosmogenic narrative, is quoted by al-Ghazālī, *Mīzān al-ʿamal*, 107.10–12 / 331.4–9.

15. *khāzinun li-anfusi khazāʾinihi*; al-Ghazālī, *Mīzān al-ʿamal*, 107.6 / 330.16–17.

16. In the twenty-first book of his *Iḥyāʾ*, 3:27.3–7 / 1380.5–9, al-Ghazālī says that the creator writes His plan for creation on the "well-guarded tablet" just as the architect writes his plan for a house on paper (see Nakamura, "Ghazālī's Cosmology Reconsidered," 35–36). Frank, *Al-Ghazālī and the Ashʿarite School*, 26–27, argues that the well-guarded tablet "designates the angel (separated intelligence) that is associated with the outermost celestial sphere." That intelligence, however, exists even beyond the outermost sphere.

17. Craig, *Kalām Cosmological Argument*, 12, 15, 150–51; idem, *The Cosmological Argument*, 56, 58.

18. See p. 142.

19. Baneth, "Jehuda Hallewi und Gazali," 35; idem, "Rabbi Yehudah ha-Levi we-Algazzali," 320.

20. See above pp. 253–60.

21. See above pp. 256–57.

22. Baljon, "The 'Amr of God' in the Koran," 15–16. On modern Western as well as Ismāʿīlite and Muslim philosophical interpretations of what the word *amr* stands for in the Qur'an, see Wakelnig, *Feder, Tafel, Mensch*, 159–62.

23. See above pp. 260–64.

24. At least four years lie between the composition of al-Ghazālī's *Faḍāʾiḥ al-bāṭiniyya*—our main source for his knowledge of Ismāʿīlite cosmology,—and his *Mishkāt*

al-anwār. They allow for an improvement of al-Ghazālī's understanding of Ismāʿīlite theology.

25. On the non-Fāṭimid Ismāʿīlite background of the *Rasāʾil Ikhwān al-ṣafāʾ*, see Wilferd Madelung in his article "Karmaṭī," in *EI2*, 4:663a

26. See above pp. 199–201.

27. In the early parts of the Veil Section in *Mishkāt al-anwār*, al-Ghazālī draws on material from the long forty-second epistle in the *Rasāʾil Ikhwān al-ṣafāʾ*, "The Different Beliefs and Religions" (*Fī l-ārāʾ wa-l-diyānāt*); see Landolt, "Ghazālī and Religionswissenschaft," 29–31.

28. Al-Ghazālī, *al-Munqidh*, 33.19–22.

29. Stern, "Authorship of the Epistles," 368.

30. Diwald, *Arabische Philosophie und Enzyklopädie*, 313–14, connects al-Ghazālī's division of the two worlds to a similar one in *Rasāʾil Ikhwān al-ṣafāʾ*, 3:282.3–7 / 3:293.19–24. The resemblance, however, remains general and unspecific and takes no account of the third Ghazalian realm, *ʿālam al-jabarūt*.

31. Ibn Taymiyya, "Sharḥ al-ʿaqīda al-iṣfahāniyya," 117–18. Ibn Taymiyya rejects al-Māzarī's view that there was also an influence from Abū Ḥayyān al-Tawḥīdī. Cf. Laoust, *Essai sur les doctrines*, 82–84.

32. Ibn Rushd, *al-Kashf ʿan manāhij al-adilla*, 183.*ult*.–184.3; see also *Tahāfut al-tahāfut*, 117.6–8. In the notes to his Hebrew translation of al-Ghazālī's *Maqāṣid al-falāsifa*, the Jewish Averroist Issac Albalag (fl. *c*. 1290) discusses Ibn Rushd's and al-Ghazālī's positions on the relationship between the mover of the *primum mobile* and God. See Vajda, *Isaac Albalag*, 31–32, 95–98; and Steinschneider, *Die hebraeischen Übersetzungen*, 1:303. Throughout his career as a writer, Ibn Rushd held different opinions about whether the mover of the highest sphere is identical to God or whether God is the creator of this mover; see Kogan, "Averroës and the Theory of Emanation," 396–97.

33. Ibn Ṭufayl, *Ḥayy ibn Yaqẓān*, 17.10–18.3. Ibn Ṭufayl did not share this view.

34. (. . .) *wa-hādhā min jinsi kalāmi l-bāṭiniyya*; Ibn al-Jawzī, *Talbīs Iblīs*, 166.3–7. Cf. Ibn Taymiyya, *Minhāj al-sunna*, 4:149.19–20.

35. Frank, *Al-Ghazālī and the Ashʿarite School*, 87, 101.

36. Frank, "The Non-Existent and the Possible in Classical Ashʿarite Teaching," 16–17.

37. See, for instance, al-Juwaynī, *Irshād*, 110.3.

38. Ibid., 210.3–4.

39. Al-Ghazālī, *Munqidh*, 23.11–13. "Elemental natures" (*al-ṭabāʾiʿ*) seems to refer to the four prime elements (*usṭuqusāt*).

40. Al-Ghazālī, *al-Munqidh*, 45.3–9.

41. Al-Ghazālī, *Iḥyāʾ*, 1:115–16 / 142.34–37.

42. Gianotti, *Al-Ghazālī's Unspeakable Doctrine of the Soul*, 168.

43. Al-Ghazālī, *Mishkāt al-anwār*, 42.2–3 / 120.8–9, 51–52 / 133.7–13, 60.2–3 / 133.11–12, 67.15–16 / 153.3–4. On emanation in al-Ghazālī, particularly in the *Mishkāt al-anwār*, see Davidson, *Alfarabi, Avicenna, and Averroes, on Intellect*, 135–36, 151; and Frank, *Creation*, 83.

44. Lazarus-Yafeh, *Studies in Al-Ghazzali*, 307–12. Gairdner, "Al-Ghazālī's Mishkāt al-Anwār," 138–39, had developed a similar argument based on the use of prepositions by al-Ghazālī.

45. In the often-quoted passage from *Mīzān al-ʿamal*, 161–64 / 405–9, in which al-Ghazālī describes three different levels of outspokenness a scholar might have with regard to his teachings, he actually rejects the described attitude. He refers to a group of scholars who express one set of teachings in public disputations, another group who

express their teachings while instructing their students, and a third group who keep it secret between themselves and God. Al-Ghazālī contrasts this attitude with the position to thoroughly investigate the subject in question, to develop *one* position, and to teach that in all circumstances to all people. The latter is the attitude to be favored. Ibn Ṭufayl's remark (*Ḥayy ibn Yaqẓān*, 16.2–6) that the attitude described in this section is the root of some apparent contradictions in al-Ghazālī's works has had a very misleading influence on many later interpreters.

46. Al-Ghazālī, *Iḥyāʾ*, 1:44–61 / 49–70.

47. See, for instance, Strauss, *Persecution and the Art of Writing*, 38–78.

BIBLIOGRAPHY

1. Only the first part of the *Naṣīḥat al-mulūk* (pp. 1–79 in Humāʾī's edition) is authored by al-Ghazālī; the second part (pp. 81–287) is by another, unknown author of the sixth/twelfth century. See Crone, "Did al-Ghazāli Write a Mirror for Princes?"; Hillenbrand, "Islamic Orthodoxy or Realpolitik?" 92; and Humāʾī's introduction to his edition of *Naṣīḥat al-mulūk*, 65–80.

Bibliography

WORKS BY AL-GHAZĀLĪ

Though trying to be comprehensive, the following list of al-Ghazālī's works can only be a guide, aiming to point the reader to his most relevant texts in the most reliable editions. In their respective catalogues of manuscripts preserved in the libraries of Turkey, Ali Riza Karabulut (3:1370–77) and Ramazan Şeşen (633–39) list works ascribed to al-Ghazālī that were not available to me and thus do not appear in this list. Some of these works, such as the *Taḥṣīn al-maʾākhiẓ* (MS Istanbul, Topkapı Serayı Müzesi Kütüphanesi, Ahmet III 1099; Karatay #4451), a mid-size book on the substantive rulings (*furūʿ*) of Islamic law, seem to be quite significant.

If a text reference in the notes of this book is followed by more than one page reference (usually divided by a slash), such as al-Ghazālī, *Fayṣal al-tafriqa*, 187/47, the numbers refer to the pages in the edition that is first mentioned in this list followed by the one mentioned second. For a more precise identification of the works that are cited in more than one edition see p. 17 of this book.

Texts of al-Ghazālī are in Arabic unless noted otherwise. Modern translations of works by al-Ghazālī are referred to under the name of the translator in the section on works by other authors.

ON THEOLOGY, PHILOSOPHY, ETHICS, AND THEIR METHODS

Al-Ghazālī. *Al-Adab fī l-dīn*. In idem. *Jawāhir al-ghawālī min rasāʾil al-Imām Ḥujjat al-Islām al-Ghazālī*. Edited by Muḥyī al-Dīn Ṣabrī al-Kurdī. Cairo: Maṭbaʿat al-Saʿāda, 1353/1934. 41–58.

———. *Kitāb al-Arbaʿīn fī uṣūl al-dīn*. Edited by Muḥyī al-Dīn Ṣabrī al-Kurdī. Cairo: al-Maṭbaʿa al-ʿArabiyya, 1344 [1925].

———. *Al-Arbaʿīn fī uṣūl al-dīn.* Edited by Muḥammad M. Jābir. Cairo: Maktabat al-Jundī, n.d. [1964].

———. *Ay farzand* (in Persian). In *Makātīb-i fārisī-yi Ghazzālī be-nām-i Fażāʾil al-anām min rasāʾil Ḥujjat al-Islām.* Edited by ʿAbbās Iqbāl Āshtiyānī. Tehran: Kitābfurūsh-i Ibn Sīnā, 1333 [1954]. 91–112.

———. *Ayyuhā l-walad = Lettre au disciple (Ayyuhā 'l-walad).* Edition and French translation by Toufic Sabbagh. 2nd ed. Beirut: Commission internationale pour la traduction des chefs-d'œuvre, 1959.

———. *Ayyuhā l-walad.* MS Berlin, Orientabteilung der Staatsbibliothek Preussischer Kulturbesitz, Sprenger 1968, foll. 51b–60a.

———. *Bidāyat al-hidāya wa-tahdhīb al-nufūs bi-l-ādāb al-sharʿiyya.* Edited by Muḥammad Saʿūd Maʿīnī. Baghdad: Sāʿadat Jāmiʿat Baghdād, 1988.

———. *Bidāyat al-hidāya.* MS Princeton, University Library, Yahuda 459. foll. 1a–23a.

———. *Al-Durra al-fākhira fī kashf ʿulūm al-ākhira = Ad-Dourra al-Fâkhira. La Perle Precieuse de Ghazâlî. Traité d'eschatologie musulmane publié d'après les manuscits de Leipzig, de Berlin, de Paris et d'Oxford et une lithographie orientale (. . .).* Edited by Lucien Gauthier. Geneva/Basle/Lyon: H. Georg, 1878.

———. *Faḍāʾiḥ al-bāṭiniyya wa-faḍāʾil al-Mustaẓhiriyya.* Edited by ʿAbd al-Raḥmān Badawī. Cairo: Dār al-Qawmiyya, 1383/1964.

———. *Fayṣal al-tafriqa bayna l-Islām wa-l-zandaqa.* Edited by Sulaymān Dunyā. Cairo: ʿĪsā al-Bābī al-Ḥalabī, 1381/1961.

———. *Fayṣal al-tafriqa bayna l-Islām wa-l-zandaqa.* Edited by Maḥmūd Bījū. Damascus: Maḥmūd Bījū, 1413/1993.

———. *[Fayṣal] al-Tafriqa bayna l-Islām wa-l-zandaqa.* MS Istanbul, Süleymaniye Kütüphanesi, Şehit Ali Paşa 1712, foll. 57a–70b.

———. *Ḥimāqat-i ahl-i ibāḥat* (in Persian). In *Streitschrift des Ġazālī gegen die Ibāḥīja.* Edited and German translation by Otto Pretzl. Munich: Bayerische Akademie der Wissenschaften, 1933.

———. *Ḥimāqat-i ahl-i ibāḥat* (in Persian). In Pūrjavādī. *Dō mujaddid. Pazhūhish-hā-yi dar bāra-yi Muḥammad-i Ghazzālī va-Fakhr-i Rāzī / Two Renewers of Faith.* 153–210.

———. *Iḥyāʾ ʿulūm al-dīn.* 5 vols. Cairo: Muʾassasat al-Ḥalabī wa-Shurakāʾhu, 1387/1967–68.

———. *Iḥyāʾ ʿulūm al-dīn.* 16 parts in 5 vols. Cairo: Lajnat Nashr al-Thaqāfa al-Islāmiyya, 1356–57 (1937–39). Photographic reprint in 6 vols, Beirut: Dār al-Kitāb al-ʿArabī, n.d. [*c.*1990].

———. *Iljām al-ʿawāmm.* Edited by Aḥmad al-Bābī al-Ḥalabī. Cairo: al-Maṭbaʿa al-Maymaniyya, 1309 [1891]. (Published in a volume together with *al-Munqidh min al-ḍalāl, al-Maḍnūn bihi ʿalā ghayri ahlih,* and *al-Maḍnūn al-ṣaghīr.*)

———. *Iljām al-ʿawāmm ʿan ʿilm al-kalām.* Edited by Muḥammad al-Muʿtaṣim bi-Llāh al-Baghdādī. Beirut: Dār al-Kitāb al-ʿArabī, 1406/1985.

———. *Iljām al-ʿawāmm ʿan ʿilm al-kalām.* MS Istanbul, Süleymaniye Kütüphanesi, Şehit Ali Paşa 1712, foll. 1a–32b.

———. *Al-Imlāʾ fī ishkālāt al-Iḥyāʾ.* In al-Ghazālī, *Iḥyāʾ ʿulūm al-dīn.* 5 vols. Cairo: Muʾassasat al-Ḥalabī wa-Shurakāʾhu, 1387/1967–68. 5:18–58.

———. *Al-Imlāʾ fī ishkālāt al-Iḥyāʾ* In al-Ghazālī, *Iḥyāʾ ʿulūm al-dīn.* 16 parts in 5 vols. Cairo: Lajnat Nashr al-Thaqāfa al-Islāmiyya, 1356–57 (1937–39). 16:3036–95.

———. *Al-Imlāʾ ʿalā l-Iḥyāʾ.* MS Yale University, Beinecke Library, Landberg 428.

———. *Al-Iqtiṣād fī l-iʿtiqād.* Edited by Ibrahim Agâh Çubukçu and Hüseyin Atay. Ankara: Nur Matbaasi, 1962.

———. "Jawāb al-masāʾil al-arbaʿ allatī saʾalahā al-bāṭiniyya bi-Hamadān." *al-Manār* (Cairo) 11.8 (1326): 601–8 (published 29 Shaʿbān 1326 / 25 September 1908).
———. *Jawāhir al-Qurʾān*. Edited by Muḥyī l-Dīn Ṣabrī al-Kurdī. Cairo: Maṭbaʿat Kurdistān al-ʿIlmiyya, 1329 [1911].
———. *Jawāhir al-Qurʾān*. MS Escorial 1130, foll. 8a–18b. (Incomplete.)
———. *Al-Kashf wa-l-tabyīn fī ghurūr al-khalq ajmaʿīn*. In ʿAbd al-Wahhāb ibn Aḥmad al-Shaʿrānī, *Tanbīh al-mughtarrīn*. Cairo: Muṣṭafā l-Bābī al-Ḥalabī wa-Awlāduhu, 1356/1937. 182–95.
———. *Kīmyāʾ al-saʿāda* [in Arabic]. In idem. *Jawāhir al-ghawālī min rasāʾil al-Imām Ḥujjat al-Islām al-Ghazālī*. Edited by Muḥyī al-Dīn Ṣabrī al-Kurdī. Cairo: Maṭbaʿat al-Saʿāda, 1353/1934. 5–19.
———. *Kīmyā-yi saʿādat* [in Persian]. Edited by Ḥusayn Khadīvjam. 2 vols. Tehran: Intishārāt-i ʿIlmī va-Farhangī 1382 [2003].
———. *Al-Lubāb min al-Iḥyāʾ*. Printed on the margins of Ibn Abī l-Munā, ʿAbd al-Malik al-Bābī al-Ḥalabī. *Nuzhat al-nāẓirīn fī tafsīr āyāt min kitāb rabb al-ʿālamīn*. Cairo: Dār al-Kutub al-ʿArabiyya al-Kubrā–Muṣṭafā l-Bābī al-Ḥalabī wa-Akhawayhi, 1328 [1910].
———. *Al-Lubāb min al-Iḥyāʾ= Mukhtaṣar Iḥyāʾʿulūm al-dīn*. Edited by Shaʿbān Muḥammad Ismāʿīl. Cairo: Maktabat Naṣīr, 1978.
———. *Al-Maʿārif al-ʿaqliyya*. Edited by ʿAlī Idrīsī. Safaqis (Tunisia): al-Taʿāḍudiyya al-ʿAmāliyya li-l-Ṭibāʿa wa-l-Nashr: 1988.
———. (?) *Maʿārij al-quds fī madārij maʿrifat al-nafs*. Edited by Muḥyī l-Dīn Ṣabrī al-Kurdī. Cairo: Maṭbaʿat al-Saʿāda, 1346/1927.
———. (?) *Al-Maḍnūn bihi ʿalā ghayr ahlih*. Edited by Aḥmad al-Bābī al-Ḥalabī. Cairo: al-Maṭbaʿa al-Maymaniyya, 1309 [1891]. (Published in a volume together with *Iljām al-ʿawāmm, al-Munqidh min al-ḍalāl*, and *al-Maḍnūn al-ṣaghīr*.)
———. (?) *Al-Maḍnūn bihi ʿalā ghayr ahlih*. MS Berlin, Orientabteilung der Staatsbibliothek Preussischer Kulturbesitz, Petermann II 35, foll. 32a–54b.
———. (?) *Al-Maḍnūn al-ṣaghīr wa-huwa al-mawsūm bi-l-Ajwiba al-Ghazāliyya fī l-masāʾil al-ukhrawiyya*. Edited by Aḥmad al-Bābī al-Ḥalabī. Cairo: al-Maṭbaʿa al-Maymaniyya, 1309 [1891]. (Published in a volume together with *Iljām al-ʿawāmm, al-Munqidh min al-ḍalāl*, and *al-Maḍnūn al-kabīr*.)
———. *Maktūb-i Ghazālī dar bāra-yi ahl-i ibāḥah* (in Persian). In Pūrjavādī. *Dō mujaddid. Pazhūhish-hā-yi dar bāra-yi Muḥammad-i Ghazzālī va-Fakhr-i Rāzī / Two Renewers of Faith*, 139–45.
———. *Maqāṣid al-falāsifa*. Edited by Muḥyī al-Dīn Ṣabrī al-Kurdī. 3 parts. Cairo: al-Maṭbaʿa al-Maḥmūdiyya al-Tijāriyya, 1936.
———. *Maqāṣid al-falāsifa = Muqaddimat Tahāfut al-falāsifa al-musammāt Maqāṣid al-falāsifa*. Edited by Sulaymān Dunyā. 2nd ed. Cairo: Dār al-Maʿarif, n.d. [1960].
———. *Al-Maqṣad al-asnā fi sharḥ maʿānī asmāʾ Allāh al-ḥusnā*. Edited by Fadlou A. Shehadi. 2nd ed. Beirut: Dār al-Mashriq, 1982.
———. *Miḥakk al-naẓar fī l-manṭiq*. Edited by Muḥammad Badr al-Dīn al-Naʿsānī and Muṣṭafā al-Qabbānī. Cairo: al-Maṭbaʿa al-Adabiyya, n.d. [1925].
———. (?) *Minhāj al-ʿābidīn ilā jannat rabb al-ʿālamīn*. Edited by Maḥmūd Muṣṭafā Ḥalāwī. Beirut: Muʾassasat al-Risāla, 1409/1989.
———. *Mishkāt al-anwār*. Edited by Abū l-ʿIlā ʿAfīfī. Cairo: al-Dār al-Qawmiyya, 1383/1964.
———. *Mishkāt al-anwār wa-miṣfāt al-asrār*. Edited by Abd al-ʿAzīz ʿIzz al-Dīn al-Sayrawān. Beirut: ʿĀlam al-Kutub, 1407/1986.

———. *Mishkāt al-anwār fī tawḥīd al-jabbār.* MS Escurial 1130, foll. 73a–87a.
———. *Mishkāt al-anwār wa-miṣfāt al-asrār.* MS American University of Beirut, Jafet Memorial Library, MS 297.3: G41 iA.
———. *Miʿyār al-ʿilm fī fann al-manṭiq.* Edited by Muḥyī al-Dīn Ṣabrī al-Kurdī. Cairo: al-Maṭbāʿa al-ʿArabiyya, 1346/1927.
———. *Miʿyār al-ʿilm fī fann al-manṭiq.* MS Vatican, Bibliotheca Apostolica, Ebr. 426, foll. 102a–176a. (Judeo-Arabic fragment corresponding to pp. 61–222 in Ṣabrī al-Kurdī's edition.)
———. *Mīzān al-ʿamal.* Edited by Muḥyī al-Dīn Ṣabrī al-Kurdī. Cairo: al-Maṭbaʿa al-ʿArabiyya, 1342 [1923].
———. *Mīzān al-ʿamal.* Edited by Sulaymān Dunyā. Cairo: Dār al-Maʿārif, 1964.
———. MS London, British Library, Or. 3126.
———. *Al-Murshid al-amīn ilā mawʿiẓat al-muʾminīn min Iḥyāʾ ʿulūm al-dīn.* Cairo: Muṣṭafā al-Bābī al-Ḥalabī 1389/1969.
———. (?) *Nafkh al-rūḥ wa-l-taswiya.* Edited by Aḥmad Ḥijāzī al-Saqqā. Cairo: Maktabat al-Madīna al-Munawwara, 1399/1979.
———. *Rawḍat al-ṭālibīn wa-ʿumdat al-sālikīn.* Edited by Muḥammad Bakhīt. Beirut: Dār al-Nahḍa al-Ḥadītha, n.d. [1966].
———. *Risāla Fī bayān maʿrifat Allāh.* In al-Ghazālī, *Talāth rasāʾil fī l-maʿrifa lam tunshar min qabl.* Edited by Maḥmūd Ḥamdī Zaqzūq. Cairo: Maktabat al-Azhar, 1399/1979. 15–22.
———. (?) *Risāla Fī l-ʿilm al-ladunī = al-Risāla al-Laduniyya.* Edited by Muḥyī al-Dīn Ṣabrī al-Kurdī. Cairo: Maktabat Kurdistān al-ʿIlmiyya, 1327 [1909].
———. (?) *Risāla Fī l-ʿilm al-ladunī = Risāla Fī bayān al-ʿilm al-ladunī.* MS Berlin, Orientabteilung der Staatsbibliothek Preussischer Kulturbesitz, Sprenger 1968. foll. 39a–51b.
———. *Risāla Fī madhāhib ahl al-salaf.* MS Berlin, Orientabteilung der Staatsbibliothek Preussischer Kulturbesitz, Petermann II 690.
———. *Risāla Fī l-maʿrifa.* In al-Ghazālī, *Talāth rasāʾil fī l-maʿrifa lam tunshar min qabl.* Edited by Maḥmūd Ḥamdī Zaqzūq. Cairo: Maktabat al-Azhar, 1399/1979. 31–65.
———. *Risāla Fī maʿrifat al-nafs.* In al-Ghazālī, *Talāth rasāʾil fī l-maʿrifa lam tunshar min qabl.* Edited by Maḥmūd Ḥamdī Zaqzūq. Cairo: Maktabat al-Azhar, 1399/1979. 73–117.
———. *Risāla Fī uṣūl al-dīn.* MS Munich, Staatsbibliothek, no. 855. foll. 23b–40b.
———. *Risāla ilā Abū l-Fatḥ al-Damīmī = Risālat al-waʿẓ wa-l-iʿtiqād.* In *Fayṣal al-tafriqa bayna l-Islām wa-zandaqa maʿa l-Risāla al-waʿẓiyya wa-Kitāb Mishkāt al-anwār wa-Risāla al-ʿAqāʾid wa-l-waʿẓ ilā Malik Shāh (. . .)* Edited by Muḥammad Badr al-Dīn al-Naʿsānī al-Ḥalabī. Cairo: Aḥmad Nājī al-Jamālī and Muḥammad Amīn al-Khānjī, 1325/1907. 27–30.
———. *Risāla ilā Abū l-Fatḥ al-Damīmī.* MS Berlin, Orientabteilung der Staatsbibliothek Preussischer Kulturbesitz, Petermann II 8, pp. 120–26.
———. (?) *Risālat al-Ṭayr.* In idem. *Jawāhir al-ghawālī min rasāʾil al-Imām Ḥujjat al-Islām al-Ghazālī.* Edited by Muḥyī al-Dīn Ṣabrī al-Kurdī. Cairo: Maṭbaʿat al-Saʿāda, 1353/1934. 147–51.
———. *Al-Qānūn al-kullī fī l-taʾwīl.* Edited by Muḥammad Zāhid al-Kawtharī. Distributed as a small pamphlet with the title *Qānūn al-taʾwīl* as a supplement (*hadiyya*) to *Majallat al-Azhar* (Cairo) 58.4 (Rabīʿ II 1406 / Dec. 1985–Jan. 1986).
———. *Qawāṣim al-bāṭiniyya.* = Ahmed Ateş. "Gazâlî'nin Batinîlerin Belini Kiran Delliler'i Kitâb Kavâṣim al-Bâṭinîya." *Ilâhiyât Fakültesi Dergisi Ankara Üniversitesi* 3 (1954): 23–54.

———. *Al-Qisṭās al-mustaqīm*. Edited by Victor Chelhot. Beirut: Imprimerie Catholique, 1959.

———. *Tahāfut al-falāsifa*. = *Tahâfot al-Falâsifat ou «Incohérence des Philosophes.»* Edited by Maurice Bouyges. Beirut: Imprimerie Catholique, 1927. Reprint Frankfurt: Institute for the History of Arabic-Islamic Science, 1999.

———. *Tahāfut al-falāsifa*. = *The Incoherence of the Philosophers / Tahāfut al-falāsifa*. A parallel English-Arabic text. Edited and translated by Michael E. Marmura. 2nd. ed. Provo (Utah): Brigham Young University Press, 2000.

———. *Zād-i ākhirat* (in Persian). Edited by Murād Awrang. Tehran: Kitāb-khānah-i Millī, 1352 [1973].

ON ISLAMIC LAW AND ITS METHOD (INCLUDING *FATĀWĀ*)

———. *Asās al-qiyās*. Edited by Fahd ibn Muḥammad al-Sadiḥān. Riyad: Maktabat al-ʿUbaykān, 1413/1993.

———. *Al-Basīṭ fī l-furūʿ fī madhhab al-Shāfiʿī*. MS Damascus, Asad-Library, Ẓāhiriyya-Collection, nos. 2111, 2112, 2113, 2114 (parts 1, 4, 5, and 6).

———. *Bayān ghāyat al-ghawr fī dirāyat al-dawr*. MS London, British Library, Or. 3102, foll. 2a–4a.

———. *Fatāwā l-Imām al-Ghazālī / The Fatāwā of Imam al-Ghazzālī*. Edited by Mustafa Mahmoud Abu-Sway. Kuala Lumpur: International Institute of Islamic Thought and Civilization (ISTAC), 1996.

———. *Al-Fatāwā li-l-Imām Ḥujjat al-Islām Muḥammad ibn Muḥammad al-Ghazālī*. Edited by ʿAlī Muṣṭafā l-Ṭassah. Beirut: Dār al-Yāmāma, 1425/2004.

———. *Fatwā ʿAlā man laʿana muslim*[an]. In Badawī. *Muʾallafāt al-Ghazālī*. 47–48.

———. *Fatwā ʿAlā man istafāḍa min amwāl ribāṭ ṣūfiyya*. In Pūrjavādī. *Dō mujaddid. Pazhūhish-hā-yi dar bāra-yi Muḥammad-i Ghazzālī va-Fakhr-i Rāzī / Two Renewers of Faith*. 96–100.

———. *Fatvā Dar bāra-yi ʿahd-i "a-lastu. . .?"* (in Persian). In Pūrjavādī. *Dō mujaddid. Pazhūhish-hā-yi dar bāra-yi Muḥammad-i Ghazzālī va-Fakhr-i Rāzī / Two Renewers of Faith*. 70–75.

———. *Fatvā Dar bāra-yi amvāl-i khānqāh* (in Persian). In Pūrjavādī. *Dō mujaddid. Pazhūhish-hā-yi dar bāra-yi Muḥammad-i Ghazzālī va-Fakhr-i Rāzī / Two Renewers of Faith*. 87–91.

———. *Fatvā Dar bāra-yi asrār-i ʿibādāt* (in Persian). In Pūrjavādī. *Dō mujaddid. Pazhūhish-hā-yi dar bāra-yi Muḥammad-i Ghazzālī va-Fakhr-i Rāzī / Two Renewers of Faith*. 113–14.

———. *Fatvā Dar bāra-yi samāʿ* (in Persian). In Pūrjavādī. *Dō mujaddid. Pazhūhish-hā-yi dar bāra-yi Muḥammad-i Ghazzālī va-Fakhr-i Rāzī / Two Renewers of Faith*. 13–18.

———. *Fatvā Dar bāra-yi takālīf-i sharʿī va-luzūm-i riʿāyat-i ānhā* (in Persian). In Pūrjavādī. *Dō mujaddid. Pazhūhish-hā-yi dar bāra-yi Muḥammad-i Ghazzālī va-Fakhr-i Rāzī / Two Renewers of Faith*. 103–7.

———. *Fatvā Dar bāra-yi taklīf va-qurbat* (in Persian). In Pūrjavādī. *Dō mujaddid. Pazhūhish-hā-yi dar bāra-yi Muḥammad-i Ghazzālī va-Fakhr-i Rāzī / Two Renewers of Faith*. 109–11.

———. *Fatvā Dar bāra-yi vājib būdan nemāz dar hameh-i aḥvāl* (in Persian). In Pūrjavādī. *Dō mujaddid. Pazhūhish-hā-yi dar bāra-yi Muḥammad-i Ghazzālī va-Fakhr-i Rāzī / Two Renewers of Faith*. 116–17.

———. *Fatwā ilā Yūsuf ibn Tāshifīn bi-sharʿiyyat jihādihi*. In Ibn al-ʿArabī. "Shawāhid al-jilla wal-aʿyān fī mashāhid al-Islam wa-l-buldān." 302–5.

———. *Fatvā-yi Maḥabbat qadīm va-maḥabbat muḥdas̱ ast?* (in Persian). In Pūrjavādī. *Dō mujaddid. Pazhūhish-hā-yi dar bāra-yi Muḥammad-i Ghazzālī va-Fakhr-i Rāzī / Two Renewers of Faith.* 77–78.
———. *Ḥaqīqat al-qawlayn,* MS Princeton, University Library, Yahuda 4358.
———. *Al-Mankhūl min ta'līqāt al-uṣūl.* Edited by Muḥammad Ḥusayn Haytū. 3rd ed. Beirut/Damascus: Dār al-Fikr / Dār al-Fikr al-Mu'āṣir, 1419/1998.
———. *Al-Muntakhal fī l-jadal.* Edited by 'Alī ibn 'Abd al-'Azīz al-'Umayrīnī. Beirut/Riyad/Damascus: Dār al-Warrāq, 1424/2004.
———. *Al-Mustaṣfā min 'ilm al-uṣūl.* Edited by Ḥamza ibn Zuhayr Ḥāfiẓ. 4 vols. Medina (Saudi Arabia): al-Jāmi'a al-Islāmiyya—Kulliyyat al-Sharī'a, 1413 [1992–93].
———. *Al-Mustaṣfā min 'ilm al-uṣūl.* 2 vols. Būlāq: al-Maṭba'a al-Amīriyya: 1322–24 [1904–7]. Photographic reprint, Beirut: Dār al-Fikr, n.d. [*c.* 1985].
———. *Shifā' al-ghalīl fī bayān al-shubah wa-l-mukhīl wa-masālik al-ta'līl.* Edited by Ḥamid 'Ubayd al-Kubaysī. Bagdad: Maṭba'at al-Irshād, 1390/1971.
———. *Al-Wajīz fī fiqh madhhab al-Imām al-Shāfi'ī.* Edited by 'Alī Mu'awwaḍ and 'Ādil 'Abd al-Mawjūd. 2 vols. Beirut: Dār al-Arqam ibn Abī l-Arqam: 1418/1997.
———. *Al-Wasīṭ fī l-madhhab wa-bi-hāmishihi al-Tanqīṭ fī sharḥ al-Wasīṭ li-Muḥyī l-Dīn ibn Sharaf al-Nawawī; Sharḥ mushkil al-Wasīṭ li-Abī 'Amr 'Uthmān ibn al-Ṣalāḥ; Sharḥ mushkilāt al-Wasīṭ li-Muwaffaq al-Dīn Ḥamza ibn Yūsuf al-Ḥamawī; Ta'līqa mūjaza 'alā l-Wasīṭ li-Ibrāhīm ibn 'Abdallāh ibn Abī l-Dam.* Edited by Aḥmad Maḥmūd Ibrāhīm and Muḥammad Muḥammad Tāmir. 7 vols. Cairo: Dār al-Salām, 1407/1997.

ON POLITICS AND THE GUIDANCE OF RULERS

———. *Naṣīḥat al-mulūk* (in Persian). Edited by Jalāl al-Din Humā'ī. Tehrān: Anjuman-i Āṣār-i Millī, 1352 [1973].[1]
———. (?) *Pandnāmah* (in Persian). In Pūrjavādī. *Dō mujaddid. Pazhūhish-hā-yi dar bāra-yi Muḥammad-i Ghazzālī va-Fakhr-i Rāzī / Two Renewers of Faith.* 425–49.
———. *Risāla ilā l-sulṭān Muḥammad [Tapar] ibn Malikshāh fī l-'aqā'id.* In *Fayṣal al-tafriqa bayna l-Islām wa-zandaqa ma'a l-Risāla al-wa'ẓiyya wa-Kitāb Mishkāt al-anwār wa-Risāla al-'Aqā'id wa-l-wa'ẓ ilā Malik Shāh (. . .)* Edited by Muḥammad Badr al-Dīn al-Na'sānī al-Ḥalabī. Cairo: Aḥmad Nājī al-Jamālī and Muḥammad Amīn al-Khānjī, 1325/1907. 61–79.
———. *Al-Tibr al-masbūk fī naṣīḥat al-mulūk.* Edited by Muḥammad Aḥmad Damaj. 2nd ed. Beirut: Mu'assasat 'Izz al-Dīn, 1416/1996.
———. *Al-Tibr al-masbūk fī naṣīḥat al-mulūk wa-l-wuzarā' wa-l-wulāt.* MS Beirut, Jafet Library of the American University, no. 172.2:G4ltbA.
———. (?) *Tuḥfat al-mulūk* (in Persian). In Pūrjavādī. *Dō mujaddid. Pazhūhish-hā-yi dar bāra-yi Muḥammad-i Ghazzālī va-Fakhr-i Rāzī / Two Renewers of Faith.* 345–412.

ON EDUCATION

———. *Fātiḥat al-'ulūm.* Edited by Muḥammad Amīn al-Khānjī. Cairo: al-Maṭba'a al-Ḥusayniyya, 1322 [1904–5].

AUTOBIOGRAPHICAL WRITINGS AND LETTERS

———. *Fażā'il al-anām = Makātīb-yi fārisī-yi Ghazzālī be-nām-i Fażā'il al-anām min rasā'il Ḥujjat al-Islām.* (in Persian). Edited by 'Abbās Iqbāl Āshtiyānī. Tehran: Kitābfurūsh-i Ibn Sīnā, 1333 [1954].

———. *Al-Munqidh min al-ḍalāl / Erreur et délivrance*. Edited and French translation by Farid Jabre. 3rd ed. Beirut: Commission libanaise pour la traduction des chefs-d'œuvre, 1969.
———. *Al-Munqidh min al-ḍalāl wa-l-mufṣiḥ bi-l-aḥwāl*. MS Istanbul, Süleymaniye Kütüphanesi, Şehit Ali Paşa 1712, foll. 33a–56b.
———. *Risāla ilā Yūsuf ibn Tāshifīn*. In Ibn al-ʿArabī. "Shawāhid al-jilla wal-aʿyān fī mashāhid al-Islam wa-l-buldān." 306–14.

WORKS BY OTHER AUTHORS INCLUDING TRANSLATED WORKS BY AL-GHAZĀLĪ

ʿAbbās, Iḥsān. "Al-Jānib al-siyāsī min riḥlat Ibn al-ʿArabī ilā l-Mashriq." *Al-Abḥāth* (Beirut) 16 (1963): 217–36.
———. "Riḥlat Ibn al-ʿArabī ilā l-Mashriq kamā ṣawwarahā Qānūn al-taʾwīl." *Al-Abḥāth* (Beirut) 21 (1968): 59–91.
Al-ʿAbd, ʿAbd al-Laṭīf Muḥammad. *Al-Insān fī fikr Ikhwān al-ṣafā*. Cairo: Maktabat al-Angelo, 1976.
Abd-el-Jalil, J. M. "Autour de la sincerité d'al-Ghazālī." In *Melanges Louis Massignon*. 3 vols. Damascus: Institut français de Damas, 1956–58. vol. 1, 57–72.
Al-Abīwardī, Muḥammad ibn Aḥmad. *Dīwān al-Abīwardī*. Edited by ʿUmar al-Asʿad. 2 vols. Damascus: Majmaʿ al-Lugha al-ʿArabiyya, 1394–95/1974–75.
Abrahamov, Binyamin. "Al-Ghazālī's Theory of Causality." *Studia Islamica* 67 (1988): 75–98.
———. "Ibn Sīnā's Influence on al-Ghazālī's Non-Philosophical Works." *Abr Nahrain* 29 (1991): 1–17.
———. "A Re-examination of al-Ashʿarī's Theory of kasb according to *Kitāb al-Lumaʿ*." *JRAS* (1989): 210–21.
Abū l-Futūḥ al-Ghazālī, Aḥmad ibn Muḥammad. *Majālis. Taqrīrāt-i Aḥmad al-Ghazālī. ʿārif-i mutavaffā-yi 520 H.* Arabic edition with Persian translation by Aḥmad Mujāhid. Tehran: Intishārāt-i Dānishgāh, 1998.
———. *Majmūʿah-yi āṣār-i fārisī-yi Aḥmad Ghazzālī. ʿārif-i mutavaffā-yi 520 H.Q.* Edited by Aḥmad Mujāhid. 3rd ed. Tehran: Intishārāt-i Dānishgāh, 1376 [1997].
———. *Kitāb al-Savāniḥ fī l-ʿishq* (in Persian). = *Aphorismen über die Liebe*. Edited by Hellmut Ritter. Istanbul: Staatsdruckerei, 1942.
———. *Gedanken über die Liebe*. German translation of the *Kitāb al-Savāniḥ fī l-ʿishq* by Richard Gramlich. Wiesbaden: F. Steiner: 1976.
———. (?) *Ghāyat al-imkān fī dirāyat al-makān*. MS London, British Library, Or. 7721, foll. 119a–140a.
———. *Al-Tajrīd fī kalimat al-tawḥīd* (in Arabic). Edited by Aḥmad Mujāhid. Tehran: Intishārāt-i Dānishgāh, 1384 [2005].
———. *Der reine Gottesglaube. Das Wort des Einheitsbekenntnisses. Aḥmad al-Ġazzālīs Schrift at-Taǧrīd fī kalimat at-tawḥīd*. German translation by Richard Gramlich. Wiesbaden: F. Steiner, 1983.
Abū l-Ḥusayn al-Baṣrī, Muḥammad ibn ʿAlī. *Taṣaffuḥ al-adilla. The Extant Parts*. Edited by Wilferd Madelung and Sabine Schmidtke. Wiesbaden: Harrassowitz, 2006.
Abu-Sway, Mustafa Mahmoud. "Al-Ghazālī's 'Spiritual Crisis' Reconsidered." *Al-Shajara* (Kuala Lumpur) 1 (1996): 77–94.
Adams, Marilyn McCord. *William Ockham*. 2 vols. Notre Dame (Ind.): University of Notre Dame Press, 1987.
Adamson, Peter. "The Arabic Sea Battle: Al-Fārābī on the Problem of Future Contingents." *Archiv für Geschichte der Philosophie* 88 (2006): 163–88.

———. "Miskawayh's Psychology." In *Classical Arabic Philosophy: Sources and Reception.* Edited by Peter Adamson. London/Turin: Warburg Institute / Nino Aragno Editore, 2007. 39–54.

———. "On Knowledge of Particulars." *Proceedings of the Aristotelian Society* 105 (2005): 273–94.

Ahlwardt, Wilhelm. *Die Handschriften-Verzeichnisse der Königlichen Bibliothek zu Berlin. Verzeichnis der arabischen Handschriften.* 10 vols. Berlin: A. W. Schade, 1887–99. Photomechanic reprint, Hildesheim: Georg Olms, 1980.

Akasoy, Anna Ayşe. *Philosophie und Mystik in der späten Almohadenzeit. Die Sizilianischen* Fragen *des Ibn Sabʿīn.* Leiden: Brill, 2006.

Al-Akiti, M. Afifi. "The Three Properties of Prophethood in Certain Works of Avicenna and al-Ġazālī." In *Interpreting Avicenna. Science and Philosophy in Medieval Islam. Proceedings of the Second Conference of the Avicenna Study Group.* Edited by Jon McGinnis. Leiden: Brill, 2004. 189–212.

Alon, Ilai. *Al-Fārābī's Philosophical Lexicon / Qāmūs al-Fārābī l-falsafī.* 2 vols. London: The E. J. W. Gibb Memorial Trust, 2002.

———. "Al-Ghazālī on Causality." *JAOS* 100 (1984): 397–405.

Al-ʿĀmirī, Muḥammad ibn Yūsuf. *Kitāb al-Fuṣūl.* In Saḥbān Khalīfāt. *Rasāʾil Abī l-Ḥasan al-ʿĀmirī wa-shadharātuhu al-falsafiyya.* Amman: al-Jāmiʿa al-Urduniyya, 1988. 363–79.

———. *Kitāb al-Fuṣūl fī l-maʿālim al-ilāhiyya.* In Elvira Wakelnig, *Feder Tafel Mensch. Al-ʿĀmirīs Kitāb al-Fuṣūl fī l-maʿālim al-ilāhīya und die arabische Proklos-Rezeption im 10 Jh.* Leiden: Brill, 2006. 82–123.

Ansari, Abdul Haq. "The Doctrine of Divine Command: A Study in the Development of Ghazālī's View of Reality." *Islamic Studies* (Islamabad) 21 (1982): 1–47.

Al-Anṣārī, Sulaymān ibn Nāṣir. *Al-Ghunya fī l-kalām.* MS Istanbul, Topkapı Sarayı Müzesi Kütüphanesi, Ahmet III, no. 1916.

———. *Sharḥ al-irshād fī uṣūl al-iʿtiqād.* MS Princeton, University Library, Yahuda 634.

Antes, Peter. *Prophetenwunder in der Ašʿarīya bis al-Ġazālī (Algazel).* Freiburg (Germany): K. Schwarz, 1970.

Al-Asad wa-l-ghawwāṣ. Ḥikāya ramziyya ʿarabiyya min al-qarn al-khāmis al-hijrī. Edited by Riḍwān al-Sayyid. Beirut: Dār al-Ṭalīʿa, 1978.

Al-Aʿsam, ʿAbd al-Amīr. *al-Faylasūf al-Ghazālī. Iʿādat taqwīm li-munḥanā taṭawwurihi al-rūḥī.* 3rd ed. Beirut: Dār al-Andalus, 1981.

ʿĀṣī, Ḥasan. *al-Tafsīr al-Qurʾānī wa-l-lugha al-ṣūfiyya fī falsafat Ibn Sīnā.* Beirut: al-Muʾassasa al-Jāmiʿiyya li-l-Dirāsāt, 1403 / 1983.

Asín Palacios, Miguel. *La espiritualidad de Algazel y su sentido christiano.* 4 vols. Madrid and Granada: Imprenta de Estanislao Maestre, 1934–41.

———. "Un faqîh siciliano, contradictor de Al Ġazzâlî (Abû ʿAbd Allâh de Mâzara)." In *Centenario della nascita de Michele Amari.* 2 vols. Palermo: Stabilimento Tipografico Virzì, 1910. 216–44.

Avicenna = Ibn Sīnā, al-Ḥusayn ibn ʿAbdallāh.

Averroes = Ibn Rushd, Muḥammad ibn Aḥmad.

ʿAyn al-Quḍāt al-Hamadhānī, ʿAbdallāh ibn Muḥammad. *Nāmah-hā-yi ʿAyn al-Qużāt.* Edited by ʿAlī Naqī Munzavī and ʿAfīf Usayrān. 2 vols. Tehran: Intishārāt-i Bunyad-i Farhang-i Īrān, 1969.

———. *Risālat Shakwā l-gharīb ʿan al-awṭān ilā ʿulamāʾ al-buldān.* Edited by ʿAfīf Usayrān. Tehran: Chāpkhānah-yi Dānishgāh, 1962. (Part 3 of *Muṣannafāt-i ʿAyn al-Qużāt-i Hamadānī.*)

———. *Tamhīdāt*. Edited by ʿAfīf Usayrān. Tehran: Chāpkhānah-yi Dānishgāh, 1341/1962. (Part 2 of *Muṣannafāt-i ʿAyn al-Quḍāt-i Hamadānī*.)

———. *Zubdat al-ḥaqāʾiq*. Edited by ʿAfīf Usayrān. Tehran: Maṭbaʿat-i Jāmiʿat-i Tehrān, 1341/1961. (Part 1 of *Muṣannafāt-i ʿAyn al-Quḍāt-i Hamadānī*.)

Bäck, Allan. "Avicenna's Conception of the Modalities." *Vivarium* 30 (1992): 217–55.

———. "Avicenna on Existence." *Journal of the History of Philosophy* 25 (1987): 351–67.

Badawī, ʿAbd al-Raḥmān. *Muʾallafāt al-Ghazālī*. 2nd ed. Kuwait: Wikālat al-Maṭbūʿāt, 1977.

Baffioni, Carmela. "An Essay on Terminological Research in Philosophy: The 'Friends of God' in the *Rasāʾil Iḫwān al-Ṣafāʾ*." *Studi Magrebini* 25 (1993–97): 23–42.

———. "From Sense Perception to the Vision of God: A Path Towards Knoweldge According to the Iḫwān al-Ṣafāʾ." *Arabic Sciences and Philosophy* 8 (1998): 213–31.

———. "Contrariety and Similarity in God According to al-Fārābī and al-Kirmānī: A Comparison." In *Classical Arabic Philosophy: Sources and Reception*. Edited by Peter Adamson. London/Turin: Warburg Institute / Nino Aragno Editore, 2007. 1–20.

Al-Baghdādī, ʿAbd al-Laṭīf ibn Yūsuf. *Kitāb al-Nāṣiḥatayn*. MS Bursa, Hüseyin Çelebi 823, foll. 62a–100b.

Al-Baghdādī, ʿAbd al-Qāhir ibn Ṭāhir. *Uṣūl al-dīn*. Istanbul: Maṭbaʿat al-Dawla: 1346/1928.

Bagley, F. R. C. *Ghazālī's Book of Council for Kings (Naṣīḥat al-Mulūk)*. Translated by F. R. C. Bagley. London: Oxford University Press, 1964.

Bahlul, Raja. "Miracles and Ghazali's First Theory of Causation." *Philosophy and Theology* 5 (1990–91): 137–50.

Baljon, J. M. S. "The 'Amr of God' in the Koran." *Acta Orientalia* 24 (1959): 7–18.

Baneth, David Z. "Jehuda Hallewi und Gazali." In *Korrespondenzblatt des Vereins zur Gründung und Erhaltung einer Akademie für die Wissenschaft des Judentums* (Berlin) 5 (1924): 27–45. Reprint in *Wissenschaft des Judentums im deutschen Sprachbereich. Ein Querschnitt*. Edited by Kurt Wilhelm. 2 vols. Tübingen: Mohr, 1967. 2: 371–89.

———. "Rabbi Yehudah ha-Levi we-Algazzali." *Kneseth* (Jerusalem) 7 (5702/1941–42): 311–29.

Al-Baqarī al-Anṣārī, ʿAbd al-Dāʾim. *Iʿtirāfāt al-Ghazālī aw kayfa arrakha l-Ghazālī nafsahu*, Cairo: Dār al-Kutub al-Ahliyya, 1943.

Al-Bāqillānī, Muḥammad ibn al-Ṭayyib. *al-Tamhīd fī l-radd ʿalā l-mulḥida wa-l-muʿaṭ-ṭila wa-l-rāfiḍa wa-l-khawārij wa-l-muʿtazila*. Edited by Richard J. McCarthy. Beirut: Librairie Orientale, 1957.

Bauer, Hans. *Die Dogmatik al-Ghazālīs nach dem II. Buche seines Hauptwerkes*. Halle: Buchdruckerei des Waisenhauses, 1912.

Bausani, Alessandro. "Some Observations on Three Problems of the Anti-Aristotelian Controversy Between al-Bīrūnī and Ibn Sīnā." In *Akten des VII. Kongresses für Arabistik und Islamwissenschaft. Göttingen, 15. bis 22. August 1974. [7th Congress of the Union Européenne d'Arabisants et d'Islamisants; UEAI.]* Edited by Albert Dietrich. Göttingen: Vandenhoeck & Ruprecht, 1976. 75–85

Al-Bayhaqī, ʿAlī ibn Zayd (Ẓahīr al-Dīn ibn Funduq). *Tatimmat Ṣiwān al-ḥikma*. Edited by Rafīq al-ʿAjm. Beirut: Dār al-Fikr al-Lubnānī, 1994.

———. *Taʾrīkh-i Bayhaq*. Edited by Aḥmad Bahmanyār. Tehran: Kitabfrūshā-yi Furūghī, n.d. [*c.* 1965].

Belo, Caterina. *Chance and Determinism in Avicenna and Averroes*. Leiden: Brill, 2007.

Berkey, Jonathan. *The Formation of Islam. Religion and Society in the Near East, 600–1800*. Cambridge: Cambridge University Press, 2003.

Bernand, Marie. "La critique de la notion de nature (ṭabʿ) par le kalām." *Studia Islamica* 5 (1980): 59–105.

Bertolacci, Amos. "The Doctrine of Material and Final Causality in the *Ilāhiyyāt* of Avicenna's *Kitāb al-Shifāʾ*." *Quaestio* 2 (2002): 25–54.

———. *The Reception of Aristotele's Metaphysics in Avicenna's Kitāb al-Shifāʾ. A Milestone of Western Metaphysical Thought*. Leiden: Brill, 2006.

Bieberstein, Klaus, and Hanswulf Bloedhorn. *Jerusalem. Grundzüge der Baugeschichte vom Chalkolithikum bis zur Frühzeit der osmanischen Herrschaft*. 3 vols. Wiesbaden: Reichert 1994.

Binder, Leonard. "Al-Ghazālī's Theory of Islamic Government." *Muslim World* 45 (1955): 229–41.

Black, Deborah L. "Avicenna on the Ontological and Epistemological Status of Fictional Beings." *Documenti e studi sulla tradizione filosofica medievale* 8 (1997): 425–53.

Bosworth, Clifford E. "The Ismaʿilis of Quhistān and the Maliks of Nimrūz or Sistān." In *Medieval Ismāʿīlī History and Thought*. Edited by Farhad Daftary. Cambridge: Cambridge University Press, 1996. 221–229.

Bousquet, Georges-Henri. *Ih'ya ʿouloûm ad-dîn ou vivication des sciences de la foi. Analyse et index*. Paris: Librairie Max Besson, 1955.

Bouyges, Maurice. *Essai de chronologie des œuvres de al-Ghazali (Algazel)*. Edited by Michel Allard. Beirut: Imprimerie Catholique, 1959.

Brockelmann, Carl. *GAL = Geschichte der arabischen Litteratur*. 2nd. ed. 2 vols. Leiden: Brill, 1943–49.

———. *GAL Suppl. = Geschichte der arabischen Litteratur. Supplementbände*. 3 vols. Leiden: Brill, 1937–42.

Brown, Jonathan A. C. "The Last Days of al-Ghazzālī and the Tripatriate Division of the Sufi World: Abū Ḥāmid al-Ghazzālī's Letter to the Seljuq Vizier and Commentary." *Muslim World* 96 (2006): 89–113.

Buchman, David. *Al-Ghazālī: The Niche of Lights / Mishkāt al-anwār*. A parallel English-Arabic text. Edited and translated by David Buchman. Provo (Utah): Brigham Young University Press, 1998.

Bulliet, Richard W. *The Patricians of Nishapur; A Study in Medieval Islamic Social History*. Cambridge (Mass.): Harvard University Press, 1972.

Al-Bundarī, al-Fatḥ ibn ʿAlī. *Zubdat al-nuṣra wa-nukhbat al-ʿuṣra*. Vol. 2 of Martijn Th. Houtsma. *Recueil de texts relatifs à l'histoire des Seldjoucides*. 4 vols. Leiden: Brill, 1889.

Burgoyne, Michael H. *Mamluk Jerusalem. An Architectural Study with Additional Historical Research by D. S. Richards*. London: Published on Behalf of the British School of Archeology in Jerusalem, 1987.

Burrell, David B. *Freedom and Creation in Three Traditions*. Notre Dame (Ind.): Notre Dame University Press, 1993.

———. and Nazih Daher. *Al-Ghazālī. The Ninety-Nine Beautiful Names of God. Al-Maqṣad al-asnā fī sharḥ asmāʾ Allāh al-ḥusnā*. Translated by David Burrell and Nazih Daher. Cambridge: Islamic Text Society, 1992.

Cabanelas, Darió. "Un opusculo inédito de Algazel: El libro le las intuicones intelectuales." *al-Andalus* 21 (1956): 19–58.

The Cambridge History of Iran. Vol. 5. The Saljuq and Mongol Periods. Edited by J. A. Boyle. Cambridge: Cambridge University Press, 1968.

Cook, Michael. *Commanding Right and Forbidding Wrong in Islamic Thought.* Cambridge: Cambridge University Press, 2000.

Corbin, Henry. *Avicenne et le récit visionaire.* 2nd ed. 2 vols. Tehran: Département d'iranologie de l'Institut franco-iranien, 1954.

———. "The Ismāʿīlī Response to the Polemic of Ghazālī." In *Ismāʿīlī Contributions to Islamic Culture.* Edited by Seyyed Hossein Nasr. Tehran: Imperial Iranian Academy of Philosophy, 1398/1977. 67–98.

Courtenay, William J. "The Critique on Natural Causality in the Mutakallimun and Nominalism." *Harvard Theological Review* 66 (1973): 77–94.

Craemer-Ruegenberg, Ingrid. "'Ens est quod primum cadit in intellectu'–Avicenna und Thomas von Aquin." In *Gottes ist der Orient / Gottes ist der Occident. Festschrift für Abdoldjavad Falaturi zum 65. Geburtstag.* Edited by Udo Tworuschka. Cologne: Böhlau, 1991. 133–42.

Craig, William Lane. *The Cosmological Argument from Plato to Leibniz.* London and Basingstoke: Macmillan Press, 1980.

———. *The Kalām Cosmological Argument.* London and Basingstoke: Macmillan Press, 1979.

Crone, Patricia. "Did al-Ghazālī Write a Mirror for Princes? On the Authorship of *Naṣīḥat al-mulūk.*" *Jerusalem Studies of Arabic and Islam* 10 (1987): 167–91.

Dabashi, Hamid. *Truth and Narrative. The Untimely Thoughts of ʿAyn al-Quḍāt al-Hamadhānī.* Richmond: Curzon, 1999.

Al-Dabbāgh, ʿAbd al-Raḥmān ibn Muḥammad, and Abū l-Qāsim ibn ʿĪsā al-Nājī. *Maʿālim al-īmān fī maʿrifat ahl al-Qayrawān.* Edited by Ibrāhīm Shabbūḥ and Muḥammad Māḍūd. 3 vols. Cairo: Maktabat al-Khānjī, and Tunis: al-Maktaba al-ʿAtīqa, 1978.

Daftary, Farhad. *The Ismāʿīlīs. Their History and Doctrine.* Cambridge: Cambridge University Press, 1990.

Dallal, Ahmad. "Ghazālī and the Perils of Interpretation." *JAOS* 122 (2003): 773–87.

Davidson, Herbert A. *Alfarabi, Avicenna, and Averroes, on Intellect. Their Cosmologies, Theories of the Active Intellect, and Theories of Human Intellect.* New York: Oxford University Press, 1992.

———. "Arguments from the Concept of Particularization in Arabic Philosophy." *Philosophy East and West* 18 (1968): 299–314.

———. "Avicenna's Proof of the Existence of God as a Necessarily Existent Being." In *Islamic Philosophical Theology.* Edited by Parviz Morewedge. Albany: State University of New York Press, 1979. 165–87.

———. *Proofs for Eternity, Creation and the Existence of God in Medieval Islamic and Jewish Philosophy.* New York: Oxford University Press, 1987.

Dawlatshāh ibn ʿAlāʾ al-Dawla Samarqandī. *Taẕkirat al-shuʿarāʾ ("Memoirs of the Poets").* Edited by Edward G. Browne. London/Leiden: Luzac & Co./Brill, 1901.

De Boer, Tjitze J. *Geschichte der Philosophie im Islam.* Stuttgart: F. Frommann, 1901.

———. *The History of Philosophy in Islam.* Translated by Edward R. Jones. London: Luzac, 1903.

De Smet, Daniel. *La quiétude de l'intellect. Néoplatonisme et gnose ismaélienne dans l'œuvre de Ḥamîd ad-Dîn al-Kirmânî (Xe/XIe s.)* Leuven: Peeters, 1995.

Al-Dhahabī, Muḥammad ibn Aḥmad. *Siyar aʿlām al-nubalāʾ.* Edited by Shuʿayb al-Arnaʾūt. 25 vols. Beirut: Muʾassasat al-Risāla, 1981–88.

———. *Taʾrīkh al-Islām wa-wafayāt al-mashāhīr wa-l-aʿlām.* Edited by ʿUmar ʿA. Tadmurī. Beirut: Dār al-Kitāb al-ʿArabi, 1407– / 1987–.

Dhanani, Alnoor. *The Physical Theory of Kalām. Atoms, Space, and Void in Basrian Muʿtazilī Cosmology*. Leiden: Brill, 1994.

D'Herbelot, Barthélemy. *Bibliotheque orientale, ou, Dictionaire universel contenant tout ce qui regarde la connaisance de Peuples de l'Orient (. . .)*. 2nd ed. 4 vols. The Hague: J. Neaulme & Van Daalen, 1777–79. Photographic reprint, Frankfurt/ Main: Institute for the History of Arabic-Islamic Science, 1995.

Dictionaire des sciences philosophique. Par une société des professeurs de philosophie. Edited by Adolphe Franck. 6 vols. Paris: Librairie A. Franck, 1844–52.

Diez, Ernst. *Die Kunst der islamischen Völker*. Berlin-Neubabelsberg: Akademische Verlagsgesellschaft Athenaion, 1915.

Al-Dimyāṭī, Aḥmad ibn Aybak. *Al-Mustafād min Dhayl Taʾrīkh Baghdād*. Edited by Qaysar Abū Faraḥ. Hyderabad: Maṭbaʿat Majlis Dāʾirat al-Maʿārif al-ʿUthmāniyya, 1399/1979.

Diwald, Susanne. *Arabische Philosophie und Wissenschaft in der Enzyklopädie. Kitāb Iḫwān aṣ-ṣafāʾ (III). Die Lehre von Seele und Intellekt*. Wiesbaden: Otto Harrassowitz, 1975.

Donaldson, Dwight M. "A Visit to the Grave of al-Ghazzali." *Moslem World* 8 (1918): 137–40.

Dreher, Josef. *Das Imamat des islamischen Mystikers Abūlqāsim Aḥmad ibn al-Ḥusain ibn Qasī (gest. 1151). Eine Studie zum Selbstverständnis des Autors des "Buchs vom Ausziehen der beiden Sandalen" (Kitāb Ḫalʿ an-naʿlain)*. Ph.D. diss., Rheinische Friedrich-Wilhelms Universität Bonn (Germany), 1985.

Druart, Thérèse-Anne. "Al-Fārābī's Causation of the Heavenly Bodies." In *Islamic Philosophy and Mysticism*. Edited by Parviz Morewedge. Delmar (N.Y.): Caravan Books, 1981. 35–45.

———. "Al-Fārābī: Metaphysics." In *Encyclopaedia Iranica*. Edited by Ihsan Yarshater. Vol. 9. New York: Bibliotheca Persica, 1999. 216–9.

———. "Al-Ghazālī's Conception of the Agent in the *Tahāfut* and the *Iqtiṣād*: Are People Really Agents?" In *Arabic Theology, Arabic Philosophy. From the Many to the One. Essays in Celebration of Richard M. Frank*. Edited by James E. Montgomery. Leuven: Peeters, 2006. 427–40.

Dutton, Blake D. "Al-Ghazālī on Possibility and the Critique of Causality." *Medieval Philosophy and Theology* 10 (2001): 23–46.

EI2 = *Encyclopaedia of Islam. New Edition*. Edited by an editorial committee consisting of H.A.R. Gibb et al. 11 vols. Leiden and London: Luzac and Brill, 1954–2003.

EI3 = *Encyclopaedia of Islam, THREE*. Edited by Gudrun Krämer, Denis Matringe, John Nawas, and Everett Rowson. Brill Online 2007–. URL <http://www.encislam.brill.nl>

EIran = *Encyclopaedia Iranica*. Edited by Ehsan Yarshater. London, New York, and Cosa Mesa (Calif.): Routledge & Kegan, Mazda, and Bibliotheca Persica, 1982–.

Einhard. "Annales. 741–829" In *Monumenta Germaniae Historica. 500–1500*. Vol. 1. Hannover: Hahn, 1826. Reprint, Stuttgart: A. Hiersemann, 1976. 135–218.

Elamrani-Jamal, Abdelali. "De la multiplicité des modes de la prophetie chez Ibn Sīnā." In *Etudes sur Avicenne*. Edited by Jean Jolivet and Roshdi Rashed. Paris: Belles Lettres, 1984. 125–42.

El-Rouayheb, Khaled. "Sunni Muslim Scholars on the Status of Logic, 1500–1800." *Islamic Law and Society* 11 (2004): 213–32.

———. "Was There a Revival of Logical Studies in Eighteenth-Century Egypt?" *Welt des Islams* 45 (2005): 1–19.

Endress, Gerhard. "Reading Avicenna in the Madrasa: Intellectual Genealogies and Chains of Transmission of Philosophy and the Sciences in the Islamic East." In *Arabic Theology, Arabic Philosophy. From the Many to the One. Essays in Celebration of Richard M. Frank*. Edited by James E. Montgomery. Leuven: Peeters, 2006. 371–422.

EQ = *Encyclopedia of the Qur'ān*. General editor Jane Dammen McAuliffe. 6 vols. Leiden: Brill, 2001–6.

Fakhry, Majid. *Islamic Occasionalism and Its Critique by Averroës and Aquinas*. London: Allan & Unwin, 1958.

Al-Fārābī, Muḥammad ibn Muḥammad. *Kitāb Bārī armīniyās ay al-ʿibāra*. In Mübahat Türker-Küyel. *Fârâbî'nin Peri Hermeneias Muhtasarı*. Ankara: Atatürk Kültür Merkezi, 1990. 21–71.

———. (?) *Al-Daʿāwa al-qalbiyya*. Hyderabad: Maṭbaʿat Majlis Dāʾirat al-Maʿārif al-ʿUthmāniyya, 1349 [1930].

———. *Fuṣūṣ al-ḥikam*. Edited by Muḥammad Ḥusayn Āl Yāsīn. Qom (Iran): Intishārāt-i Bīdār, 1405 (1984–85).

———. *Mabādiʾ arāʾ ahl al-madīna al-fāḍila / Al-Farabi on the Perfect State*. Edited and translated by Richard Walzer. Oxford and New York: Clarendon Press and Oxford University Press, 1985.

———. *Sharḥ al-Fārābī li-kitāb Arisṭutālīs fī l-ʿibāra*. Edited by Wilhelm Kutsch and Stanley Marrow. Beirut: Libraririe Orientale. 1960.

———. *Al-Siyāsa al-madaniyya*. Edited by Fawzī M. Najjār. Beirut: Imprimerie Catholique, 1964.

Faris, Nabih Amin. *The Book of Knowledge. Being a Translation with Notes of the Kitāb al-ʿIlm of al-Ghazzālī's Iḥyāʾ ʿUlūm al-Dīn*. 4th ed. Lahore: Muhammad Ashraf, 1974.

Fakhr al-Dīn al-Rāzī, Muḥammad ibn ʿUmar. *Al-Munāẓarāt fī bilād mā waraʾa l-nahr*. In Fathallah Kholeif. *A Study on Fakhr al-Dīn al-Rāzī and His Controversies in Transoxania*. 2nd ed. Beirut: Dār al-Mashriq, 1987. Arabic Pagination 7–63.

———. *Al-Maṭālib al-ʿāliya min al-ʿilm al-ilāhī*. Edited by Aḥmad Ḥijāzi al-Saqqā. 9 parts in 5 vols. Beirut: Dār al-Kitāb al-ʿArabī, 1987.

———. *Kitāb al-Muḥaṣṣal wa-huwa Muḥaṣṣal afkār al-mutaqaddimīn wa-l-mutaʾakhkhirīn min al-ḥukamāʾ wa-l-mutakallimīn*. Edited by Hüseyin Atay. Cairo: Maktabat Dār al-Turāth, 1411/1991.

———. *Al-Tafsīr al-kabīr aw-Mafātīḥ al-ghayb*. Edited by Muḥammad Muḥyī al-Dīn ʿAbd al-Ḥamīd. 32 vols. Cairo: al-Maṭbāʿa al-Bahiyya al-Miṣriyya, 1352/1933.

Fayḍ al-Kāshānī, al-Muḥsin Muḥammad ibn al-Murtaḍā. *Al-Maḥajja al-bayḍā fī tahdhīb al-Iḥyāʾ*. Edited by ʿAlī Akbar al-Ghaffārī. 8 vols. Tehran: Maktabat al-Ṣadūq, 1339–42 [1960–64].

Al-Fayyūmī, Aḥmad ibn Muḥammad. *Miṣbāḥ al-munīr fī gharīb al-Sharḥ al-kabīr*. Edited by ʿAbd al-ʿAẓīm al-Shannawī. Cairo: Dār al-Maʿārif, 1977.

Field, Claud. *The Alchemy of Happiness*. Translated from Urdu by Claud Field. Edited by Elton L. Daniel. Armonk (N.Y.): M.E. Sharpe, 1991.

Forest, Aimé. *La structure métaphysique du concret selon Saint Thomas d'Aquin*. Paris: J. Vrin, 1931.

Frank, Richard M. "The Ašʿarite Ontology: 1. Primary Entities." *Arabic Sciences and Philosophy* 9 (1999): 163–231. Reprint in idem. *Classical Islamic Theology. The Ashʿarites*. Text ix.

———. *Creation and the Cosmic System. Al-Ghazâlî & Avicenna*. Heidelberg: Carl Winter, 1992.

———. "Currents and Countercurrents." In *Islam. Essays on Scripture, Thought and Society*. Edited by Peter G. Riddell and Tony Street. Leiden: Brill, 1997. 113–34. Reprint in Frank, *Philosophy, Theology, and Mysticism in Medieval Islam*. Text viii.

———. *Al-Ghazālī and the Ash'arite School*. Durham: Duke University Press, 1994.

———. "Al-Ghazālī's Use of Avicenna's Philosophy." *Revue des etudes islamiques* 55–57 (1987–89): 271–85. Reprint in idem. *Philosophy, Theology, and Mysticism in Medieval Islam*. Text xi.

———. "Al-Ghazālī on *Taqlīd*. Scholars, Theologians, and Philosophers." *Zeitschrift für die Geschichte der arabisch-islamischen Wissenschaften* 7 (1991–92): 207–52. Reprint in idem. *Philosophy, Theology, and Mysticism in Medieval Islam*. Text x.

———. "Knowledge and *Taqlīd*. The Foundation of Religious Belief in Classical Ash'arism." *Journal of the American Oriental Society* 109 (1989): 258–78. Reprint in idem. *Classical Islamic Theology. The Ash'arites*. Text vii.

———. "The Non-Existent and the Possible in Classical Ash'arite Teaching." *Mélanges d'institut dominicain d'études orientales* 24 (2000): 1–37. Reprint in idem. *Classical Islamic Theology. The Ash'arites*. Text viii.

———. "The Structure of Created Causality According to al-Ash'arī. An Analysis of the *Kitâb al-Luma'*, §§ 82–164." *Studia Islamica* 25 (1966): 13–75. Reprint in idem. *Early Islamic Theology. The Mu'tazilites and al-Ash'arī*. Text vii.

———. "Al-Ustādh Abū Isḥāq: An *'Akīda* Together with Selected Fragments." *Mélanges de l'Institut Dominicain d'Études Orientales du Caire* 19 (1989): 129–202. Reprint in idem. *Classical Islamic Theology. The Ash'arites*. Text xiv.

———. *Philosophy, Theology, and Mysticism in Medieval Islam*. Edited by Dimitri Gutas. Aldershot (Hampshire, UK): Ashgate, 2005.

———. *Early Islamic Theology. The Mu'tazilites and al-Ash'arī*. Edited by Dimitri Gutas. Aldershot (Hampshire, UK): Ashgate, 2007.

———. *Classical Islamic Theology. The Ash'arites*. Edited by Dimitri Gutas. Aldershot (Hampshire, UK): Ashgate, 2009.

Frye, Richard. *The Histories of Nishapur*. London: Mouton. 1965.

Gairdner, W. H. T. "Al-Ghazālī's Mishkāt al-Anwār and the Ghazālī-Problem." *Der Islam* 5 (1914): 121–53.

———. *Al-Ghazzālī's Mishkāt al-Anwār ("The Niche of Lights"). A Translation with Introduction*. London: Royal Asiatic Society, 1924.

Garden, Kenneth. *Al-Ghazālī's Contested Revival. Iḥyā' 'ulūm al-dīn and its Critics in Khorasan and the Maghrib*. Ph.D. diss., University of Chicago, 2005.

Gardet, Louis. *Dieu et la destinée de l'homme*. Paris: J. Vrin, 1967.

———, and Georges C. Anawati. *Introduction à la théologie musulmane. Essai de théologie comparée*. Paris: J. Vrin, 1948.

Gätje, Helmut. "Logisch-semasiologische Theorien bei al-Ġazzālī." *Arabica* 21 (1974): 151–82.

Gavison, Abraham ben Ya'qov. *Sefer 'Omer ha-shikheḥah*. Livorno: Abraham Meldola, 1748. Photographic reprint, Brooklyn (N.Y.): Ch. Reich, 1993.

Gāzurgāhī, Kamāl al-Dīn Ḥusayn ibn Ismā'īl. *Majālis al-'ushshāq: taẕkirah-yi 'urafā'*. Edited by Ghulām-Riżā Ṭabāṭabā'ī-Majd. Tehran: Intishārāt-i Zarrīn, 1375 [1996].

Gianotti, Timothy J. *Al-Ghazālī's Unspeakable Doctrine of the Soul. Unveiling the Esoteric Psychology and Eschatology of the Iḥyā'*. Leiden: Brill, 2002.

Giese, Alma. "Zur Erlösungsfunktion des Traumes bei den Iḫwān aṣ-Ṣafā': Aus der *Risāla al-khāmisa min al-'ulūm an-nāmūsiyya wa-š-šar'iyya*." In *Gott ist schön und Er liebt die Schönheit / God is Beautiful and He Loves Beauty*. Edited by Alma Giese and J. Christoph Bürgel. Bern: Peter Land, 1994. 191–207.

Gimaret, Daniel. *La doctrine d'al-Ash'arī*. Paris: Cerf, 1990.

———. *Le livre de Bilawhar et Būḏāsaf selon la version arabe ismaélienne*. Geneva/Paris: Droz, 1971.

———. *Les noms divins en Islam. Exégèse lexicographique et théologique*. Paris: Cerf, 1988.

———. *Théories de l'acte humain en théologie musulmane*. Paris: Vrin, 1980.

Ghurāb, Sa'd (Saâd Ghrab). "Ḥawla iḥrāq al-Murābiṭīn li-Iḥyā' al-Ghazālī." *Actas del IV Coloquio Hispano-Tunecino (Palma de Mallorca, 1979)*. Madrid: Instituto Hispano-Arabe de Cultura, 1983. 133–63.

Gil'adi, Avner. "On the Origin of Two Key-Terms in al-Ġazzālī's Iḥyā' 'ulūm al-dīn." *Arabica* 36 (1989): 81–92.

Glassen, Erika. *Der mittlere Weg. Studien zur Religionspolitik und Religiösität der späten Abbasiden-Zeit*. Wiesbaden: F. Steiner, 1981.

Goichon, Amélie-Marie. *La distinction de l'essence et de l'existence d'après Ibn Sīnā (Avicenne)*. Paris: Desclée de Brouwer, 1937.

———. *Lexique de la langue philosophique d'Ibn Sīnā (Avicenne)*. Paris: Desclée de Brouwer, 1938.

Goldziher, Ignaz. *Gesammelte Schriften*. Edited by Joseph Desomogyi. 6 vols. Hildesheim: Olms, 1967–73.

———. *Die islamische und die jüdische Philosophie des Mittelalters*. In Wilhelm Wundt et al. *Allgemeine Geschichte der Philosophie*. 2nd ed. Berlin/Leipzig: B. G. Teubner, 1913. 301–337.

———. "Materialien zur Kenntnis der Almohadenbewegung." *ZDMG* 41 (1887): 30–140.

———. *Die Richtungen der islamischen Koranauslegung*. Leiden: Brill, 1920.

———. "Stellung der alten islamischen Orthodoxie zu den antiken Wissenschaften." *Abhandlungen der königlich preussischen Akademie der Wissenschaften* 8 (1915–16): 3–46.

———. *Streitschrift des Ġazālī gegen die Bāṭinijja-Sekte*. Leiden: Brill, 1916.

Goodman, Lenn E. *Avicenna*. London: Routledge, 1992.

———. "Did al-Ghazâlî Deny Causality?" *Studia Islamica* 47 (1978): 83–120.

———. "Ghazâlî's Argument from Creation." *IJMES* 2 (1971): 67–85, 168–188.

Goodrich, David Raymond. *A Ṣūfī Revolt in Portugal. Ibn Qasī and his Kitāb khal' al-na'layn*. Ph.D. diss., Columbia University, 1978.

Gramlich, Richard. *Muḥammad al-Ġazzālīs Lehre von den Stufen zur Gottesliebe. Die Bücher 31–36 seines Hauptwerkes eingeleitet, übersetzt und kommentiert*. Wiesbaden: F. Steiner, 1984.

———. *Die Nahrung der Herzen. Abū Ṭālib al-Makkīs Qūt al-qulūb eingeleitet, übersetzt und kommentiert*. 4 vols. Stuttgart: F. Steiner, 1992–95.

Griffel, Frank. "Al-Ghazālī's Cosmology in the Veil Section of His *Mishkāt al-Anwār*." In *Avicenna and His Heritage. A Golden Age of Science and Philosophy*. Edited by Tzvi Langermann. Turnhout (Belgium): Brepols, forthcoming.

———. "Al-Ghazālī or al-Ghazzālī? On a Lively Debate Among Ayyūbid and Mamlūk Historians in Damascus." In *Islamic Thought in the Middle Ages. Studies in Transmission and Translation in Honour of Hans Daiber*. Edited by Anna Ayşe Akasoy and Wim Raven. Leiden: Brill, 2008. 101–112.

———. *Al-Ġazālī: Über Rechtgläubigkeit und religiöse Toleranz. Eine Übersetzung der Schrift Das Kriterium in der Unterscheidung zwischen Islam und Gottlosigkeit (Fayṣal at-tafriqa bayn al-Islām wa-z-zandaqa)*. German translation by Frank Griffel. Zurich: Spur Verlag, 1998.

———. "Al-Ġazālī's Concept of Prophecy: The Introduction of Avicennan Psychology into Aš'arite Theology." *Arabic Sciences and Philosophy* 14 (2004): 101–44.

———. "Ibn Tūmart's Rational Proof for God's Existence and Unity of and His Connection to the Niẓāmiyya *madrasa* in Baghdad." In *Los Almohades: problemas y perspectivas*. Edited by Patrice Cressier, Maribel Fierro, and Luis Molina. 2 vols. Madrid: Consejo Superior de Investigationes Científicas, 2005. 2;753–813

———. "MS London, British Library Or. 3126: An Unknown Work by al-Ghazālī on Metaphysics and Philosophical Theology." *Journal of Islamic Studies* 17 (2006): 1–42.

———. "On Fakhr al-Dīn al-Rāzī's Life and the Patronage He Received." *Journal of Islamic Studies* 18 (2007): 313–44.

———. *Toleranz und Apostasie im Islam. Die Entwicklung zu al-Ġazālīs Urteil gegen die Philosophie und die Reaktionen der Philosophen*. Leiden: E. J. Brill, 2000.

———. "Toleration and Exclusion: al-Shāfiʿī and al-Ghazālī on the Treatment of Apostates." *Bulletin of the School of Oriental and African Studies* 64 (2001): 339–54.

———. "*Taqlīd* of the Philosophers. Al-Ghazālī's Initial Accusation in His *Tahāfut*." In *Ideas, Images, and Methods of Portrayal. Insights into Classical Arabic Literature and Islam*. Edited by Sebastian Günther. Leiden: Brill, 2005. 273–296.

———. "The Relationship Between Averroes and al-Ghazālī as It Presents Itself in Averroes' Early writings, Especially in his Commentary on al-Ghazālī's *al-Mustaṣfā*." In *Medieval Philosophy and the Classical Tradition in Islam, Judaism, and Christianity*. Edited by John Inglis. Richmond: Curzon Press, 2002. 51–63.

Gutas, Dimitri. *Avicenna and the Aristotelian Tradition. Introduction to Reading Avicenna's Philosophical Works*. Leiden: Brill, 1988.

———. "Avicenna's maḏhab. With an Appendix on the Question of His Date of Birth." *Quaderni di Studia Arabi* 5–6 (1987–88): 323–336.

Gwynne, Rosalind W. "Al-Jubbāʾī, al-Ashʿarī and the Three Brothers: The Uses of Fiction." *Muslim World* 75 (1985): 132–61.

———. *Logic, Rhetoric, and Legal Reasoning in the Qurʾān. God's Arguments*. London: RoutledgeCurzon, 2004.

Gyekye, Kwame. "Al-Ghazālī on Action." In *Ghazâlî, la raison et le miracle. Table ronde UNESCO, 9–10 decembre 1985*. Paris: Maisonneuve et Larose, 1987. 83–91.

Al-Ḥaddād, Abū ʿAbdallāh Maḥmūd ibn Muḥammad. *Takhrīj aḥādīth Iḥyāʾ ʿulūm al-dīn li-l-ʿIraqī (725–806) wa-bn al-Subkī (727–771) wa-l-Zabīdī (1145–1205)*. 7 vols. Riyad: Dār al-ʿĀṣima, 1408/1987.

Ḥājjī Khalīfa KātibČelebī, Muṣṭafā ibn ʿAbdallāh. *Kashf al-ẓunūn ʿan asāmī l-kutub wa-l-funūn = Lexicon bibliographicum et encyclopaedicum, a Mustapfa ben Abdallah, Katib Jelebi dicto et nomine Haji Khalfa celebrato compositum; ad codicum Vindobonensium, Parisiensium et Berolinensis fidem primum edidit Latine vertit et commentario indicibusque instruxit Gustavus Fluegel*. 7 vols. Leipzig/London: Published for the Oriental Translation Fund by R. Bentley, 1835–58.

Hakami, Nasrine. *Pèlerinage de l'Emâm Rezâ. Étude Socio-économique*. Tokyo: Institute for the Study of Languages and Cultures of Asia and Africa, 1989.

Halm, Heinz. *Die Ausbreitung der šāfiʿitischen Rechtsschule von den Anfängen bis zum 8./14. Jahrhundert*, Wiesbaden: L. Reichert, 1974.

———. *The Fatimids and Their Tradition of Learning*. London: I.B. Tauris, 1997.

———. *Kosmologie und Heilslehre der frühen Ismāʿīlīya. Eine Studie zur islamischen Gnosis*. Wiesbaden: F. Steiner, 1978.

Hasnawi, Ahmad. "Fayḍ." In *Encyclopedie philosophique universelle*. Edited by André Jacob. 4 vols. Paris: Presses universitaires de France, 1989–98. 2:966–72.

Hasse, Dag N. *Avicenna's De Anima in the West. The Formation of a Peripatetic Philosophy of the Soul 1160–1300*. London/Turin: The Warburg Institute / Nino Aragno Editore, 2000.

———. "Avicenna on Abstraction." In *Aspects of Avicenna*. Edited by Robert Wisnovsky. Princeton: Marcus Wiener, 2001. 39–72.

Heath, Peter. "Reading al-Ghazālī: The Case of Psychology." In *Reason and Inspiration in Islam. Theology, Philosophy and Mysticism in Muslim Thought. Essays in Honor of Hermann Landolt*. Edited by Todd Lawson. London: I.B. Tauris, 2005. 185–99.

Heemskerk, Margaretha T. *Suffering in the Muʿtazilite Theology. ʿAbd al-Jabbār's Teaching on Pain and Divine Justice*. Leiden: Brill, 2000.

Heer, Nicholas L. "Abū Ḥāmid al-Ghazālī's Esoteric Exegesis of the Koran." In *Classical Persian Sufism. From Its Origins to Rumi*. Edited by Leonard Lewisohn. London/New York: Khaniqahi Nimatullahi Publications, 1993. 235–57.

———. "Moral Deliberation in al-Ghazālī's *Iḥyāʾ ʿulūm al-dīn*." In *Islamic Philosophy and Mysticism*. Edited by Parviz Morewedge. Delmar (N.Y.): Caravan Books, 1981. 163–76.

Heinen, Anton M. *Islamic Cosmology. A Study of as-Suyūṭī's al-Hayʾa as-sanīya fī l-hayʾa as-sunnīya*. Beirut/Wiesbaden: F. Steiner, 1982.

Hill, Donald R. *Arabic Water-Clocks*. Aleppo (Syria): University of Aleppo, Intitute for the History of Arabic Science, 1981.

———. "The Toledo Water-Clock of c. 1075." *History of Technology* 16 (1994): 62–71.

Hillenbrand, Carole. "1092: A Murderous Year. " *Arabist. Budapest Studies in Arabic* 15–16 (1995): 281–296. = Proceedings of the fourteenth Congress of the Union Européenne d'Arabisants et Islamisants in Budapest 1988. Vol. 2.

———. "Islamic Orthodoxy or Realpolitik? Al-Ghazālī's Views on Government." *Iran. Journal of the British Institute of Persian Studies* 26 (1988): 81–94.

———."A Little Known Mirror for Princes by al-Ghazālī." In *Words, Texts, and Concepts Cruising the Mediterranean Sea. Studies on the Sources, Contents and Influences of Islamic Civilization and Arabic Philosophy and Science. Dedicated to Gerhard Endress on His Sixty-fifth Birthday*. Edited by Rüdiger Arnzen and J. Thielmann. Leuven: Peeters, 2004. 592–601.

———. "The Power Struggle Between the Saljuqs and the Ismāʿīlis of Alamūt, 487–518 / 1094–1124: The Saljuq Perspestive." In *Medieval Ismāʿīlī History and Thought*. Edited by Farhad Daftary. Cambridge: Cambridge University Press, 1996. 205–220.

Hintikka. Jaakko. *Time & Necessity. Studies in Aristotle's Theory of Modalities*. Oxford: Clarendon, 1973.

Horst, Heribert. *Die Staatsverwaltung der Großselǧuqen und Ḫorazmšāhs (1038–1231). Eine Untersuchung nach Urkundenformularen der Zeit*. Wiesbaden: F. Steiner, 1964.

Hourani, George F. "The Chronology of Ghazālī's Writings." *JAOS* 79 (1959): 225–33.

———. "A Revised Chronology of Ghazālī's Writings." *JAOS* 104 (1984): 289–302.

———. "Ibn Sīnā's 'Essay on the Secret of Destiny.'" *BSOAS* 29 (1966): 25–48.

———. "Ibn Sina on Necessary and Possible Existence." *Philosophical Forum* 4 (1972): 74–86.

Humāʾī, Jalāl al-Dīn. *Ghazzālī-nāmah*. 2nd ed. Tehran: Kitābfurūshī-yi Furūghī, 1342 [1963].

HWdP = *Historisches Wörterbuch der Philosophie*. Edited by Joachim Ritter, Karlfried Gründer, and Gottfried Gabriel. 13 vols. Basle: Schwabe Verlag, 1971–2007.

Ibn Abī Uṣaybiʿa, Aḥmad ibn al-Qāsim. *ʿUyūn al-anbāʾ fī ṭabaqāt al-aṭṭibāʾ*. Edited by August Müller. 2 vols. Cairo: al-Maṭbaʿa al-Wahbiyya, 1299/1882.

Ibn al-ʿAdīm, ʿUmar ibn Aḥmad. *Bughyat al-ṭalab fī taʾrīkh Ḥalab*. Edited by Suhayl Zakkār. 12 vols. Beirut: Dār al-Fikr, 1988.

Ibn al-ʿArabī, Abū Bakr Muḥammad ibn ʿAbdallāh. *ʿĀriḍat al-aḥwadhī bi-sharḥ Ṣaḥīḥ al-Tirmidhī*. Edited by Jamāl Marʿashlī. 14 parts in 8 vols. Beirut: Dār al-Kutub al-ʿIlmiyya, 1997.

———. *Al-ʿAwāṣim min al-qawāṣim = al-Naṣṣ al-kāmil li-kitāb al-ʿAwāṣim min al-qawāṣ im*. Edited by ʿAmmār Ṭālibī. Cairo: Maktabat Dār al-Turāth, 1417/1997.

———. *Qānūn al-taʾwīl*. Edited by Muḥammad Sulaymānī. Beirut: Dār al-Gharb al-Islāmī, 1990.

———. "Shawāhid al-jilla wal-aʿyān fī mashāhid al-Islam wa-l-buldān." In *Tres textos árabes sobre beréberes en el occidente Islámico*. Edited by Muḥammad Yaʿlā. Madrid: CSIC, 1996. 273–333.

Ibn ʿAsākir, ʿAlī ibn al-Ḥasan. *Tabyīn kadhib al-muftarī fī-mā ansaba al-Imām Abī l-Ḥasan al-Ashʿarī*. Damascus: Maṭbaʿat al-Tawfīq, 1347/1928.

———. *Taʾrīkh Dimashq wa-dhikr faḍlihā wa-tasmiyat man ḥallahā*. Edited by Muḥibb al-Dīn ʿUmar al-ʿAmrawī and ʿAlī Shīrī. 80 vols. Beirut: Dār al-Fikr al-Lubnānī, 1415–22 / 1995–2001.

Ibn al-Athīr, ʿIzz al-Dīn ʿAlī ibn Muḥammad. *al-Kāmil fī taʾrīkh = Chronicon quod perfectissimus inscribitur*. Edited by Carl J. Tornberg. 14 vols. Leiden and Uppsala: E. J. Brill, 1851–76.

Ibn al-ʿImād, ʿAbd al-Ḥayy ibn Aḥmad. *Shadharāt al-dhahab fī akhbār man dhahab*. 8 vols. Cairo: Maktabat al-Qudsī, 1351 [1931–32]. Photomechanic reprint, Beirut: al-Maktab al-Tijārī, 1966.

Ibn Bājja, Muḥammad ibn Yaḥyā. *Sharḥ al-samāʿ al-ṭabīʿī*. In *Shurūḥāt al-samāʿ al-ṭabīʿī li-bn Bājja al-Andalusī*. Edited by Maʿan Ziyāda. Beirut: Dār al-Fikr and Dār al-Kindī, 1398/1978. 11–178.

Ibn Fūrak, Muḥammad ibn al-Ḥasan. *Mujarrad maqālāt al-Ashʿarī*. Edited by Daniel Gimaret. Beirut: Librairie Orientale, 1986.

———. *Kitāb Mushkil al-ḥadīth wa-bayānih*. Hyderabad (India): Maṭbaʿat Jamʿiyyat Dāʾirat al-Maʿārif al-ʿUthmāniyyah, 1362 [1943].

Ibn Ghaylān al-Balkhī, ʿUmar ibn ʿAlī. *Ḥudūth al-ʿālam*. Edited by Mahdī Muḥaqqiq. Tehran: Muʾassasat-i Muṭālaʿāt-i Islāmī, 1377 [1998].

Ibn al-Jawzī, ʿAbd al-Raḥmān ibn ʿAlī. *al-Muntaẓam fī taʾrīkh al-mulūk wa-l-umam*, vols. 5–10. Hyderabad (India): Maṭbaʿat Majlis Dāʾirat al-Maʿārif al-ʿUthmāniyya, 1358 [1938–39].

———. *Talbīs Iblīs*. Edited by Muḥammad Munīr al-Dimashqī. Cairo: Idārat al-Ṭibāʿa al-Munīriyya, 1368 / 1950. Photographic reprint, Beirut: Dār al-Kutub al-ʿIlmiyya, n.d. [1975].

Ibn Jubayr al-Kinānī, Muḥammad ibn Aḥmad. *Tadhkira bi-l-akhbār ʿan ittifāqāt al-asfār*. Published unter the title: *Riḥlat ibn Jubayr*. Edited by Muḥammad Zaynhum. Cairo: Dār al-Maʿārif, 2000.

Ibn Kathīr, Ismāʿīl ibn ʿUmar. *Al-Bidāya wa-l-nihāya fī l-taʾrikh*. 14 vols. Cairo: al-Maṭbaʿa al-Saʿāda, 1351/1932.

———. *Ṭabaqāt al-fuqahāʾ al-shāfiʿiyīn*. Edited by Aḥmad ʿUmar Hāshim and Muḥammad Z. M. ʿAzab. 2 vols. Cairo: Maktabat al-Thaqāfa al-Dīniyya, 1413/1993.

Ibn Khaldūn, ʿAbd al-Raḥmān ibn Muḥammad. *Al-Muqaddima*. Edited by ʿAbd al-Salām al-Shaddādī. 3 vols. Casablanca: Bayt al-Funūn wa-l-ʿUlūm wa-l-Ādāb, 2005.

———. *The Muqaddimah. An Introduction to History*. Translated from the Arabic by Franz Rosenthal. 2nd ed. 3 vols. Princeton: Princeton University Press, 1967.

———. *Kitāb al-ʿIbar*. 7 vols. Beirut: Dār al-Kitāb al-Lubnānī, 1956–61.

Ibn Khallikān, Aḥmad ibn Muḥammad. *Wafayāt al-aʿyān wa-anbāʾ abnāʾ al-zamān*. Edited by Iḥsān ʿAbbās. 8 vols. Beirut: Dār Ṣādir, 1968–72.

Ibn al-Malāḥimī, Maḥmūd ibn Muḥammad. *Kitāb al-Muʿtamad fī uṣul al-dīn*. Edited by Martin McDermott and Wilferd Madelung. London: Al-Hoda, 1991.
Ibn al-Muqaffaʿ, ʿAbdallāh. *La version arabe de Kalîlah et Dimnah. D'après le plus ancien manuscrit arabe daté*. Edited by Louis Cheikho. Beirut: Imprimerie Catholique, 1905. Photographic reprint, Amsterdam: Philo Press, 1981.
———. *Kalīla wa-Dimna*. Edited by ʿAbd al-Wahhāb ʿAzzām. Cairo: Maṭbaʿat al-Maʿārif wa-Maktabatuhā, 1941. Photographic reprint, Cairo: Dār al-Maʿārif, 1980.
Ibn al-Najjār, Muḥammad ibn Maḥmūd. *Dhayl Taʾrīkh Baghdād*. Edited by Sharaf al-Dīn Aḥmad, Sayyida Mahr al-Nisāʾ, and Ceasar E. Farah. 5 vols. Hyderabad (India): Maṭbaʿat Majlis Dāʾirat al-Maʿārif al-ʿUthmāniyya, 1978–86.
Ibn al-Qalānisī, Ḥamza ibn Asad. *Dhayl Taʾrīkh Dimashq = History of Damascus 363–555* AH. Edited by Henry F. Amedroz. Leiden: Brill, 1908.
Ibn Qasī, Aḥmad ibn al-Ḥusayn. *Kitāb Khalʿ al-naʿlayn wa-iqtibās al-nūr min mawḍiʿ al-quddamayn*. Edited by Muḥammad Amrānī. Safi (Morocco): IMBH, 1418/1997.
Ibn Rushd, Muḥammad ibn Aḥmad. *Faṣl al-maqāl wa-taqrīr mā bayna l-sharīʿa wa-l-ḥikma min al-ittiṣāl*. Edited by George F. Hourani. Leiden: Brill, 1959.
———. *Al-Kashf ʿan manāhij al-adilla fī ʿaqāʾid al-milla = Manāhij al-adilla fī ʿaqāʾid al-milla*. Edited by Maḥmūd Qāsim. 2nd ed. Cairo: Maktabat Angelo, 1964.
———. *Tahāfut al-tahāfut = Averroès Tahafot at-Tahafot (L'Incohérence de l'incohérence)*. Edited by Maurice Bouyges. 2nd ed. Beirut: Dar el-Mashreq, 1987.
———. *Averroes' Tahafut al-Tahafut (The Incoherence of the Incoherence*. Translated by Simon van den Bergh. 2 vols. Oxford and London: Oxford University Press and Luzac & Co., 1954.
Ibn Sabʿīn, ʿAbd al-Ḥaqq ibn Ibrāhīm. *Budd al-ʿārif*. Edited by Jurj Katūrah. Beirut: Dār al-Andalus / Dār al-Kindī, 1978.
Ibn Ṣāḥib al-Ṣalāt, ʿAbd al-Malik ibn Muḥmmad. *Taʾrīkh al-mann bi-l-imāma*. Edited by Abd al-Hādī al-Tāzī. Beirut: Dār al-Andalus, 1964.
Ibn al-Ṣalāḥ al-Shahrazūrī, ʿUthmān ibn ʿAbd al-Raḥmān. *Ṭabaqāt al-fuqahāʾ al-shāfiʿiyya*. In the recension of al-Nawawī. Edited by Muḥyī al-Dīn ʿAlī Najīb. 2 vols. Beirut: Dār al-Bashāʾir al-Islāmiyya, 1413/1992.
Ibn Sīnā, al-Ḥusayn ibn ʿAbdallāh. *Aḥwāl al-nafs. Risāla Fī l-nafs wa-baqāʾihā wa-maʿādihā*. Edited by Aḥmad Fuʾād al-Ahwānī. Cairo: ʿĪsā al-Bābī al-Ḥalabī, 1371/1952.
———. *Dānishnāmah-yi ʿAlāʾ-i. Ilāhiyyāt*. Edited by Muḥammad Muʿīn. Tehran: Intishārāt-i Anjuman-i Āsār-i Millī, 1331 [1952].
———. *Fawāʾid wa-nukat*. MS Istanbul, Nuruosmaniye Kütüphanesi, 4894, foll. 242b–243a.
———. *Al-Ḥikma al-ʿarshiyya* = al-*Risāla al-ʿArshiyya fī tawḥīdihi taʿālā wa-ṣifātihi*. Hyderabad (India): Maṭbaʿat Dāʾirat al-Maʿārif al-ʿUthmāniyya, 1353 [1934–35].
———. *Al-Ishārāt wa-l-tanbīhāt*. Edited by Jacob Forget. Leiden: Brill, 1892. Reprint Frankfurt: Institute for the History of Arabic-Islamic Science, 1999.
———. *Lettre au vizir Abû Saʿd. Editio princeps d'après le manuscript de Bursa*. Edited and French translation by Yahya Michot. Beirut: Les Éditions Albouraq, 1421/2000.
———. *Kitāb al-Najāt*. Edited by Muḥyī l-Dīn Ṣabrī al-Kurdī. 2nd ed. Cairo: Maṭbaʿat al-Saʿāda, 1357/1938.
———. *Al-Najāt min al-gharq fī baḥr al-ḍalālāt*. Edited by Muḥammad Taqī Dānishpazhūh. Tehran: Intishārāt-i Dānishgāh-i Tehrān, 1364/1985.
———. *Al-Risāla al-Aḍḥawiyya fī l-maʿād. = Epistola sulla vita futura*. Edited and Italian translation by Francesca Lucchetta. Padua: Editrice Antenore, 1969.

———. *Réfutation de l'astrologie*. Edited, translated into French, and introduced by Yahya Michot. Beirut: Albouraq, 1427/2006.
———. *Risālat al-Ḥukūma fī-l-ḥujaj al-muthbitīn li-l-māḍī mabda'an zamaniyyan*. In Ibn Ghaylān al-Balkhī. *Ḥudūth al-ʿālam*. Edited by Mahdī Muḥaqqiq. Tehran: Muʾassasat-i Muṭālaʿāt-i Islāmī, 1377 [1998]. 131–53.
———. *Al-Shifāʾ, al-Manṭiq, al-Madkhal*. Edited by G. S. Qanawātī, Maḥmūd al-Khuḍayrī, and M. Fuʾād al-Ahwānī. Cairo: al-Maṭbaʿa al-Amīriyya, 1371/1953.
———. *Al-Shifāʾ, al-Manṭiq, al-Qiyās*. Ibrāhīm Madkūr and Saʿīd Zāyid. Cairo: al-Hayʾa al-ʿĀmma li-Shuʾūn al-Maṭābiʿ al-Amīriyya, 1384/1964.
———. *Al-Shifāʾ, al-Manṭiq, al-Burhān*. Edited by ʿAbd al-Raḥmān Badawī. 2nd ed. Cairo: Dār al-Nahḍa al-ʿArabiyya, 1966.
———. *Al-Shifāʾ, al-Ṭabīʿiyyāt, al-Samāʿ al-ṭabīʿī*. Edited by Saʿīd Zāyid. Cairo: al-Hayʾa al-Miṣriyya al-ʿĀmma li-l-Kitāb, 1983.
———. *Al-Shifāʾ, al-Ṭabīʿiyyāt, al-Nafs. = Avicenna's De Anima (Arabic Text) Being the Psychological Part of Kitāb al-Shifāʾ*. Edited by Fazlur Rahman. London: Oxford University Press, 1959.
———. *Al-Shifāʾ, al-Ilāhiyyāt. = The Metaphysics of the Healing*. A parallel English-Arabic text. Edited and translated by Michael E. Marmura. Provo (Utah): Brigham Young University Press, 2005.
———. *Al-Taʿlīqāt*. Edited by ʿAbd al-Raḥmān Badawī. Cairo: al-Hayʾa al-Miṣriyya al-ʿĀmma li-l-Kitāb, 1973.
———. *Kitāb al-Taʿlīqāt*. Edited by Ḥasan Majīd al-ʿUbaydī. Baghdad: Bayt al-Ḥikma, 2002.
———. *Avicenna latinus. Liber de philosophia prima sive scientia divina*. Edited by Simone van Riet. 3 vols. Louvain/Leiden: Peeters / Brill, 1977–83.
Ibn Taymiyya, Aḥmad ibn ʿAbd al-Ḥalīm. *Darʾ taʿāruḍ al-ʿaql wa-l-naql*. Edited by M. Rashād Sālim. 11 vols. Beirut: Dār al-Kunūz al-Adabiyya, n.d. [1980].
———. *Majmūʿ fatāwā Shaykh al-Islām Aḥmad ibn Taymiyya*. Edited by ʿAbd al-Raḥmān ibn Muḥammad ibn Qāsim al-ʿĀṣimī. 35 vols. Riyad: Maṭābiʿ al-Riyāḍ, 1381–86 [1961–67].
———. *Minhāj al-sunna al-nabawiyya fī naqd kalām al-shīʿa wa-l-qadariyya*. 4 vols. Būlāq: al-Maṭbaʿa al-Kubrā al-Amīriyya, 1321–22 [1903–5]. Photographic reprint, Beirut: Dār Ṣādir, 1973.
———. "Sharḥ al-ʿaqīda al-iṣfahāniyya." In *Majmūʿat fatāwā Shaykh al-Islām Taqī l-Dīn Ibn Taymiyya*. Edited by Farajallāh Zakī al-Kurdī. 5 vols. Cairo: Maṭbaʿat Kurdistān al-ʿIlmiyya, 1326–29 [1908–11]. Vol. 5. 3rd text.
Ibn Ṭufayl, Muḥammad ibn ʿAbd al-Malik. *Risālat Ḥayy ibn Yaqẓān fī asrār al-ḥikmā al-mushriqīya [sic] / Hayy ben Yaqdhân. Roman philosophique d'Ibn Thofaïl*. Edited and translated into French by Léon Gauthier. 2nd ed. Beirut: Imprimerie catholique, 1936.
Ibn Tūmart, Amghār ibn ʿAbdallāh. *Sifr fīhi jamīʿ taʿāliq al-Imām al-maʿṣūm al-Mahdī al-maʿlūm (. . .) mimma amlaʾhu Sayyiduna al-Imām al-Khalīfa Amir al-Muʾminīn Abū Muḥammad ʿAbd al-Muʾmin ibn ʿAlī = Aʿazz mā yuṭlab*. Edited by ʿAmmār Ṭālibī. Algiers: al-Muʾassasat al-Waṭaniyya li-l-Kitāb, 1985.
Ibn Ṭumlūs, Yūsuf ibn Muḥammad. *Madkhal li-sinaʿat al-manṭiq = Introducción al arte de la lógica por Abentomlús de Alcira*. Edited by Miguel Asín Palacios. Madrid: Centro de Estudios Históricos, 1916.
Ibn al-Walīd, ʿAlī ibn Muḥammad. *Dāmigh al-bāṭil wa-ḥatf al-munāḍil*. Edited by Muṣṭafā Ghālib. 2 vols. Beirut: Muʾassasat ʿIzz al-Dīn, 1403/1982.

Iqbāl Āshtiyānī, ʿAbbās. *Vizārāt dar ʿahd-i salāṭīn-i buzurg-i saljūqī. Az tārīkh-i tashkīl-i īn silsila tā marg-i Sulṭān Sanjar.* Tehran: Intishārāt-i Dānishgāh-i Tehrān, 1338 [1959].

Irwin, Robert. "Islamic Science and the Long Siesta. Did Scientific Progress in the Islamic World Really Grind to a Halt After the Twelfth Century?" *Times Literary Supplement* (January 25, 2008); 8.

Al-Isfarāʾīnī, Ibrāhīm ibn Muḥammad. "al-ʿAqīda." In Richard M. Frank, "Al-Ustādh Abū Isḥāq: An ʿAqīda Together with Selected Fragments." *Mélanges d'institut dominicain d'études orientales* 19 (1989): 129–202.

İskenderoğlu, Muammar. *Fakhr al-Dīn al-Rāzī and Thomas Aquinas on the Question of the Eternity of the World.* Leiden: Brill, 2002.

Al-Isnawī, ʿAbd al-Raḥmān ibn al-Ḥasan. *Ṭabaqāt al-shāfiʿiyya.* Edited by ʿAbdallāh al-Jibūrī. 2 vols. Baghdad: Riʾāsat Dīwān al-Awqāf, 1390–1391 [1970–71].

Ivry, Alfred L. "Destiny Revisited: Avicenna's Concept of Determinism." In *Islamic Theology and Philosophy. Studies in Honor of George F. Hourani.* Edited by Michael Marmura. Albany: State University of New York Press, 1984. 160–71.

Izutsu, Toshihiko. "Mysticism and the Linguistic Problem of Equivocation in the Thought of ʿAyn al-Quḍāt Hamadānī." *Studia Islamica* 31 (1970): 153–70.

———. "Creation and the Timeless Order of Things: A Study in the Mystical Philosophy of ʿAyn al-Quḍāt." *Philosophical Forum* 4 (1972): 124–40.

Jabre, Farid. "La Biographie et l'œuvre de Ghazali reconsiderées a la lumiere des *Tabaqat* de Sobki." *Mélanges d'institut dominicain d'études orientales* 1 (1954): 73–102.

———. *Essai sur le lexique de Ghazali. Contribution à l'étude de la terminologie de Ghazali dans ses principaux ouvrages à l'exception du Tahāfut.* Beirut: Librairie Orientale, 1985.

Jackson, Sherman A. *On the Boundaries of Theological Tolerance in Islam. Abū Ḥāmid al-Ghāzalī's Fayṣal al-Tafriqa bayna al-Islam wa al-zandaqa.* Karachi: Oxford University Press, 2002.

Jāmiʿ al-badāʾiʿ. Majmūʿ (. . .) yaḥtawī ʿalā thamāniya ʿashra risāla (. . .) li-(. . .) Ibn Sīnā wa- (. . .) ʿUmar al-Khayyām (. . .). Edited by Muḥyī al-Din Ṣabrī al-Kurdī. Cairo: Muḥyī al-Din Ṣabrī al-Kurdī 1335/1917.

Janssens, Jules. "Creation and Emanation in Ibn Sīnā." *Documenti e studi sulla tradizione filosofica medievale* 8 (1997): 455–77. Reprint in idem. *Ibn Sīnā and His Influence.* Text iv.

———. "'Experience' (*tajriba*) in Classical Arabic Philosophy (al-Fārābī—Avicenna)." *Quaestio* 4 (2004): 45–62.

———. "Al-Ghazzālī and His Use of Avicennian Texts." In *Problems in Arabic Philosophy.* Edited by Miklós Maróth. Piliscaba (Hungary): Avicenna Institute of Middle East Studies, 2003. 37–49. Reprint in idem. *Ibn Sīnā and His Influence.* Text xi.

———. "Ibn Sīnā (Avicenne): Un projet 'religieux' de philosophie?" In *Was ist Philosophie im Mittelalter?* Edited by Jan A. Aertzen and Andreas Speer. Berlin: Walter de Gruyter, 1998. 863–70. Reprint in idem. *Ibn Sīnā and His Influence.* Text v.

———. "Ibn Sīnā, and His Heritage in the Islamic World and the Latin West." In idem. *Ibn Sīnā and His Influence.* Text i.

———. *Ibn Sīnā and His Influence on the Arabic and Latin World.* Aldershot (Hampshire, UK): Ashgate, 2006.

———. "Filosofische elementen in de mystieke leer van al-Ghazali (*Iḥyâ, 31–36*)," *Tijdschrift voor filosofie* (Leuven) 50 (1988): 334–342.

———. "Le *Dânesh-Nâmeh* d'Ibn Sînâ: Un text à revoir?" *Bulletin de philosophie médiévale* 28 (1986): 163–77. Reprint in idem. *Ibn Sīnā and His Influence.* Text vii.

———. "Le Ma'ârij al-quds fî madârij ma'rifat al-nafs: Un élément-clé pour le dossier Ghazzâli-Ibn Sînâ?" *Archive d'histoire doctrinale et littéraire du Moyen Age* 60 (1993): 27–55. Reprint in idem, *Ibn Sīnā and His Influence*. Text viii.

———. "The Problem of Human Freedom in Ibn Sînâ." *Actes del Simposi Internacional de l'Edat Mitjana*. Edited by Paloma Llorente et al. Vic (Spain): Patronat d'Estudis Osonencs, 1996. 112–18. Reprint in idem. *Ibn Sīnā and His Influence*. Text iii.

Jaouiche, Habib. *The Histories of Nishapur. 'Abdalġāfir al-Fārisī. Siyāq Ta'rīḫ Naisābūr. Register der Personen- und Ortsnamen*. Mit einem Vorwort von Josef van Ess. Wiesbaden: Ludwig Reichert, 1984.

Al-Jazarī, Ismā'īl ibn al-Razzāz. *The Book of Knowledge of Ingenious Mechanical Devices (Kitāb Fī ma'rifat al-ḥiyal al-handasiyya)*. Translated and annotated by Donald R. Hill. Dortrecht (Holland): D. Reidel, 1974.

Al-Juwaynī, 'Abdalmalik ibn 'Abdallāh. *Al-'Aqīda al-Niẓāmiyya*. Edited by Muḥammad Zāhid al-Kawtharī. Cairo: Maktabat al-Khānjī, 1367/1948.

———. *Al-Burhān fī uṣūl al-fiqh*. Edited by 'Abd al-Aẓīm al-Dīb. 2nd ed. 2 vols. Cairo: Tawzī' Dār al-Anṣār, 1400 [1979/80].

———. *Al-Irshād ilā qawāṭī l-adilla fī uṣūl al-i'tiqād*. Edited by Muḥammad Yūsuf Mūsā and 'Alī 'A. 'Abd al-Ḥamīd. Cairo: Maktabat al-Khānjī, 1369/1950.

———. *Luma' fī qawā'id ahl al-sunna wa-l-jamā'a*. In *Textes apologétiques de Ǧuwainī (m. 478/1085)*. Edited and translated by Michel Allard. Beirut: Dar El-Machreq, 1968. 118–77.

———. *A Guide to Conclusife Proos for the Principles of Belief. Kitāb al-irshād ilā qawāṭi' al-adilla fī uṣūl al i'tiqād [by] Imām al-Ḥaramayn al-Juwaynī*. Translated by Paul E. Walker. Reading [U.K.]: Garnet Publishing, 2000.

———. *Al-Shāmil fī uṣūl al-dīn*. Edited by 'Alī S. al-Nashshār, Fayṣal B. 'Awn, and Suhayr M. Mukhtār. Alexandria: Manshū'at al-Ma'ārif, 1969.

———. *Al-Shāmil fī uṣūl al-dīn. Some Additional Portions of the Text*. Edited by Richard M. Frank. Tehran: Intishārāt-i Mu'assasah-yi Muṭāla'āt-i Islāmī-yi Dānishgāh-i McGill, 1401/1981.

Juvaynī, 'Aṭā-Malik ibn Muḥammad. *Ta'rīkh-i Jahāngushāy*. Edited by Muḥammad ibn 'Abd al-Vahhāb Qazvīnī. 3 vols. London/Leiden: Luzac and Brill, 1912–37.

Kaplony, Andreas. *The Ḥaram of Jerusalem 324–1099. Temple, Friday Mosque, Area of Spritual Power*. Suttgart: F. Steiner, 2002.

Karabulut, Ali Riza (also: 'Alī Riḍā Qara-Bulūṭ). İstanbul ve Anadolu Kütüphanelerinde Mevcut El Yazmasi Eserler Ansiklopedisi. 3 vols. Kayseri: Akabe Kitabevi Sahabiye Medreseci, n.d. [2005].

Karatay, Fehmi Edhem. *Topkapı Serayı Müzesi Kütüphanesi Arapça Yazmalar Kataloğu*. 4 vols. Istanbul: Topkapı Serayı Müzesi, 1962–69.

Kasā'ī, Nūrallāh. *Madāris-i niẓāmiyyah va-tā'sīrāt-i 'ilmī va-ijtimā'ī-yi ān*. Tehran: Intishārāt-yi Amīr-i Kabīr, 1374 [1995].

Kemal, Salim. *The Philosophical Poetics of Alfarabi, Avicenna and Averroës. The Aristotelian Reception*. London: RoutledgeCourzon, 2003.

Al-Kiyā' al-Harrāsī, 'Alī ibn Muḥammad. *Uṣūl al-dīn*. MS Cairo, Dār al-Kutub al-Miṣriyya, *tawḥīd* 295.

Al-Khāzinī, 'Abd al-Raḥmān. *Mīzān al-ḥikma*. Hayderabat (India): Maṭba'at Dā'irat al-Ma'ārif al-'Uthmāniyya, 1359 [1940].

Al-Khayyām, 'Umar ibn Ibrāhīm. *Dānishnāmah-yi Khayyāmī. Majmū'ah-yi rasā'il-i 'ilmī, falsafī, va-adabī*. Edited by Raḥīm Riżāzādah Malik. Tehran: 'Ilm va-Humar, 1377 [1998].

Al-Khwānsārī, Muḥammad Bāqir ibn Muḥammad Naqī. *Rawḍat al-jannāt fī aḥwāl al-ʿulamāʾ wa-l-sādāt.* 8 vols. Beirut: Dār al-Islāmiyya, 1411/1991.

Al-Kirmānī, Aḥmad ibn ʿAbdallāh. *Rāḥat al-ʿaql.* Edited by Muṣṭafā Ghālib. Beirut: Dār al-Andalus, 1983.

Klausner, Carla L. *The Seljuk Vezirate. A Study in Civil Administration 1055–1194.* Cambridge (Mass.): Center for Middle Eastern Studies, 1973.

Klein-Franke, Felix. *Die klassische Antike in der Tradition des Islam.* Darmstadt: Wissenschaftliche Buchgesellschaft, 1980.

Kleinknecht, Angelika. "Al-Qisṭās Al-Mustaqīm: Eine Ableitung der Logik aus dem Koran." In *Islamic Philosophy and the Classical Tradition.* Edited by Salomon M. Stern, Albert Hourani, and Vivian Brown. Columbia: University of South Carolina Press, 1972. 159–87.

Knuuttila, Simo. *Modalities in Medieval Philosophy.* London and New York: Routledge, 1993.

———. "Plentitude, Reason and Value: Old and New in the Metaphyscis of Nature." In *Nature and Lifeworld. Theoretical and Practical Metaphysics.* Edited by Carsten Bengt-Pedersen and Niels Thomassen. Odense (Denmark): Odense University Press, 1998. 139–51.

Kogan, Barry S. "Averroës and the Theory of Emanation." *Mediaeval Studies* 43 (1981): 384–403.

———. "The Philosophers al-Ghazālī and Averroes on Necessary Connection and the Problem of the Miraculous." In *Islamic Philosophy and Mysticism.* Edited by Parviz Morewedge. Delmar (N.Y.): Caravan Books, 1981. 113–132.

———. "Some Reflections on the Problem of Future Contingency in Alfarabi, Avicenna, and Averroes." In *Divine Omniscience and Omnipotence in Medieval Philosophy. Islamic, Jewish, and Christian Prespectives.* Edited by Tamar Rudavsky. Dordrecht (Holl.): D. Reidel, 1984. 95–101.

Krawulsky, Dorothea. *Briefe und Reden des Abū Hāmid Muhammad al-Gazzālī.* Freiburg (Germany): Klaus Schwarz, 1971.

Kukkonen, Taneli. "Causality and Cosmology: The Arabic Debate." In *Infinity, Causality and Determinism. Cosmological Enterprises and their Preconditions.* Edited by Eeva Martikainen. Frankfurt/Main: Peter Lang, 2002. 19–43.

———. "Plentitude, Possibility, and the Limits of Reason: A Medieval Arabic Debate on the Metaphyscis of Nature." *Journal of the History of Ideas* 61 (2000): 539–60.

———. "Possible Worlds in the *Tahâfut al-Falâsifa.* Al-Ghazâlî on Creation and Contingency." *Journal of the History of Philosophy* 38 (2000): 479–502.

Lagardère, Vincent. "A propos d'un chapitre du *Nafḫ wal-taswiya* attribué à Ġazālī." *Studia Islamica* 60 (1984): 119–36.

Landolt, Hermann. "Ghazālī and 'Religionswissenschaft' Some Notes on the Mishkāt al-Anwār." *Asiatische Studien. Zeitschrift der Schweizer Gesellschaft für Asienkunde* (Bern) 45 (1991): 19–72.

———. "Two Types of Mystical Thought in Muslim Iran: An Essay on Suhrawardī *Shaykh al-Ishrāq* and ʿAynulquẓāt-i Hamadānī." *Muslim World* 68 (1978): 187–204; and 70 (1980): 83–84.

Laoust, Henri. *Essai sur les doctrines sociale et politiques de Takī-d-Dīn Aḥmad b. Taimīya.* Cairo: Institut Français d'Archéologie Orientale, 1939.

———. *La politique de Ġazālī.* Paris: Paul Geuthner, 1970.

———. "La survie de Ġazālī d'après Subkī. "*Bulletin d'études orientales* 25 (1972): 153–172.

Al-Lawkarī, Faḍl ibn Muḥammad. *Bayān al-ḥaqq bi-ḍimān al-ṣidq. al-Manṭiq. 1– al-Madkhal.* Edited by Ibrāhīm Dībājī. Tehran: Intishārāt-yi Amīr-i Kabīr, 1364 [1986].

———. *Bayān al-ḥaqq bi-ḍimān al-ṣidq. al-ʿIlm al-ilāhī*. Edited by Irāhīm Bībājī. Tehran: International Institute of Islamic Thought and Civilization, 1373/1995.
———. *Sharḥ-i qaṣīdah-yi asrār al-ḥikmah*. Edited by Ilahah Rūḥī-Dil under the direction of Muḥammad-Rasūl Daryāgasht and Riżā Pūrjavādī. Tehran: Markaz-i Nashr-i Dānishgāhī, 1382 [2003].
Lazarus-Yafeh, Hava. *Studies in al-Ghazzālī*. Jerusalem: Magnes Press, 1975.
Leaman, Oliver. "God's Knowledge of the Future in the Philosophy of al-Fārābī." *Occasional Papers of the School of Abbasid Studies* 1 (1986): 23–29.
Leo Africanus = al-Ḥasan ibn Muḥammad al-Wazzān. "Libellus de viris quibusdam illustribus apud Arabes." In Johann Heinrich Hottinger. *Bibliothecarius quadripartitus*. Zurich: Melchior Stauffacher, 1664. 246–94.
Le Strange, Guy. *Baghdad during the Abbasid Caliphate*. Oxford: Clarendon 1900.
———. *The Lands of the Eastern Caliphate. Mesopotamia, Persia, and Central Asia from the Moslem Conquest to the Time of Timur*. Cambridge: University Press, 1905.
———. *Palestine under the Moslems. A Description of Syria and the Holy Land from A.D. 650 to 1500*. London: Palestine Exploration Fund, 1890.
Lettinck, Paul. *Aristotle's Physics and Its Perception in the Arabic World with an Edition of the Unpublished Parts on Ibn Bājja's Commentary on the Physics*. Leiden: Brill, 1994.
Lizzini, Olga. "Occasionalismo e causalità filosofica: la discussione della causalità in al-Ġazâlî." *Quaestio* 2 (2002): 155–83.
Lumbard, Joseph E. B. *Aḥmad al-Ghazālī (d. 517/1123 or 520/1127) and the Metaphysics of Love*. Ph.D. diss., Yale University, 2003.
Macdonald, Duncan B. "The Life of al-Ghazzālī, with special reference to his religious experience and opinions." *JAOS* 20 (1899): 71–132.
———. "The Name al-Ghazzālī." *JRAS* 1901: 18–22.
Mach, Rudolph. *Catalogue of Arabic Manuscripts (Yahuda Section) in the Garrett Collection. Princeton University Library*. Princeton: Princeton University Press, 1977.
Madelung, Wilferd. "Abū l-Ḥusayn al-Baṣrī's Proof for the Existence of God." In *Arabic Theology, Arabic Philosophy. From the Many to the One. Essays in Celebration of Richard M. Frank*. Edited by James E. Montgomery. Leuven: Peeters, 2006. 273–80.
———. "The Late Muʿtazila and Determinism: The Philosopher's Trap." In *Yād-Nāma. In Memoria di Alessandro Bausani*. Edited by Biancamaria Scarcia Amoretti and Lucia Rostagnio. 2 vols. Rome: Bardi Editore, 1991. 1:245–57.
———. "Ar-Rāġib al-Iṣfahānī und die Ethik al-Ġazālīs." In *Islamkundliche Abhandlungen. Fritz Meier zum sechzigsten Geburtstag*. Edited by Richard Gramlich. Wiesbaden: F. Steiner, 1974. 152–163.
Mafākhir al-barbar = Fragments historiques sur les Berbères au moyen age. Extraits inédits d'un recuils anonyme compilé en 712–1312 et intitulé Kitāb Mafākhir al-Barbar. Edited by Évariste Lévi-Provençal. Rabat: F. Moncho, 1934.
Mafākhir al-barbar. In *Tres textos árabes sobre beréberes en el occidente Islámico*. Edited by Muḥammad Yaʿlā. Madrid: CSIC, 1996. 123–383
Mahdavī, Yaḥyā. *Fihrist-i nuskhat-hā-yi muṣannafāt-i Ibn Sīnā*. Tehran: Intishārāt-i Dānishgāh-yi Tehrān, 1333/1954.
Maimonides, Mūsā ibn Maymūn. *Dalālat al-ḥāʾirīn*. Edited by Solomon Munk and Issacher Joel. Jerusalem: J. Junovitch, 1929.
———. *The Guide of the Perplexed*. Translated by Shlomo Pines. 2 vols. Chicago: University of Chicago Press, 1963.
Majmūʿah-yi falsafī-yi Marāgha / A Philosophical Anthology from Maragha. Containing Works by Abū Ḥāmid Ghazzālī, ʿAyn al-Quḍāt al-Ḥamadānī, Ibn Sīnā, ʿUmar ibn

Sahlān Sāvi, Majduddīn Jīlī and others. Facsimile edition with introductions in Persian and English by Nasrollah Pourjavady. Tehran: Markaz-i Nashr-I Dānishgāh, 1380/2002.

Makdisi, George. "Ash'ari and Ash'arites in Islamic Religious History." *Studia Islamica* 17 (1962): 37–80; and 18 (1963): 19–39.

———. *Ibn ʿAqīl et la resurgence de l'Islam traditionaliste au XIe siècle (Ve siècle de l'Hegire).* Damascus: Institut français de Damas, 1963.

———. "Muslim Institutions of Learning in Eleventh-Century Baghdad." *BSOAS* 24 (1961): 1–56.

———. "The Non-Ashʿarite Shafiʿism of Abū Ḥāmid al-Ghazzālī." *Revue des Études Islamiques* 54 (1986): 239–57.

———. *The Rise of Colleges. Institutions of Learning in Islam and the West.* Edinburgh: Edinburgh University Press, 1981.

Al-Makkī, Muḥammad ibn ʿAlī. *Qūt al-qulūb fī muʿāmalat al-maḥbūb wa-waṣf ṭarīq al-murīd ilā maqām al-tawḥīd.* 2 vols. Cairo: al-Maṭbaʿa al-Maymaniyya, 1310 [1892–3]. Photographic reprint, Cairo: Tawzīʿ Maktabat al-Mutanabbī, n.d. [*c.* 1985].

Malter, Heinrich. *Die Abhandlung des Abû Hâmid al-Ġazzâlî. Antworten auf Fragen, die an ihn gerichtet wurden nach mehreren Handschriften ediert.* Frankfurt/Main: J. Kauffmann, 1896.

Marmura, Michael E. "Avicenna on Causal Priority." In *Islamic Philosophy and Mysticism.* Edited by Parviz Morewedge. Delmar (N.Y.): Caravan Books, 1981. 64–83.

———. "Avicenna's Proof from Contingency in the *Metaphysics* of his *al-Shifāʾ.*" *Mediaeval Studies* 42 (1980): 34–56. Reprint in idem. *Probing.* 131–48.

———. "Divine Omniscience and Future Contingents in Alfarabi and Avicenna." In *Divine Omniscience and Omnipotence in Medieval Philosophy. Islamic, Jewish and Christian Prespectives.* Edited by Tamar Rudavsky. Dordrecht (Holl.): D. Reidel, 1984. 81–94. Reprint in Marmura. *Probing.* 375–89.

———. "Al-Ghazālī." In *The Cambridge Companion to Arabic Philosophy.* Edited by Peter Adamson and Richard C. Taylor. Cambridge: Cambridge University Press, 2005. 137–54.

———. "Ghazali and Ashʿarism Revisited." *Arabic Sciences and Philosophy* 12 (2002): 91–110.

———. "Ghazali's Attitude to the Secular Sciences and Logic." In *Essays on Islamic Philosophy and Science.* Edited by George F. Hourani. Albany: State University of New York Press, 1975. 100–111.

———. "Al-Ghazālī on Bodily Resurrection and Causality in Tahafut and The Iqtisad." *Aligarh Journal of Islamic Thought* 2 (1989): 46–75. Reprint in idem. *Probing.* 273–99.

———. "Ghazālī and Demonstrative Science." *Journal of the History of Philosophy* 3 (1965): 183–204. Reprint in idem. *Probing.* 231–60.

———. "Ghazālian Causes and Intermediaries." *JAOS* 115 (1995): 89–100.

———. "Ghazali's Chapter on Divine Power in the *Iqtiṣād.*" *Arabic Sciences and Philosophy* 4 (1994): 279–315. Reprint in idem. *Probing.* 301–34.

———. "Ghazālī on Ethical Premises." *The Philosophical Forum* 1 (1968–69): 393–403. Reprint in idem. *Probing.* 261–65.

———. "Ghazali's *al-Iqtisad fi al-I'tiqad.* Its Relation to *Tahafut al-Falasifa* and to *Qawa'id al-Aqa'id.*" *Aligarh Journal of Islamic Philosophy* 10 (2004): 1–12.

———. "Al-Ghazālī's Second Causal Theory in the 17th Discussion of His Tahāfut." In *Islamic Philosophy and Mysticism.* Edited by Parviz Morewedge. Delmar (N.Y.): Caravan Books, 1981. 85–112.

———. "The Metaphysics of Efficient Causality in Avicenna (Ibn Sīnā)." In *Islamic Theology and Philosophy. Studies in Honor of George F. Hourani*. Edited by Michael Marmura. Albany: State University of New York Press, 1984. 172–87.

———. "Some Aspects of Avicenna's Theory of God's Knowledge of Particulars," *Journal of the American Oriental Society* 83 (1962): 299–312.

———. *Probing in Islamic Philosophy. Studies in the Philosophies of Ibn Sīnā, al-Ghazālī and Other Major Muslim Thinkers*. Binghampton (N.Y.): Global Academic Publishing, 2005.

Marquet, Yves. *La philosophie des Iḫwān al-Ṣafāʾ. Nouvelle édition augmentée*. Paris/Milano: S.É.H.A / Archè: 1999.

———. "Révélation et vision véridique ches les Ikhwān al-ṣafā." *Revue des Études Islamiques* 32 (1964): 27–44.

Al-Masʿūdī, Muḥammad ibn Masʿūd. *Al-Shukūk wa-l-shubah ʿalā l-Ishārāt*. MS Istanbul, Süleymaniye Kütüphanesi, Hamidiye 1452, foll. 109b–150a.

Mayer, Tobias. "Ibn Sīnāʾs *Burhān al-Ṣiddiqīn*." *Journal of Islamic Studies* 12 (2001): 18–39.

Māyil Hirāwī, Najīb. *Khāṣṣiyyat-i āyinagī. Naqd-i ḥāl, guzārah-yi ārā va-guzīdah-yi āsār-i fārisī-yi ʿAyn al-Qużāt-i Hamadānī*. Tehran: Nashr-i Nay 1374 [1995].

McCarthy, Richard J. *Al-Ghazali: Deliverance from Error. Five Key Texts Including His Spriritual Autobiography, al-Munqidh min al-Dalal*. Translated and annotated by R.J. McCarthy. Louisville (Ky.): Fons Vitae: 2000.

McDermott, Martin J. "Abū l-Ḥuṣayn al-Baṣrī on God's Volition." In *Culture and Memory in Medieval Islam. Essays in Honor of Wilferd Madelung*. Edited by Farhad Daftary and Josef W. Meri. London: I.B. Tauris, 2003. 86–93.

McGinnis, Jon. "Making Abstraction Less Abstract: The Logical, Psychological, and Metaphysical Dimensions of Avicenna's Theory of Abstraction." *Proceedings of the American Philosophical Association* 80 (2006): 169–83.

———. "Occasionalism, Natural Causation and Science in al-Ghazālī." In *Arabic Theology, Arabic Philosophy. From the Many to the One. Essays in Celebration of Richard M. Frank*. Edited by James E. Montgomery. Leuven: Peeters, 2006. 441–63.

———. "Scientific Methodologies in Medieval Islam." *Journal of the History of Philosophy* 41 (2003): 307–27.

Menn, Stephen. "The *Discourse on the Method* and the Tradition of Intellectual Autobiography." In *Hellenistic and Early Modern Philosophy*. Edited by Jon Miller and Brad Inwood. Cambridge: Cambridge University Press, 2003. 141–91.

Michot, Yaḥyā (also: Jean). *La destinee de l'homme selon Avicenne. Le retour à Dieu (maʿād) et l'imagination*. Leuven: Peeters, 1986.

———. "La pandémie Avicennienne au VIe/XIIe siècle. Kitāb ḥudūth al-ʿālam d'Ibn Ghaylan al-Balkhī." *Arabica* 40 (1993): 287–344.

Minorsky, Vladimir. *A History of Sharvān and Darband in the 10th–11th Centuries*. Cambridge: Heffer, 1958.

Miskawayh, Aḥmad ibn Muḥammad. *Kitāb al-Fawz al-aṣghar*. Edited by Sāliḥ ʿUḍayma. French translation by Roger Arnaldez. Tunis: Dār al-ʿArabiyya li-l-Kitāb, 1987.

Moosa, Ebrahim. *Ghazālī and the Poetics of Imagination*. Chapel Hill: University of North Carolina Press, 2005.

Müller August. *Ibn Abi Useibia*. Königsberg: Selbstverlag, 1884.

Munk, Solomon. *Mélanges de philosophie juive et arabe*. Paris: Alophe Franck, 1859.

Muntajab al-Dīn Badīʿ Juvaynī. *ʿAtabat al-kataba fī bayān taʿlīm al-kitāba wa-l-inshāʾ = Kitāb-i ʿAtabat al-katabah. Majmūʿah-yi murāsalāt-i dīvān-i Sultān-i Sanjar*. Edited by Muḥammad Qazvīnī and ʿAbbās Iqbāl Āshtiyānī. Tehran: Sihāmī-yi Chāp, 1329 [1950].

Nagel, Tilman. *Die Festung des Glaubens. Triumph und Scheitern des islamischen Rationalismus im 11. Jahrhundert.* Munich: C.H. Beck, 1988.

———. *Im Offenkundigen das Verborgene. Die Heilszusage des sunnitischen Islams.* Göttingen: Vandenhoeck und Ruprecht, 2002.

Nakamura, Kojiro. "An Approach to Ghazzālī's Conversion." *Orient* (Tokyo) 21 (1985): 46–59.

———. "Imām Ghazālī's Cosmology Reconsidered with Special Reference to the Concept of *Jabarūt*." *Studia Islamica* 80 (1994): 29–46.

Nasr, Seyyed Hossein. *An Introduction to Islamic Cosmological Doctrines.* Rev. ed. Albany: State University of New York Press, 1993.

Al-Nawawī, Yaḥyā ibn Sharaf. *Mukhtaṣar Ṭabaqāt al-fuqahāʾ.* Edited by ʿĀdil ʿAbd al-Mawjūd and ʿAlī Mūʿawwaḍ. Beirut: Muʾassasat al-Kutub al-Thaqāfiyya, 1416/1995.

———. *Rawḍat al-ṭālibīn wa-ʿumdat al-muftiyīn.* Edited by ʿAbdallāh ʿUmar al-Bārādī. 10 vols. Beirut: Dār al-Fikr, 1995.

Nemoy, Leon. *Arabic Manuscripts in the Yale University Library.* New Haven: Yale University Press, 1956.

Niẓām al-Mulk, al-Ḥasan ibn ʿAlī. *Siyāsat-nāmah.* Edited by Mehmet A. Köymen. Ankara: Ankara Üniversitesi, 1976.

Niẓāmī ʿArūḍī Samarqandī, Aḥmad ibn ʿUmar. *Chahār Maqāla ("The Four Discourses").* Edited by Mírzá Muḥammad. London/Leiden: Luzac & Co./Brill, 1910.

Niẓāmī Ganjāvī, Ilyās ibn Yūsuf. *Khamsah = Makhzan al-asrār, Khusraw va-Shīrīn, Laylī va-Majnūn, Haft paykar, Sharafnāmah, Iqbālnāmah.* Edited by Vaḥīd Dastgirdī. 6 vol. Tehran: Armaghān, 1313–17 [1934–38].

Normore, Calvin G. "Duns Scotus's Modal Theory." In *The Cambridge Companion to Duns Scotus.* Edited by Thomas Williams. Cambridge: Cambridge University Press, 2003. 129–60.

Obermann, Julian. "Das Problem der Kausalität bei den Arabern." *Wiener Zeitschrift für die Kunde des Morgenlandes* 29 (1915): 323–50 and 30 (1918): 37–90.

———. *Der philosophische und religiöse Subjektivismus Ghazālīs. Ein Beitrag zum Problem der Religion.* Vienna and Leipzig: W. Braunmüller: 1921.

Opwis, Felicitas. "Islamic Law and Legal Change: The Concept of *Maṣlaḥa* in Classical and Contemporary Islamic Legal Theory." In *Shari'a. Islamic Law in the Contemporary Context.* Edited by Abbas Amanat and Frank Griffel. Stanford (Calif.): Stanford University Press, 2007. 62–82, 203–211.

Ormsby, Eric L. *Theodicy in Islamic Thought. The Dispute over al-Ghazâlî's "Best of All Possible Worlds."* Princeton: Princeton University Press, 1984.

Perler, Dominik, and Ulrich Rudolph. *Occasionalismus. Theorien der Kausalität im arabisch-islamischen und im europäischen Denken.* Göttingen: Vandenhoeck & Ruprecht, 2000.

Peskes, Esther. *Al-ʿAidarūs und seine Erben. Eine Untersuching zu Geschichte und Sufismus einer ḥaḍramatischen Sāda-Gruppe vom fünfzehnten bis zum achtzehnten Jahrhundert.* Stuttgart: F. Steiner, 2005.

Pines, Shlomo (also: Salomon). "Quelques notes sur les rapports de l'*Iḥyāʾ ʿulūm al-dīn* d'al-Ghazâlî avec la pensée d'Ibn Sînâ." In *Ghazâlî, la raison et le miracle. Table ronde UNESCO, 9–10 decembre 1985.* Paris: Maisonneuve et Larose, 1987. 11–16.

———. "La conception de la conscience de soi chez Avicenne et chez Abu'l-Barakât al-Baghdâdî." In *Studies in Abu'l-Barakât al-Baghdâdî. Physcis and Metaphysics.* Jerusalem/Leiden: Magnes Press / Brill, 1979. 181–258.

———. "Some Problems of Islamic Philosophy." *Islamic Culture* 11 (1937): 66–80.

Poonawala, Ismail K. *Biobibliography of Ismāʿīlī Literature.* Malibu (Calif.): Undena, 1977.

Poggi, Vincenzo M. *Un classico della spiritualità musulmana. Saggio monografico sul "Munqiḏ" di al-Ġazālī.* Rome: Libreria dell'Università Gregoriana, 1967.

Pūrjavādī, Naṣrallāh (also: Pourjavady, Nasrollah). *ʿAyn al-Qużāt va-ustāẕān-i ū.* Tehran: Intishārāt-i Asāṭīr, 1374 [1995].

———. "Chahār aṣar-i kutāh-i fārisī az Abū Ḥāmid Ghazzālī." *Maʿārif* 7 (1369 [1990]): 3–19.

———. "Dō maktūb-i fārisī az Imām Muḥammad-i Ghazzālī." *Maʿārif* 8 (1370 [1991]): 3–36.

———. *Dō mujaddid. Pazhūhish-hā-yi dar bāra-yi Muḥammad-i Ghazzālī va-Fakhr-i Rāzī / Two Renewers of Faith. Studies on Muhammad-i Ghazzālī and Fakhruddīn-i Rāzī.* Tehran: Markaz-i Nashr-i Dānishgāhī, 1381 [2002].

———. "Faḫr-e Rāzī und Ġazzālīs Miškāt al-anwār (Lichternische)." *Spektrum Iran* (Bonn, Germany) 2 (1989): 49–70.

Al-Qushayrī, ʿAbd al-Karīm ibn Hawāzin. *al-Risāla al-Qushayriyya.* Edited by ʿAbd al-Ḥalīm Maḍmūd and Maḍmūd ibn al-Sharīf. 2 vols. Cairo: Riḍā Tawfīq ʿAfīfī, 1972.

Al-Rāfiʿī, ʿAbd al-Karīm ibn Muḥammad. *Al-Fatḥ al-ʿazīz fī sharḥ al-Wajīz = al-ʿAzīz sharḥ al-Wajīz al-maʿrūf bi-l-sharḥ al-kabīr.* Edited by ʿAlī Muḥammad Muʿawwaḍ and ʿĀdil Aḥmad ʿAbd al-Mawjūd. 14 vols. Beirut: Dār al-Kutub al-ʿIlmiyya, 1997.

Al-Rahim, Ahmed H. "The Twelver-Sīʿī Reception of Avicenna in the Mongol Period." In *Before and After Avicenna. Proceedings of the First Conference of the Avicenna Study Group.* Edited by David C. Reisman and Ahmed H. al-Rahim. Leiden: Brill, 2003. 219–231.

Rahman, Fazlur. *Prophecy in Islam. Philosophy and Orthodoxy.* London: Allan & Unwin, 1958.

Rasāʾil Ikhwān al-ṣafā wa-khillān al-wafāʾ. Edited by Khayr al-Dīn Ziriklī. 4 vols. Cairo: al-Maktaba al-Tijāriyya al-Kubrā, 1347/1928. Reprint, Frankfurt: Institute for the History of Arabic-Islamic Science, 1999.

———. 4 vols. Beirut: Dār Ṣādir, n.d. [1957].

Rashīd al-Dīn Ṭabīb, Fażlallāh ibn ʿImād al-Dawla. *Jāmiʿ al-tavārīkh.* Edited by Muḥammad Rawshan Muṣṭafā Mūsavī. 4 vols. Tehran: Nashr-i Alburz, 1373 [1994].

———. *The History of the Seljuq Turks from the Jāmiʿ al-Tawārīkh. An Ilkhanid Adaptation of the Saljuq-nāma of Ẓahīr al-Dīn Nīshāpūrī.* Translated and annotated by Kenneth Allin Luther. Edited by C. Edmund Bosworth. Richmond (U.K.): Curzon, 2001.

Rayan, Sobhi. "Al-Ghazali's Use of the Terms 'Necessity' and 'Habit' in His Theory of Natural Causality." *Theology and Science* 2 (2004): 255–68.

Reichmuth, Stefan. "Murtaḍā az-Zabīdī (d. 1791) in Biographical and Autobiographical Accounts. Glimpses of Islamic Scholarship in the 18th Century." *Welt des Islams* 39 (1999): 64–102.

———. *The World of Murtada al-Zabidi.* Cambridge: E. J. W. Gibb Memorial Trust, forthcoming.

Reisman, David C. *The Making of the Avicennan Tradition. The Transmission, Contents, and Structure of Ibn Sīnā's* al-Mubāḥaṯāt *(The Discussions).* Leiden: Brill, 2002.

———. "Al-Fārābī and the Philosophical Curriculum." *The Cambridge Companion to Arabic Philosophy.* Edited by Peter Adamson and Richard C. Taylor. Cambridge: Cambridge University Press, 2005. 52–71.

Renan, Ernest. *Averroès et l'averroïsme. Essai historique.* Paris: Librairie Auguste Durand, 1852.

Rescher, Nicholas. *The Development of Arabic Logic*. Pittsburgh: University of Pittsburgh Press, 1964.
———. "The Concept of Existence in Arabic Logic and Philosophy." In idem. *Studies in Arabic Philosophy*. Pittsburgh: University of Pittsburgh Press, n.d. [1967]. 69–80.
———. *Temporal Modalities in Arabic Logic*. Dortrecht (Holland): D. Reidel, 1967.
Rosenthal, Franz. "Die arabische Autobiographie." *Analecta Orientalia* 14 (1937): 3–40.
———. *The Technique and Approach of Muslim Scholarship*. Rome: Pontificium Institutum Biblicum, 1947.
Rotter, Gernot. *Löwe und Schakal*. German translation of the Arabic *al-Asad wa-l-ghawwāṣ*. Tübingen and Basel: Horst Erdmann, 1980.
Routledge Encyclopedia of Philosophy. General editor Edward Craig. 10 vols. London and New York: Routledge, 1998.
Rudolph, Ulrich. *Die Doxographie des Pseudo-Ammonios. Ein Beitrag zur neuplatonischen Überlieferung im Islam*. Wiesbaden: F. Steiner, 1989.
———. "La preuve de l'existence de Dieu chez Avicenne et dans la théologie musulmane." In *Langages et philosophie. Hommage à Jean Jolivet*. Edited by Alain de Libera, Abdelali Elamrani-Jamal, and Alain Galonnier. Paris: J. Vrin, 1997. 339–46.
———. "Die Neubewertung der Logik durch al-Ġazālī." In *Logik und Theologie. Das Organon im arabischen und lateinischen Mittelalter*. Edited by Dominik Perler and Ulrich Rudolph. Leiden: Brill, 2005. 73–97.
———. see also under Perler, Dominik.
Rūmī, Jalāl al-Dīn ibn Bahāʾ al-Dīn. *Maṣnavī = The Mathnawí of Jalálu'ddín Rúmí*. Edited and translated by Reynolds A. Nicholson. 8 vols. London/Leiden: Luzac & Co. / Brill, 1925–40.
Ṣabāt, Khalīl, *Taʾrīkh al-ṭibāʿa fī l-sharq al-ʿarabī*. 2nd ed. Cairo: Dār al-Maʿārif, 1966.
Sabra, Abdelhamid I. "*Kalām* Atomism as an Alternative Philosophy to Hellenizing *Falsafa*." In *Arabic Theology, Arabic Philosophy. From the Many to the One. Essays in Celebration of Richard M. Frank*. Edited by James E. Montgomery. Leuven: Peeters, 2006. 199–271.
———. "The Appropriation and Subsequent Naturalization of Greek Sciences in Medieval Islam: A Preliminary Statement." *History of Science* 25 (1987): 223–43.
Al-Ṣafadī, Khalīl ibn Aybak. *Al-Wāfī bi-l-wafayāt*. Edited by Helmut Ritter et al. Istanbul/Beirut/Wiesbaden: Orient Institut der DMG, 1931–.
Safi, Omid. *The Politics of Knowledge in Premodern Islam. Negotiating Ideology and Religious Inquiry*. Chapel Hill: University of North Carolina Press, 2006.
Saflo, Mohammad Moslem Adel. *Al-Juwaynī's Thought and Methodology. With a Translation and Commentary on Lumaʿ al-Adilla*. Berlin: K. Schwarz, 2000.
Saleh, Walid A. "The Last of the Nishapuri School of Tafsīr: Al-Wāḥidī (d. 468/1076) and His Significance in the History of Qurʾanic Exegesis." *Journal of the American Oriental Society* 126 (2006): 223–43.
Saliba, George. "The Ashʿarites and the Science of the Stars." In *Religion and Culture in Medieval Islam*. Edited by Richard G. Hovannisian and Georges Sabagh. Cambridge: Cambridge University Press, 1999. 79–92.
Al-Samʿānī, ʿAbd al-Karīm ibn Muḥammad. *Kitāb al-Ansāb*. 13 vols. Hayderabat (India): Maṭbaʿat Majlis Dāʾirat al-Maʿārif al-ʿUthmāniyya, 1382–1402/1962–83.
———. *Al-Taḥbīr fī l-muʿjam al-kabīr*. Edited by Munīra Nājī Sālim. 2 vols. Bagdad: Riʾāsat Dīwān al-Awqāf, 1395/1975.
Al-Ṣarīfīnī, Ibrāhīm ibn Muḥammad. *al-Muntakhab min al-Siyāq li-Taʾrīkh Nīsābūr*. Edited by Muḥammad Kāẓim al-Maḥmūdī. Qum: Jamāʿat al-Mudarrisīn fī l-Ḥawza al-ʿIlmiyya, 1362 [1983].

Sarkīs, Yūsuf Alyān (also: Sarkis, Joseph Elain). *Muʿjam al-maṭbūʿāt al-ʿarabiyya wa-l-muʿarraba / Dictionaire encyclopédique de bibliographie arabe.* Cairo: Maṭbaʿat Sarkīs, 1346/1928.

Sawwaf, Abdul-Fattah. *Al-Ghazzali. Etude de la réforme ghazzalienne dans l'histoire de son devéloppement.* Ph.D. diss., Université de Fribourg (Switzerland), 1962.

Schöck, Cornelia. "Möglichkeit und Wirklichkeit menschlichen Handelns. 'Dynamis' (*qūwa/qudra/istiṭāʿa*) in der islamischen Theology." *Traditio* 59 (2004): 79–128.

Schmölders, August. *Essai sur les écoles philsophiques chez les Arabes et notamment sur la doctrine d'Algazzali.* Paris: Firmin Didot freres, 1842.

Schwarz, Michael. "'Acquisition' (kasb) in Early Kalām." In *Islamic Philosophy and the Classical Tradition.* Edited by Salomon M. Stern, Albert Hourani, and Vivian Brown. Columbia: University of South Carolina Press, 1972. 355–387.

Serrano Ruano, Delfina. "Why Did the Scholars of al-Andalus Distrust al-Ghazālī? Ibn Rushd al-Jadd's *Fatwā* on *Awliyāʾ Allāh." Der Islam* 83 (2006): 137–56.

Şeşen, Ramazan (also: Ramaḍān Shishin). *Mukhtārāt min al-makhṭūṭāt al-ʿarabiyya al-nādira fī maktabāt Turkiyā.* Edited by Ekmeleddin Ihsanoğlu. Istanbul: İSAR, 1997.

Al-Shahrastānī, Muḥammad ibn ʿAbd al-Karīm. *Al-Milal wa-l-niḥal = Book of Religious and Philosophical Sects.* Edited by William Cureton. 2 vols. London: Printed for the Society for the Publication of Oriental Texts, 1842–46.

———. *Livre des religions et des sects.* French translation by Daniel Gimaret, Jean Jolivet, and Guy Monnot. 2 vols. Louvain/Paris: Peeters/UNESCO, 1986–93.

Al-Shahrazūrī, Muḥammad ibn Maḥmūd. *Nuzhat al-arwāḥ wa-rawḍat al-afrāḥ fī tawārīkh al-ḥukamāʾ al-mutaqaddimīn wa-l-mutaʾakhkhirīn.* 2 vols. Hayderabat (India): Maṭbaʿat Majlis Dāʾirat al-Maʿārif al-ʿUthmāniyya, 1396/1976.

Al-Shīrāzī, Ibrāhīm ibn ʿAlī. *Ṭabaqāt al-fuqahāʾ.* Edited by Iḥsān ʿAbbās. Beirut: Dār al-Rāʾid al-ʿArabī: 1970.

Al-Shushtarī, al-Qāḍī Nūrallāh ibn ʿAbdallāh. *Majālis al-muʾminīn.* Edited by Aḥmad ʿAbd-Manāfī. 2 vols. Tehran: Kitābfurūsh-i Islāmiyya, 1365 [1986].

Shihadeh, Ayman. "From al-Ghazālī to al-Rāzī: 6th/12th Century Developments in Muslim Philosophical Theology." *Arabic Sciences and Philosophy,* 15 (2005): 141–79.

———. *The Teleological Ethics of Fakhr al-Dīn al-Rāzī.* Leiden: Brill, 2006.

Sibṭ ibn al-Jawzī, Yūsuf ibn Qizoghlu. *Mirʾāt al-zamān fī taʾrīkh al-aʿyān.* Part 8. 2 vols. Hyderabad (India): Maṭbaʿat Majlis Dāʾirat al-Maʿārif al-ʿUthmāniyya, 1370–77/1951–52.

———. *Mirʾāt al-zamān fī taʾrīkh al-aʿyān. 481–517 h. / 1077–1123 m.* Edited by Musfir Sālim al-Ghāmidī. 2 vols. Mecca: Markaz Iḥyāʾ al-Turāth al-Islāmī, 1407/1987.

Smith, Gerard. "Avicenna and the Possibles." *New Scholasticism* 17 (1943): 340–57.

Smith, Margaret. *Al-Ghazālī the Mystic.* London: Luzac & Co., 1944.

Smith, Wilfred Cantwell. "Faith as Taṣdīq." In *Islamic Philosophical Theology.* Edited by Parviz Morewedge. Albany: State University of New York Press, 1979. 96–119.

Steel, Carlos. "Avicenna and Thomas Aquinas on Evil." In *Avicenna and His Heritage. Acts of the International Colloquium, Leuven—Louvain-la-Neuve September 8–September 11, 1999.* Edited by Jules Janssens and Daniel de Smet. Leuven: Leuven Universty Press. 171–96.

Steinschneider, Moritz. *Die hebraeischen Übersetzungen des Mittelalters und die Juden als Dolmetscher.* 2 vols. Berlin: Kommissionsverlag des Bibliographischen Bureaus, 1893. Photographic reprint, Graz: Akademische Druck- und Verlagsanstalt, 1956.

———. "Typen von M. Steinschneider. II. Waage und Gewichte." *Jeschurun. Zeitschrift für die Wissenschaft des Judentums* (Bamberg) 9 (1873): 65–97.

Stern, Samuel M. "The Authorship of the Epistles of the Ikhwān-aṣ-Ṣafā'." *Islamic Culture* 20 (1946): 367–72; and 21 (1947): 403–4.
Strauss, Leo. *Persecution and the Art of Writing.* Glencoe (Ill.): The Free Press, 1952.
Street, Tony. "Faḫraddīn ar-Rāzī's Critique of Avicennan Logic." In *Logik und Theologie. Das Organon im arabischen und lateinischen Mittelalter.* Edited by Dominik Perler and Ulrich Rudolph. Leiden: Brill, 2005. 101–16.
Soucek, Priscilla P. "Niẓāmī on Painters and Painting." in *Islamic Art in the Metropolitan Museum of Art.* Edited by Richard Ettinghausen. New York: Metropolitan Museum, 1972. 9–21.
Al-Subkī, 'Abd al-Wahhāb ibn 'Alī. *Ṭabaqāt al-shāfi'iyya al-kubrā.* Edited by Maḥmūd M. al-Ṭanāḥī and 'Abd al-Fattāh M. al-Ḥilw. 10 vols. Cairo: 'Īsā al-Bābī al-Ḥalabī, 1964–76.
Sublet, Jacqueline. "Un itinéraire du *fiqh* šāfi'ite d'apres al-Ḫaṭīb al-'Uthmānī." *Arabica* 11 (1964): 188–95.
Al-Ṭabarī, Muḥammad ibn Jarīr. *Jāmi' al-bayān 'an tafsīr al-Qur'ān.* Edited by Maḥmūd M. Shākir and Aḥmad M. Shākir. 16 vols. Cairo: Dār al-Ma'ārif, 1954–69.
———. *Jāmi' al-bayān fī tafsīr al-Qur'ān.* 30 vols. Būlāq: al-Maṭba'a al-Kubrā al-Amīriyya, 1323–30 [1905–12]. Photographic reprint, Beirut: Dār al-Ma'rifa, 1406/1986.
Ṭālibī, 'Ammār: *Arā' Abī Bakr ibn al-'Arabī l-kalāmiyya.* 2 vols. Algiers: al-Sharika al-Waṭaniyya li-l-Nashr wa-l-Tawzī', n.d. (1974).
Al-Tawḥīdī, 'Alī ibn Muḥammad. *al-Imtā' wa-l-mu'ānasa.* Edited by Aḥmad Amīn and Aḥmad al-Zayn. 3 vols. Cairo: Lajnat al-Ta'līf wa-l-Tarjama wa-l-Nashr, 1939–44.
Terkan, Fehrullah. "Does Zayd Have the Power *Not* to Travel Tomorrow? A Preliminary Analysis of al-Fārābī's Discussion on God's Knowledge of Future Human Acts." *Muslim World* 94 (2004): 45–64.
Tibawi, Abdel Latif. "Al-Ghazālī's Sojourn in Damascus and Jerusalem." *Islamic Quarterly* 9 (1965): 65–122.
Treiger, Alexander. "Monism and Monotheism in al-Ghazālī's *Mishkāt al anwār.*" *Journal of Qur'anic Studies* 9 (2007): 1–27.
Al-Ṭurṭūshī, Muḥammad ibn al-Walīd. *Kitāb al-Ḥawādith wa-l-bida' (El libro de las novedades y las innovaciones).* Spanish translation and study by Maribel Fierro. Madrid: CSIC, 1993.
———. *Risāla ilā 'Abdallāh ibn Muẓaffar.* In Sa'd Ghurāb. "Ḥawla iḥrāq al-Murābiṭūn li-Iḥyā' al-Ghazālī." *Actas del IV Coloquio Hispano-Tunecino (Palma de Mallorca, 1979).* Madrid: Instituto Hispano-Arabe de Cultura, 1983. 133–63, 158–63.
Al-'Ulaymī, 'Abd al-Raḥmān ibn Muḥammad. *Al-Uns al-jalīl bi-ta'rīkh al-Quds wa-l-Khalīl.* Edited by Muḥammad Baḥr al-'Ulūm. 2 vols. Najaf: al-Maṭba'a al-Ḥaydariyya: 1388/1968. Reprint, Amman: Maktabat al-Muktasib, 1973.
'Uthmān, 'Abd al-Karīm. *Sīrat al-Ghazālī wa-aqwāl al-mutaqaddimīn fīhi.* Edited by Aḥmad Fu'ād al-Ahwānī. Damascus: Dār al-Fikr, n.d. [*c.* 1960].
Vajda, Georges. *Isaac Albalag, averroïste juif, traducteur et annotateur de'l-Ghazâlî.* Paris: J. Vrin, 1960
Van den Bergh, Simon. "Ghazali on 'Gratitude Towards God' and Its Greek Sources." *Studia Islamica* 7 (1957): 77–98.
Van Ess, Josef. *Die Erkenntnislehre des 'Aḍudaddīn al-Īcī. Übersetzung und Kommentar des ersten Bandes seiner Mawāqif,* Wiesbaden: F. Steiner, 1966.
———. "Neuere Literatur zu Ġazzālī. "*Oriens* 20 (1967): 299–308.
———. "Quelques remarques sur le Munqiḏ min aḍ-ḍalāl." In *Ghazâlî, la raison et le miracle. Table ronde* UNESCO, *9–10 decembre 1985.* Paris: Maisonneuve et Larose, 1987. 57–68.

———. *Theologie und Gesellschaft im 2. und 3. Jahrhundert Hidschra. Eine Geschichte des religiösen Denkens im frühen Islam.* 6 vols. Berlin: Walter de Gruyter, 1991–97.

———. *Zwischen Ḥadīṯ und Theologie. Studien zum Entstehen prädestinatianischer Überlieferung.* Berlin: Walter de Gruyter, 1975.

Wakelnig, Elvira. *Feder Tafel Mensch. Al-ʿĀmirīs Kitāb al-Fuṣūl fī l-Maʿālim al-ilāhīya und die arabische Proklos-Rezeption im 10 Jh.* Leiden: Brill, 2006.

Walker, Paul E. *Early Philosophical Shiism. The Ismaili Neoplatonism of Abū Yaʿqūb al-Sijistānī.* Cambridge: Cambridge University Press, 1993.

———. *Ḥamīd al-Dīn al-Kirmānī. Ismaili Thought in the Age of al-Ḥākim.* London: I.B. Tauris, 1999.

———. "The Ismāʾīlīs." In *The Cambridge Companion to Arabic Philosophy.* Edited by Peter Adamson and Richard C. Taylor. Cambridge: Cambridge University Press, 2005. 72–91.

———. "The Ismaili Vocabulary of Creation." *Studia Islamica* 40 (1974): 75–85.

Watt, William Montgomery: "The Authenticity of Works Ascribed to al-Ghazālī." *JRAS* (1952): 24–45.

———. "A Forgery in al-Ghazālī's *Mishkāt*?" *JRAS* (1949): 5–22.

———. *The Faith and Practice of al-Ghazālī.* London: Allen & Unwin, 1953.

———. *The Formative Period of Islamic Thought.* Edinburgh: University Press, 1973.

———. *Free Will and Predestination in Early Islam.* London: Luzac & Co., 1948.

———. *Islamic Philosophy and Theology.* Edinburgh: Edinburgh University Press, 1962.

———. "Die islamische Theologie 950–1850." In W. M. Watt and Michael Marmura. *Der Islam II. Politische Entwicklungen und theologische Konzepte.* Stuttgart: W. Kohlhammer, 1985. 393–487.

Weinberg, Steven. "A Deadly Certitude." *Times Literary Supplement* (January 19, 2007); 5–6.

———. *The First Three Minutes. A Modern View of the Origin of the Universe.* 2nd ed. New York: BasicBooks, 1993.

Wensinck, Arent Jan. "On the Relationship Between al-Ghazālī's Cosmology and His Mysticism." *Mededeelingen der Koninklijke Akademie van Wetenschappen, Afdeeling Letterkunde, Deel 75, Serie A* (1933): 183–209.

———et al. (eds.): *Concordance et indices de la tradition musulmane.* 8 vols. Leiden: Brill, 1936–88.

Weiss, Bernard. "Knowledge of the Past: The Theory of *tawātur* According to Ghazālī." *Studia Islamica* 61 (1985): 81–105.

Wehr, Hans. *Al-Ġazzālī's Buch vom Gottvertrauen. Das 35. Buch des Iḥyāʾ ʿulūm ad-dīn übersetzt und mit Anmerkungen versehen.* Halle: Niemeyer, 1940.

Whittingham, Martin. *Al-Ghazālī and the Qurʾān. One Book, Many Meanings.* London and New York: Routledge, 2007.

Winter, Timothy J. *Al-Ghazālī. The Remembrance of Death and the Afterlife. Kitāb dhikr al-mawt wa mā baʿdahu. Book XL of the Revival of the Religious Sciences.* Cambridge: Islamic Text Society, 1989.

Wisnovsky, Robert. *Avicenna's Metaphysics in Context.* London: Gerald Duckworth & Co., 2003.

———. "Avicenna and the Avicennian Tradition." In *The Cambridge Companion to Arabic Philosophy.* Edited by Peter Adamson and Richard C. Taylor. Cambridge: Cambridge University Press, 2005. 92–136.

———. "Final and Efficient Causality in Avicenna's Cosmology and Theology." *Quaestio* 2 (2002): 97–123.

———. "The Nature and Scope of Arabic Philosophical Commentary in Post-Classical (ca. 1100–1900 AD) Islamic Intellectual History: Some Preliminary Observations." In *Philosophy, Science and Exegesis in Greek, Arabic, and Latin Commentaries*. Edited by Peter Adamson, Han Baltussen, and M.W.F. Stone. London: Institute of Classical Studies, 2004. 149–91.

Wörterbuch der klassischen arabischen Sprache. Edited by the Deutsche Morgenländische Gesellschaft. Wiesbaden: Harrassowitz, 1970–.

Al-Yāfiʿī, ʿAbdallāh ibn Asʿad. *Mirʾāt al-jinān wa-ʿibrāt al-yaqẓān fī maʿrifat mā yuʿtabir min ḥawādith al-zamān*. 2nd ed. 4 vols. Beirut: Muʾassasat al-Aʿlāmī: 1390/1970.

Yāqūt al-Ḥamawī al-Rūmī: *Muʿjam al-buldān = Jacut's Geographisches Wörterbuch*. Edited by Ferdinand Wüstenfeld. 6 vols. Leipzig: F.A. Brockhaus, 1866–73.

———. *Muʿjam al-udabāʾ. Irshād al-arīb ilā maʿrifat al-adīb*. Edited by Iḥsān ʿAbbās. 7 vols. Beirut: Dār al-Gharb al-Islāmī, 1993.

Al-Zabīdī, al-Murtaḍā Muḥammad ibn Muḥammad, *Itḥāf al-sāda al-muttaqīn bi-sharḥ Iḥyāʾ ʿulūm al-dīn*. 10 vols. Cairo: al-Maṭbaʿa al-Maymaniyya, 1311 [1894]. Photographic reprint, Beirut: Iḥyāʾ al-Turāth al-ʿArabī, *c*. 1975 and 1994.

———. *Al-Muʿjam al-mukhtaṣṣ (. . .): yaḥtawī ʿalā tarājim akthar min sitmiʾa min aʿyān al-qarn al-thānī ʿashar al-Hijrī (. . .)*. Edited by Niẓām M. Ṣ. Yaʿqūbī and Muḥammad ibn Nāṣir al-ʿAjamī. Beirut: Dār al-Bashāʾir al-Islāmiyya, 1427/2006.

Zarrīnkūb, ʿAbd al-Ḥusayn. *Firār az madrasah. Dar bāra-yi zendegi-yi va-andīshah-yi Abū Ḥāmid-i Ghazzālī*. Tehran: Anjuman-i Āsār-i Milli, 1353 [1974].

Zedler, Beatrice H. "Another Look at Avicenna." *New Scholasticism* 55 (1981): 113–30.

———. "Why Are the Possibles Possible?" *New Scholasticism* 50 (1976): 504–21.

Ziriklī, Khayr al-Dīn. *Al-Aʿlām. Qāmūs tarājim li-ashhar al-rijāl wa-l-nisāʾ min al-ʿArab wa-l-mustaʿribīn wa-l-mustashriqīn*. 10th ed. 8 vols. Beirut: Dār al-ʿIlm li-l-Malāyīn, 1992.

Zimmermann, F.W. *Al-Farabi's Commentary and Short Treatise on Aristotle's De Interpretatione*. Oxford: Oxford University Press, 1981.

General Index

ʿAbbās, Iḥsān, 70, 309 n. 64
ʿAbd al-Ghāfir al-Fārisī, 9, 20–23, 27–29, 31, 34, 43–44, 49–54, 56–57, 266
ʿAbd al-Muʾmin ibn ʿAlī (Almohad ruler), 78
al-Abīwardī, 58
Abraham, the Prophet, 8, 24, 44, 48, 51, 100–1, 157, 246–51, 253–54, 284
Abrahamov, Binjamin, 179–80
Abū l-Fatḥ Naṣr al-Maqdisī al-Nabulūsī, xii, 43, 44, 46–49, 71
Abū-l Futūḥ Aḥmad al-Ghazālī, 26–27, 42, 58, 62, 81, 82
Abū l-Ḥusayn al-Baṣrī, 134, 170, 189, 323, 332–33, 336, 343, 345
Abū Naṣr al-Ismāʿīlī, 27–28
Abū l-Qāsim ʿAbdallāh ibn ʿAlī, 27
Abū l-Qāsim Ismāʿīl ibn Masʿada al-Ismāʿīlī, 295 n. 69
Abū Ṭāhir al-Shabbāk, 42
Abū Yazīd al-Bisṭāmī, 40
accident (ʿ*araḍ*), 70–71, 109, 125–27, 129, 133, 140–41, 159, 189, 202, 209–212, 223, 229, 249–50, 281, 284–85
Acre, 63
Aḥmad al-Ghazālī. *See* Abū-l Futūḥ Aḥmad al-Ghazālī
Alamūt, 40
al-Āmidī, 72
Almohads, 77–81, 187
Almoravids, 63, 80–81, 85
Alp Arslan, xi, 27
al-amr ("the divine command"), 237, 243, 252–53, 256–59, 274, 280 (*see also* laws of nature)
al-ʿĀmirī, 48, 256, 283, 356 n. 155, 358 n. 22
al-Andalus, 44, 61–63, 67, 77–81, 85, 199, 275
al-Anṣārī, Abū l-Qāsim, 57, 130, 134
Antioch, 48
Anūshirwān ibn Khālid, 40, 314 n. 161
apostasy, 7, 85–86, 103–6, 108–9, 111, 119
Aristotle, 3, 29, 30, 97, 99, 116–17, 121, 134–36, 139, 163, 166–70, 199–200 208–9, 230, 248–52, 262–63, 276, 278, 282
Aristotelianism, 6–7, 29, 81, 98, 101, 116, 120, 122, 134–35, 144, 147–48, 150–53, 158–59, 167–68, 175, 177, 181, 184, 199, 208, 210, 212–13, 246, 249–50, 261–62, 278
al-ʿarsh ("the throne"), 118, 123, 243, 268, 280

As'ad al-Mayhanī, 27–28, 71–74, 77, 83, 86
Ascalon, 63
Asín Palacios, Miguel, 9
al-Ash'arī, 29, 42, 125–30, 169, 188, 195, 198, 217
Ash'arism, 6, 10–12, 21–22, 27, 29–30, 54, 67, 69, 75, 77, 80, 105–6, 124–28, 152, 159, 167, 170–72, 177–78, 181–82, 188–91, 195, 197–98, 202, 204, 216–17, 219, 225–26, 230, 224, 226, 274, 249–50, 262–63, 266, 271
atomism, 71, 125–27, 159, 202–04, 249–50, 284–85
Averroes (Ibn Rushd), 4, 6, 19–20, 61, 81 95, 118–19, 178, 180, 212, 246, 277, 284
Avicenna (Ibn Sīnā), 6–7, 10–12, 14, 19, 29–30, 35, 61, 71–72, 79, 98–99, 104–6, 108, 117–19, 133–41, 147, 161–73, 175–78, 180, 182, 184–85, 189, 193–201, 208–12, 219, 223, 225–26, 229–30, 233–34, 242, 249, 251–67, 271–73, 276–83, 286
  on causality, 134–37, 143–44, 150–53
  on prophecy and the quasi-prophetical faculties of *awliyā'*, 67–69
  on the necessity of God's creation, 141–43
  proof of God's existence, 77, 83–85, 135, 251–52
awliyā' ("friends of God"), 67, 69, 86, 179, 198, 199, 200, 264
al-'Aydarūs, 16
'Ayn al-Quḍāt al-Hamadhānī, 62, 81–86

al-Bābī al-Ḥalabī Family, 16
Baghdad, xi–xiii, 8–9, 20, 24–29, 31–32, 34–37, 39–45, 48–52, 56–59, 63–65, 67, 73—74, 77, 80, 86, 97, 128, 204
al-Baghdādī, 'Abd al-Qāhir, 128, 189, 262, 349 n. 97, 355
al-Baghdādī, 'Abd al-Laṭīf, 72
Bauer, Hans, 15–16
Bausani, Alessandro, 6
al-Baqarī, 'Abd al-Dā'im, 20
al-Bayhaqī, Ẓāhir al-Dīn ibn Funduq, 22, 71–72, 81
Berkey, Jonathan, 4
Berk-Yaruq, xii–xiii, 36–38, 40, 298 n. 132
Bījū, Maḥmūd, 15
Bougie (*also* Béjaïa, in Algeria), 63
Bouyges, Maurice, 13, 17, 20, 35
Brethren of Purity, 199–201, 269, 282–83
Brockelmann, Carl, 13, 306 n. 7
al-Bundarī, 314
Burrell, David B., 259

Cairo, 15–16, 40, 70
causality, 12, 83, 127–28, 147–62, 175, 179, 182, 203–3, 206, 211–12, 257, 277–79
  secondary, 10, 122, 128, 132–37, 143–46, 149, 151, 153, 172–73, 178–83, 201, 204–5, 215, 217, 221, 223, 235–36, 239, 243, 262, 264, 274, 276–79, 284
chance, coincidence (*ittifāq*), 139–40, 143, 195, 200, 209–10, 233, 244–45
choice between alternatives (*ikhtiyār*), 11–12, 124–25, 127, 129–32, 139–43, 150, 153, 160, 173, 184–85, 192–93, 217–19, 231, 234, 263
creation, (*see also* world)
  ex nihilo, 118–19, 261
  of human acts (*See* power, *kasb*, and al-Ghazālī)
Columbus, Christopher, 3
conditions (*shurūṭ*) for God's creation, 84, 222–36, 245
contingency, 30, 79, 84, 139–40, 143, 167–73, 194, 251–52, 258, 272–73 (*see also* modalities)
Cordoba, 80
Crusades, 46, 48, 301 n. 192, 307 n. 28

Damascus, xii, 15, 22, 25, 42–49, 51, 63–64, 76–77, 239
Dānishmand, 42, 64–65, 71
al-Dargazinī, 86
Dawlatshāh Samarqandī, 73–74
Daylam, 36, 40
de Boer, Tjitze J., 4, 6
demonstration (*burhān*), 29, 70–71, 73, 99–100, 107, 112–13, 115–16, 120–22, 176, 205, 232, 268, 278, 319 n. 20

determination, 30, 80, 91, 130–31, 137–39, 143, 187–94, 219–20, 234, 239, 241, 259–60, 273, 277–79, 286 (*see also* predestination *and* God, the foreknowledge of)
al-Dhahabī, 22, 26, 64
dhawk ("tasting, direct experience"), 41, 197, 254, 351 n. 50
dhikr ("mystical practice"), 67
al-Dimashqī, Abū l-Barakāt Naṣr, 314 n. 167
Ḍiyā᾿ al-Mulk ibn Niẓām al-Mulk, 51, 57
Dunyā, Sulaymān, 14–15
Dutton, Blake, 165

Egypt / Egyptians, 3, 14–15, 20, 59, 63–64, 81
Elburz Mountains, 36
evil, harm, 90–91, 93, 188, 229–31, 245, 281
experience, experimentation (*tajriba*), 68, 205–211 (*see also* dhawq)

falsafa, *falāsifa*. (*see* Aristotelianism, Avicenna, and al-Fārābī)
fā῾il ("efficient cause, maker"), 12, 79, 83, 132–37, 141, 144–47, 150–53, 160, 169, 172, 181–85, 217, 221, 232, 235, 240–41, 244, 247, 261, 277, 279
Fakhr al-Mulk ibn Niẓām al-Mulk, xiii, 53–55, 57
al-Fārābī, 35, 98, 145–46, 155, 167, 169, 180, 199, 218, 242, 247, 252–53, 256, 258–60, 264–66, 271, 279, 283
  the cosmology of, 136–37
  on future contingencies, 139–41
al-Fāramadhī, Abū ῾Alī, 9, 26, 52
Fāṭimids, 36, 40, 260, 282
fatwā ("legal opinion"), 49–50, 55, 67, 101–3, 106, 276
al-Fayyūmī, 59
Fès, 64
fiqh, 27–29, 34, 48, 72–74, 77, 81
Frank, Richard M., 8, 10–11, 105, 128, 132, 180–82, 202, 212, 221, 233–34, 242, 245, 255, 259, 266, 271, 275, 284
furū῾ ("substantial rulings in Islamic law"), 48, 75–76

Gabriel (archangel), 256
Gairdner, W. H. T., 9, 180, 283
Galilei, Galileo, 3
Garden, Kenneth, 304 n. 244
Gavison, Abraham, 19
al-Ghazālī, *passim*
  and politics, 38–40, 44, 85, 102–4, 114–15
  date of birth of, 23–25
  life of, 25–59
  name of, 22
  on the creation of human acts, 216–21, 276–76
  on ethics, 48, 73, 92–93, 193, 215, 285–86
  on logics, 7, 35, 116, 212–13, 272
  on miracles, 156–57, 173, 176, 182–83, 185, 194–201, 258, 282
  on prophecy, 67–70, 100–1, 114, 194–201, 226, 244
  on the human soul, 7, 42, 66–71, 93, 98, 101, 224, 269–71, 283, 285–86
  pseudo-epigrahic works of, 13–14
  works of. *See* separate index
Ghazna (in Afghanistan), 72
Ghosheh, Muhammad, 301 n. 193
Gimaret, Daniel, 128, 132
Glassen, Erika, 37
God
  attributes of, 103, 114, 130, 170, 189–92, 219–20, 230–32, 234, 247, 250, 258, 263, 276, 284
  foreknowledge of, 139–43, 187–94, 218–19, 234, 270
  proofs for the existence of, 77–80, 83–84, 135, 144, 170–71, 220–21, 251–52, 282.
Goodman, Lenn E., 159
Goldziher, Ignaz, 4–5, 254, 355
Gramlich, Richard, 62, 225, 290 n. 46, 318 n. 9, 353 n. 91
Granada, 81
Gurgan, 27–28

habit (*῾āda, sunna*), 69, 127, 130, 149, 154–57, 160, 162, 173–76, 179, 181, 185–87, 193–95, 198, 201–7, 221, 228, 243, 257–58, 270–71, 276–77
ḥadīth, *ḥadīth*-studies, 56–57, 68, 70–71, 73, 104, 106–8, 115, 148, 188, 205, 220, 230, 237, 246, 266–68

ḥads ("intellectual intuition"), 68 (*see also* ilhām)
Ḥāfiẓ, Ḥamza, 15
Ḥājjī Khalīfa Kātib Čelebī, 13, 306 n. 7
Hamadan, 36, 72, 81, 85
ḥaqīqa ("truth of the matter, essence"), 53, 66–68, 107, 264
al-ḥaqq ("that what is due"), 225, 228, 233
Hārūn al-Rashīd, 59, 239
Hārūniyya (monument in Ṭūs), 58–59, 305 n. 261
al-Ḥārith al-Muḥāsibī, 40, 67
Ḥasan ibn al-Ṣabbāḥ, 40
Hebron, xii, 8, 24–25, 42, 44, 47–49
Hegel, G. W., 11
Herat (in Afghanistan), 29, 76, 94, 265
Hillenbrand, Carole, 37
Hourani, George F., 20, 35
Humā'ī, Jalāl al-Dīn, 20

Ibn al-ʿAdīm, 28
Ibn ʿArabī, Muḥyī l-Dīn, 200, 314 n. 156
Ibn al-ʿArabī, Abū Bakr, 9, 20, 23, 38, 42–43, 48, 62–71, 77, 199
Ibn al-ʿArabī, Abū Muḥammad, 63–64
Ibn ʿAsākir, 21–22, 45
Ibn al-Athīr, 22, 38, 44
Ibn Bājja, 95, 249, 352
Ibn Fūrak, 128
Ibn Ghaylān al-Balkhī, 62, 116–120, 122, 199
Ibn al-ʿImād, 59
Ibn Jahīr, ʿAmīd al-Dawla, 64
Ibn al-Jawzī, 22–23, 32, 34, 38, 40, 44, 50, 71–72, 95, 282, 284
Ibn Jubayr, 41
Ibn Kathīr, 22, 38, 72
Ibn Khallikān, 22–23, 291 n. 11
Ibn Khaldūn, 4, 133–34
Ibn al-Malāḥimī, 323 n. 51, 332 n. 130
Ibn al-Muqaffaʿ, 17, 87, 315 n. 168
Ibn Muẓaffar, 64, 66
Ibn al-Najjār, 26–28
Ibn Qasī, 85
Ibn al-Razzāq, Abū Manṣūr, 297 n. 99
Ibn Rushd. *See* Averroes
Ibn Rushd al-Jadd, 67
Ibn Sabʿīn, 200, 289 n. 35,
Ibn al-Ṣaffār, Abū Bakr, 76
Ibn al-Ṣalāḥ al-Shahrazūrī, 46, 72, 76–77, 81, 132, 201, 250
Ibn al-Samḥ, Abū ʿAlī, 352 n. 68
Ibn Sīnā. *See* Avicenna
Ibn Taymiyya, 4, 66, 200–1, 283, 314 n. 157
Ibn Ṭufayl, 81, 90, 95, 180, 275, 284
Ibn Tūmart, 77–81, 83, 187, 190–91, 220–21, 254
ilhām ("inspiration"), 101, 199 (*see also* ḥads)
impossibility, 30, 79, 148, 154, 157–72, 178, 196, 218–19, 233–34, 259–60, 268–71, 277, 282 (*see also* modalities)
intermediation, 10, 143–46, 151, 178, 180–82, 187, 255, 260, 262, 266, 274, 277–78 (*see also* causality, secondary)
Iraq, 39, 55, 62–64, 77
al-Irāqī, 16
Irwin, Robert, 357 n. 4
Isfahan, 32, 34, 36–38, 40, 74
al-Iṣfahānī, ʿImād al-Dīn, 314 n. 161
Isfara'in, 76
al-Isfarā'īnī Abū Isḥāq, 128, 188, 322 n. 15, 323 n. 50
Ismā'īlites, 32, 35–340, 53, 55, 66, 86, 102, 106, 113–14, 116, 121, 200, 260–64, 282, 284
al-Isnawī, 291 n. 11
Izutsu, Toshihiko, 86, 313–14

Janza (*also* Ganja, in Azerbaijan), 75
Jeddah, 15
Jerusalem, xii, 42, 45–47, 51, 63
jihād ("holy war", *also* "strong inner effort"), 48–50, 302 n. 199
Junayd, 40
al-Juwaynī, Abū l-Maʿālī, xi–xii, 21, 27–32, 34–35, 37, 42, 52, 57, 75, 77–79, 128–34, 139–71, 189–91, 195, 198, 220, 284–85
  position in Ashʿarite theology, 29–31
  on human actions, 128–33
Juvaynī, ʿAṭā'-Malik, 37, 297 n. 116, 298 n. 121, 310 n. 84, 311 n. 106

al-Kaʿbī, 189, 191
kalām, mutakallimūn, 6–9, 13, 19, 54, 70–71, 78–79, 83–85, 99, 119,

125–26, 132–34, 159, 167, 170, 172, 182, 188, 192, 201, 204, 212, 249–50, 266–67, 275, 278, 285 (*see also* Ash'arism and Mu'tazilism)
karamāt ("wonderous deeds of saints"), 69, 179, 198
kasb, iktisāb ("acquisition"), 199, 217–219 (*see also* power)
al-Kiyā' al-Harrāsī, xiii, 29, 57, 77, 134
al-Khāzinī, Abū l-Fatḥ, 239
khānqāh ("Sufi convent"), 49–51, 55, 73–74, 95
  khānqāh of Abū Sa'd in Baghdad, 48, 64–65
al-Khawāfī, Aḥmad, 75
Khorasan, 20–23, 27, 29, 39, 40–59, 64, 72–73, 76–77, 80, 90, 117, 239
al-Kirmānī, Ḥamīd al-Dīn, 261–63, 282
Krawulsky, Dorothea, 21, 56
Kukkonen, Taneli, 16
al-Kurdī, Muḥyī al-Dīn Ṣabrī, 289 n. 37
al-kursī ("the pedestal"), 356 n. 155, 357 n. 180

Landolt, Hermann, 86, 246, 260
Laoust, Henri, 61, 297 n. 120, 298, 299 n. 158
al-Lawkarī, 72
laws of nature, 12, 127, 173, 175–76, 198, 201, 205, 225, 237, 243, 252–53, 256–59, 274, 276 277, 280
Lazarus of Bethany, 195–96
Lazarus-Yafeh, Hava, 9–10, 286
Leo Africanus, 19

madrasa ("seminary"), 4, 27, 295 n. 65 (*see* also Niẓāmiyya and *zāwiya*)
  Tājiyya madrasa in Baghdad, 42
  Bayhaqī madrasa in Nishapur, 57
Mahdiyya (in Tunisia), 63
māhiyya ("quiddity, essence"), 163, 168, 248, 252, 257, 279
Maḥmūd ibn Muḥammad Tapar, 72, 86
Maimonides, 179, 249, 275, 321–322, 352, 385
Majd al-Mulk, 39
Makdisi, George, 4, 295 n. 65
al-Makkī, Abū Ṭālib, 40, 48, 221, 226–28, 278
al-Malik al-Mu'aẓẓam al-Ayyūbī, 46
Malikshāh, xi–xii, 27, 32–33, 36–38, 40
Marmura, Michael E., 10–11, 113, 147, 155, 181–83, 213, 240, 277
Marrakesh, 78
al-Mas'ūdī, Sharaf al-Dīn, 61, 117
Mayhana, 72–74
al-Māzarī al-Imām, 200, 283, 303 n. 232, 316 n. 2, 339 n. 138
al-Māzarī al-Dhakī, 48
McGinnis, Jon, 136, 182, 210
Mecca, xii, 29, 41, 47
Medina, xii, 29, 47
Melissos, 249
Merw, 72, 76, 117
Miskawayh, 35, 48, 256, 283, 319 n. 12
modalities, epistemological character of, 30, 166–72, 203, 206, 276, 278
Mongols, 59, 77,
Morocco, 77–78
motive to act (*dā'in, dā'iya*), 131–32, 142–43, 216–18, 220, 223, 232, 236, 242, 279
mover, movers
  celestial, 242, 247–48, 250–53, 256, 279, 281
  the unmoved, 251–52, 262–63, 266, 281, 284
Mu'ayyad al-Mulk, 39
Mujīr al-Dīn, 39–40, 298 n. 121
Muḥammad, the Prophet of Islam, 42, 49, 56, 70–71, 80, 91, 106–8, 112, 115, 193, 204, 208, 220–21
Muḥammad ibn Yaḥyā al-Janzī, 74–77
Muḥammad Tapar, xiii, 40, 51, 57, 73–74 86
Mullā Ṣadrā Shīrāzī, 6, 293 n. 38
Munk, Solomon, 5, 147
al-Muqtadī, Caliph, xii, 36–39
al-Murdār, Abū Mūsā, 104
al-Murtaḍā al-Zabīdī, 16, 23, 59, 64, 191
musabbib al-asbāb ("the one who makes the causes function as causes"), 132, 220–21, 236–37,
Muslim Brotherhood, 16
al-Mustaẓhir, Caliph, xii, 64
al-muṭā' ("the obeyed one"), 252–53, 255–59, 262–63, 268, 274, 280–81 (*see also al-'arsh*, and Gabriel)

Muʿtazilism, Muʿtaziltes, 6, 103–5, 124–29, 133–34, 139, 151, 170, 172–73, 185, 188–89, 191, 193, 204, 220, 239–40, 246–47, 266, 284

Nabulus, 43
al-Nasafī, 262–63, 282
al-Nawawī, Yaḥyā, 22, 76, 81
necessity (*see also* Modalities)
  by itself versus through something else, 30, 139–41, 155, 163, 169, 245, 251, 270–71, 276
  of causal connections, 126, 135–37, 148–50, 153, 172–73, 178, 201–204, 215, 257, 261, 279–80
  of human actions, 218, 222,
  of empirical judgments, 176, 178, 206, 208–13, 212, 218
  of God to act in a certain way, 11, 141–43, 158, 185, 225, 231–34, 255, 261–63, 271–74, 281
  of the judgments in logics, 116, 136, 167, 212–13
Neoplatonism, 97, 133–35, 199, 260, 282–83
Nishapur, 21, 24–27, 29–32, 48, 50–51, 53–58, 74–77, 117, 128, 236, 239
Nīshābūrī, Ẓāhir al-Dīn, 298 n. 122
Niẓām al-Mulk, xi–xii, 27–30, 32, 34, 36–40, 51, 53, 64, 129
Niẓāmī Ganjawī, 75, 94, 264–65
Niẓāmiyya madrasa, 27, 29, 51
  in Baghdad, 8, 20, 24, 29, 31, 34, 40–42, 44, 48, 51, 57, 72–74, 77, 80, 86, 97, 204
  in Merw, 72, 117
  in Nishapur, 24–27, 29–32, 48, 51, 53–58, 75–77, 117, 236
Niẓāmiyya (political party), 36–40, 299 n. 139
nominalism, 71, 97, 163, 166, 176–77, 205, 211, 285

Obermann, Julian, 160–62
occasionalism, 10, 122, 124–28, 132, 149, 153–56, 180–83, 186–87, 284–85
orthodoxy, 5–6, 9, 103, 105–9, 111
Oğuz Turks (*ghuzz*), 4, 56, 76, 95

Parmenides, 249
Pharao, 107, 247–48, 327 n. 3
Philoponus, John (Yaḥyā l-Naḥwī), 164, 352 n. 68
Pines, Shlomo, 6
Plato, 99, 121, 243, 257 (*see also* Neoplatonism)
power
  of God, 10, 124–5, 127, 140–41, 148–49, 153–56, 157–58, 160, 162–63, 167, 172, 176, 178–79, 191–92, 223–24, 232–34, 241, 247, 255, 270–71, 280, 285–86
  of humans to act, 128–33, 217–19. 224, 276–77, 285
  ordained versus absolute, 155 (*see also* necessity, by itself versus through something else)
  the things' active and passive, 136, 182, 208–9, 211, 277
predestination, 12, 91, 137–43, 188–90, 192–93, 220, 222, 227, 237, 240–41, 258, 270, 280 (*see also* determination *and* God, foreknowledge of)
primum mobile (*falak al-aflāk*), 12, 137, 251, 263, 281, 284
providence, divine (*ʿināya*), 142, 226, 229, 245
Ptolemy, 12, 136, 251, 263
Pythagoras, 200, 250

Qazwin, 76
qiyās ("syllogism," *also* "legal analogy"), 5, 7, 68, 99–100, 116, 120, 206–13, 279
Quhistan (Province in Iran), 40
Qur'an, 73, 75, 78, 99, 100, 102, 104, 106–8, 111–13, 155–16, 118–20, 123–24, 129, 146, 148, 188, 193–200, 204–05, 217, 220, 237, 246, 248, 254, 256–57, 261–63, 266–71, 274, 281, 282, 284 (*see also* the separate index of verses in the Qur'an)
al-Qushayrī, Abū l-Qāsim, xi, 21, 67, 75, 90
Quṭṭā al-ʿAdawī, ʿAbd al-Raḥmān, 16

al-Rādhakānī, Aḥmad, 27–29
al-Rādhakānī, Abū Saʿd ʿAbd al-Malik, 27–29

al-Rāfiʿī, Abū l-Qāsim, 76
al-Rāghib al-Iṣfahānī, 48, 283
Rahman, Sayeed, 318 n. 9
Rashīd al-Dīn Ṭabīb, 33, 37, 53, 298 n. 122
Rāwandī, Muḥammad ibn ʿAlī, 298 n. 122
al-Rāzī, Abū Bakr, 117
al-Rāzī, Fakhr al-Dīn 6, 59, 75–76, 116–20, 122, 131–32, 321 n. 64, 336 n. 88, 338, n. 125, 343 n. 3. 344 n. 22
realism, epistemological, 161, 177
Renan, Ernest, 5, 61
Roscelin, 177
Rudolph, Ulrich, 149, 155, 158–59
Rūmī, Jalāl al-Dīn, 264

al-Sāʿātī, Muḥammad, 239
Sabra, Abdelhamid, 6
Sabzawar, 20
al-Ṣafadī, 22
Safi, Omid, 298
Ṣalāh al-Dīn (Saladin), 46
al-Sāmirī ("the Samaritan"), 197
al-Samʿānī, Abū Saʿd, 21–22, 26, 28, 75, 86
al-Samʿānī, Abū l-Muẓaffar, 72
Samarqand, 117
Sanjar, xiii, 23–25, 28, 30, 32, 35, 39–40, 50, 53–57, 59, 73–74, 76, 86, 90, 94–95, 239
al-Ṣayrafī, Abū l-Ḥasan, 77
al-Sayyid, Riḍwān, 87
seclusion (*ʿuzla*), 24, 42, 50–51, 53, 67, 95
Seville, 63–65
Shāfiʿism, 23, 25, 27, 34, 43, 52, 57, 61, 72–77, 104
al-Shahrastānī, 72, 77, 132, 250, 310 n. 92,
Sharīʿa, 48, 93, 129, 236 (*see also fiqh*)
al-Shāshī, Abū Bakr, 77
al-Shāṭibī, 81
shawka ("political and military power"), 39
al-Shiblī, 40
Shihāb al-Islām ʿAbd al-Razzāq, 28
al-Shīrāzī, Abū Isḥāq, 295 n. 69, 307 n. 34
al-Shīrāzī, Abū Muḥammad al-Fāmī, 297 n. 100
Sibṭ ibn al-Jawzī, 22
al-Sijistānī, Abū Yaʿqūb, 260–63, 282
al-Simnānī, al-Kamāl, 75
Smith, Gerard, 259
spheres (*aflāk* or *kurāt*), 10, 12, 79, 84–85, 101, 105, 123, 135–41, 145–46, 151, 171, 180, 184, 200, 230, 237, 240, 242–44, 247–53, 256–57, 262–63, 266, 268–69, 274, 279–84, 286
al-Subkī, Tāj al-Dīn, 13, 21, 23, 26–28, 45, 64, 71–72
substance (*jawhar*), 68, 70–71, 93, 101, 125–27, 139, 148, 159, 166, 178, 223–24, 248, 250, 269, 285
al-Suhrawardī, Shihāb al-Din ʿUmar, 16
al-Suhrawardī, Shihāb al-Din Yaḥyā "al-Maqtūl," 6, 95
al-Suyūṭī, Jalāl al-Dīn, 81, 356 n. 155
Sūs Valley (in Morocco), 77
Syria, 21, 40, 42–43, 45, 48, 51, 63–65, 75

al-Ṭabarī, Abū ʿAbdallāh, 297 n. 100
Tāj al-Mulk, 36–38
takhṣīṣ ("particularization"), 170
tarjīḥ ("preponderance"), 170
al-Tawḥīdī, Abū Ḥayyān, 200, 359 n. 31
taʿlīq, or taʿlīqa ("notes"), 28, 32, 72, 112, 297 n. 93
Terken Khātūn, xii, 36–38
Thales, 250
Ṭībāwī, ʿAbd al-Laṭīf, 43
Tlemcen, 64
transliteration, rules of, 16–17
Transoxania, 61, 117
Treiger, Alexander, 255, 294 n. 51
Toghril-Bey, xi, 29
Tunis, 64, 78
al-Ṭurṭūshī, 63–64, 66, 199–200
Ṭūs (district in Iran containing the towns or villages:)
  Fāramadh, 303 n. 223
  Nūqān, 25
  Rādhakān, 27
  Sanābādh / Meshed, 25, 302 n. 208, 305 n. 261
  Ṭābarān, xi, xiii, 22–23, 25, 27, 49, 51–53, 55–59, 76, 117, 236
  Turūq, 304 n. 242
al-Ṭūsī, Naṣīr al-Dīn, 6
al-Ṭūsī, Abū Manṣūr Muḥammad, 311 n. 94

al-Ṭūsī, Abū l-Fatḥ Muḥammad, 311 n. 94
Tyros, 47

al-ʿUlaymī, Mujīr al-Dīn, 45–47

volition (*irāda*), 129, 182, 184, 190, 216–220, 223–25, 227, 232–33, 279
al-Wāḥidī, Abū l-Ḥasan, 75
al-Wāsiṭī, Abū ʿAbdallāh, 23, 71, 293 n. 36, 304 n. 249
waterclock (*ṣandūq al-sāʿāt*), 13, 80, 236–43, 245, 257–59
Watt, William M., 5, 9
Weinberg, Steven, 289 n. 32, 357 n. 4
Wensinck, Arent J., 9, 257, 283
William of Ockham, 177
Wisnovsky, Robert, 133–34
Wolff, Christian, 123
world, worlds
  best of all possible, 70, 201, 225–43, 273, 280–81
  eternity of, 5, 30, 78, 83, 85–86, 101–04, 117–22, 164–65, 189–90, 270–71, 280
  temporary creation of, 5, 30, 78, 83, 117–19, 126, 184–5, 258, 280

al-Yāfiʿī, 293 n. 36
Yāqūt, 22, 23, 291 n. 11, 293 n. 27
Yūsuf ibn Tāshifīn, 63–64

zāʾid ʿalā l-dhāt ("additional to God's essence"), 142, 191, 272, 234, 276
zandaqa ("clandestine apostasy"), 103–5, 108–9, 111, 151
Zarrīnkūb, ʿAbd-Ḥusayn, 49
zāwiya ("small, private madrasa"), 44–46, 49–51, 53, 55, 57, 75
Zedler, Beatrice H., 259

Index of Works by al-Ghazālī

al-Arbaʿīn fī uṣūl al-dīn / Book of Forty, 180, 197, 236–241

al-Basīṭ fī l-furūʿ fī madhhab al-Shāfiʿī / The Extended One, 34, 75
Bidāyat al-hidāya / Beginning of Guidance, 43–44

Faḍāʾiḥ al-bāṭiniyya / The Scandals of the Esoterics, 35, 83, 102, 262, 297 n. 113,
Fayṣal al-tafriqa bayna l-Islām wa-l-zandaqa / Decisive Criterion for Distinguishing Islam from Clandestine Apostasy, 11, 14, 57, 101, 105–14, 119, 248, 256, 264, 267

Hujjat al-ḥaqq fī-l radd ʿalā l-bāṭiniyya / Proof of Truth Responding to the Ismāʿīlites, 32

Iḥyāʾ ʿulūm al-dīn / Revival of the Religious Sciences, 13, 15–16, 23, 26, 44–45, 48–49, 56, 62, 64, 67, 71, 78, 80–81, 84–85, 92–93, 102, 105–7, 114, 121, 146, 151, 179, 180–81, 184, 186, 190–99, 204–5, 215–37, 239–45, 254, 264, 269, 278, 283, 285–86
Iljām al-ʿawāmm ʿan ʿilm al-kalām / Restraining the Ordinary People from the Science of Kalām, 13, 57–58, 180, 264–74, 277
al-Imlāʾ fī ishkālāt al-Iḥyāʾ / The Dictation on Difficult Passages in the Revival, 16, 227, 230–33, 347 n. 64
al-Iqtiṣād fī l-iʿtiqād / The Balanced Book on What-To-Belief, 9, 14, 35, 79, 83, 102, 154–55, 170, 179–82, 192, 202–4, 218, 226, 234, 271–72, 285, 288

Jawāhir al-Qurʾān / Jewels of the Qurʾān, 31, 269

Kīmyā-yi saʿādat / Alchemy of Happiness, 93

al-Lubāb min al-Iḥyāʾ / The Kernels of the Revival 62

al-Maʿārif al-ʿaqliyya / Intellectual Insights, 221
(?) *Maʿārij al-quds fī madārij maʿrifat al-nafs = The Stairs of Jerusalem*

in the Steps Leading to Knowledge on the Soul, 45, 280, 289 n. 35, 314 n. 156, 338 n. 117, 339 n. 140, 353 n. 101

(?) *al-Maḍnūn bihi ʿalā ghayr ahlih) / The Book to Be Withheld from Those For Whom it is Not Written*, 14

al-Mankhūl min taʿlīqāt al-uṣūl / The Sifted among the Notes on the Methods of Jurisprudence, 32

Maqāṣid al-falāsifa / The Intentions of the Philosophers, 39, 66, 98, 143, 144

al-Maqṣad al-asnā fi sharḥ maʿānī asmāʾ Allāh al-ḥusnā / The Highest Goal in Explaining the Beautiful Names of God, 14, 84, 177, 180, 230, 236–46, 264, 269

Miḥakk al-naẓar fī l-manṭiq / Touchstone of Reasoning in Logic, 13, 35, 116, 121, 205–07, 212, 243, 269

Mishkāt al-anwār / Niche of Lights, 9, 13, 54, 62, 83, 85–86, 113, 180–81, 245–260, 262, 264, 266, 274, 281, 284, 286

Miʿyār al-ʿilm fī fann al-manṭiq / Standard of Knowledge in Logics, 13, 35, 201, 205, 271–72

Mīzān al-ʿamal / Scale of Action, 26, 264, 280, 359 n. 45

MS London, British Library, Or. 3126, 143–46, 226, 251, 272, 282, 308 n. 56, 313 n. 141–43, 316, 319 n. 12, 326–27, 338 n. 117, 346, 352 n. 76, 353 n. 93, 357, 364

al-Munqidh min al-ḍalāl / Deliverer from Error, 7–9, 14, 19–21, 23–25, 29, 35, 40–43, 45, 49–53, 57, 75, 83, 97, 100–1, 120, 144, 157, 184, 195–98, 200, 208, 282, 285

al-Mustaṣfā min ʿilm al-uṣūl / The Choice Essentials of the Science of the Methods [of Jurisprudence], 14–15, 117, 216, 255

(?) *Nafkh al-Rūḥ wa-l-taswiya / Breathing of the Spirit and the Shaping*, 14, 70

Naṣīḥat al-mulūk / Council for Kings, 39, 55, 93, 298

al-Qānūn al-kullī fī l-taʾwīl / The Universal Rule in Interpreting Revelation, 70

Qawāṣim al-bāṭiniyya. / Weak Positions of the Esoterics, 35

al-Qisṭās al-mustaqīm / The Straight Balance, 36, 116, 120

Risāla Fī l-ʿilm al-ladunī / The Epistle on Intimate Knowledge, 14

al-Risāla al-Qudsiyya / Letter for Jerusalem, 45, 191

Shifāʾ al-ghalīl fī bayān al-shubah wa-l-mukhīl wa-masālik al-taʿlīl / The Quenching of Thirst in Explaining Difficult Questions, 39, 296 n. 92

Tahāfut al-falāsifa / The Incoherence of the Philosophers, 5, 7, 11, 13–14, 19–20, 35–36, 66, 83, 86, 97–103, 101, 106, 115–16, 119, 122–23, 144, 147–73, 175–79, 181–87, 194–98, 203, 205–07, 213, 222, 225, 234, 258, 262, 266 272, 275–76

(?) *Tuḥfat al-mulūk / Gift for Kings*, 48

al-Wajīz fī fiqh madhhab al-Imām al-Shāfiʿī / The Succinct One, 75–76

al-Wasīṭ fī l-madhhab / The Middle One, 34, 75–76

Index of Manuscripts

Ankara, Dil ve Tarih-Cografya Fakültesi Kütüphanesi
  İsmail Saib 1544, 292 n. 15

Beirut, Jafet Library of the American University
  no. 172.2:G4ltbA, 366
  no. 297.3: G41 iA, 307 n. 22, 364
Berlin, Staatsbibliothek Preussischer Kulturbesitz—Orientabteilung
  Petermann II 8 (Ahlwardt 10070.2), 304 n. 256, 306, n. 5, 315 n. 171–73, 364
  Petermann II 35 (Ahlwardt 1721), 363, 364
  Petermann II 545 (Ahlwardt 1714), 347–48
  Petermann II 690 (Ahlwardt 2301), 356–57, 364
  Wetzstein 99 (Ahlwardt 1708), 306 n. 7
  Wetzstein II 19 (Ahlwardt 1680), 290 n. 46
  Wetzstein II 1806 (Ahlwardt 2075), 290 n. 38, 315 n. 168, 320 n. 40
  Wetzstein II 1807 (Ahlwardt 1707), 306, n. 7
  Sprenger 763 (Ahlwardt 1715), 349 n. 1
  Sprenger 941 (Ahlwardt 1716), 349 n. 1
  Sprenger 1968 (Ahlwardt 3210, 3976), 362, 364
Bursa (Turkey)
  Hüseyin Çelebi 823, 369

Cairo, Dār al-Kutub al-Miṣriyya
  majāmiʿ 180, 309 n. 66
  tawḥīd 295, 382

Damascus, Maktabat al-Asad al-Waṭaniyya, Ẓāhiriyya Collection
  nos. 2111–14, 365
  no. 6469, 290 n. 39
  no. 6595, 290 n. 39
Dublin, Chester Beatty Library
  no. 3372, 289 n. 34

Edinburgh, University Library
  Arab 20, 33, 37, 53
Escurial Collection, Madrid
  no. 1130, 290 n. 38, 296 n. 88, 363, 364

Istanbul, Nuruosmaniye Kütüphanesi
  4894, 326 n. 97, 379

Istanbul, Süleymaniye Kütüphanesi
Hamidiye 1452, 289 n. 34, 386
Şehit Ali Paşa 1712, 290 n. 38, 318 n. 58, 356 n. 140, 356 n.143, 362, 363, 367
Istanbul, Topkapı Serayı Müzesi Kütüphanesi
Ahmet III 1099, 361
Ahmet III 1916, 367

London, British Library
Or. 3102, 365
Or. 3126. *See* index of works by al-Ghazālī
Or. 6810, 94
Or. 7721, 367
Or. 11837, 82

Munich, Bayrische Staatsbibliothek
no. 855, 364

New York, The Metropolitan Museum of Art
Gift of Alexander Smith Cochran, 13.228.3, 265

Oxford, Bodleien Library
Pococke 400, 88, 89

Paris, Bibliothèque Nationale
fonds arabe 1331, 293 n. 39
5639 (Fonds Archinard), 309 n. 61

Patna (India), Khuda Bakhsh Oriental Public Library (Bankipore)
no. 1825, 314 n. 167
Princeton, University Library
Yahuda 364, 368
Yahuda 459, 362
Yahuda 838, 306 n. 7
Yahuda 3714, 306 n. 7
Yahuda 3893, 349 n. 1
Yahuda 4358, 366
Yahuda 4374, 349 n. 1

Rabat, Bibliothèque Générale
1275 *kāf*, 307, n. 19
Rome, Vatican, Bibliotheca Apostolica
Ebr. 426, 340–43, 364

Vienna, Österreichische Nationalbibliothek
no. 1656, 290 n. 46

Yale University, Beinecke Rare Books and Manuscripts Library
Arab suppl. 425, 349 n. 1
Landberg 98, 314 n. 167
Landberg 318, 305 n. 263
Landberg 428, 347–49, 362
Salisbury 38, 306, n. 7

Index of Verses in the Qur'an

2:7, 124
2:23, 195
2:81, 217
2:134, 217
2:235, 237
2:255, 123

3:7, 83
3:39, 195
3:47, 124
3:52, 124
3:39, 124
3:133, 104

4:11–12, 123
4:93, 108
4:113, 124

5:38, 217
5:110, 195, 196

6:2, 188
6:18, 268
6:73, 124, 252
6:75–79, 248–54, 352 n. 71
6:135, 344 n. 26

7:34, 188
7:54, 123, 268

9:129, 123

10:3, 268
10:38, 195
10:61, 190

11:3, 188
11:6, 188

12:76, 112

13:2, 256
13:26, 188

14:10, 188

15:16, 124
15:26, 124
15:29, 70
15:26, 124

16:50, 268
16:61, 188

17:36, 274

20:9–36, 113
20:12, 85, 113
20:24, 107
20:43, 107

21:30, 124
21:33, 123
21:68, 157

23:12, 124

26:23, 247

28:88, 254

29:24, 157

30:30, 198

32:9, 70

33:62, 198, 270, 277

35:42, 198, 270, 277
35:43, 257

36:40, 123

37:06, 124
37:97, 157

38:72, 70, 200

39:42, 188

40:7, 123

41:9, 124
41:12, 123

48:6, 108
48:23, 198, 270, 277

54:1, 148

55:14, 124
55:15 124
55:19–20, 124

58:14, 108

64:1, 124

65:12, 124

66:8, 124

67:3, 123

71:16, 123

78:13, 123

81:19–21, 256

85:16, 221

89:16, 188

89:22, 123

112:1–4, 206

CPSIA information can be obtained at www.ICGtesting.com
Printed in the USA
BVOW03s1415160813

328628BV00002B/3/P

9 780199 773701